Methods in Molecular Biology

Series Editor
John M. Walker
School of Life and Medical Sciences
University of Hertfordshire
Hatfield, Hertfordshire, UK

For further volumes:
http://www.springer.com/series/7651

For over 35 years, biological scientists have come to rely on the research protocols and methodologies in the critically acclaimed *Methods in Molecular Biology* series. The series was the first to introduce the step-by-step protocols approach that has become the standard in all biomedical protocol publishing. Each protocol is provided in readily-reproducible step-by-step fashion, opening with an introductory overview, a list of the materials and reagents needed to complete the experiment, and followed by a detailed procedure that is supported with a helpful notes section offering tips and tricks of the trade as well as troubleshooting advice. These hallmark features were introduced by series editor Dr. John Walker and constitute the key ingredient in each and every volume of the *Methods in Molecular Biology* series. Tested and trusted, comprehensive and reliable, all protocols from the series are indexed in PubMed.

The Epithelial-to Mesenchymal Transition

Methods and Protocols

Edited by

Kyra Campbell

Department of Biomedical Science and Bateson Centre, University of Sheffield, Sheffield, UK

Eric Theveneau

Centre de Biologie du Développement (CBD), Centre de Biologie Intégrative (CBI), Université de Toulouse, Toulouse, France

Humana Press

Editors
Kyra Campbell
Department of Biomedical Science
and Bateson Centre
University of Sheffield
Sheffield, UK

Eric Theveneau
Centre de Biologie du Développement (CBD)
Centre de Biologie Intégrative (CBI)
Université de Toulouse
Toulouse, France

ISSN 1064-3745 ISSN 1940-6029 (electronic)
Methods in Molecular Biology
ISBN 978-1-0716-0778-7 ISBN 978-1-0716-0779-4 (eBook)
https://doi.org/10.1007/978-1-0716-0779-4

This Humana imprint is published by the registered company Springer Science+Business Media, LLC, part of Springer Nature.
The registered company address is: 1 New York Plaza, New York, NY 10004, U.S.A.

Preface

The epithelial-to-mesenchymal transition (EMT) describes a cellular process during which cells transition to a mesenchymal cell state. Reversible in nature, EMTs not only play crucial roles during development, their inappropriate activation is associated with a number of pathologies, including cancer metastasis and fibrosis. Long thought of as a binary transition, the realization that cells activate the EMT program to different extents, and often stop at varying points along the way to becoming mesenchymal (or epithelial), has revolutionized thinking in the field. Furthermore, it is increasingly clear that the way that the program is implemented is very context-dependent, which can lead to difficulties extrapolating from results "in a dish." In this book of methods for studying *The Epithelial-to-Mesenchymal Transition*, we have first explored the complexity of EMT research, by asking a diverse range of EMT researchers to write personal perspectives on their opinion of the field, which has led to a number of thought-provoking pieces. The book provides a comprehensive range of methods for imaging EMT/MET in in vivo systems, as well as methods to leverage these systems to dissect the underlying mechanisms. We hope that these will aid future studies, deepening our understanding of context-dependent features, as well as cross-species commonalities. Finally, we finish with a series of innovative methods for studying different features of epithelial-mesenchymal plasticity, at multiple levels, from metabolism to single-cell level. As two researchers whose science has been largely driven by the simple excitement of watching cells moving and changing behaviors in vivo, we hope that this combination of perspectives and methods in EMT will help to unite and drive research in this exciting field forward!

Sheffield, UK   *Kyra Campbell*
Toulouse, France   *Eric Theveneau*

Contents

Preface .... *v*
Contributors .... *xi*

Part I Perspectives on the Past and Future of EMT Research

1 A Spectrum of Cell States During the Epithelial-to-Mesenchymal Transition 3
Erica J. Hutchins and Marianne E. Bronner

2 Perspective on Epithelial-Mesenchymal Transitions in Embryos 7
David R. McClay

3 EMT, One of Many Morphological Transitions in Cellular Phase Space 13
Denise J. Montell

4 Are You Interested or Afraid of Working on EMT? 19
M. Angela Nieto

5 Partial EMT/MET: An Army of One 29
Sofiane Hamidi, Hiroki Nagai, and Guojun Sheng

6 EMT: An Update 35
Jean Paul Thiery

Part II Review on Mesenchymal-to-Epithelial Transitions

7 Mesenchymal-to-Epithelial Transitions in Development and Cancer 43
John-Poul Ng-Blichfeldt and Katja Röper

Part III Methods to Study EMT in In Vivo Systems

8 Live Imaging of Epithelial-Mesenchymal Transition in Mesoderm Cells of Gastrulating *Drosophila* Embryos 65
Lingkun Gu and Mo Weng

9 Zebrafish Neural Crest: Lessons and Tools to Study In Vivo Cell Migration 79
Zain Alhashem, Macarena Alvarez-Garcillan Portillo, Mint Ravinand Htun, Anton Gauert, Luis Briones Montecinos, Steffen Härtel, and Claudia Linker

10 Live Imaging of the Neural Crest Cell Epithelial-to-Mesenchymal Transition in the Chick Embryo 107
Mary Cathleen McKinney and Paul M. Kulesa

11 Exploiting *Drosophila melanogaster* Wing Imaginal Disc Eversion to Screen for New EMT Effectors 115
Sofia Golenkina, Rosemary Manhire-Heath, and Michael J. Murray

12 Visualizing Mouse Embryo Gastrulation Epithelial-Mesenchymal Transition Through Single Cell Labeling Followed by Ex Vivo Whole Embryo Live Imaging 135
Wallis Nahaboo, Bechara Saykali, Navrita Mathiah, and Isabelle Migeotte

13 Fluorescence Recovery After Photobleaching to Study the Dynamics of Membrane-Bound Proteins In Vivo Using the *Drosophila* Embryo 145
Joshua Greig and Natalia A. Bulgakova

14 Methods to Generate and Assay for Distinct Stages of Cancer Metastasis in Adult *Drosophila melanogaster* 161
Jamie Adams, Andreu Casali, and Kyra Campbell

PART IV IN VIVO IMAGING OF MET

15 4D Live Imaging and Analysis of Chick Embryo Somites 173
Gi Fay Mok, James McColl, and Andrea Münsterberg

16 In Vivo Analysis of the Mesenchymal-to-Epithelial Transition During Chick Secondary Neurulation 183
Elena Gonzalez-Gobartt, Guillaume Allio, Bertrand Bénazéraf, and Elisa Martí

17 Multiscale In Vivo Imaging of Collective Cell Migration in *Drosophila* Embryos 199
Gordana Scepanovic, Alexandru Florea, and Rodrigo Fernandez-Gonzalez

PART V EX VIVO/IN VITRO METHODS TO STUDY EMT/MET

18 Methods to Generate Tube Micropatterns for Epithelial Morphogenetic Analyses and Tissue Engineering 227
Minerva Bosch-Fortea and Fernando Martín-Belmonte

19 CAFs and Cancer Cells Co-Migration in 3D Spheroid Invasion Assay 243
Sefora Conti, Takuya Kato, Danielle Park, Erik Sahai, Xavier Trepat, and Anna Labernadie

20 Using Xenopus Neural Crest Explants to Study Epithelial-Mesenchymal Transition 257
Nadège Gouignard, Christian Rouvière, and Eric Theveneau

21 *Xenopus* Deep Cell Aggregates: A 3D Tissue Model for Mesenchymal-to-Epithelial Transition 275
Hye Young Kim and Lance A. Davidson

22 Molecular Tension Microscopy of E-Cadherin During Epithelial-Mesenchymal Transition 289
Helena Canever, Pietro Salvatore Carollo, Romain Fleurisson, Philippe P. Girard, and Nicolas Borghi

PART VI METHODS TO STUDY EPITHELIAL-MESENCHYMAL PLASTICITY AT THE SINGLE CELL, SUBCELLULAR, AND MOLECULAR LEVEL

23 Methodologies for Following EMT In Vivo at Single Cell Resolution 303
Abdull J. Massri, Geoffrey R. Schiebinger, Alejandro Berrio, Lingyu Wang, Gregory A. Wray, and David R. McClay

24 Isolation and Identification of EMT Subtypes 315
Robert J. Norgard and Ben Z. Stanger

25 Studying the Metabolism of Epithelial-Mesenchymal Plasticity Using the Seahorse XFe96 Extracellular Flux Analyzer 327
Sugandha Bhatia, Erik W. Thompson, and Jennifer H. Gunter

26 Inducing Sequential Cycles of Epithelial-Mesenchymal and Mesenchymal-Epithelial Transitions in Mammary Epithelial Cells 341
Cecile Davaine, Eva Hadadi, William Taylor, Annelise Bennaceur-Griscelli, and Hervé Acloque

27 Analysis of Cellular EMT States Using Molecular Biology and High Resolution FISH Labeling 353
Noémie Kempf, Fatima Moutahir, Isabelle Goiffon, Sylvain Cantaloube, Kerstin Bystricky, and Anne-Claire Lavigne

28 Mathematical Modeling of Plasticity and Heterogeneity in EMT 385
Shubham Tripathi, Jianhua Xing, Herbert Levine, and Mohit Kumar Jolly

29 Control of Cell Migration Using Optogenetics 415
Leo Valon and Simon de Beco

Index *427*

Contributors

HERVÉ ACLOQUE • *Inserm, UMRS935 ESTeam Paris Sud, Malignant and Therapeutic Stem Cell Models, Villejuif, France; GABI, INRA, AgroParisTech, Université Paris-Saclay, Jouy-en-Josas, France*

JAMIE ADAMS • *Department of Biomedical Science and Bateson Centre, The University of Sheffield, Sheffield, UK*

ZAIN ALHASHEM • *Randall Centre for Cell and Molecular Biophysics, King's College, London, UK*

GUILLAUME ALLIO • *Centre de Biologie du Développement (CBD), Centre de Biologie Inté grative (CBI), Université de Toulouse, CNRS, UPS, Toulouse, France*

BERTRAND BÉNAZÉRAF • *Centre de Biologie du Développement (CBD), Centre de Biologie Inté grative (CBI), Université de Toulouse, CNRS, UPS, Toulouse, France*

ANNELISE BENNACEUR-GRISCELLI • *Inserm, UMRS935 ESTeam Paris Sud, Malignant and Therapeutic Stem Cell Models, Villejuif, France; Service d'hématologie, APHP, GHU Paris Sud, Villejuif, France; University Paris Sud, University Paris-Saclay, UFR de Médecine Kremlin Bicêtre, Le Kremlin-Bicêtre, France*

ALEJANDRO BERRIO • *Department of Biology, Duke University, Durham, NC, USA*

SUGANDHA BHATIA • *Faculty of Health, Institute of Health and Biomedical Innovation and School of Biomedical Sciences, Queensland University of Technology, Brisbane, QLD, Australia; Translational Research Institute, Brisbane, QLD, Australia*

NICOLAS BORGHI • *Institut Jacques Monod, Université de Paris, CNRS, Paris, France*

MINERVA BOSCH-FORTEA • *Program of Tissue and Organ Homeostasis, Centro de Biología Molecular Severo Ochoa (CSIC-UAM), Madrid, Spain; Institute of Bioengineering and School of Engineering and Materials Science, Queen Mary University of London, London, UK*

MARIANNE E. BRONNER • *Department of Biology and Biological Engineering, California Institute of Technology, Pasadena, CA, USA*

NATALIA A. BULGAKOVA • *Department of Biomedical Science and Bateson Centre, The University of Sheffield, Sheffield, UK*

KERSTIN BYSTRICKY • *Center for Integrative Biology (CBI), Laboratoire de Biologie Molé culaire des Eucaryotes (LBME), University of Toulouse, UPS, CNRS, Toulouse, France*

KYRA CAMPBELL • *Department of Biomedical Science and Bateson Centre, The University of Sheffield, Sheffield, UK*

HELENA CANEVER • *Institut Jacques Monod, Université de Paris, CNRS, Paris, France*

SYLVAIN CANTALOUBE • *Center for Integrative Biology (CBI), Laboratoire de Biologie Molé culaire des Eucaryotes (LBME), University of Toulouse, UPS, CNRS, Toulouse, France*

PIETRO SALVATORE CAROLLO • *Institut Jacques Monod, Université de Paris, CNRS, Paris, France*

ANDREU CASALI • *Serra Húnter fellow, Department of Basic Medical Sciences, Institut de Recerca Biomèdica de Lleida (IRBLleida), University of Lleida, Lleida, Spain*

SEFORA CONTI • *Institute for Bioengineering of Catalonia (IBEC), The Barcelona Institute for Science and Technology (BIST), Barcelona, Spain*

CECILE DAVAINE • *Inserm, UMRS935 ESTeam Paris Sud, Malignant and Therapeutic Stem Cell Models, Villejuif, France*

LANCE A. DAVIDSON • *Department of Bioengineering, Swanson School of Engineering, University of Pittsburgh, Pittsburgh, PA, USA; Department of Developmental Biology, School of Medicine, University of Pittsburgh, Pittsburgh, PA, USA; Department of Computational and Systems Biology, School of Medicine, University of Pittsburgh, Pittsburgh, PA, USA*

SIMON DE BECO • *Cell Adhesion and Mechanics, Institut Jacques Monod, CNRS UMR7592, Paris Diderot University, Paris, France*

RODRIGO FERNANDEZ-GONZALEZ • *Department of Cell and Systems Biology, University of Toronto, Toronto, ON, Canada; Ted Rogers Centre for Heart Research, University of Toronto, Toronto, ON, Canada; Institute of Biomaterials and Biomedical Engineering, University of Toronto, Toronto, ON, Canada; Developmental and Stem Cell Biology Program, The Hospital for Sick Children, Toronto, ON, Canada*

ROMAIN FLEURISSON • *Institut Jacques Monod, Université de Paris, CNRS, Paris, France*

ALEXANDRU FLOREA • *Ted Rogers Centre for Heart Research, University of Toronto, Toronto, ON, Canada; Institute of Biomaterials and Biomedical Engineering, University of Toronto, Toronto, ON, Canada*

ANTON GAUERT • *Department of Pediatrics Oncology and Hematology, Charité—Universitätsmedizin Berlin, corporate member of Freie Universität Berlin, Humboldt—Universität zu Berlin, and Berlin Institute of Health, Berlin, Germany*

PHILIPPE P. GIRARD • *Institut Jacques Monod, Université de Paris, CNRS, Paris, France; Faculty of Basic and Biomedical Sciences, Université de Paris, Paris, France*

ISABELLE GOIFFON • *Center for Integrative Biology (CBI), Laboratoire de Biologie Molé culaire des Eucaryotes (LBME), University of Toulouse, UPS, CNRS, Toulouse, France*

SOFIA GOLENKINA • *School of BioSciences, University of Melbourne, Parkville, VIC, Australia*

ELENA GONZALEZ-GOBARTT • *Instituto de Biología Molecular de Barcelona, CSIC, Parc Científic de Barcelona, Barcelona, Spain*

NADÈGE GOUIGNARD • *Centre de Biologie du Développement (CBD), Centre de Biologie Inté grative (CBI), Université de Toulouse, CNRS, UPS, Toulouse Cedex 09, France*

JOSHUA GREIG • *Department of Biomedical Science and Bateson Centre, The University of Sheffield, Sheffield, UK*

LINGKUN GU • *School of Life Sciences, University of Nevada, Las Vegas, NV, USA*

JENNIFER H. GUNTER • *Faculty of Health, Institute of Health and Biomedical Innovation and School of Biomedical Sciences, Queensland University of Technology, Brisbane, QLD, Australia; Translational Research Institute, Brisbane, QLD, Australia; Australian Prostate Cancer Research Centre, Queensland University of Technology, Brisbane, QLD, Australia*

EVA HADADI • *Inserm, UMRS935 ESTeam Paris Sud, Malignant and Therapeutic Stem Cell Models, Villejuif, France*

SOFIANE HAMIDI • *International Research Centre for Medical Sciences, Kumamoto University, Kumamoto, Japan*

STEFFEN HÄRTEL • *Program of Anatomy and Developmental Biology, BNI, CIMT, CENS, ICBM, Faculty of Medicine, University of Chile, Santiago, Chile*

MINT RAVINAND HTUN • *Randall Centre for Cell and Molecular Biophysics, King's College, London, UK; The Francis Crick Institute, London, UK*

ERICA J. HUTCHINS • *Department of Biology and Biological Engineering, California Institute of Technology, Pasadena, CA, USA*

MOHIT KUMAR JOLLY • *Centre for BioSystems Science and Engineering, Indian Institute of Science, Bangalore, Karnataka, India*

TAKUYA KATO • *The Francis Crick Institute, London, UK*
NOÉMIE KEMPF • *Center for Integrative Biology (CBI), Laboratoire de Biologie Moléculaire des Eucaryotes (LBME), University of Toulouse, UPS, CNRS, Toulouse, France*
HYE YOUNG KIM • *Department of Bioengineering, Swanson School of Engineering, University of Pittsburgh, Pittsburgh, PA, USA; Center for Vascular Research, Institute for Basic Science (IBS), Daejeon, Republic of Korea*
PAUL M. KULESA • *Stowers Institute for Medical Research, Kansas City, MO, USA; Department of Anatomy and Cell Biology, University of Kansas School of Medicine, Kansas City, KS, USA*
ANNA LABERNADIE • *Institute for Bioengineering of Catalonia (IBEC), The Barcelona Institute for Science and Technology (BIST), Barcelona, Spain*
ANNE-CLAIRE LAVIGNE • *Center for Integrative Biology (CBI), Laboratoire de Biologie Molé culaire des Eucaryotes (LBME), University of Toulouse, UPS, CNRS, Toulouse, France*
HERBERT LEVINE • *Center for Theoretical Biological Physics, Rice University, Houston, TX, USA; Department of Physics and Department of Bioengineering, Northeastern University, Boston, MA, USA*
CLAUDIA LINKER • *Randall Centre for Cell and Molecular Biophysics, King's College, London, UK*
ROSEMARY MANHIRE-HEATH • *School of BioSciences, University of Melbourne, Parkville, VIC, Australia*
ELISA MARTÍ • *Instituto de Biología Molecular de Barcelona, CSIC, Parc Científic de Barcelona, Barcelona, Spain*
FERNANDO MARTÍN-BELMONTE • *Program of Tissue and Organ Homeostasis, Centro de Biología Molecular Severo Ochoa (CSIC-UAM), Madrid, Spain*
ABDULL J. MASSRI • *Department of Biology, Duke University, Durham, NC, USA*
NAVRITA MATHIAH • *IRIBHM, Université Libre de Bruxelles, Brussels, Belgium*
DAVID R. MCCLAY • *Department of Biology, Duke University, Durham, NC, USA*
JAMES MCCOLL • *School of Biological Sciences, University of East Anglia, Norwich, UK; Chemistry Department, University of Cambridge, Cambridge, UK*
MARY CATHLEEN MCKINNEY • *Stowers Institute for Medical Research, Kansas City, MO, USA*
ISABELLE MIGEOTTE • *IRIBHM, Université Libre de Bruxelles, Brussels, Belgium; WELBIO, Wavre, Belgium*
GI FAY MOK • *School of Biological Sciences, University of East Anglia, Norwich, UK*
LUIS BRIONES MONTECINOS • *Program of Anatomy and Developmental Biology, BNI, CIMT, CENS, ICBM, Faculty of Medicine, University of Chile, Santiago, Chile*
DENISE J. MONTELL • *Molecular, Cellular, and Developmental Biology Department, University of California, Santa Barbara, Santa Barbara, CA, USA*
FATIMA MOUTAHIR • *Center for Integrative Biology (CBI), Laboratoire de Biologie Molé culaire des Eucaryotes (LBME), University of Toulouse, UPS, CNRS, Toulouse, France*
ANDREA MÜNSTERBERG • *School of Biological Sciences, University of East Anglia, Norwich, UK*
MICHAEL J. MURRAY • *School of BioSciences, University of Melbourne, Parkville, VIC, Australia*
HIROKI NAGAI • *International Research Centre for Medical Sciences, Kumamoto University, Kumamoto, Japan*
WALLIS NAHABOO • *IRIBHM, Université Libre de Bruxelles, Brussels, Belgium*

JOHN-POUL NG-BLICHFELDT • *MRC Laboratory of Molecular Biology, Francis Crick Avenue, Cambridge Biomedical Campus, Cambridge, UK*
M. ANGELA NIETO • *Instituto de Neurociencias (CSIC-UMH), Alicante, Spain*
ROBERT J. NORGARD • *Abramson Family Cancer Research Institute, Perelman School of Medicine at the University of Pennsylvania, Philadelphia, PA, USA*
DANIELLE PARK • *The Francis Crick Institute, London, UK*
MACARENA ALVAREZ-GARCILLAN PORTILLO • *Randall Centre for Cell and Molecular Biophysics, King's College, London, UK*
KATJA RÖPER • *MRC Laboratory of Molecular Biology, Francis Crick Avenue, Cambridge Biomedical Campus, Cambridge, UK*
CHRISTIAN ROUVIÈRE • *Centre de Biologie du Développement (CBD), Centre de Biologie Inté grative (CBI), Université de Toulouse, CNRS, UPS, Toulouse Cedex 09, France*
ERIK SAHAI • *The Francis Crick Institute, London, UK*
BECHARA SAYKALI • *IRIBHM, Université Libre de Bruxelles, Brussels, Belgium*
GORDANA SCEPANOVIC • *Department of Cell and Systems Biology, University of Toronto, Toronto, ON, Canada; Ted Rogers Centre for Heart Research, University of Toronto, Toronto, ON, Canada*
GEOFFREY R. SCHIEBINGER • *Department of Mathematics, University of British Columbia, Vancouver, BC, Canada*
GUOJUN SHENG • *International Research Centre for Medical Sciences, Kumamoto University, Kumamoto, Japan*
BEN Z. STANGER • *Abramson Family Cancer Research Institute, Perelman School of Medicine at the University of Pennsylvania, Philadelphia, PA, USA; Department of Medicine, Perelman School of Medicine at the University of Pennsylvania, Philadelphia, PA, USA; Department of Cell and Developmental Biology, Perelman School of Medicine at the University of Pennsylvania, Philadelphia, PA, USA*
WILLIAM TAYLOR • *Inserm, UMRS935 ESTeam Paris Sud, Malignant and Therapeutic Stem Cell Models, Villejuif, France*
ERIC THEVENEAU • *Centre de Biologie du Développement (CBD), Centre de Biologie Inté grative (CBI), Université de Toulouse, CNRS, UPS, Toulouse Cedex 09, France*
JEAN PAUL THIERY • *Guangzhou Regenerative Medicine and Health, Guangdong Laboratory, Guangzhou, China*
ERIK W. THOMPSON • *Faculty of Health, Institute of Health and Biomedical Innovation and School of Biomedical Sciences, Queensland University of Technology, Brisbane, QLD, Australia; Translational Research Institute, Brisbane, QLD, Australia*
XAVIER TREPAT • *Institute for Bioengineering of Catalonia (IBEC), The Barcelona Institute for Science and Technology (BIST), Barcelona, Spain; Unitat de Biofisica i Bioenginyeria, Facultat de Medicina, Universitat de Barcelona, Barcelona, Spain; Institució Catalana de Recerca i Estudis Avançats (ICREA), Barcelona, Spain; Center for Networked Biomedical Research on Bioengineering, Biomaterials and Nanomedicine (CIBER-BBN), Barcelona, Spain*
SHUBHAM TRIPATHI • *PhD Program in Systems, Synthetic, and Physical Biology, Rice University, Houston, TX, USA; Center for Theoretical Biological Physics, Rice University, Houston, TX, USA*
LEO VALON • *Department of Developmental and Stem Cell Biology, Institut Pasteur, UMR3738, CNRS, Paris, France*
LINGYU WANG • *Department of Biology, Duke University, Durham, NC, USA*
MO WENG • *School of Life Sciences, University of Nevada, Las Vegas, NV, USA*

GREGORY A. WRAY • *Department of Biology, Duke University, Durham, NC, USA*
JIANHUA XING • *Department of Computational and Systems Biology, University of Pittsburgh, Pittsburgh, PA, USA*

Part I

Perspectives on the Past and Future of EMT Research

Chapter 1

A Spectrum of Cell States During the Epithelial-to-Mesenchymal Transition

Erica J. Hutchins and Marianne E. Bronner

Abstract

The epithelial-to-mesenchymal transition (EMT) encompasses a complex cascade of events through which a cell transits to reduce its epithelial characteristics and become migratory. Classically, this transition has been considered complete upon loss of molecular markers characteristic of an "epithelial" state and acquisition of those associated with "mesenchymal" cells. Recently, however, evidence from both developmental and cancer EMT contexts suggest that cells undergoing EMT are often heterogeneous, concomitantly expressing both epithelial and mesenchymal markers to varying degrees; rather, cells frequently display a "partial" EMT phenotype and do not necessarily require full "mesenchymalization" to become migratory. Here, we offer a brief perspective on recent important advances in our fundamental understanding of the spectrum of cellular states that occur during partial EMT in the context of development and cancer metastasis.

Key words Epithelial-to-mesenchymal transition, Partial EMT, Cancer, Post-transcriptional regulation, Heterogeneity

1 Perspective

The complex process of the epithelial-to-mesenchymal transition (EMT) has reiterative roles throughout embryonic morphogenesis to generate the myriad of tissues and organ systems needed to form the adult organism. Developmental EMT is analogous in many ways to the EMT that occurs during cancer metastasis and invasion, requiring the same signaling pathways as well as often the same effector genes [1–3]. The process of EMT (both developmentally and during metastasis) requires cells to coordinately orchestrate a complex series of intra- and extra-cellular events to achieve separation from the epithelium and become migratory [3, 4]. In traditional descriptions of EMT, the transition was thought to have a specific, intrinsic end-point—the point at which a cell has completely lost all "epithelial" markers (e.g., E-cadherin) and gained a cadre of "mesenchymal" markers (e.g., N-cadherin,

Kyra Campbell and Eric Theveneau (eds.), *The Epithelial-to Mesenchymal Transition: Methods and Protocols*, Methods in Molecular Biology, vol. 2179, https://doi.org/10.1007/978-1-0716-0779-4_1,

vimentin) (reviewed in [1, 5]), thus completing its mesenchymalization. Recently, however, our perception of what constitutes "complete" EMT has expanded from this former all-or-nothing view to one that instead exists as a spectrum [3, 6].

During developmental EMT, mesenchymalization requires the loss of stable basement membrane adhesions and, subsequently, a loss of apico-basal polarity. Beyond this delamination event, there exists a wide array of "mesenchymal" phenotypes (based on stereotypical marker expression and morphology) across organisms and migratory cell types (reviewed in [1, 7]). Neural crest cells are a classic example of the ranges of developmental EMT; in the chick, these cells lose epithelial characteristics to delaminate from the neuroepithelium and acquire mesenchymal characteristics to migrate largely as individual cells. In *Xenopus,* neural crest cells undergo collective migration, wherein cells retain some epithelial characteristics and cell–cell connections as they migrate (reviewed in [8, 9]). Interestingly, the collective migration of lateral line cells in zebrafish and border cells in *Drosophila* exhibit even greater retention of epithelial characteristics (reviewed in [6]). This span of migration and EMT phenotypes observed across development supports the notion that there exists a sliding scale of EMT, where partial mesenchymalization may be a normal end-point for EMT in certain contexts.

The notion that this "partial" EMT is observed in developmental contexts is interesting in light of recent discoveries describing a similar partial EMT phenomenon in cancer cells [10–12]. Single-cell transcriptomics have uncovered tremendous heterogeneity in tumor cells, and importantly, revealed partial EMT programs through which these cells transit to become invasive [13]. Cancer cells that undergo partial EMT often exhibit collective migration [14], which imbues them with increased therapy resistance and generates tumor cells that, based on molecular signatures, retain greater plasticity and ability to transdifferentiate (reviewed in [2]). Thus, from a biomedical perspective, complete understanding of the spectrum of tumor EMT states, and the molecular factors driving these transitions, will likely provide novel and improved targets for cancer therapies.

The mechanisms underlying EMT have previously been largely attributed to "classical" transcription-dependent cellular changes (reviewed in [1, 15]); though undoubtedly important, there are additional modes of regulation beyond transcription—from post-transcriptional to post-translational—that are emerging as critical regulators of EMT (reviewed in [2, 3, 12]). Direct modification to or localization of EMT-related proteins, from transcription factors to adhesion molecules, has critical roles in regulating the progression of EMT programs [11, 16, 17]. In addition, post-transcriptional regulatory mechanisms mediated by miRNAs, lncRNAs, and RNA-binding proteins are increasingly implicated in the regulation of key transcription factors that drive both developmental and cancer EMT [2, 15, 18–22]. Furthermore, post-transcriptional regulation is often coordinated for functionally related targets; for example, multiple miRNAs recently have been shown to cooperate to combinatorially regulate epithelial and mesenchymal markers to facilitate EMT [23]. With these many modes of EMT regulation beyond transcription, it will become increasingly important, as we continue to define contexts of partial EMT, to fully understand how regulation beyond transcriptional control participates in the dynamics of partial EMT.

As the impact of partial EMT in the context of both development and disease becomes clearer, new questions begin to emerge: Are these partial EMT states controlled cell autonomously? Is partial EMT a "halted" program or an alternative program? What molecular signatures underlie distinct states within the spectrum of partial EMT? Mathematical modeling approaches are providing insight into the factors likely to be involved in partial EMT [24, 25]. However, these approaches are limited by the knowledge of which combination of factors, most of which are currently unknown, define the spectrum of partial EMTs. Single-cell RNA-sequencing (scRNA-seq) is currently the best technology to resolve the heterogeneity of states in between an epithelial and a mesenchymal cell state [26]. The use of scRNA-seq in both developmental and cancer EMT contexts, in combination with multiplex fluorescent in situ hybridization for spatial information, will provide new molecular markers for the various partial EMTs. Understanding the molecular mechanisms underpinning the transitional states across partial EMT will begin to illuminate how these states are intrinsically controlled. Furthermore, understanding the molecular signatures defining intermediate states across partial EMT will enhance modeling approaches to better predict the dynamics of transition states [27], which will better define the partial EMT programs.

The importance of partial EMT in both development and cancer is becoming increasingly well established. As technologies and methods emerge to characterize and define the heterogeneity of states comprising the spectrum of partial EMT states, this information will provide new insights into important and complex developmental processes, as well as provide novel targets for cancer therapies.

Acknowledgments

We thank members of the Bronner lab for helpful discussions. The authors are supported by the National Institutes of Health K99 DE028592 (E.J. Hutchins), R01DE027538 and P01 HD037105 (M.E. Bronner).

References

1. Thiery JP, Acloque H, Huang RY et al (2009) Epithelial-mesenchymal transitions in development and disease. Cell 139:871–890
2. Aiello NM, Kang Y (2019) Context-dependent EMT programs in cancer metastasis. J Exp Med 216:1016–1026
3. Nieto MA, Huang RY, Jackson RA et al (2016) Emt: 2016. Cell 166:21–45
4. Thiery JP, Sleeman JP (2006) Complex networks orchestrate epithelial-mesenchymal transitions. Nat Rev Mol Cell Biol 7:131–142
5. Kalluri R, Weinberg RA (2009) The basics of epithelial-mesenchymal transition. J Clin Invest 119:1420–1428
6. Campbell K, Casanova J (2016) A common framework for EMT and collective cell migration. Development 143:4291–4300
7. Campbell K (2018) Contribution of epithelial-mesenchymal transitions to organogenesis and cancer metastasis. Curr Opin Cell Biol 55:30–35
8. Theveneau E, Mayor R (2012) Neural crest delamination and migration: from epithelium-to-mesenchyme transition to collective cell migration. Dev Biol 366:34–54
9. Hutchins EJ, Kunttas E, Piacentino ML et al (2018) Migration and diversification of the vagal neural crest. Dev Biol 444(Suppl 1): S98–S109
10. Pastushenko I, Blanpain C (2019) EMT transition states during tumor progression and metastasis. Trends Cell Biol 29:212–226
11. Aiello NM, Maddipati R, Norgard RJ et al (2018) EMT subtype influences epithelial plasticity and mode of cell migration. Dev Cell 45 (681–695):e684
12. Lu W, Kang Y (2019) Epithelial-mesenchymal plasticity in cancer progression and metastasis. Dev Cell 49:361–374
13. Pastushenko I, Brisebarre A, Sifrim A et al (2018) Identification of the tumour transition states occurring during EMT. Nature 556:463–468
14. Grigore AD, Jolly MK, Jia D et al (2016) Tumor budding: the name is EMT. Partial EMT. J Clin Med 5:E51
15. Lamouille S, Xu J, Derynck R (2014) Molecular mechanisms of epithelial-mesenchymal transition. Nat Rev Mol Cell Biol 15:178–196
16. Chang R, Zhang Y, Zhang P et al (2017) Snail acetylation by histone acetyltransferase p300 in lung cancer. Thorac Cancer 8:131–137
17. Chen A, Wong CS, Liu MC et al (2015) The ubiquitin ligase Siah is a novel regulator of Zeb1 in breast cancer. Oncotarget 6:862–873
18. Sanchez-Vasquez E, Bronner ME, Strobl-Mazzulla PH (2019) Epigenetic inactivation of miR-203 as a key step in neural crest epithelial-to-mesenchymal transition. Development 146:dev171017
19. Ma L, Young J, Prabhala H et al (2010) miR-9, a MYC/MYCN-activated microRNA, regulates E-cadherin and cancer metastasis. Nat Cell Biol 12:247–256
20. Li J, He M, Xu W et al (2019) LINC01354 interacting with hnRNP-D contributes to the proliferation and metastasis in colorectal cancer through activating Wnt/beta-catenin signaling pathway. J Exp Clin Cancer Res 38:161
21. Zhou Y, Chang R, Ji W et al (2016) Loss of scribble promotes snail translation through translocation of HuR and enhances cancer drug resistance. J Biol Chem 291:291–302
22. Nagaoka K, Fujii K, Zhang H et al (2016) CPEB1 mediates epithelial-to-mesenchyme transition and breast cancer metastasis. Oncogene 35:2893–2901
23. Cursons J, Pillman KA, Scheer KG et al (2018) Combinatorial targeting by MicroRNAs co-ordinates post-transcriptional control of EMT. Cell Syst 7(77–91):e77
24. Bocci F, Jolly MK, Tripathi SC et al (2017) Numb prevents a complete epithelial-mesenchymal transition by modulating notch signalling. J R Soc Interface 14:20170512
25. He P, Qiu K, Jia Y (2018) Modeling of mesenchymal hybrid epithelial state and phenotypic transitions in EMT and MET processes of cancer cells. Sci Rep 8:14323
26. Maclean AL, Hong T, Nie Q (2018) Exploring intermediate cell states through the lens of single cells. Curr Opin Syst Biol 9:32–41
27. Jolly MK, Tripathi SC, Somarelli JA et al (2017) Epithelial/mesenchymal plasticity: how have quantitative mathematical models helped improve our understanding? Mol Oncol 11:739–754

Chapter 2

Perspective on Epithelial-Mesenchymal Transitions in Embryos

David R. McClay

Abstract

The epithelial-mesenchymal transition (EMT) is a key process required for building the early body plan of metazoa. It involves coordinated and precisely timed changes in multiple cell processes such as de-adhesion, motility, invasion, and cell polarity. While much has been learned about how embryos deploy epithelial-mesenchymal transitions since Betty Hay named the process decades ago, a number of things are still not well understood. Here I will discuss some of the big questions that remain, including how is all of this controlled, how does each of the cell biological events work, and how are they so nicely coordinated with one another?

Key words Gastrulation, Epithelial-mesenchymal transition, Morphogenesis, Sea urchin

1 Introduction

Morphogenetic movements at gastrulation reshape the embryos of multicellular animals. Among those movements, an epithelial-mesenchymal transition (EMT) removes mesoderm cells from an epithelium and places them in between the ectoderm and endoderm. Other embryonic cell types also go through EMTs at various stages of development in many animals. The EMT process is thus essential for building the early body plan of metazoa, and because of this it is important to learn how the process works at a molecular level.

A number of properties of developmental EMTs are not well understood. As part of the EMT, the cells become motile while still in the epithelium. That motility results in cell shape changes, helps the cell penetrate through the basement membrane, and mechanically contributes to the loss of adhesion as the cell leaves the adherens junction. The invasion through the basement membrane involves at least a partial remodeling of the basement membrane plus a determined polarized movement of the cell. The cell

Kyra Campbell and Eric Theveneau (eds.), *The Epithelial-to Mesenchymal Transition: Methods and Protocols*, Methods in Molecular Biology, vol. 2179, https://doi.org/10.1007/978-1-0716-0779-4_2,

de-adheres from the adherens junction. Movies of the process show that to de-adhere, the cells stretch, mechanically, pulling away from the adherens junction before that junctional connection is lost. Release from the adherens junction occurs late in the EMT, and as soon as the cell "tail" is released, the membrane containing cadherin is endocytosed while membrane bearing mesenchymal determinants are inserted via exocytosis. The cell, now in the interstitial spaces, rapidly demonstrates a mesenchymal phenotype and expresses very different cell surface markers relative to its earlier state as an epithelial cell. The big questions then are to ask how all of this is controlled, how does each of the cell biological events work, and how they are so nicely coordinated with one another. The approach of my lab over the years has changed with the technologies available but with each iteration, the question has always been: How does it work?

Our first observation was made 35 years ago. We had begun a study of sea urchin skeletogenic cells which can be seen to undergo an EMT beginning about 9 h after fertilization. The sea urchin embryo is transparent, and the skeletogenic cells are the first cells to engage in gastrulation movements so the EMT is easy to see. Methods had been developed to separate the skeletogenic cells from the other cells of the embryo, and we noticed that at 9 h post fertilization the skeletogenic precursor cells, now in culture, began to move. They changed shape and began to crawl on the substrate. In other words, they behaved as if they were still part of the embryo and they began those movements autonomously, at the same time as skeletogenic cells in vivo started EMT. We developed a quantitative adhesion assay using a centrifuge to measure the force needed to remove the cells from a substrate. Rachel Fink, now a professor at Mt. Holyoke, and I observed that during the 45 min period of the EMT the skeletogenic cells lost their affinity for other cells and gained an affinity for extracellular matrix [1]. In other words, the EMT was accompanied by a dramatic and quantifiable adhesion change.

The next obvious question was to ask what molecules were responsible for those adhesion changes? At the time the field was busy identifying cell-cell and cell substrate adhesion molecules. Masatoshi Takeichi had discovered cadherins [2], and Richard Hynes, Clayton Buck, and others had identified integrins [3, 4], so our efforts shifted to identifying as many of those molecules as possible. The field also was embracing molecular biology so all of my graduate students from that time onwards became familiar with molecular technologies. We identified cadherins, integrins, a number of basement membrane proteins and put each of them to the test in an effort to learn which molecules participated in the EMT change. Those efforts were productive and showed that as the skeletogenic cells went through EMT they lost an adhesive affinity to cadherin and gained an affinity for extracellular matrix components via integrins. Cadherin was removed from the cell surface as the skeletogenic cell went through EMT and was immediately endocytosed [5, 6].

While these efforts were productive, we were bothered by the approach. It seemed as if we were stuck playing hunches about molecules and it did not seem like we were gaining a broad understanding of how the EMT process was coordinately regulated. We did not understand the detailed cell biology behind the changes, and we did not understand what was directly controlled by the regulatory apparatus. Productive genetic approaches were being taken in other embryonic systems, most notably by Maria Leptin in *Drosophila* [7], but the sea urchin lacked the use of genetics as a tool for discovery. So, my choices were either to switch to a more tractable genetic model, or to find a way to move forward using the sea urchin system.

At about that time new discoveries in the lab sent us down a different path. We began identifying genes that regulated developmental specification. Because of that, Eric Davidson at Caltech, with a longstanding focus on mechanisms of gene regulation, asked if I might be interested in joining him and his lab in identifying the gene regulatory network (GRN) that established early specification of the sea urchin embryo. I thought about it overnight and realized that if we were able to identify the regulatory network governing specification we would also be generating the regulatory apparatus controlling the EMT. With that goal in mind I called Eric back and eagerly entered into an entirely new direction for the lab. Other members of the sea urchin community quickly joined in and over the next 10 years our understanding of the complexity of early specification grew enormously. The skeletogenic cell lineage led the way with a detailed regulatory sequence [8].

With a fairly detailed GRN model in hand we returned to the EMT problem. Could we use the GRN to understand how the EMT is regulated and coordinated? Our strategy was fairly simple. Since we knew when the EMT began we started by asking what regulatory events occurred in the 2 h prior to launching the EMT. We systematically knocked down each of the transcription factors in the GRN that were activated within that two-hour time period and asked whether, in the absence of that transcription factor, was the EMT crippled? To assess the outcome, we designed several simple assays, and used time-lapse microscopy to watch the cell behavior during the EMT. To focus on the EMT only, the assays employed fluorescently tagged skeletogenic cells in embryos that were unlabeled. We could also assess whether each EMT event was cell autonomous or if there might be a non-autonomous component to the process. Lindsay Saunders, a graduate student, and I did the analysis [9]. We learned that at least 10 transcription factors were involved in the EMT regulation. Some regulated only the motility. Others regulated the invasive process, and still others regulated the de-adhesion component. No single transcription factor was involved in all of the EMT behaviors we scored. Some of those behaviors were most interesting. For example, we found that FoxN2/3 was involved in acquisition of motility. FoxN2/3 KO cells remained in the epithelium but continued to remodel the basement membrane as part of the invasive function. Without motility, however, the FoxN2/3 KO cells failed to take advantage of their own remodeled matrix.

With at least a partial understanding of the EMT transcriptional control, the next goal was to ask what genes were controlled by each contributing transcription factor. Those downstream effector genes, we hypothesized, encoded the proteins that initiated and conducted the EMT mechanics. We knew this strategy would not yield all proteins involved in the EMT. For example, actin and myosin are present constitutively in the cell and surely are participants in the motility component, but other proteins, controlled by the motility transcriptional sub-circuit, we hypothesized, were the drivers of actin and myosin cytoskeletal contractions. Those proteins were our targets. We decided to first use RNA-seq and did a temporal profile starting with sampling 2 h before, then during and after the EMT. We also did the same profile with embryos in which *twist* or *snail* had been knocked down. The database was huge, but we could eliminate constitutively expressed RNAs since we wanted to find the genes activated by *twist* and *snail*. We were able to segment that population of genes into clusters that were activated at the predicted times relative to the EMT, and we could identify which of those genes were putative targets of Twist or Snail regulation. That allowed us to narrow the search from over 16,000 genes to fewer than 50, a manageable group to then work with. Upon further narrowing we arrived at a small number of proteins that we could functionally confirm as participants in the EMT.

While we were successful in pulling out a few genes involved in the de-adhesion phase of the EMT, and those studies continue, we wanted to access proteins in all component processes. A handicap with the RNA-seq profiles, however, is noise. Even though the skeletogenic cells went through EMT at a similar time, they were not perfectly synchronous, and also there were a number of cells in the database that were not involved in the EMT. With improvements in single cell-sequencing (sc-RNAseq) it is now possible to interrogate each cell of an embryo. For that reason, we launched a sc-RNAseq project, the methods of which are described in this Methods book. It still is too early to fill in the details about that approach, but it offers a way to identify a substantial fraction of the RNAs present in a cell at any given time and has the very nice advantage of being able to computationally project a temporal sequence of RNA appearance and disappearance. This along with the known GRN has enabled us to pry much more deeply into the workings of the EMT than we ever thought possible. Still, there are many questions to address.

The future holds many remaining questions. Along with gaining details of the de-adhesion, motility, invasion, and cell polarity changes, it is important to know how uniform or diverse the EMT process is in different tissues. As morphogenesis progresses in the sea urchin, five different cell types undergo an EMT hours apart from one another. In time-lapse movies, those EMTs look similar but from what little we have learned, each EMT is quite different from the others at a molecular level. What is needed is an in-depth analysis of many of these EMTs to learn whether in fact each EMT is indeed unique in its molecular control, both transcriptionally and with effector proteins, or, are there proteins that are universally and uniquely involved in EMTs? That universality seems to be the target of cancer EMT studies because it would provide a focal point for attempts to inhibit that EMT that initiates metastasis. In embryos, such a protein would also be valuable for many reasons, ranging from evolutionary mechanisms to targeted perturbations in studies of morphogenesis.

Another big question in morphogenesis is how the EMT is timed and coordinated. As indicated above, five different EMTs occur during gastrulation. Each starts at a fairly precise time, and each is coordinated such that all the cell biological changes of that EMT occur harmoniously. An ability to follow events in single cells as outlined above will help uncover these mechanisms. From the time-lapse movies of the process it appears as though there are a number of possibilities for mechanical sensing, and if that is the case there will be opportunities for gaining an understanding of how such sensing relays information to other components of the EMT.

Even though the remaining questions are quite numerous, much has been learned about how embryos deploy epithelial-mesenchymal transitions since Betty Hay named the process decades ago. Since then progress has paralleled advances in molecular technologies. With many new technologies being introduced at a rapid pace, the next decade should be rich with discoveries about how EMTs work in embryos.

Acknowledgments

Thanks to the many students and postdocs in the McClay lab who contributed so much over the years to advance our understanding of the EMT. Thanks also to the NIH for supporting this work: RO1 HD14483 (to DRM) and PO1 HD 37105 (Project 2 to DRM).

References

1. Fink RD, McClay DR (1985) Three cell recognition changes accompany the ingression of sea urchin primary mesenchyme cells. Dev Biol 107:66–74
2. Takeichi M (1988) The cadherins: cell-cell adhesion molecules controlling animal morphogenesis. Development 102:639–655
3. Hynes RO (1987) Integrins: a family of cell surface receptors. Cell 48:549–554
4. Damsky CH, Richa J, Solter D, Knudsen K, Buck CA (1983) Identification and purification of a cell surface glycoprotein mediating intercellular adhesion in embryonic and adult tissue. Cell 34:455–466
5. Miller JR, McClay DR (1997) Characterization of the role of cadherin in regulating cell adhesion during sea urchin development. Dev Biol 192 (2):323–339
6. Hertzler PL, McClay DR (1999) alphaSU2, an epithelial integrin that binds laminin in the sea urchin embryo. Dev Biol 207(1):1–13
7. Leptin M (1995) Drosophila gastrulation: from pattern formation to morphogenesis. Annu Rev Cell Dev Biol 11:189–212
8. Oliveri P, Tu Q, Davidson EH (2008) Global regulatory logic for specification of an embryonic cell lineage. Proc Natl Acad Sci U S A 105 (16):5955–5962. https://doi.org/10.1073/pnas.0711220105
9. Saunders LR, McClay DR (2014) Sub-circuits of a gene regulatory network control a developmental epithelial-mesenchymal transition. Development 141(7):1503–1513. https://doi.org/10.1242/dev.101436

Chapter 3

EMT, One of Many Morphological Transitions in Cellular Phase Space

Denise J. Montell

Abstract

The epithelial to mesenchymal transition (EMT) is an enticingly simple mechanism that converts stationary epithelial cells into migratory mesenchymal cells. EMT is meant to provide a unified explanation for phenomena as complex as gastrulation and metastasis. However, cell movements turn out to be diverse, and many are collective. Cells commonly migrate in clusters, strands, sheets, elongating tubes, or in fluid-like masses. Moreover, plenty of cells move without activating the EMT program. Here I propose that EMT can be understood as one of many types of transitions in a broader landscape—or phase space—of cell morphologies and behaviors. Throughout biology, and at multiple scales, complexity arises from the combinatorial deployment of simple, modular components. I propose that diversity of cell shapes and behaviors similarly arises from combinatorial use of modular biomechanical properties.

Key words Epithelial to mesenchymal transition, Collective cell migration, Cell forces, Mechanical properties, Cell shape and behavior

1 The Enticing EMT

Simple ideas are enticing. Take the DNA double helix. Just like that, the simple concept of base pairing explained all kinds of puzzling phenomena such as replication of genetic information, built-in redundancy, and the known but perplexing perfect match between the number of adenine and thymine bases and between guanine and cytosine. Base pairing neatly explains it all. Or consider action potentials. The electricity that generates our ideas, thoughts, and feelings boils down to the mathematical simplicity of the Nernst equation. Simple ideas that explain complex phenomena are the pinnacle in science.

The epithelial to mesenchymal transition (EMT) is an enticingly simple idea meant to explain some of the most complex phenomena in all of biology, like gastrulation and tumor metastasis. This simplicity has been the source of the EMT's popularity over

Kyra Campbell and Eric Theveneau (eds.), *The Epithelial-to Mesenchymal Transition: Methods and Protocols*, Methods in Molecular Biology, vol. 2179, https://doi.org/10.1007/978-1-0716-0779-4_3,

the past 30 years. In its strongest (simplest) form, the EMT hypothesis is that a simple transcriptional switch drives a dramatic change in cellular morphology and behavior, from a stationary epithelium into individual, motile, mesenchymal cells [1]. EMT transcription factors include Twist and Snail (first identified and named based on their Drosophila gastrulation phenotypes), and Zebs 1 and 2. The wholesale EMT includes loss of apicobasal polarity in favor of leading/lagging migratory polarity, complete loss of cell-cell junctions, expression of mesenchymal instead of epithelial intermediate filament proteins and adhesion molecules, expression of matrix metalloproteinases, and resistance to apoptosis.

Since cancer cells coopt normal developmental mechanisms, an elegant explanation for the acquisition of migratory and invasive characteristics by carcinoma cells (cancers of epithelial origin) is that they reactivate the embryonic EMT program. This simple idea was satisfying, and a great deal of evidence accumulated in support of it [1–4]. Excitement grew that inhibition of EMT would prevent metastasis and thereby render most cancers curable.

2 Unwelcome Complexity

Cracks began to appear in the foundation of this simple line of thinking. For example, pathologists rarely if ever observe individual, mesenchymal cells in or around tumors. In principle this might be because EMT is a transient state that stimulates motility, making it hard to catch in fixed samples. But technical improvements in live imaging of embryonic and organ development also revealed that the majority of cell movements do not involve complete EMT. Rather, many movements are collective, meaning that some degree of cell-cell contact and apicobasal polarity are maintained [5, 6]. Cells commonly migrate in clusters, strands, sheets, elongating tubes, or in fluid-like masses. Plenty of migratory cells move without activating the EMT program, and in some cases "EMT transcription factors" such as Twist promote collective types of motility independently of EMT [7]. Tumor cells, like normal cells, are found in all of kinds of arrangements too. Moreover, some circulating tumor cell clusters are 60–100 times more effective than single cells at seeding metastases in animal models [8–10]. Mother Nature seems determined to defy our zealous pursuit of simplicity.

Out of necessity then, the EMT concept morphed into partial EMT, transient EMT, and partial/transient EMT. The idea emerged that a spectrum of metastable intermediate states exists along a continuum between the extremes of epithelial and mesenchymal morphologies [4, 11]. The need to account for diverse cell morphologies and behaviors led to a somewhat nebulous concept of epithelial plasticity [12, 13].

The more we look, the more diverse cell morphologies and migration modes appear. Single cells can move not only in a fibroblast-like, mesenchymal mode but also in a blebbing manner sometimes referred to as amoeboid or myeloid. Cells can transition between these different modes as well. To the EMT, we must add the epithelial to amoeboid transition, the amoeboid to mesenchymal transition, the mesenchymal to amoeboid transition, etc. [14]. Some collective cell migrations, such as border cell migration, which we study in my laboratory, do not involve any form of an EMT [15]. These six epithelial cells do not express Twist or Snail. They maintain apicobasal polarity even while acquiring leading/lagging migratory polarity. They require increased rather than decreased E-cadherin expression to move. They remain connected to one another by adherens and gap junctions. One might be tempted to dismiss these cells as a quirk of fly biology, except that they look an awful lot like the tumor cell clusters that seed metastases [8, 10].

3 Moving Forward

So how should we think about cell motility in normal development and in cancer? It is more complex than we originally envisioned. But is it infinitely complex and incomprehensible? I don't think so. Everywhere we look, nature builds complexity from simple, modular components. To return to DNA, our remarkably complex genetic information, for example, is encoded by just four nucleotides, used in combination. Genes themselves are built from modular exons. The byzantine regulation of gene expression that drives embryonic development is accomplished by combinatorial use of simple, modular enhancers. Tens of thousands of proteins are built from just 20 amino acids. Complex organs and animals are built from simpler organizational units like the nephron or vertebrae. Throughout biology, and at multiple scales, complexity arises from the combinatorial deployment of simple, modular components. Kind of like Legos.

Perhaps then, nature also builds complexity and diversity in cellular morphology and behavior from combinations of modules. But what modules? Ultimately the shapes and behaviors of cells are determined by forces. So one idea is that cellular morphologies and behaviors are produced by combinations of simple and modular forces [16](Fig. 1). Consider four force-generating properties of cells: protrusion, contractility, cell-cell cohesion, and cell-matrix adhesion. I propose that the shape of a single cell and the arrangement of a group of cells are determined in large measure by the relative strengths and localizations of these four forces as illustrated in Fig. 1.

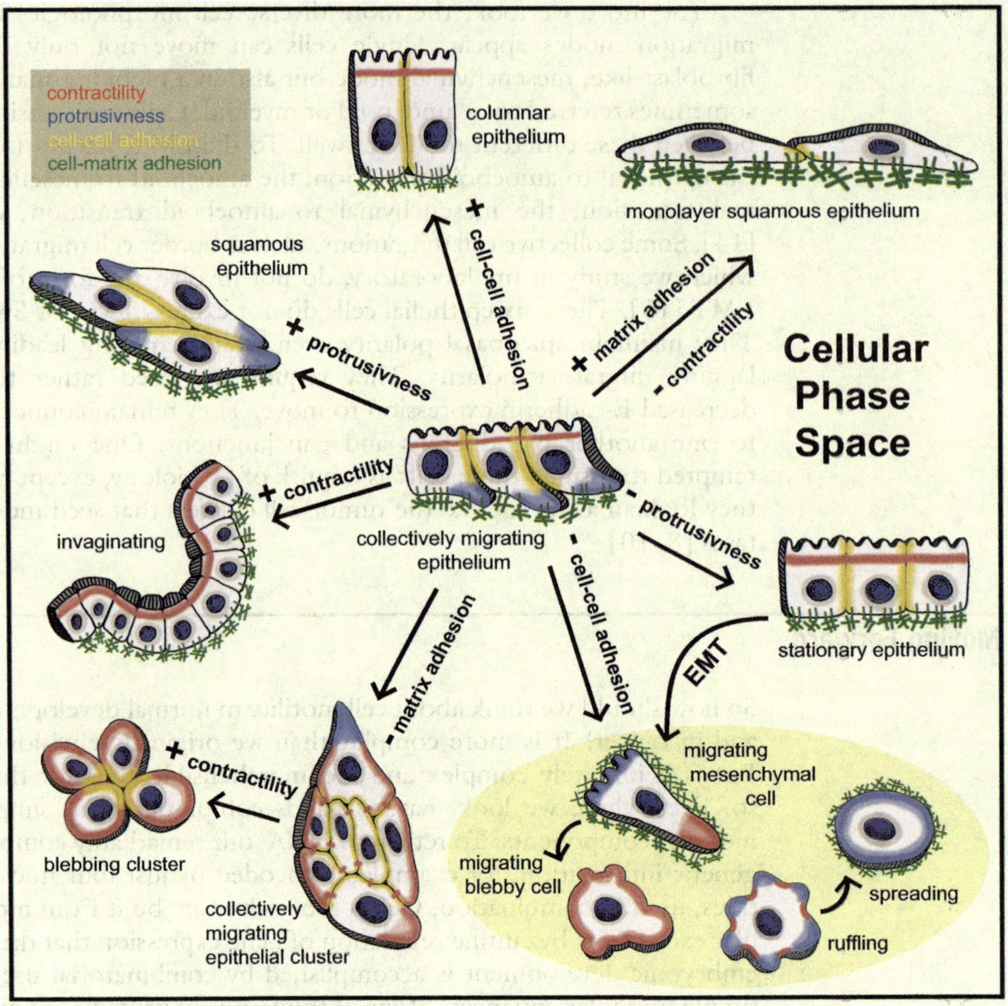

Fig. 1 The shape of a single cell and the arrangement of a group of cells are determined by the relative strengths and localizations of the four force-generating properties of cells: contractility, protrusiveness, cell-cell adhesion, and cell-matrix adhesion

An attractive feature of this idea is that, in principle, it can account for a great variety of cell shapes, arrangements, and behaviors. It allows for transitions between any of them, simply by tuning one or more of the four parameters. The EMT appears as one of many possible transitions within this multi-dimensional space (Fig. 1).

To test whether this qualitative idea works in practice, Hannezo et al. undertook the ambitious project of generating a mathematical formalization of this conceptual framework [17]. They generated a three-dimensional model of epithelia and found that tuning contractility, cell-cell and cell-matrix adhesion yields the predicted transitions between columnar, cuboidal, and squamous cell shapes.

Turn up apical actomyosin contractility and the model produces epithelial invagination [17]. This study did not attempt to model dynamics, or transitions of epithelia into individual cells, or cell motility, or transitions between migratory modes. Modeling of such transitions would therefore be an interesting future goal for physicists and mechanical engineers interested in biology, in order to continue to test how well this conceptual framework explains diverse cell morphologies and behaviors.

Are these mechanical properties actually modular? That is, can they be altered independently of one another? Mostly yes. The families of molecules that endow cells with these properties are known. Cell adhesion molecules such as classical cadherins provide cell-cell cohesion, which can be tuned by expression levels and/or post-translational modifications. Integrins are the dominant cell-matrix adhesion receptors, and it has long been known that an intermediate level of cell-matrix adhesion is optimal for fibroblasts to migrate [18]. Too little adhesion and cells slip, like trying to walk on ice; too much and they stick. Actomyosin endows cells with contractility, and this property is primarily tuned by levels and localization of Rho GTPase activity. Moreover, high Rho/myosin activity is sufficient to convert a variety of individual cells and cell clusters from mesenchymal to amoeboid migration modes [19–21]. This applies to tumor cells and normal cells, in vitro and in vivo. Actin polymerization is the main generator of protrusive force, usually downstream of the GTPases Rac and Cdc42. Increasing Rac activity converts cells from amoeboid to mesenchymal motility. Therefore these four mechanical properties can be tuned independently, leading to transitions in shape and behavior. Encouragingly, others have converged onto similar ideas [14, 22, 23].

Though admittedly not as simple as the original EMT hypothesis, this framework of a cellular "phase space" is useful because it synthesizes numerous, disparate observations into a unified concept. Like learning that the earth is not the center of the solar system, and that there are many solar systems and galaxies, sometimes a deeper understanding requires recognition of greater complexity. Nevertheless, recognizing that simple, modular components can build diverse and complex forms is an enticingly powerful idea.

References

1. Thiery JP et al (2009) Epithelial-mesenchymal transitions in development and disease. Cell 139:871–890
2. Lim J, Thiery JP (2012) Epithelial-mesenchymal transitions: insights from development. Development 139:3471–3486
3. Lambert AW et al (2017) Emerging biological principles of metastasis. Cell 168:670–691
4. Nieto MA et al (2016) EMT: 2016. Cell 166:21–45
5. Friedl P, Gilmour D (2009) Collective cell migration in morphogenesis, regeneration and cancer. Nat Rev Mol Cell Biol 10:445–457
6. Derynck R, Weinberg RA (2019) EMT and cancer: more than meets the eye. Dev Cell 49:313–316

7. Shamir ER et al (2014) Twist1-induced dissemination preserves epithelial identity and requires E-cadherin. J Cell Biol 204:839–856
8. Cheung KJ et al (2016) Polyclonal breast cancer metastases arise from collective dissemination of keratin 14-expressing tumor cell clusters. Proc Natl Acad Sci U S A 113: E854–E863
9. Aceto N et al (2014) Circulating tumor cell clusters are oligoclonal precursors of breast cancer metastasis. Cell 158:1110–1122
10. Cheung KJ, Ewald AJ (2016) A collective route to metastasis: seeding by tumor cell clusters. Science 352:167–169
11. Kröger C et al (2019) Acquisition of a hybrid E/M state is essential for tumorigenicity of basal breast cancer cells. Proc Natl Acad Sci U S A 116:7353–7362
12. Thiery JP, Chopin D (1999) Epithelial cell plasticity in development and tumor progression. Cancer Metastasis Rev 18(1):31–42
13. Beerling E et al (2016) Plasticity between epithelial and mesenchymal states unlinks EMT from metastasis-enhancing stem cell capacity. Cell Rep 14:2281–2288
14. Friedl P, Wolf K (2010) Plasticity of cell migration: a multiscale tuning model. J Cell Biol 188:11–19
15. Montell DJ et al (2012) Group choreography: mechanisms orchestrating the collective movement of border cells. Nat Rev Mol Cell Biol 13:631–645
16. Montell DJ (2008) Morphogenetic cell movements: diversity from modular mechanical properties. Science 322:1502–1505
17. Hannezo E et al (2014) Theory of epithelial sheet morphology in three dimensions. Proc Natl Acad Sci U S A 111:27–32
18. DiMilla PA et al (1991) Mathematical model for the effects of adhesion and mechanics on cell migration speed [published erratum appears in Biophys J 1991 Oct;60(4):983]. Biophys J 60(1):15–37
19. Yamazaki D et al (2009) Involvement of Rac and Rho signaling in cancer cell motility in 3D substrates. Oncogene 28:1570–1583
20. Panková K et al (2010) The molecular mechanisms of transition between mesenchymal and amoeboid invasiveness in tumor cells. Cell Mol Life Sci 67:63–71
21. Mishra AK et al (2019) Coordination of protrusion dynamics within and between collectively migrating border cells by myosin II. Mol Biol Cell 30(19):2490–2502. https://doi.org/10.1091/mbc.E19-02-0124
22. Lämmermann T, Sixt M (2009) Mechanical modes of "amoeboid" cell migration. Curr Opin Cell Biol 21:636–644
23. Murrell M et al (2015) Forcing cells into shape: the mechanics of actomyosin contractility. Nat Rev Mol Cell Biol 16:486–498

Chapter 4

Are You Interested or Afraid of Working on EMT?

M. Angela Nieto

Abstract

When referring to the epithelial-to-mesenchymal transition (EMT), readers are familiar with sentences alluding to its pivotal role both in embryonic development and in disease. Following that argument, usually there is a point on the importance of studying the process and the impact it has on the design of therapeutic strategies. However, it is also very common to find arguments on how the EMT is very difficult to tackle, being a somehow obscure and complex process, where the field cannot reach universal conclusions, particularly in pathological contexts. Even worse, it is sometimes defined as a process that cannot be described with universal markers, making it therefore very difficult for cancer studies, where there is a need to use optimal animal models and stratify patients for differential therapeutic strategies. In the face of all this, the question is whether you have been frightened off working on pathological EMTs, or even if you are not interested anymore and would prefer waiting till the field reaches a steady state of robust knowledge. Do not be afraid and be interested now. It only involves being more plastic, like the EMT itself.

Key words Epithelial-to-mesenchymal transition, Morphogenesis, Cancer, Fibrosis, Stemness

When Kyra Campbell and Eric Theveneau invited me to participate in this issue, I initially thought that I was not ready to write yet another review on EMT. Then, after considering that I had been working on this exciting and more than challenging process for 25 years, I thought I could write a short informal view-point, mainly as a reflection on the ins and outs that we encounter on a daily basis in the lab. I will refer to several points, which may help think about the EMT in a very positive manner, while still posing the challenges ahead. First of all, to set out the scene, refer to excellent recent reviews on EMT [1–5].

1 The Origin of the EMT: The Importance of Evolution

One way to help clarify the importance of the EMT is acknowledging its origin and function in physiology. It is a process that appeared and was fixed in evolution for cells that were born far from

Kyra Campbell and Eric Theveneau (eds.), *The Epithelial-to Mesenchymal Transition: Methods and Protocols*, Methods in Molecular Biology, vol. 2179, https://doi.org/10.1007/978-1-0716-0779-4_4,

their final destinations, a solution that was found to help cells move leaving behind their tissue of origin [6]. Importantly, all metazoans throughout evolution have used EMT-like processes as the mechanism for cell dissemination and survival. *Thus, it seems very reasonable that this mechanism is also used by cancer cells, and can be regarded as a pathological reactivation of a crucial developmental program.*

2 The Intrinsic Complexity/Heterogeneity: Do Not Be Afraid

One of the main reasons why some researchers are disillusioned and not very interested in studying EMT is the complexity of the process. Indeed, EMT is complex: it is induced by multiple extracellular cellular signals; the signals converge in a plethora of transcription factors (EMT-TFs) and how these factors implement the EMT program is not so evident. Perhaps the best way to look at it is to think of commonalities, just going back to the main purpose of the process: cell dissemination and survival.

The EMT is all about how to achieve cell dissemination in an efficient manner, with the requirement that cells reach their final destination. In other words, that cells survive till the end of the journey. How this is implemented depends on the cellular context. And here there is the second unwanted word: context-dependent. For many researchers, this may just indicate that we do not know much about the process. The answer is however much more interesting. Evolution has provided us with the richness of biology which, in this particular context, implies the existence of many EMT transcription factors that trigger the expression of a plethora of target genes. How these EMT-TFs operate is still a matter of active investigation, with both commonalities and specificities to the programs they can implement individually or in combination. This enables cells to deal with the different topologies and microenvironments that they encounter while disseminating, both in development and cancer. As a result of that, both embryonic and cancer cells express combinations of EMT-TFs, an EMT-TF code. Importantly, cell behavior will be the outcome of the specific EMT-TF code implemented by the signaling inputs. In cancer, this heterogeneity adds another level of complexity to that provided by mutations and genomic instability, with implications for the design and interpretation of experiments in the field. For instance, loss of function analysis for a single EMT factor cannot be taken as surrogate of the whole EMT process, as different developing tissues and carcinomas use different EMT programs [7, 8]. *Thus, do not be afraid, pay attention to the commonalities, but characterize the system by examining the expression code in the tissue/tumor of interest.*

3 The Lack of Universal Markers and Some Misleading Concepts

In general terms, one process is considered to be well defined when specific markers are described, and one drawback in the study of EMT is the lack of universal specific markers. During EMT, epithelial genes (E) are downregulated and mesenchymal markers (M) are activated. Their expression or better, that of the proteins they encode, is used as markers of EMT. The problem is that, in many studies, only one marker is used as representatives of the E or M phenotype, E-cadherin, and vimentin, respectively. And it is a problem because E-cadherin and vimentin are not universal or exclusive markers of the E and M states and also because the E and M phenotypes are not static entities. At this point it is worth going back to the actual term, EMT, the epithelial-to-mesenchymal *transition* and discuss some terms that can be misleading.

1. *Transition* versus *binary transformation*. The EMT is a transition, and it is not implemented as a binary decision. In agreement with that, the EMT community agreed already in 2003 at the first meeting of the Epithelial-Mesenchymal transition International Association, (TEMTIA), to use *transition* rather than the original term *transformation* [9]. Transition implies the sequential nature of the process and, although not that evident at that time, the existence of intermediate states (Fig. 1). Even though the latter is easy to understand, *in the field we have failed to convey this message of intermediate states*, leading to false expectations and misinterpretations, some of which I will discuss below.
2. *Cell behavior, not cell fate*. During embryonic development, cell behavior and fate occur in parallel, but they are independently regulated. The EMT is a program for cell behavior, not fate. When Snail, one of the most potent EMT-TFs was first identified in *Drosophila*, it was described as a mesodermal inducer [10], but it was later found in the majority of vertebrate tissues where cell delamination occurs in the developing embryo, regardless of their embryonic origin [11]. As such, *the role of EMT-TFs is related to changes in cell shape and motility rather than to specific cell fates*, and this needs to be considered in all physiological and pathological contexts.
3. *Kinetics and hierarchy*. In the search for EMT markers, temporal aspects and functional hierarchy are frequently not considered, and this can have an impact in the definition of the process. We need to consider at least three hierarchical levels: (1) the inducing signals (non-cell autonomous), (2) the EMT-TFs as gene expression regulators (cell autonomous), and (3) the final effectors: structural molecules and enzymes that altogether include those involved in adhesion, cytoskeletal

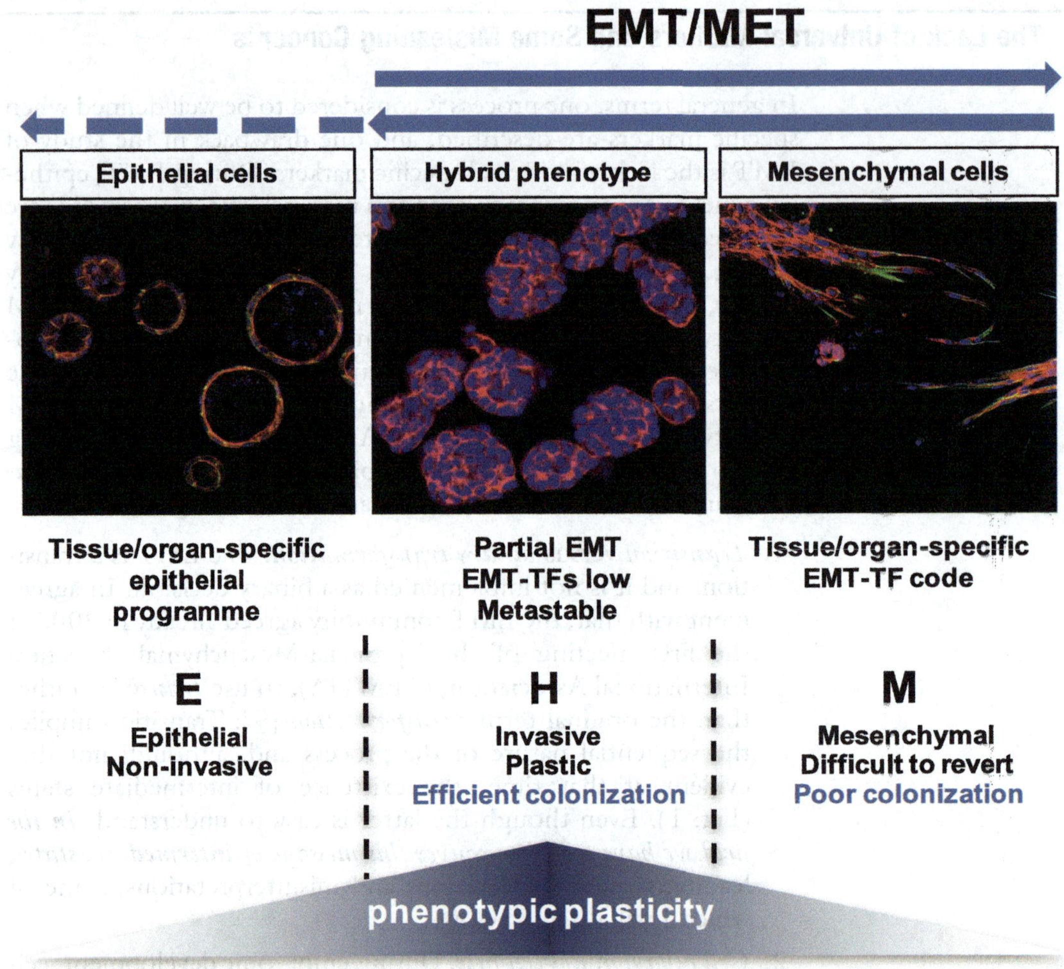

Fig. 1 The EMT and associated cellular traits. The cellular properties associated to the EMT and shown in the figure can be applied to both physiological (developmental) and pathological (cancer progression) scenarios. It is worth noting that cell plasticity after EMT does not necessarily take the exact reverse pathway. *E* Epithelial, *H* Hybrid, *M* Mesenchymal

organization and basement membrane or extracellular matrix degradation. None of the extracellular signals or the effectors are specific for the EMT program, and this is why it can be misleading when considered EMT markers. The EMT-TFs are more specific, but nevertheless, the EMT is better defined in terms of cell morphology and behavior than by gene expression. In any case, *it is not appropriate to consider EMT-TFs and effector molecules as equivalent readouts of the EMT*, as the former are regulators that can implement the whole program through the activation of multiple target (effector) genes.

4. *The pure E and M phenotypes.* One important misleading message is conveyed by the standard illustrations of the EMT. In the text book image that we all have in mind, often there is a

representation of the full epithelial (E) and the full mesenchymal (M) states, usually represented on one hand by a beautiful intestinal epithelial cell with microvilli and on the other hand by a fibroblast; black and white, without the richness of the gray shades, and biology is all about richness. In the vast majority of EMT processes, both in physiology and pathology, in development and in cancer, EMT phenotypes are intermediate, just meaning that they do not correspond to the pure E or pure M states as they appear in textbooks and reviews (Fig. 1). One paradigmatic example is that observed during the progression of renal fibrosis, where damaged epithelial cells undergo a partial EMT. This EMT program does not include the activation of invasive properties; cells downregulate markers of epithelial character, gain mesenchymal markers, activate the EMT-TFs and cannot work as renal epithelial cells. However, they do not delaminate from the renal tubules. Interestingly, these damaged cells release signals to promote both fibrogenesis and inflammation, the two hallmarks of fibrotic disease progression in a paracrine manner [12, 13]. Thus, in summary, and valid for all EMT contexts, *we should not expect to observe individual migratory cells to invoke the activation of an EMT program*.

5. *E-Cadherin loss* versus *E-Cadherin downregulation*. Another misleading message is that we have classically described the EMT as a process where E-cadherin is lost. As a matter of fact, in the vast majority of cases, if not all, E-Cadherin expression decreases during EMT, but cells maintain some degree of expression. Thus, cells are formally kept in an "E-cadherin positive state," which should not be considered as a failure to undergo EMT. The maintenance of *certain E-cadherin level is important to establish cell-cell contacts that are albeit frequently transient, help collective migration of both embryonic and cancer cells and, importantly, colonization* [14–18]. Conversely, it is worth noting that E-cadherin loss is not equivalent to an EMT, as it does not imply changes in mesenchymal markers and importantly, is not associated to crucial subprograms that are frequently accompanying EMT, including the activation of survival signaling cascades (see below). As such, E-cadherin loss in epithelial cells frequently leads to cell death, which would not happen when E-cadherin is downregulated in the course of an EMT, due to the concomitant acquired survival properties. This is in agreement with recent findings [18] and old observations suggesting that the formation of cadherin-mediated clusters helps survival of cancer cells in the circulation [19].
6. *EMT does not need to be massive*. Another confounding issue has been the difficulty in finding cells with EMT features in many

cancer studies. The main reason is that only a small population of cancer cells undergoes EMT-like processes. This is important, as bulk RNAseq for whole tumors may fail to detect an EMT program. The phenomenon may be focal, not massive, but importantly, *focal EMT is enough to fulfill the role of the program, cell dissemination*. EMT occurs mainly at the invasion front, and as a result of the interaction with the stroma. This is again very similar to the situation in developing tissue in the embryo, such as the delamination of the neural crest, occurring only from a small region of the neural tube.

7. *The confusing terms*. In the last few years, reviews and textbooks refrain from showing the two extreme E and M phenotypes alluded above. However, together with the acceptance of the transition process, a few terms associated with the non-extreme phenotypes have also led to some confusion. When referred to EMT, "metastable" alludes to the *stability/plasticity of a particular state*; "hybrid/intermediate" refers to the *actual cell phenotype*, with simultaneous expression of both epithelial and mesenchymal traits; and "partial" denotes the *EMT process*, alluding to a partial conversion in the spectrum from E to M or vice versa.

4 EMT Subprograms or Compatibility with Other Programs

The EMT is not only the downregulation of epithelial properties and the upregulation of mesenchymal character, as it also includes the ability to delaminate, move, and invade. Thus, the secretome of the cells changes dramatically to be able to break the basement membrane and the metabolism also changes. In addition, the EMT has been associated with other fundamental cellular programs, including survival, proliferation, and stemness [6]. If we think of the EMT as a global program to move cells from their place of origin to their destinations, it is easy to understand that cells need to arrive to their destination and therefore, be resistant to the death signals that they may encounter in their way. Similarly, to improve migration, cells attenuate proliferation, and the EMT has indeed been associated with a decrease in cell division.

1. *Survival*. The activation of survival signaling cascades associated with EMT has many implications in physiology and pathology. As mentioned above, the main purpose of developmental EMTs is cell dissemination and survival. Similarly, in cancer progression, the completion of the program is dependent on the survival of the disseminated cells to be able to form distant metastases. Importantly, the activation of resistance to cell death has enormous implications in therapeutic strategies,

as *the survival properties of cells that undergo EMT also make them resistant to chemo- and radiotherapy* [20].

2. *Stemness.* The EMT has been proposed to endow cells with stem cell-like properties, which again is fundamental in the embryo for cells to differentiate into different tissues once they reach their destination. In cancer, disseminated cells need to maintain the ability to form new tumors at distant sites. Out of the three associated properties, survival, low proliferation, and stemness, the latter has been more difficult to analyze and somehow also subjected to debate. After the seminal discovery of the relationship between EMT and stemness [21, 22], the idea was that there was a correlation between the mesenchymal phenotype and stemness, and this is not correct. The general idea is that the full mesenchymal phenotype is incompatible with stemness, in agreement with (1) embryonic stem cells (ESCs) being epithelial, and (2) the requirement for fibroblasts to undergo the EMT reverse process (MET) during reprogramming to iPSCs [23, 24]. This is also compatible with the finding the EMT-TFs are downregulated during colonization and metastasis formation [17, 25–28]. In summary, *stemness is better manifested at intermediate states in the EMT spectrum* and is not a fixed property of cells, but rather relies on cell plasticity. As such, EMT is important for delamination and local invasion, but it is not sufficient for metastatic colonization, that requires a certain degree of reversibility towards a more epithelial phenotype compatible with stemness. All this explains why *correlation between EMT and metastasis may be not direct, as it is plasticity that really matters. See* Ref. 29 for a discussion on cell heterogeneity in EMT and stemness.

5 The Challenge: Designing and Interpreting Optimal Models

One of the big challenges in the EMT in cancer field is the design of optimized animal models that can recapitulate the different steps of tumor progression while allowing cell plasticity and behavior to be followed throughout the metastatic cascade. Needless to say, and as mentioned above, the models need to take into account the complexity of the process, and among other sources of heterogeneity, the tissue-specific EMT programs. Thus, it is important to analyze the experimental system for expression codes and phenotypic plasticity before designing the reporters and/or the players to target. Better reporter systems that can simultaneously detect cell plasticity and cell lineages should be very helpful, together with optimized protocols to recover the full complement of circulating tumor cells (CTCs). Fortunately, several reports presented at the IX TEMTIA meeting held in Kumamoto in November 2019 showed improved animal models, which should be of tremendous help on this matter.

Importantly, the models should also be able to reproduce the timing of events, not only to be able to more accurately follow the progression to the metastatic disease but also to prevent misleading interpretations. For instance, in animal models, the silencing of EMT-TFs is usually performed from the very early stages of tumorigenesis and tumor progression. Under these experimental conditions, the lack of these EMT regulators leads to a well-documented decrease in metastatic burden. This is usually interpreted as a requirement of EMT for metastasis formation, and suggests that inhibiting EMT should be beneficial in the clinic. However, the most likely explanation in the aforementioned experimental setting is that the decrease in metastatic burden results from a much lower degree of invasion and dissemination. It is worth noting here that the situation is very different in the clinical setting, as the vast majority of cancer patients are diagnosed at the time when dissemination has already occurred. As it is accepted that disseminated cancer cells need to revert towards a more epithelial phenotype for metastatic colonization (see above), a strategy aimed at inhibiting EMT under these conditions could be counterproductive, as it may promote colonization of already disseminated cancer cells. Thus, a note of caution is essential when extrapolating from data obtained in animal models, and this is important because it may have a direct impact on the decision of therapeutic strategies.

Another source of putative misleading conclusions is due to (1) the focal nature of the EMT and (2) the low level of expression of the EMT-TFs. These two factors may affect the results and interpretations of whole genome transcriptome analyses, preventing the detection of an EMT program. The limited areas of EMT within a tumor may be diluted in bulk RNA sequencing. Although this should not be a problem in single cell RNA sequencing (scRNAseq) experiments, the low expression levels of EMT-TFs may as well prevent their detection in scRNAseq studies.

6 The Benefits

Are you still afraid of the EMT? Think of it in a conceptual manner, integrate it into the biology of the system and consider that the progress in our knowledge of the molecular and cellular properties of the different EMT programs together with the emergence of commonalities should prove very useful in the design of optimized animal models. This is all for the benefit of knowledge and essentially, for the opportunity to design better therapeutic strategies for the benefit of the patients. Cancer and degenerative organ diseases including fibrosis progress through loss of epithelial homeostasis and EMT. Aging is the biggest risk factor for both cancer and fibrosis, and thus, with the increase in the aging population, the associated social and economic burden is increasing dramatically, and so too will the benefit.

Acknowledgments

Work in the lab is funded by the Spanish Ministry of Science, Innovation and Universities (MICIU RTI2018-096501-B-I00), Generalitat Valenciana (PROMETEOII/2017/150), and Spanish Association against Cancer (AECC) to M.A.N. The first two grants are co-financed by the European Regional Development Fund, ERDF.

References

1. Brabletz T et al (2018) EMT in cancer. Nat Rev Cancer 18(2):128–134
2. Derynck R, Weinberg RA (2019) EMT and cancer: more than meets the eye. Dev Cell 49 (3):313–316
3. Nieto MA et al (2016) Emt: 2016. Cell 166 (1):21–45
4. Williams ED et al (2019) Controversies around epithelial-mesenchymal plasticity in cancer metastasis. Nat Rev Cancer 19(12):716–732
5. Lu W, Kang Y (2019) Epithelial-mesenchymal plasticity in cancer progression and metastasis. Dev Cell 49(3):361–374
6. Thiery JP et al (2009) Epithelial-mesenchymal transitions in development and disease. Cell 139(5):871–890
7. Stemmler MP et al (2019) Non-redundant functions of EMT transcription factors. Nat Cell Biol 21(1):102–112
8. Nieto MA (2017) Context-specific roles of EMT programmes in cancer cell dissemination. Nat Cell Biol 19(5):416–418
9. Hay ED (1995) An overview of epithelio-mesenchymal transformation. Acta Anat (Basel) 154(1):8–20
10. Alberga A et al (1991) The snail gene required for mesoderm formation in drosophila is expressed dynamically in derivatives of all three germ layers. Development 111 (4):983–992
11. Nieto MA et al (1994) Control of cell behavior during vertebrate development by slug, a zinc finger gene. Science 264(5160):835–839
12. Lovisa S et al (2015) Epithelial-to-mesenchymal transition induces cell cycle arrest and parenchymal damage in renal fibrosis. Nat Med 21(9):998–1009
13. Grande MT et al (2015) Snail1-induced partial epithelial-to-mesenchymal transition drives renal fibrosis in mice and can be targeted to reverse established disease. Nat Med 21 (9):989–997
14. Montero JA et al (2005) Shield formation at the onset of zebrafish gastrulation. Development 132(6):1187–1198
15. Scarpa E et al (2015) Cadherin switch during EMT in neural crest cells leads to contact inhibition of locomotion via repolarization of forces. Dev Cell 34(4):421–434
16. Campbell K, Casanova J (2015) A role for E-cadherin in ensuring cohesive migration of a heterogeneous population of non-epithelial cells. Nat Commun 6:7998
17. Pastushenko I et al (2018) Identification of the tumour transition states occurring during EMT. Nature 556(7702):463–468
18. Padmanaban V et al (2019) E-cadherin is required for metastasis in multiple models of breast cancer. Nature 573(7774):439–444
19. Takeichi M (1993) Cadherins in cancer: implications for invasion and metastasis. Curr Opin Cell Biol 5(5):806–811

20. van Staalduinen J et al (2018) Epithelial-mesenchymal-transition-inducing transcription factors: new targets for tackling chemoresistance in cancer? Oncogene 37(48):6195–6211
21. Mani SA et al (2008) The epithelial-mesenchymal transition generates cells with properties of stem cells. Cell 133(4):704–715
22. Morel AP et al (2008) Generation of breast cancer stem cells through epithelial-mesenchymal transition. PLoS One 3(8): e2888
23. Samavarchi-Tehrani P et al (2010) Functional genomics reveals a BMP-driven mesenchymal-to-epithelial transition in the initiation of somatic cell reprogramming. Cell Stem Cell 7 (1):64–77
24. Li R et al (2010) A mesenchymal-to-epithelial transition initiates and is required for the nuclear reprogramming of mouse fibroblasts. Cell Stem Cell 7(1):51–63
25. Ocana OH et al (2012) Metastatic colonization requires the repression of the epithelial-mesenchymal transition inducer Prrx1. Cancer Cell 22(6):709–724
26. Tsai JH, Yang J (2013) Epithelial-mesenchymal plasticity in carcinoma metastasis. Genes Dev 27(20):2192–2206
27. Tran HD et al (2014) Transient SNAIL1 expression is necessary for metastatic competence in breast cancer. Cancer Res 74 (21):6330–6340
28. Beerling E et al (2016) Plasticity between epithelial and mesenchymal states unlinks EMT from metastasis-enhancing stem cell capacity. Cell Rep 14(10):2281–2288
29. Celia-Terrassa T, Jolly MK (2019) Cancer stem cells and epithelial-to-mesenchymal transition in cancer metastasis. Cold Spring Harb Perspect Med

Chapter 5

Partial EMT/MET: An Army of One

Sofiane Hamidi, Hiroki Nagai, and Guojun Sheng

Abstract

As our understanding of Epithelial Mesenchymal Transition (EMT) increases, the original binary concept of E versus M no longer fits with experimental evidence. Re-definition of the EMT paradigm as spectral transitions between a full epithelium and a full mesenchyme suggests the existence of a virtual infinity of intermediate cellular states. The new challenge is to develop technical tools needed to contextualize each of these states and identify biologically significant cellular mechanisms that could be targeted in combatting EMT-related diseases.

Key words Epithelial-to-mesenchymal transition, Partial EMT/MET, Morphogenesis, Cancer

EMTs (epithelial mesenchymal transitions) and its reverse METs (mesenchymal epithelial transitions) are normal morphogenetic processes in animal development and tissue homeostasis [1]. EMT/METs are also known to play important roles in abnormal morphogenetic processes such as cancer and fibrosis [2, 3]. The field of EMT/MET research has been growing exponentially in the last decade, with over 5000 publications in 2018 alone and with more papers published in the last 5 years (2015–2019) than in all the previous 35 years combined (Fig. 1). No research field, however, could sustain exponential growth forever. It is therefore timely to examine what has made the EMT/MET concept attractive to a diverse group of researchers and where this field is heading in the next decade.

Functional compartmentalization in multicellular organisms is achieved through cell lineage specialization and epithelial barrier establishment. EMT/METs are involved in the development of all mesendoderm-derived and some ectoderm-derived cell lineages that form epithelial barriers in a vertebrate body [1]. Failure in maintaining their epithelial organization characterizes over 90% of all human cancers (2014 World Cancer Report). For example, the top eight human cancer types, together causing 2/3 of all cancer

Kyra Campbell and Eric Theveneau (eds.), *The Epithelial-to Mesenchymal Transition: Methods and Protocols*, Methods in Molecular Biology, vol. 2179, https://doi.org/10.1007/978-1-0716-0779-4_5,

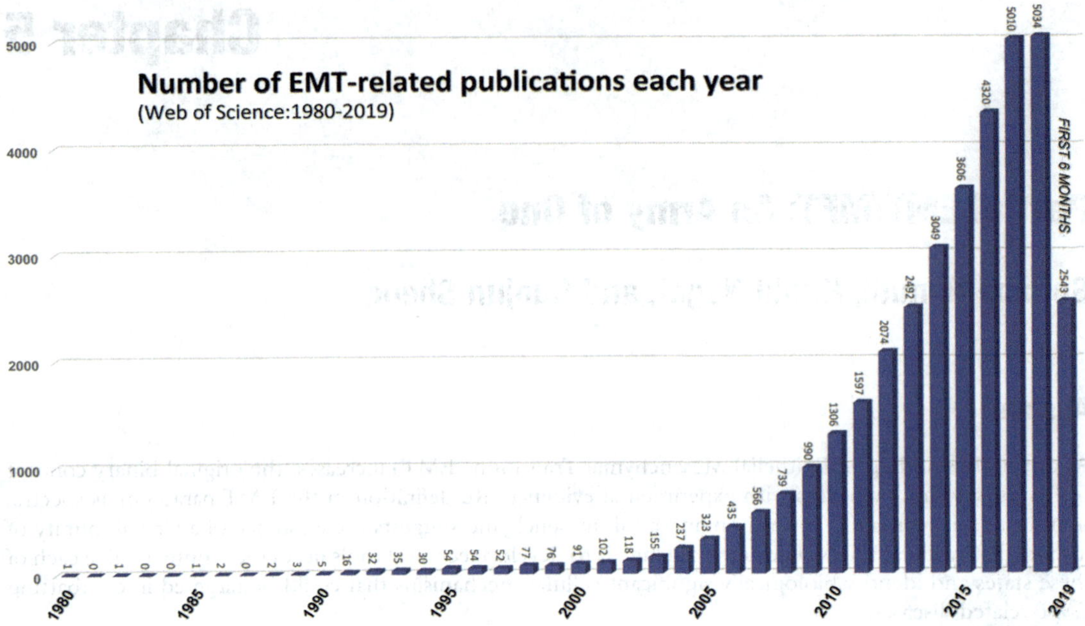

Fig. 1 Number of EMT-related publications by year, from 1980 to June 2019, based on Web of Science search

deaths, originate from the lung, liver, stomach, colorectum, breast, esophagus, pancreas, and prostate (2014 World Cancer Report), all of which requiring EMT/METs in their developmental ontogeny and being derived (except for the mammary gland) from the endoderm germ layer [1], suggesting that human cancer incidence/mortality preferentially affects endoderm-derived epithelial structures. It is therefore not surprising that both developmental and cancer biologists have become increasingly interested in applying the EMT/MET concept in their respective research and in trying to uncover common regulators for both physiological (developmental) and pathological (disease) EMT/METs [2, 3].

A common theme emerged from such studies in recent years concerns the issue of EMT/MET diversity as it has become increasingly clear that in both development and cancer, a complete transition of E to M or M to E is rare [4]. Most EMT/MET phenomena involve transitions that partially overlap with the full spectrum between full epithelium and full mesenchyme (Fig. 2). With such observations came questions of whether diverse EMT/METs could have a standardized definition with a conserved set of regulators and, if not, whether comparison between developmental and cancer EMT/METs (or between any two EMT/METs) is meaningful [5].

Answer to these questions will come from advancement in related fields of cellular, molecular, biophysical, bioinformatic, and computational research that will place precise quantitative values to each EMT/MET descriptor [6, 7]. For example, both epithelial and mesenchymal cells are polarized, and their polarity is

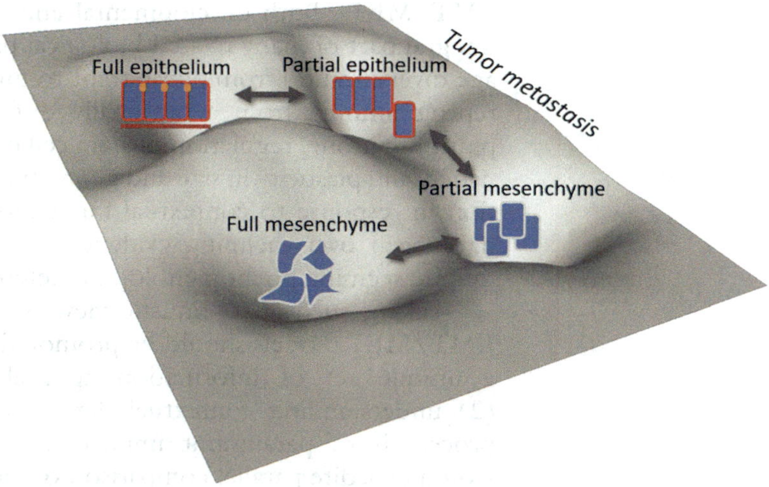

Fig. 2 Schematic representation of EMT/MET and partial EMT/MET features. Tumor metastasis preferentially occurs in partial epithelial status

regulated by a set of evolutionarily conserved molecules [8]. An epithelium is organized by cells that have uniform cellular polarity and form relatively stable intercellular junctions and cell–matrix interactions, whereas a mesenchyme is a collection of non-uniformly polarized cells with more dynamic intercellular and cell–matrix interactions. The value of any molecular or cellular descriptor of EMT/MET, therefore, would depend on how it can effectively and faithfully distinguish the mesenchymal and epithelial populations. For example, the presence or absence of E-Cadherin (E-cad) staining is not a good indicator of an epithelium, but its sub-cellular localization to the apical adherens junctions and the lateral membrane is. Such descriptive value is further enhanced by differentiating E-cad from P-Cadherin (P-cad), which is a closely related cadherin with distinct expression patterns but is recognized by most E-cad antibodies [9, 10]. Similarly, predictive value of most EMT/MET markers can be substantially increased if attention is paid to these (and other similar) cellular and molecular details. It is reasonable to expect that, for any given full or partial EMT/MET, a set of descriptors could be extracted in the near future by means of bioinformatics and computational biology.

With increased precision in EMT/MET description, one can start to address the fundamental question of whether a limited set of molecular descriptors would be sufficient to characterize a myriad of partial EMT/METs and whether all partial EMT/METs could fit in a 1- or 2-dimensional continuum between the two end states (fully epithelial and fully mesenchymal) (Fig. 2). With extreme diversity in physiological circumstances underlying various normal and disease EMT/METs, one would be tempted to believe otherwise. However, a more optimist view would emerge if we view

EMT/METs, both developmental and pathological, as being an integral part of many related biological phenomena, e.g., the lineage diversification from pluripotent to uni-functional cells, cellular reprogramming from a terminally differentiated fate back to a pluripotent one, regulation of stem cell differentiation and maintenance, and plasticity in size and geometry in multicellular organization in response to contextual fluctuations, in all of which there have been overwhelming evidence for mechanistic conservation despite seeming phenomenological heterogeneity [11].

Taking such an optimistic view would mean that (1) more EMT/MET models should be promoted because each can give us a unique set of information applicable to all EMT/METs; (2) understanding contextual details of any given EMT/MET process is of paramount importance because such knowledge would expedite parallel comparison of mechanistic insights gained from different EMT/MET models; (3) each normal or disease EMT/MET model, with its unique advantages and limitations, should be compared with its closest "EMT/MET" relatives because a limited number of EMT/MET contextual variations would suggest the possibility of establishing an EMT/MET "phylogenetic tree"; and (4) most importantly, potential applications of EMT/MET concept in cancer diagnosis and prevention can be facilitated by integration of normal and disease EMT/MET research.

In conclusion, we believe that a key element for conceptual breakthroughs in EMT/MET research is to treat each partial EMT/MET process as "an army of one," with their unique cellular history and tissue circumstance and ready to offer us mechanistic insight of general importance. By doing so, we will be getting ever closer to winning the battle to conquer cancer, the loftiest goal of the EMT/MET research [12].

References

1. Nakaya Y, Sheng G (2013) EMT in developmental morphogenesis. Cancer Lett 341 (1):9–15. https://doi.org/10.1016/j.canlet.2013.02.037
2. Thiery JP, Acloque H, Huang RY, Nieto MA (2009) Epithelial-mesenchymal transitions in development and disease. Cell 139 (5):871–890. https://doi.org/10.1016/j.cell.2009.11.007
3. Derynck R, Weinberg RA (2019) EMT and cancer: more than meets the eye. Dev Cell 49 (3):313–316. https://doi.org/10.1016/j.devcel.2019.04.026
4. Sha Y, Haensel D, Gutierrez G, Du H, Dai X, Nie Q (2019) Intermediate cell states in epithelial-to-mesenchymal transition. Phys Biol 16(2):021001. https://doi.org/10.1088/1478-3975/aaf928
5. Yang J, Antin P, Berx G, Blanpain C, Brabletz T, Bronner M, Campbell K, Cano A, Casanova J, Christofori G, Dedhar S, Derynck R, Ford HL, Fuxe J, Garcia de Herreros A, Goodall GJ, Hadjantonakis AK, Huang RJY, Kalcheim C, Kalluri R, Kang Y, Khew-Goodall Y, Levine H, Liu J, Longmore GD, Mani SA, Massague J, Mayor R, McClay D, Mostov KE, Newgreen DF, Nieto MA, Puisieux A, Runyan R, Savagner P, Stanger B, Stemmler MP, Takahashi Y, Takeichi M, Theveneau E, Thiery JP, Thompson EW, Weinberg RA, Williams ED, Xing J, Zhou BP, Sheng G, Association EMTI (2020) Guidelines and definitions for research on epithelial-mesenchymal transition. Nat Rev

Mol Cell Biol 21 (6):341–352. doi:https://doi.org/10.1038/s41580-020-0237-9

6. Xing J, Tian XJ (2019) Investigating epithelial-to-mesenchymal transition with integrated computational and experimental approaches. Phys Biol 16(3):031001. https://doi.org/10.1088/1478-3975/ab0032
7. Burger GA, Danen EHJ, Beltman JB (2017) Deciphering epithelial-mesenchymal transition regulatory networks in cancer through computational approaches. Front Oncol 7:162. https://doi.org/10.3389/fonc.2017.00162
8. Hamidi S, Sheng G (2018) Epithelial-mesenchymal transition in haematopoietic stem cell development and homeostasis. J Biochem 164(4):265–275. https://doi.org/10.1093/jb/mvy063
9. Ribeiro AS, Paredes J (2014) P-cadherin linking breast cancer stem cells and invasion: a promising marker to identify an "intermediate/metastable" EMT state. Front Oncol 4:371. https://doi.org/10.3389/fonc.2014.00371
10. Vieira AF, Paredes J (2015) P-cadherin and the journey to cancer metastasis. Mol Cancer 14:178. https://doi.org/10.1186/s12943-015-0448-4
11. Liao TT, Yang MH (2017) Revisiting epithelial-mesenchymal transition in cancer metastasis: the connection between epithelial plasticity and stemness. Mol Oncol 11 (7):792–804. https://doi.org/10.1002/1878-0261.12096
12. Li C, Balazsi G (2018) A landscape view on the interplay between EMT and cancer metastasis. NPJ Syst Biol Appl 4:34. https://doi.org/10.1038/s41540-018-0068-x

Chapter 6

EMT: An Update

Jean Paul Thiery

Abstract

Epithelial Mesenchymal Transition (EMT) initially discovered as a key developmental mechanism is now shown to be indirectly involved in fibrosis and is contributing to the progression of carcinomas. Additionally, to transcription factors driving the morphological transition, novel mechanisms are now described to modulate the different features of the transition. The debate as to whether EMT is essential for the dissemination of carcinoma cells from the primary tumors is likely to be resolved soon, considering that EMT is not a linear transition from an epithelial to a mesenchymal state. Multiple intermediate states can be reached without involving the presence of some of known transcription factors initially described as indispensable for the acquisition of mesenchymal-like phenotypes.

Key words Epithelial-to-mesenchymal transition, Morphogenesis, Gastrulation, Cancer, Fibrosis, EMT spectrum, Ribosome biogenesis, Stemness

Epithelial Mesenchymal Transition (EMT) designates a fundamental process driving morphogenesis and organogenesis in multicellular organisms. During gastrulation, the primitive epithelium undergoes a drastic morphological transition, engaging presumptive mesodermal and endodermal cells into a migratory behavior through the transient or permanent loss of apico-basal polarity of epithelial cells and the acquisition of front-rear polarity in newly formed mesenchymal cells. The reverse mechanism, Mesenchymal Epithelial Transition (MET), is critically important to resume an epithelial state during subsequent stages of morphogenesis and in organogenesis. MET is a mandatory step in the newly formed mesenchyme to generate somites, precursors of the vertebrae, and the kidney. Heart development also involves cycles of EMT and MET [1].

EMT is actively studied in embryonic development. Gene regulatory networks during gastrulation in *Drosophila melanogaster*, the sea urchin, and in the neural crest of vertebrates have been partially elucidated [2, 3]. Mechanisms driving gastrulation and neural crest ontogeny share transcriptional regulation, particularly

Kyra Campbell and Eric Theveneau (eds.), *The Epithelial-to Mesenchymal Transition: Methods and Protocols*, Methods in Molecular Biology, vol. 2179, https://doi.org/10.1007/978-1-0716-0779-4_6,

with regard to SNAIL (SNAI), known as SNAI1 and SNAI2 in vertebrates. However, it is clear that the pathways involved are quite complex and not well conserved in controlling cell shape changes, the initiation of migration, and the engagement into differentiation. For instance, in the sea urchin, none of the contributing pathways alone can complete gastrulation [4]. Thus, one cannot infer that one EMT transcription factor is sufficient to execute the full gastrulation process although, when overexpressed in some cell lines in vitro, such a factor can lead to morphological transitions. These findings point to the intricate nature of EMT signal transduction that is established during evolution, a process that appears to ensure the robustness of this fundamental event.

Our understanding of EMT has considerably improved over the past 10 years [5], particularly following the discovery of the contributions of epigenetics [6], splicing, [7], posttranslational modifications [8], metabolism [9], and redox states [10]. Notably, EMT is now shown to require ribosome biogenesis, which occurs at the G1/S restriction point [11] in cell lines and in vivo during neural crest cell dissociation from the neural fold, as well as in their early migratory phase. Snail1 recruited to rDNA operons displaces the repressive nucleolar chromatin complex and becomes associated with the rRNA along with the core components of polymerase 1 and rictor, a component of the mammalian target of rapamycin. This mechanism also contributes to the maintenance of the dedifferentiated state and the promotion of invasion and metastasis. These events can be abrogated by CX-5461, a drug that interferes with rRNA biogenesis. Some recent efforts have been made to better understand the intricate feedback loop controlling EMT and MET. A hysteretic mechanism has now been revealed to ensure the robustness of the EMT process [12]. The mechanism—driven by the mir-200/Zeb1 double-negative feedback loop—ensures that the level of E-cadherin remains high in a concentration range of the TGFβ inducer if the cells are in an epithelial state, and low in a mesenchymal state. This so-called bi-stability of E-cadherin is lost if the Zeb1 binding site on the E-cadherin promoter is deleted, leading to a gradual progression of EMT. Such high sensitivity and memory of hysteretic EMT allows cells to accomplish the full metastatic cascade upon just a short exposure to the inducer.

Several EMT scoring methods have been published recently, albeit with different degrees of robustness [13, 14]. The scoring methods more precisely position cells in distinct intermediate stages; however, these scores fail to tell us how these intermediate stages are achieved. We now require a matrix of scores that provides "n"-dimensional computations to position each of the mechanisms that contribute to these intermediate stages. The debate surrounding whether EMT is dispensable for metastasis is directly linked to this issue [15–20]. It is likely that the reporters used to depict the E and M states in these studies are activated in a wide range of EMT values rather than strictly at one of the two extremes of the spectrum. Clearly, using only one reporter, such as FSP-1, to quantify the EMT status is far from sufficient to describe the position of carcinoma cells on the EMT spectrum. A single- or double knockout does not guarantee that these carcinoma cells have not retained mesenchymal-like features, even if they do express E-cadherin [21]. A recent study [22] provides additional support to the importance of intermediate EMT states, with distinct subpopulations of intermediate EMT-staged carcinoma cells exhibiting either tumor initiation and expansion or invasive and metastatic properties. These different populations are localized in distinct microenvironments in primary tumors. Cells with intermediate EMT values are more prone to invasion and metastasis than their epithelial counterpart, indicative of a role for EMT in the progression of carcinoma.

However, it remains to be determined precisely when cells acquire stemness and how stable is the phenotype [23]. One way to approach this critical issue is to decipher how somatic cells can be reprogrammed into stem cells. To achieve reprogramming, somatic cells must be brought into a fully mesenchymal state, during which a dramatic reorganization of the chromatin renders the cells susceptible to stem cell transcription factors. The acquisition of stemness appears to occur during the MET process, initiated through the combined actions of EMT inhibitors and MET inducers [24, 25]. Stemness is likely acquired at an intermediate stage during reversal to the epithelial state. It remains unclear whether both normal and carcinoma cells can acquire stem-like properties at similar or distinct intermediates stages and the stability of these phenotypes in carcinoma cells. Carcinoma cell plasticity is also exploited to induce the transdifferentiation of breast carcinoma cells into adipocytes [26]. Interestingly, the reprogramming of breast carcinoma cells into adipocytes will prevent their invasion and metastasis.

The question remains: Are intermediate phenotypes seen during carcinoma progression? There is accumulating evidence for the existence of intermediate-staged EMT in human primary tumors and in circulating tumor cells [27–29]. The EMT spectrum seen among all tumor types also points to intermediate stages in primary tumors, and this is further reflected in the spectrum of EMT noted in corresponding carcinoma cell lines [14]. Future studies will address in depth the mechanisms driving carcinoma cell plasticity in relation to the microenvironment for the design of better therapeutic strategies.

References

1. Lim J, Thiery JP (2012) Epithelial-mesenchymal transitions: insights from development. Development 139(19):3471–3486. https://doi.org/10.1242/dev.071209
2. Erkenbrack EM, Davidson EH, Peter IS (2018) Conserved regulatory state expression controlled by divergent developmental gene regulatory networks in echinoids. Development 145(24). https://doi.org/10.1242/dev.167288
3. Martik ML, Bronner ME (2017) Regulatory logic underlying diversification of the neural crest. Trends Genet 33(10):715–727. https://doi.org/10.1016/j.tig.2017.07.015
4. Saunders LR, McClay DR (2014) Sub-circuits of a gene regulatory network control a developmental epithelial-mesenchymal transition. Development 141(7):1503–1513. https://doi.org/10.1242/dev.101436
5. Dongre A, Weinberg RA (2019) New insights into the mechanisms of epithelial-mesenchymal transition and implications for cancer. Nat Rev Mol Cell Biol 20(2):69–84. https://doi.org/10.1038/s41580-018-0080-4
6. Tam WL, Weinberg RA (2013) The epigenetics of epithelial-mesenchymal plasticity in cancer. Nat Med 19(11):1438–1449. https://doi.org/10.1038/nm.3336
7. Climente-Gonzalez H, Porta-Pardo E, Godzik A, Eyras E (2017) The functional impact of alternative splicing in cancer. Cell Rep 20(9):2215–2226. https://doi.org/10.1016/j.celrep.2017.08.012
8. De Craene B, Berx G (2013) Regulatory networks defining EMT during cancer initiation and progression. Nat Rev Cancer 13(2):97–110. https://doi.org/10.1038/nrc3447
9. Shaul YD, Freinkman E, Comb WC, Cantor JR, Tam WL, Thiru P, Kim D, Kanarek N, Pacold ME, Chen WW, Bierie B, Possemato R, Reinhardt F, Weinberg RA, Yaffe MB, Sabatini DM (2014) Dihydropyrimidine accumulation is required for the epithelial-mesenchymal transition. Cell 158(5):1094–1109. https://doi.org/10.1016/j.cell.2014.07.032
10. Krishnan V, Chong YL, Tan TZ, Kulkarni M, Bin Rahmat MB, Tay LS, Sankar H, Jokhun DS, Ganesan A, Chuang LSH, Voon DC, Shivashankar GV, Thiery JP, Ito Y (2018) TGFbeta promotes genomic instability after loss of RUNX3. Cancer Res 78(1):88–102. https://doi.org/10.1158/0008-5472.CAN-17-1178
11. Prakash V, Carson BB, Feenstra JM, Dass RA, Sekyrova P, Hoshino A, Petersen J, Guo Y, Parks MM, Kurylo CM, Batchelder JE, Haller K, Hashimoto A, Rundqivst H, Condeelis JS, Allis CD, Drygin D, Nieto MA, Andang M, Percipalle P, Bergh J, Adameyko I, Farrants AO, Hartman J, Lyden D, Pietras K, Blanchard SC, Vincent CT (2019) Ribosome biogenesis during cell cycle arrest fuels EMT in development and disease. Nat Commun 10(1):2110. https://doi.org/10.1038/s41467-019-10100-8
12. Celia-Terrassa T, Bastian C, Liu DD, Ell B, Aiello NM, Wei Y, Zamalloa J, Blanco AM, Hang X, Kunisky D, Li W, Williams ED, Rabitz H, Kang Y (2018) Hysteresis control of epithelial-mesenchymal transition dynamics conveys a distinct program with enhanced metastatic ability. Nat Commun 9(1):5005. https://doi.org/10.1038/s41467-018-07538-7
13. George JT, Jolly MK, Xu S, Somarelli JA, Levine H (2017) Survival outcomes in cancer patients predicted by a partial EMT gene expression scoring metric. Cancer Res 77(22):6415–6428. https://doi.org/10.1158/0008-5472.CAN-16-3521
14. Tan TZ, Miow QH, Miki Y, Noda T, Mori S, Huang RY, Thiery JP (2014) Epithelial-mesenchymal transition spectrum quantification and its efficacy in deciphering survival and drug responses of cancer patients. EMBO Mol Med 6(10):1279–1293. https://doi.org/10.15252/emmm.201404208
15. Aiello NM, Brabletz T, Kang Y, Nieto MA, Weinberg RA, Stanger BZ (2017) Upholding a role for EMT in pancreatic cancer metastasis. Nature 547(7661):E7–E8. https://doi.org/10.1038/nature22963
16. Brabletz T, Kalluri R, Nieto MA, Weinberg RA (2018) EMT in cancer. Nat Rev Cancer 18(2):128–134. https://doi.org/10.1038/nrc.2017.118
17. Fischer KR, Altorki NK, Mittal V, Gao D (2017) Fischer et al. reply. Nature 547(7661):E5–E6. https://doi.org/10.1038/nature22817
18. Fischer KR, Durrans A, Lee S, Sheng J, Li F, Wong ST, Choi H, El Rayes T, Ryu S, Troeger J, Schwabe RF, Vahdat LT, Altorki NK, Mittal V, Gao D (2015) Epithelial-to-mesenchymal transition is not required for lung metastasis but contributes to chemoresistance. Nature 527(7579):472–476. https://doi.org/10.1038/nature15748

19. Ye X, Brabletz T, Kang Y, Longmore GD, Nieto MA, Stanger BZ, Yang J, Weinberg RA (2017) Upholding a role for EMT in breast cancer metastasis. Nature 547(7661):E1–E3. https://doi.org/10.1038/nature22816
20. Zheng X, Carstens JL, Kim J, Scheible M, Kaye J, Sugimoto H, Wu CC, LeBleu VS, Kalluri R (2015) Epithelial-to-mesenchymal transition is dispensable for metastasis but induces chemoresistance in pancreatic cancer. Nature 527(7579):525–530. https://doi.org/10.1038/nature16064
21. Stemmler MP, Eccles RL, Brabletz S, Brabletz T (2019) Non-redundant functions of EMT transcription factors. Nat Cell Biol 21(1):102–112. https://doi.org/10.1038/s41556-018-0196-y
22. Pastushenko I, Brisebarre A, Sifrim A, Fioramonti M, Revenco T, Boumahdi S, Van Keymeulen A, Brown D, Moers V, Lemaire S, De Clercq S, Minguijon E, Balsat C, Sokolow Y, Dubois C, De Cock F, Scozzaro S, Sopena F, Lanas A, D'Haene N, Salmon I, Marine JC, Voet T, Sotiropoulou PA, Blanpain C (2018) Identification of the tumour transition states occurring during EMT. Nature 556(7702):463–468. https://doi.org/10.1038/s41586-018-0040-3
23. Chaffer CL, Weinberg RA (2015) How does multistep tumorigenesis really proceed? Cancer Discov 5(1):22–24. https://doi.org/10.1158/2159-8290.CD-14-0788
24. Li Q, Hutchins AP, Chen Y, Li S, Shan Y, Liao B, Zheng D, Shi X, Li Y, Chan WY, Pan G, Wei S, Shu X, Pei D (2017) A sequential EMT-MET mechanism drives the differentiation of human embryonic stem cells towards hepatocytes. Nat Commun 8:15166. https://doi.org/10.1038/ncomms15166
25. Pei D, Shu X, Gassama-Diagne A, Thiery JP (2019) Mesenchymal-epithelial transition in development and reprogramming. Nat Cell Biol 21(1):44–53. https://doi.org/10.1038/s41556-018-0195-z
26. Ishay-Ronen D, Diepenbruck M, Kalathur RKR, Sugiyama N, Tiede S, Ivanek R, Bantug G, Morini MF, Wang J, Hess C, Christofori G (2019) Gain fat-lose metastasis: converting invasive breast cancer cells into adipocytes inhibits cancer metastasis. Cancer Cell 35(1):17–32. e16. https://doi.org/10.1016/j.ccell.2018.12.002
27. Sarrio D, Rodriguez-Pinilla SM, Hardisson D, Cano A, Moreno-Bueno G, Palacios J (2008) Epithelial-mesenchymal transition in breast cancer relates to the basal-like phenotype. Cancer Res 68(4):989–997. https://doi.org/10.1158/0008-5472.CAN-07-2017
28. Yadavalli S, Jayaram S, Manda SS, Madugundu AK, Nayakanti DS, Tan TZ, Bhat R, Rangarajan A, Chatterjee A, Gowda H, Thiery JP, Kumar P (2017) Data-driven discovery of extravasation pathway in circulating tumor cells. Sci Rep 7:43710. https://doi.org/10.1038/srep43710
29. Yu M, Bardia A, Wittner BS, Stott SL, Smas ME, Ting DT, Isakoff SJ, Ciciliano JC, Wells MN, Shah AM, Concannon KF, Donaldson MC, Sequist LV, Brachtel E, Sgroi D, Baselga J, Ramaswamy S, Toner M, Haber DA, Maheswaran S (2013) Circulating breast tumor cells exhibit dynamic changes in epithelial and mesenchymal composition. Science 339(6119):580–584. https://doi.org/10.1126/science.1228522

Part II

Review on Mesenchymal-to-Epithelial Transitions

Chapter 7

Mesenchymal-to-Epithelial Transitions in Development and Cancer

John-Poul Ng-Blichfeldt and Katja Röper

Abstract

The evolutionary emergence of the mesenchymal phenotype greatly increased the complexity of tissue architecture and composition in early Metazoan species. At the molecular level, an epithelial-to-mesenchymal transition (EMT) was permitted by the innovation of specific transcription factors whose expression is sufficient to repress the epithelial transcriptional program. The reverse process, mesenchymal-to-epithelial transition (MET), involves direct inhibition of EMT transcription factors by numerous mechanisms including tissue-specific MET-inducing transcription factors (MET-TFs), micro-RNAs, and changes to cell and tissue architecture, thus providing an elegant solution to the need for tight temporal and spatial control over EMT and MET events during development and adult tissue homeostasis.

Key words Epithelial morphogenesis, MET, Apical-basal polarity, Transcription factors

1 Introduction

Epithelial and mesenchymal cells represent the two major cell types within triploblastic Metazoan organisms. Epithelial cells are defined as cells forming continuous sheets with apical-basal polarity, and that possess a low degree of individual motility due to strong intercellular adhesive forces via apical-lateral tight junctions (TJs), adherens junctions (AJs), and desmosomal complexes, and basal interactions with extracellular matrix (ECM) via cell-matrix junctions [1]. Conversely, mesenchymal cells are capable of a high degree of motility due to an absence of stable intercellular and cell-matrix adhesive complexes, and possess front-back instead of apical-basal polarity. It is now recognized, however, that these definitions reflect extreme ends of a spectrum of phenotypes that includes a variety of cell states with intermediate levels of junctional complexity, and varying capabilities for either individual or collective motility [2]. Consequently, plasticity between the epithelial and mesenchymal state is an intensive area of research; both

Kyra Campbell and Eric Theveneau (eds.), *The Epithelial-to Mesenchymal Transition: Methods and Protocols*, Methods in Molecular Biology, vol. 2179, https://doi.org/10.1007/978-1-0716-0779-4_7,

epithelial-to-mesenchymal transitions (EMT) and mesenchymal-to-epithelial transitions (MET) are integral to a wide range of developmental and pathological processes. Here, we briefly discuss the molecular and transcriptional control of MET events that occur during development, and their implications for cancer metastasis.

2 Making and Breaking Epithelial Junctions

MET involves the stepwise formation of epithelial intercellular contacts and establishment of apical-basal polarity (Fig. 1). While the precise order of events is likely to be highly dependent on the cell- and tissue-context, insight has predominantly come from cell culture models in which epithelial cells establish apical-basal polarity (reviewed in [3, 4]). An early adhesion event between epithelial (–izing) cells is via the transmembrane protein nectin, which forms calcium-independent hetero-trans-dimers between adjacent cells [5]. Nectin complexes form scaffolds that recruit the actin-binding protein afadin, and E-cadherin (CDH1), a prototypical member of the cadherin family expressed by most epithelial cell types. E-cadherin forms calcium-dependent homo-trans-dimers with adjacent cells that progressively mature into AJs and are attached to the intracellular actin cytoskeleton via alpha- and beta-catenin.

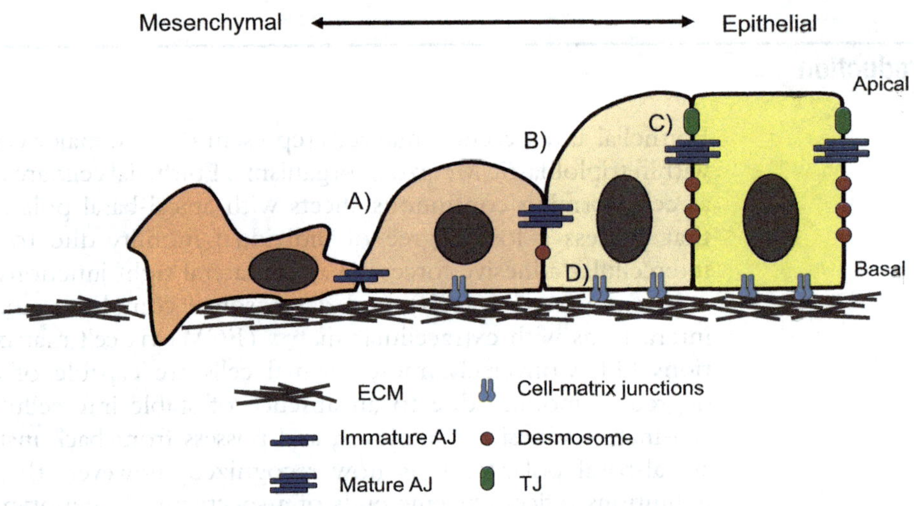

Fig. 1 Potential mechanisms by which cells establish epithelial junctions during MET. Mesenchymal cells (**a**) adhere to neighboring cells via nectin-based hetero-trans-dimers, which allow accumulation of E-cadherin, leading to establishment of immature adherens junctions (AJs). These progressively develop into mature AJs (**b**), leading to establishment of desmosomes, tight junctions (TJs), and consolidation of apical-basal polarity (**c**). Epithelializing cells progressively establish connections with the underlying extracellular matrix (ECM) through integrin-based cell-matrix junctions (**d**). The precise order of events is likely to be highly specific to the cell- and tissue-context, and is reversible in the context of EMT

Nectin also recruits partitioning defective 3 (Par3), which together with Par6 and atypical protein kinase C forms the Par polarity complex that establishes the apical domain, in part by recruiting another apical determinant, the Crumbs complex, and by displacing the basolateral determinant, the Scribble complex [6]. AJ formation promotes the formation of desmosomes, which form electron-dense plaques between neighboring cells and are attached to intracellular intermediate cytokeratin filaments. AJ formation also promotes the formation of sub-apical TJs, comprised of occludin, zona occludens (ZO) proteins, and claudins (CLDN), which are essential for restricting paracellular transport of solutes and ions and for consolidating apical-basal polarity [1, 7].

Much attention has been paid to events occurring in the dissolution of epithelial cell-cell contacts during EMT (reviewed in [3]). A general apical-to-basal "unzipping" mechanism has been proposed, initiating with disassembly of TJs and followed by loss of AJs, driven by both post-translational degradation and transcriptional downregulation of E-cadherin [3]. Subsequently, desmosomes and cell-ECM contacts are dismantled, concordant with a loss of apical-basal polarity, culminating in increased individual motility and delamination through the underlying basement membrane. However, this model, based largely on TGF-β-treated epithelial cells in vitro, is unlikely to hold universally. For example, in epiblast cell ingression during chick gastrulation, basement membrane breakdown is the first recognizable event in EMT, preceding loss of AJs, TJs, and apical-basal polarity [8]. Thus, the precise order of events is also likely to be highly dependent on the cell- and tissue-context.

3 EMT-MET in Development

In amniote embryogenesis, the first cells with epithelial characteristics are blastomeres of the late 8-cell stage embryo; these express E-cadherin, which accumulates at intercellular junctions that, together with cortical actomyosin-mediated tension, induces "compaction" in which cells flatten together and maximize interfacial contacts. Coincident with this is the emergence of apical-basal polarity, followed by the biogenesis of TJs and desmosomes [9]. "Proto-epithelial" cells are thus the origin of all subsequent cells in the organism [10]. Further cleavages lead to segregation of the trophoectoderm epithelial lineage from the inner cell mass, which gives rise to the epiblast. The first instance of EMT occurs during gastrulation with the emergence of the three primitive germ layers, where migratory epiblast cells ingress to form transient mesenchymal mesendodermal progenitors, whereas remaining epiblast cells form ectoderm [11]. Mesodermal and endodermal cells, through successive rounds of EMT-MET, ultimately generate the wide diversity of adult organs, most of which comprise a mixture of epithelial cells and supporting mesenchymal cells in close association. Thus, throughout development, epithelial cells arise from non-epithelial cell types through MET.

Mesodermal cells along the mediolateral axis undergo MET to generate the epithelial precursors of the notochord, somites, splanchnopleure, somatopleure, and nephric ducts [11]. Nephrogenesis during mammalian kidney development is one of the best-studied developmental MET events (Fig. 2, top panel). Occurring at the caudal end of the intermediate mesoderm, the ureteric bud epithelium initially invades the surrounding metanephric mesenchyme. Mesenchymal nephron progenitor cells surrounding the

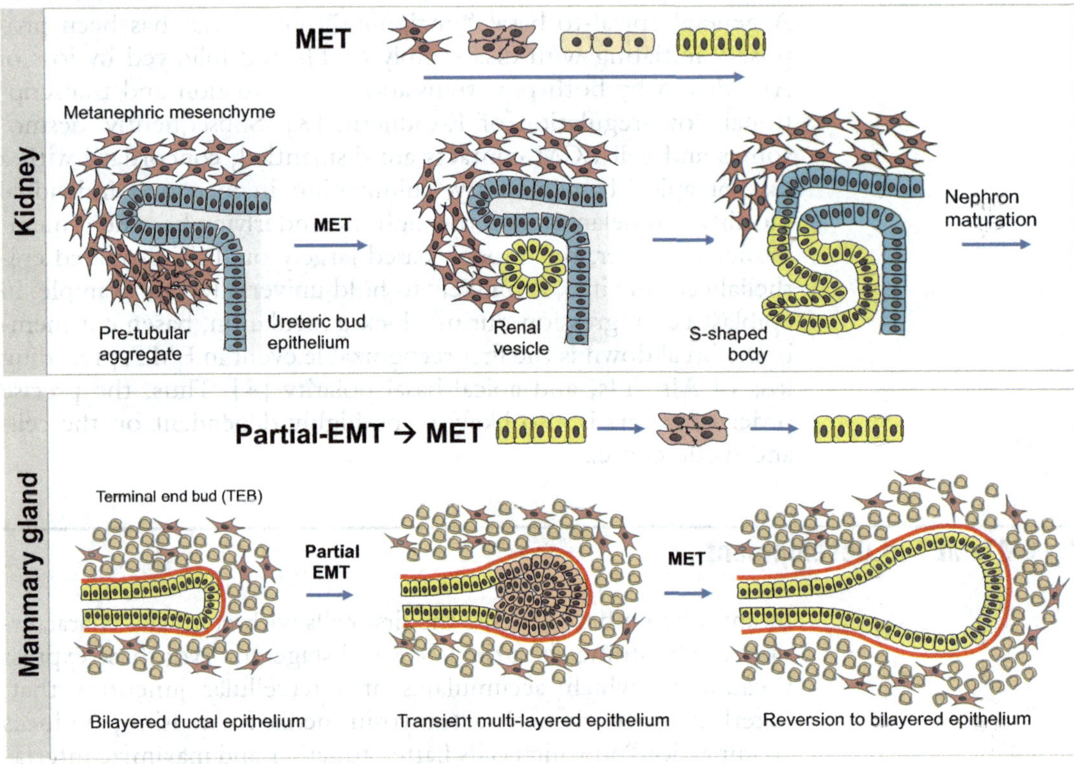

Fig. 2 Developmental MET events. (Top panel) Nephrogenesis during embryonic kidney development. Metanephric mesenchymal cells receive signals from tips of the ureteric bud (blue), inducing cell condensation to form a pre-tubular aggregate. Condensed cells undergo MET to form the renal vesicle, which elongates to form the S-shaped body and eventually fuses with the tip of the ureteric bud to form the nascent nephron. (Bottom panel) Ductal elongation during pubertal mammary morphogenesis. The developing mammary duct is a bilayered epithelium consisting of polarized luminal epithelial cells (yellow) surrounded by myoepithelium (red). During mammary morphogenesis in puberty, cells of the terminal end bud (TEB) undergo partial-EMT, losing apical-basal polarity while partially reducing intercellular adhesion to form a multi-layered, migratory epithelium that invades the surrounding fat pad (composed of adipocytes and fibroblasts) to drive ductal elongation. Subsequently, TEB cells re-establish apical-basal polarity and revert to a normal ductal bilayered epithelium

bud tips are induced to condense into a pre-tubular aggregate, and undergo MET to form the primitive epithelial renal vesicle, which then elongates and fuses to the ureteric bud tip to form the nascent nephron [12]. Another example of developmental MET occurs during mammary gland morphogenesis in puberty (Fig. 2, bottom panel). Mammary epithelial stem/progenitor cells residing at the tips of terminal end buds (TEBs) acquire migratory ability through partial-EMT, transiently losing apical-basal polarity and reducing intercellular adhesive junctions to invade the surrounding fat pad, and subsequently revert to an epithelial state through MET to drive ductal morphogenesis [13, 14].

4 Establishment and Maintenance of the Epithelial Transcriptional Program

Epithelial cells are widely believed to precede the mesenchymal state in both evolutionary and ontological terms [15]. Accordingly, the epithelial state has been hypothesized to represent a "default" state [10]. Early work on the E-cadherin promoter elucidated regulatory control consistent with a default state of transcriptional activation. The CDH1 promoter lacks a TATA-box, usually associated with tissue-specific gene expression [16], but possesses discrete positive regulatory components comprising a CCAAT-box, a GC-rich region, a CE-box, and an enhancer in the first intron [17–21]. These regions are recognized by ubiquitous transcription factors (TFs) including AP-2, c-Myc, RB, and Sp1 [22, 23]. Notably, overexpression of AP-2α was sufficient to restore E-cadherin expression in mesenchymal cells by directly binding to and activating the CDH1 promoter [24]. Moreover, AP-2 activated expression of the epidermal-specific gene KRT14 in keratinocytes [25], and AP-2γ directly controlled expression of apical-basal polarity determinant Pardb6 and TJ component CLDN4 [26], suggesting that ubiquitous TFs like AP-2 may control expression of general epithelial features. Later studies using transgenic reporter mice revealed that the proximal promoter was insufficient to drive epithelial-specific CDH1 expression during embryogenesis, and additional cis-regulatory elements in the second intron were required [27, 28], likely due to a requirement for epigenetic remodeling to enhance chromatin accessibility at the CDH1 locus [29].

The CDH1 promoter also contains E-boxes that mediate transcriptional repression in non-epithelial cell types [17, 30]. E-boxes within the CDH1 gene are recognized by specific TFs that function as transcriptional repressors. These include the zinc-finger TFs SNAIL1/Snail [31, 32] and SNAIL2/Slug [33, 34], the two-handed zinc-finger TFs delta EF1/ZEB1 [35, 36] and SIP1/ZEB2 [37], and the basic helix-loop-helix TFs Twist [38] and E12 and E47 [38]. These TFs recruit co-repressors with chromatin modifying ability to induce genome-wide epigenetic changes required to repress the epithelial phenotype (reviewed in [39]). For example, Snail repression of the CDH1 locus involves the DNA methyltransferase (DNMT) G9a [40], the histone demethylase LSD1 [41], or histone deacetylase 1 (HDAC1) and HDAC2 in complex with either mSin3a [42] or polycomb repressor complex 2 components including enhancer of zeste homolog 2 (EZH2) [43, 44]. Accordingly, ectopic expression of these TFs in epithelial cells is often sufficient to suppress the epithelial program and invoke mesenchymal features, thus leading to the designation EMT-inducing TFs (EMT-TFs).

The action of EMT-TFs is not restricted to E-cadherin. ZEB1 binds to the promoters of, and represses, the Crumbs polarity complex components Crumbs3 (Crb3) and Pals1-associated tight junction protein (PATJ), and the Scribble complex components human Lethal giant larvae homolog 2 (HUGL2) [45] and Lethal giant larvae 2 (LGL2) [46], whereas ZEB2 represses CLDN4 and ZO-3 [47]. Similarly, Snail represses expression of Crb3 [48], epithelial cytoskeletal components KRT17, KRT18 and the epithelial apical recycling endosomal component Rab25 [49] and the tight junction components occludin [50] and CLDN3, 4 and 7 [51], and Snail and Slug represses CLDN1 [52]. Therefore, EMT-TFs can suppress multiple genes required for the epithelial phenotype. Consequently, EMT-TFs induce the progressive loss of epithelial junctional integrity leading to acquisition of a mesenchymal state.

Numerous signaling pathways converge on EMT-TFs to initiate EMT, including receptor tyrosine kinase (RTK)-mediated signaling (e.g., fibroblast growth factor (FGF)-FGF receptor (FGFR) interactions), transforming growth factor beta (TGF-β) signaling, Wnt/β-catenin signaling, and vascular endothelial growth factor signaling; the precise signals are highly dependent on the cell-, tissue-, and developmental-context in which EMT occurs (reviewed in [53]).

CDH1 downregulation is not an absolute requirement for EMT. During chick gastrulation, ingressing epiblast cells retain E-cadherin despite increased migratory ability [8], and dissemination of mammary epithelial cells in 3D-culture driven by ectopic Twist expression increases migration but without loss of E-cadherin [54]. Numerous studies suggest that cells that maintain epithelial features but exhibit intermediate expression of mesenchymal markers, termed partial-EMT, retain a plasticity that better allows reversion to the epithelial state via MET compared to cells that have undergone full EMT. Loss of polarity and acquisition of mesenchymal features through EMT is a hallmark of many invasive carcinomas, and metastasis is the primary cause of death in cancer patients [55]. However, the degree to which EMT contributes to invasiveness is hotly debated [56, 57], and it is proposed that retention of epithelial features via partial-EMT and collective, rather than individual, delamination from primary tumors renders cancer cells more prone to distant metastatic colonization, consistent with the identification of partial-EMT states of circulating tumor cells and the polyclonal nature of metastatic tumors observed in patients [58, 59]. Understanding how EMT and MET programs are dysregulated during carcinoma progression is thus likely to be critical to developing better therapeutic approaches for cancer.

The identification of EMT-TFs supports the hypothesis that the evolutionary emergence of the mesenchymal phenotype was driven by the innovation of direct transcriptional repression of the epithelial program [15]; such EMT-TFs are able to overcome the activity of ubiquitous TFs that drive expression of epithelial-specific genes such as CDH1. Accordingly, EMT-TFs are expressed in a spatial-temporal pattern during vertebrate embryogenesis consistent with the appearance of migratory cells.

5 Cellular Strategies to Restrict EMT

Of critical importance to developmental EMT events is their short-lived nature, permitting re-establishment of epithelial tissues at the correct time and place for normal tissue patterning. It is therefore surprising that relatively little attention has been given to mechanisms of MET compared to EMT. Cells employ several mechanisms to inhibit the activity of EMT-TFs, providing temporally controlled de-repression of epithelial genes and reversion to an epithelial state. These include specific TFs that directly repress EMT-TFs to induce MET (which we designate MET-TFs, summarized in Table 1), micro-RNAs, and post-translational modifications of EMT-TFs, regulated in a tissue-specific manner to enable precise spatio-temporal control over EMT-MET events during development and in adult tissue maintenance.

5.1 Grainy-Head Family

TFs of the Grainy-head family, including Grainyhead (grh) in *Drosophila* and Grainy-head like (GRHL1, 2, and 3) in humans and mice, have emerged as TFs with potentially broad roles in establishing and maintaining the epithelial phenotype in developing and adult epithelial tissues. GRLH2 null mice show defects in neural tube closure, which requires establishment of epithelial cell-cell adhesion, due to disrupted expression of adhesion molecules including E-cadherin and CLDN4, 6, and 7 [60]. During kidney development, GRHL2 co-expresses with E-cadherin in the distal nephron, collecting duct, and ureteric bud epithelium [61, 62]. Notably, GRHL2 binds to enhancer elements in intron 2 of the CDH1 gene to activate promoter activity through chromatin looping [61], suggesting that tissue-specific GRHL2 expression may contribute to epithelial-specific CDH1 activation during

Table 1
Summary of proposed MET-TFs, and roles in developmental processes described to date

MET-TF	Family members	Developmental processes involved in	References
Grainy-head family	grh (Drosophila)	GRHL2 controlled neural tube closure in mice	[60]
	GRHL1, GRHL2, GRHL3 (mice and human)	GRHL2 controlled activation of CDH1 and CLDN4 in human airway epithelium	[63]
		GRHL2 controlled activation of CLDN3 and RAB25 in mouse liver epithelial cells	[64]
Ovo-like family	ovo (Drosophila)	OVOL2 restricted EMT in mammary TEB cells in mice	[70]
	OVOL1, OVOL2 (mice and human)	OVOL1/2 controlled basal epidermal proliferation in mice	[71]
		OVOL2 controlled activation of CDH1, RAB25, and CLDN4 in mouse renal epithelial cysts in vitro	[62]
		OVOL1 controlled kidney and urogenital development in mice	[62, 73]
Ets-domain containing	ELF3, ELF5 (mice and human)	ELF5 restricted EMT in mammary TEB cells in mice	[77]
		ELF3 controlled morphogenesis of intestinal epithelium in mice	[80]
Singleminded 2	Sim2 (mice and human)	Sim2 restricted EMT in mammary TEB cells in mice	[82]
SRY-box 3	Sox3 (mice and human)	Sox3 repressed Slug to define non-ingressing ectoderm during chick gastrulation	[83]

embryogenesis [27, 28]. GHRL2 also binds to and mediates activation of the CLDN4 gene in renal epithelial cells [61], and GRHL2 mediates activation of CDH1 and CLDN4 in human airway epithelium, where it is required for maintenance and establishment of epithelial barrier function [63]. Furthermore, GRHL2 mediates activation of CLDN3 and RAB25 in liver epithelial cells, which is necessary for translocation of CLDN4 to TJs [64]. Thus, GRHL TFs may function to maintain the epithelial phenotype in adult epithelial tissues. Accordingly, GRHL TFs are implicated in a number of adult human cancers (reviewed in [65]). For example, GRHL2 is downregulated in mesenchymal ("claudin-low") primary mammary tumor cells, and ectopic expression is sufficient to initiate MET in mesenchymal breast cancer cells via binding to the ZEB1 promoter, mediating its repression, and via induction of miR-200b/c which inhibits ZEB1 expression [66]. Furthermore,

GRHL2 maintains the epithelial state of ovarian cancer cells by activating expression of the miR-200 family, which in turn inhibits expression of ZEB1 [67].

These studies raise the question of how GRHL TFs can mediate both repression and activation of its various gene targets. Genome-wide analysis of Grh binding during *Drosophila* development revealed that Grh stably bound DNA of target genes, whereas activation of Grh-target genes was dynamic, suggesting a permissive role for Grh [68]. Recently, Grh was revealed to be a pioneer factor, able to bind to nucleosomal DNA and induce remodeling of epithelial gene enhancers to increase chromatin accessibility. Grh binding is not sufficient to activate gene expression, but instead primes genes to be activated or repressed by other factors, a function conserved by human GRHL1, 2, and 3 within mammary epithelial cells [29]. Interestingly, different GRHL family members have both common and distinct target genes despite high amino acid sequence similarity and identical DNA consensus sequences [69]; distinct transcriptional co-regulator binding partners may thus determine specificity, although this aspect of target gene regulation by GRHL TFs is poorly understood.

5.2 Ovo-Like Family

The zinc-finger Ovo-like (OVOL) family of transcriptional repressors has emerging roles as guardian of the epithelial phenotype. During mammary morphogenesis in mice, in which TEB cells undergo transient partial-EMT, TEB-specific loss of OVOL2 leads to increased cell invasion and ductal dysmorphia and impaired restoration of epithelial characteristics [70]. OVOL2 directly represses several EMT-inducing TFs including ZEB1, Snail, and Twist1, and represses expression of the mesenchymal marker vimentin [70]. Similarly, in the epidermal epithelium in mice, targeted OVOL1/2 double mutations lead to hyperproliferation and EMT of basal epidermal cells, together with induction of ZEB1, Slug, and vimentin [71]. Moreover, these changes are partially inhibited by concomitant deletion of ZEB1, suggesting that OVOL1/2 represses ZEB1 to establish the epithelial phenotype in epidermis. OVOL2 binds to and represses ZEB1/2 to maintain epithelial genes in corneal epithelial cells [72], and targeted disruptions in the *ovol* gene in mice cause defects in kidney and urogenital system, organs in which MET is critical during development [73]. At the chromatin level, OVOL1 represses transcription by recruiting HDAC1 to induce remodeling at target loci [74]. In turn, ZEB1 can directly bind and repress the OVOL2 promoter, indicating a ZEB1-OVOL2 mutual inhibitory circuit [75], and OVOL1 can bind and repress its own promoter, indicating negative feedback [74], providing mechanisms by which expression of OVOL1/2 can be tuned to exert spatio-temporal control of EMT.

Interestingly, OVOL2 expression is activated by GRHL2 in kidney epithelial cells, and ectopic OVOL2 expression is sufficient to rescue the expression of CDH1, RAB25, and CLDN4 after GRHL2 knockdown, restoring luminal expansion in renal epithelial cysts in vitro [62]. Thus, in contrast to the described role of OVOL2 as a transcriptional repressor [76], this study suggests that OVOL2 may co-operate with GRHL2 to induce activation of epithelial functional genes through as-yet unidentified mechanisms.

5.3 Epithelium-Specific Ets-Domain Containing TFs

Recent attention on epithelium-specific Ets (ESE) TFs, a subfamily of E26 transformation specific (Ets)-domain containing TFs, has revealed potential roles as negative regulators of EMT and enforcers of the epithelial state. Mice with targeted mutations of ESE subfamily member ELF5/ESE2 specifically in the mammary gland display impaired lobulogenesis, decreased AJs, and increased mesenchymal markers, and ELF5 represses the Slug promoter in mouse mammary cells [77]. Another ESE subfamily member, ELF3/ESE1/ESX, is a negative regulator of EMT in ovarian cancer cells, and ELF3 overexpression induces MET with downregulation of Slug [78]. Moreover, ELF3 nuclear localization correlates with epithelial gene expression in ovarian tumors and with improved survival in patients with ovarian cancer [78]. ELF3 inhibits EMT in bladder carcinoma cells concomitant with decreased CDH2 and Slug [79]. Notably, ELF3 null mice display disrupted morphogenesis of small intestinal epithelium with impaired differentiation of absorptive enterocyte and goblet cell lineages [80], and loss of ELF causes delayed epithelial repair after selective injury to airway epithelium in adult mice [81]. Another ESE family member EHF directly represses TWIST1 and ZEB2 to control epithelial characteristics of prostate epithelial cells, and loss of EHF induces EMT [82]. ELF3, ELF5, and EHF are expressed specifically in various epithelial tissues in adult humans and mice [83], supporting broad roles in establishing and maintaining epithelial characteristics (reviewed in [84]).

5.4 Singleminded 2

The bHLH TF singleminded 2 (Sim2) restricts EMT during mammary gland morphogenesis, and Sim2 null mice display invasive TEB cells in developing mammary gland tips. Furthermore, silencing Sim2 induces EMT in a mammary epithelial cell line, and Sim2 directly represses the SNAIL2 promoter [85]. Together, the above studies suggest that ELF5, OVOL2, and Sim2 may co-operate to promote MET during mammary gland morphogenesis.

5.5 SRY-Box 3

A study of chick gastrulation revealed a fundamental role for SRY-box 3 (Sox3) in defining mesenchymal tissue boundaries. Here, Sox3 represses the Slug promoter to define non-ingressing ectoderm, whereas Slug represses the Sox3 promoter in ingressing mesendoderm [86]. Sox3-Slug antagonism thus controls cell adhesion and behavior to define cell territories rather than directly inducing cell fate. This is consistent with the role of Snail in controlling cell migration during gastrulation rather than being a specific mesoderm inducer [87]. Moreover, Sox3 loss in mouse ES cells leads to de-repression of Snail and downregulation of E-cadherin, and Sox3 overexpression in human breast cancer cell lines decreases Snail and induces CDH1 [86], supporting a role for Sox3 as a MET-TF.

Together, these studies indicate the existence of TFs that are able to restrict EMT both during developmental processes and in adult epithelial homeostasis. Of note, the evolutionary origins of the Grh family appear to pre-date the emergence of the Snail family [88, 89]. Thus, the pre-existence of mechanisms to maintain epithelial features could have provided a permissive background to allow EMT-TFs to emerge, ensuring induction of EMT is temporally controlled to allow proper patterning during development.

Signaling mechanisms upstream of MET-TFs are still poorly characterized. Intriguingly, cyclic AMP (cAMP)-mediated signaling has long been recognized to promote epithelial characteristics in cultured epithelial cells. A recent study identified a cAMP-protein kinase A (PKA)-mediated pathway as able to restore epithelial characteristics in mesenchymal breast cancer cells via direct recruitment of the H3K9 demethylase PHF2, which binds to and de-represses a range of epithelial genes including CDH1, CDH3, and CLDN4 [90]. An interesting possibility is that global target gene specificity of PHF2 is mediated by MET-TFs such as those described above; however, this is yet to be formally investigated.

5.6 Micro-RNA Control of EMT-TFs

Inhibition of EMT-TFs by micro-RNAs has emerged as an additional means by which cells exert control over EMT. In early studies, the miR-200 family, consisting of two closely related sub-families (comprising miR-200a and miR-141, and miR-200b, miR-200c and miR-429), was shown to directly inhibit ZEB1/2 to maintain E-cadherin expression in epithelial human cancer cell lines [91]. Ectopic expression of miR-200a and miR-200c induces MET in mesenchymal cancer cells [91] and prevents TGF-β-induced EMT in epithelial cells [92]. In turn, ZEB1/2 represses miR-200 family members, indicating a negative feedback loop [93, 94], allowing tight temporal control over EMT. A multitude of other miRNAs has since been demonstrated to control expression of both ZEB1/ZEB2 and other EMT-TFs, providing an additional layer of complexity to control of EMT (reviewed in [95]).

6 Cell and Tissue Architecture as a Checkpoint for EMT

Numerous studies point to post-translational control of EMT-TF activity as critical in restricting EMT. GSK3β can phosphorylate Snail to promote bTrcp-mediated ubiquitination [96]. Conversely, lysyl oxidase-like 2 (LOXL2) can stabilize the Snail protein via oxidative deamination of K98/K137 residues [97]. A recent study identified a new regulation of Snail activity via the apical-basal polarity machinery of mammary epithelial cells in 3D culture, which acts to prevent EMT and consolidate the epithelial phenotype [98]. Here, the Par complex component atypical PKC (aPKC) phosphorylates S249 of Snail to promote its degradation. This is dependent on the integrity of the apical Par3-aPKC complex in polarized cells, revealing apical-basal polarity as a key checkpoint in restricting EMT in this system [98].

In gastrulating *Drosophila* embryos, ingressing mesodermal cells of the ventral furrow exhibit AJs despite co-expressing Snail. Apical myosin contractility protects AJs from Snail-mediated disassembly, whereas in the absence of myosin contractility, AJs are disassembled in Snail-expressing cells through post-transcriptional mechanisms [99]. Thus, mechanical forces may stabilize AJs despite active Snail expression. Moreover, during development of the Malpighian tubules, the *Drosophila* renal system, mesenchymal stellate cells undergo MET to integrate with principle cells to form mature tubules, which requires basolateral cues from principal cells to allow stellate cells to integrate and establish apical-basal polarity [100].

Together, these studies suggest that cell and tissue architecture serve to control EMT and MET. Notably, early studies revealed a primary role for AJs in establishing epithelial features. For example, ectopic E-cadherin expression in mouse fibroblasts causes acquisition of Ca^{2+}-dependent intercellular contacts [101], and CDH1 overexpression in sarcoma cells causes polygonal epithelial morphology with formation of adherens and gap junctions, although not of desmosomes or TJs [102]. E-cadherin overexpression in normally mesenchymal ovarian surface cells causes MET with re-localization of α-, β- and γ-catenin and F-actin to intercellular borders, and subsequent tight junction formation and cytokeratin expression [103]. The mechanism by which AJs function upstream of establishment of epithelial features may involve interactions between E-cadherin and the catenins, which are able to function both at the cell membrane to connect AJs to the actin cytoskeleton, and also in the nucleus as transcriptional co-factors [104]. For example, in confluent colon cancer cells in culture, establishment of AJs sequesters β-catenin at intercellular borders and prevents Wnt/β-catenin activation, preventing transcriptional activation of Slug, thus stabilizing E-cadherin transcription [105]. Indeed, E-cadherin has increasingly recognized roles as a key regulator of gene transcription via interactions with signaling mediators in addition to its cell adhesive functions. For example, E-cadherin directly interacts with EGF receptors (EGFR) to influence downstream MAP kinase signaling [106], and epidermal junctional tension mediated by E-cadherin controls EGFR localization to restrict signaling activity, thereby allowing polarized tight junction formation [107]. Feedback between junction assembly and downstream signaling activity may thus be central to ensuring appropriate control over establishment of the epithelial state.

7 Conclusions

MET events are likely to be as pervasive as EMT during development and adult tissue homeostasis, yet our understanding of MET mechanisms is comparably limited. This is likely in part due to a paucity of models with which to interrogate MET. It is now recognized that MET is a critical and initiating event during reprogramming of somatic cells to pluripotency [108, 109], and that sequential EMT-MET events are likely to govern differentiation of pluripotent stem cells to a variety of somatic cell types in vitro [110, 111]. Remarkably, changes in cell shape enforced by seeding on micropatterned plates induced MET-like changes in fibroblasts and enhanced reprogramming [112], indicating possible roles for mechanotransduction in determining the epithelial state. These models will help to shed further light on the various control mechanisms that cells use to suppress and establish epithelial features. Importantly, the advent of human pluripotent stem cell derived organoid models, in which cells recapitulate complex developmental events to establish diverse tissue fates, provides a wealth of opportunities for studying EMT and MET events in more complex systems [113]. Organoids retain many three-dimensional features of tissues in vivo that are likely to be critical in determining fate during differentiation. Furthermore, the emergence of high-throughput methods to investigate genome-wide epigenetic status of developing cells will shed further light on chromatin-level control of EMT and MET events. Finally, MET is frequently dysregulated in pathological processes, highlighting the need for a better understanding of this process to design better therapeutic approaches for disease, including fibrosis due to aberrant wound healing, and cancer metastasis [56, 114].

Acknowledgments

Funding: Work in the Röper lab is supported by the Medical Research Council (file reference number U105178780 and BSF30).

References

1. Rodriguez-Boulan E, Macara IG (2014) Organization and execution of the epithelial polarity programme. Nat Rev Mol Cell Biol 15(4):225–242. https://doi.org/10.1038/nrm3775
2. Campbell K, Casanova J (2016) A common framework for EMT and collective cell migration. Development 143(23):4291–4300. https://doi.org/10.1242/dev.139071
3. Huang RY, Guilford P, Thiery JP (2012) Early events in cell adhesion and polarity during epithelial-mesenchymal transition. J Cell Sci 125(Pt 19):4417–4422. https://doi.org/10.1242/jcs.099697
4. Pei D, Shu X, Gassama-Diagne A, Thiery JP (2019) Mesenchymal-epithelial transition in development and reprogramming. Nat Cell Biol 21(1):44–53. https://doi.org/10.1038/s41556-018-0195-z
5. Takai Y, Nakanishi H (2003) Nectin and afadin: novel organizers of intercellular junctions. J Cell Sci 116(Pt 1):17–27
6. Rikitake Y, Mandai K, Takai Y (2012) The role of nectins in different types of cell-cell adhesion. J Cell Sci 125(Pt 16):3713–3722. https://doi.org/10.1242/jcs.099572
7. Roignot J, Peng X, Mostov K (2013) Polarity in mammalian epithelial morphogenesis. Cold Spring Harb Perspect Biol 5(2). https://doi.org/10.1101/cshperspect.a013789
8. Nakaya Y, Sukowati EW, Wu Y, Sheng G (2008) RhoA and microtubule dynamics control cell-basement membrane interaction in EMT during gastrulation. Nat Cell Biol 10 (7):765–775. https://doi.org/10.1038/ncb1739
9. Eckert JJ, Fleming TP (2008) Tight junction biogenesis during early development. Biochim Biophys Acta 1778(3):717–728. https://doi.org/10.1016/j.bbamem.2007.09.031
10. Frisch SM (1997) The epithelial cell default-phenotype hypothesis and its implications for cancer. Bioessays 19(8):705–709. https://doi.org/10.1002/bies.950190811
11. Acloque H, Adams MS, Fishwick K, Bronner-Fraser M, Nieto MA (2009) Epithelial-mesenchymal transitions: the importance of changing cell state in development and disease. J Clin Invest 119(6):1438–1449. https://doi.org/10.1172/JCI38019
12. McMahon AP (2016) Development of the mammalian kidney. Curr Top Dev Biol 117:31–64. https://doi.org/10.1016/bs.ctdb.2015.10.010
13. Inman JL, Robertson C, Mott JD, Bissell MJ (2015) Mammary gland development: cell fate specification, stem cells and the microenvironment. Development 142 (6):1028–1042. https://doi.org/10.1242/dev.087643
14. Ewald AJ, Huebner RJ, Palsdottir H, Lee JK, Perez MJ, Jorgens DM, Tauscher AN, Cheung KJ, Werb Z, Auer M (2012) Mammary collective cell migration involves transient loss of epithelial features and individual cell migration within the epithelium. J Cell Sci 125(Pt 11):2638–2654. https://doi.org/10.1242/jcs.096875
15. Perez-Pomares JM, Munoz-Chapuli R (2002) Epithelial-mesenchymal transitions: a mesodermal cell strategy for evolutive innovation in metazoans. Anat Rec 268 (3):343–351. https://doi.org/10.1002/ar.10165
16. Carninci P, Sandelin A, Lenhard B, Katayama S, Shimokawa K, Ponjavic J, Semple CA, Taylor MS, Engstrom PG, Frith MC, Forrest AR, Alkema WB, Tan SL, Plessy C, Kodzius R, Ravasi T, Kasukawa T, Fukuda S, Kanamori-Katayama M, Kitazume Y, Kawaji H, Kai C, Nakamura M, Konno H, Nakano K, Mottagui-Tabar S, Arner P, Chesi A, Gustincich S, Persichetti F, Suzuki H, Grimmond SM, Wells CA, Orlando V, Wahlestedt C, Liu ET, Harbers M, Kawai J, Bajic VB, Hume DA, Hayashizaki Y (2006) Genome-wide analysis of mammalian promoter architecture and evolution. Nat Genet 38(6):626–635. https://doi.org/10.1038/ng1789
17. Behrens J, Lowrick O, Klein-Hitpass L, Birchmeier W (1991) The E-cadherin promoter: functional analysis of a G.C-rich region and an epithelial cell-specific palindromic regulatory element. Proc Natl Acad Sci U S A 88 (24):11495–11499. https://doi.org/10.1073/pnas.88.24.11495
18. Ringwald M, Baribault H, Schmidt C, Kemler R (1991) The structure of the gene coding for the mouse cell adhesion molecule uvomorulin. Nucleic Acids Res 19(23):6533–6539. https://doi.org/10.1093/nar/19.23.6533
19. Bussemakers MJ, Giroldi LA, van Bokhoven A, Schalken JA (1994) Transcriptional regulation of the human E-cadherin gene in human prostate cancer cell lines: characterization of the human E-cadherin gene promoter. Biochem Biophys Res Commun 203(2):1284–1290

20. Hennig G, Lowrick O, Birchmeier W, Behrens J (1996) Mechanisms identified in the transcriptional control of epithelial gene expression. J Biol Chem 271(1):595–602. https://doi.org/10.1074/jbc.271.1.595
21. Rodrigo I, Cato AC, Cano A (1999) Regulation of E-cadherin gene expression during tumor progression: the role of a new Ets-binding site and the E-pal element. Exp Cell Res 248(2):358–371. https://doi.org/10.1006/excr.1999.4438
22. Faraldo ML, Rodrigo I, Behrens J, Birchmeier W, Cano A (1997) Analysis of the E-cadherin and P-cadherin promoters in murine keratinocyte cell lines from different stages of mouse skin carcinogenesis. Mol Carcinog 20(1):33–47
23. Batsche E, Muchardt C, Behrens J, Hurst HC, Cremisi C (1998) RB and c-Myc activate expression of the E-cadherin gene in epithelial cells through interaction with transcription factor AP-2. Mol Cell Biol 18 (7):3647–3658. https://doi.org/10.1128/mcb.18.7.3647
24. Schwartz B, Melnikova VO, Tellez C, Mourad-Zeidan A, Blehm K, Zhao YJ, McCarty M, Adam L, Bar-Eli M (2007) Loss of AP-2alpha results in deregulation of E-cadherin and MMP-9 and an increase in tumorigenicity of colon cancer cells in vivo. Oncogene 26(28):4049–4058. https://doi.org/10.1038/sj.onc.1210193
25. Leask A, Byrne C, Fuchs E (1991) Transcription factor AP2 and its role in epidermal-specific gene expression. Proc Natl Acad Sci U S A 88(18):7948–7952. https://doi.org/10.1073/pnas.88.18.7948
26. Choi I, Carey TS, Wilson CA, Knott JG (2012) Transcription factor AP-2gamma is a core regulator of tight junction biogenesis and cavity formation during mouse early embryogenesis. Development 139 (24):4623–4632. https://doi.org/10.1242/dev.086645
27. Stemmler MP, Hecht A, Kemler R (2005) E-cadherin intron 2 contains cis-regulatory elements essential for gene expression. Development 132(5):965–976. https://doi.org/10.1242/dev.01662
28. Stemmler MP, Hecht A, Kinzel B, Kemler R (2003) Analysis of regulatory elements of E-cadherin with reporter gene constructs in transgenic mouse embryos. Dev Dyn 227 (2):238–245. https://doi.org/10.1002/dvdy.10301
29. Jacobs J, Atkins M, Davie K, Imrichova H, Romanelli L, Christiaens V, Hulselmans G, Potier D, Wouters J, Taskiran II, Paciello G, Gonzalez-Blas CB, Koldere D, Aibar S, Halder G, Aerts S (2018) The transcription factor grainy head primes epithelial enhancers for spatiotemporal activation by displacing nucleosomes. Nat Genet 50(7):1011–1020. https://doi.org/10.1038/s41588-018-0140-x
30. Giroldi LA, Bringuier PP, de Weijert M, Jansen C, van Bokhoven A, Schalken JA (1997) Role of E boxes in the repression of E-cadherin expression. Biochem Biophys Res Commun 241(2):453–458. https://doi.org/10.1006/bbrc.1997.7831
31. Batlle E, Sancho E, Franci C, Dominguez D, Monfar M, Baulida J, Garcia De Herreros A (2000) The transcription factor snail is a repressor of E-cadherin gene expression in epithelial tumour cells. Nat Cell Biol 2 (2):84–89. https://doi.org/10.1038/35000034
32. Cano A, Perez-Moreno MA, Rodrigo I, Locascio A, Blanco MJ, del Barrio MG, Portillo F, Nieto MA (2000) The transcription factor snail controls epithelial-mesenchymal transitions by repressing E-cadherin expression. Nat Cell Biol 2 (2):76–83. https://doi.org/10.1038/35000025
33. Bolos V, Peinado H, Perez-Moreno MA, Fraga MF, Esteller M, Cano A (2003) The transcription factor slug represses E-cadherin expression and induces epithelial to mesenchymal transitions: a comparison with snail and E47 repressors. J Cell Sci 116 (Pt 3):499–511
34. Hajra KM, Chen DY, Fearon ER (2002) The SLUG zinc-finger protein represses E-cadherin in breast cancer. Cancer Res 62 (6):1613–1618
35. Eger A, Aigner K, Sonderegger S, Dampier B, Oehler S, Schreiber M, Berx G, Cano A, Beug H, Foisner R (2005) DeltaEF1 is a transcriptional repressor of E-cadherin and regulates epithelial plasticity in breast cancer cells. Oncogene 24(14):2375–2385. https://doi.org/10.1038/sj.onc.1208429
36. Grooteclaes ML, Frisch SM (2000) Evidence for a function of CtBP in epithelial gene regulation and anoikis. Oncogene 19 (33):3823–3828. https://doi.org/10.1038/sj.onc.1203721
37. Comijn J, Berx G, Vermassen P, Verschueren K, van Grunsven L, Bruyneel E, Mareel M, Huylebroeck D, van Roy F (2001) The two-handed E box binding zinc finger protein SIP1 downregulates E-cadherin and induces invasion. Mol Cell 7(6):1267–1278

38. Yang J, Mani SA, Donaher JL, Ramaswamy S, Itzykson RA, Come C, Savagner P, Gitelman I, Richardson A, Weinberg RA (2004) Twist, a master regulator of morphogenesis, plays an essential role in tumor metastasis. Cell 117(7):927–939. https://doi.org/10.1016/j.cell.2004.06.006
39. Skrypek N, Goossens S, De Smedt E, Vandamme N, Berx G (2017) Epithelial-to-mesenchymal transition: epigenetic reprogramming driving cellular plasticity. Trends Genet 33(12):943–959. https://doi.org/10.1016/j.tig.2017.08.004
40. Dong C, Wu Y, Yao J, Wang Y, Yu Y, Rychahou PG, Evers BM, Zhou BP (2012) G9a interacts with snail and is critical for snail-mediated E-cadherin repression in human breast cancer. J Clin Invest 122 (4):1469–1486. https://doi.org/10.1172/JCI57349
41. Lin T, Ponn A, Hu X, Law BK, Lu J (2010) Requirement of the histone demethylase LSD1 in Snai1-mediated transcriptional repression during epithelial-mesenchymal transition. Oncogene 29(35):4896–4904. https://doi.org/10.1038/onc.2010.234
42. Peinado H, Ballestar E, Esteller M, Cano A (2004) Snail mediates E-cadherin repression by the recruitment of the Sin3A/histone deacetylase 1 (HDAC1)/HDAC2 complex. Mol Cell Biol 24(1):306–319. https://doi.org/10.1128/mcb.24.1.306-319.2004
43. Herranz N, Pasini D, Diaz VM, Franci C, Gutierrez A, Dave N, Escriva M, Hernandez-Munoz I, Di Croce L, Helin K, Garcia de Herreros A, Peiro S (2008) Polycomb complex 2 is required for E-cadherin repression by the Snail1 transcription factor. Mol Cell Biol 28(15):4772–4781. https://doi.org/10.1128/MCB.00323-08
44. Tong ZT, Cai MY, Wang XG, Kong LL, Mai SJ, Liu YH, Zhang HB, Liao YJ, Zheng F, Zhu W, Liu TH, Bian XW, Guan XY, Lin MC, Zeng MS, Zeng YX, Kung HF, Xie D (2012) EZH2 supports nasopharyngeal carcinoma cell aggressiveness by forming a co-repressor complex with HDAC1/HDAC2 and snail to inhibit E-cadherin. Oncogene 31(5):583–594. https://doi.org/10.1038/onc.2011.254
45. Aigner K, Dampier B, Descovich L, Mikula M, Sultan A, Schreiber M, Mikulits W, Brabletz T, Strand D, Obrist P, Sommergruber W, Schweifer N, Wernitznig A, Beug H, Foisner R, Eger A (2007) The transcription factor ZEB1 (deltaEF1) promotes tumour cell dedifferentiation by repressing master regulators of epithelial polarity. Oncogene 26 (49):6979–6988. https://doi.org/10.1038/sj.onc.1210508
46. Spaderna S, Schmalhofer O, Wahlbuhl M, Dimmler A, Bauer K, Sultan A, Hlubek F, Jung A, Strand D, Eger A, Kirchner T, Behrens J, Brabletz T (2008) The transcriptional repressor ZEB1 promotes metastasis and loss of cell polarity in cancer. Cancer Res 68(2):537–544. https://doi.org/10.1158/0008-5472.CAN-07-5682
47. Vandewalle C, Comijn J, De Craene B, Vermassen P, Bruyneel E, Andersen H, Tulchinsky E, Van Roy F, Berx G (2005) SIP1/ZEB2 induces EMT by repressing genes of different epithelial cell-cell junctions. Nucleic Acids Res 33(20):6566–6578. https://doi.org/10.1093/nar/gki965
48. Whiteman EL, Liu CJ, Fearon ER, Margolis B (2008) The transcription factor snail represses Crumbs3 expression and disrupts apico-basal polarity complexes. Oncogene 27 (27):3875–3879. https://doi.org/10.1038/onc.2008.9
49. De Craene B, Gilbert B, Stove C, Bruyneel E, van Roy F, Berx G (2005) The transcription factor snail induces tumor cell invasion through modulation of the epithelial cell differentiation program. Cancer Res 65 (14):6237–6244. https://doi.org/10.1158/0008-5472.CAN-04-3545
50. Ohkubo T, Ozawa M (2004) The transcription factor snail downregulates the tight junction components independently of E-cadherin downregulation. J Cell Sci 117 (Pt 9):1675–1685. https://doi.org/10.1242/jcs.01004
51. Ikenouchi J, Matsuda M, Furuse M, Tsukita S (2003) Regulation of tight junctions during the epithelium-mesenchyme transition: direct repression of the gene expression of claudins/occludin by snail. J Cell Sci 116 (Pt 10):1959–1967. https://doi.org/10.1242/jcs.00389
52. Martinez-Estrada OM, Culleres A, Soriano FX, Peinado H, Bolos V, Martinez FO, Reina M, Cano A, Fabre M, Vilaro S (2006) The transcription factors slug and snail act as repressors of Claudin-1 expression in epithelial cells. Biochem J 394(Pt 2):449–457. https://doi.org/10.1042/BJ20050591
53. Peinado H, Olmeda D, Cano A (2007) Snail, Zeb and bHLH factors in tumour progression: an alliance against the epithelial phenotype? Nat Rev Cancer 7(6):415–428. https://doi.org/10.1038/nrc2131
54. Shamir ER, Pappalardo E, Jorgens DM, Coutinho K, Tsai WT, Aziz K, Auer M, Tran

PT, Bader JS, Ewald AJ (2014) Twist1-induced dissemination preserves epithelial identity and requires E-cadherin. J Cell Biol 204(5):839–856. https://doi.org/10.1083/jcb.201306088

55. Chaffer CL, Weinberg RA (2011) A perspective on cancer cell metastasis. Science 331 (6024):1559–1564. https://doi.org/10.1126/science.1203543

56. Brabletz T, Kalluri R, Nieto MA, Weinberg RA (2018) EMT in cancer. Nat Rev Cancer 18(2):128–134. https://doi.org/10.1038/nrc.2017.118

57. Campbell K (2018) Contribution of epithelial-mesenchymal transitions to organogenesis and cancer metastasis. Curr Opin Cell Biol 55:30–35. https://doi.org/10.1016/j.ceb.2018.06.008

58. Cheung KJ, Padmanaban V, Silvestri V, Schipper K, Cohen JD, Fairchild AN, Gorin MA, Verdone JE, Pienta KJ, Bader JS, Ewald AJ (2016) Polyclonal breast cancer metastases arise from collective dissemination of keratin 14-expressing tumor cell clusters. Proc Natl Acad Sci U S A 113(7):E854–E863. https://doi.org/10.1073/pnas.1508541113

59. Aceto N, Bardia A, Miyamoto DT, Donaldson MC, Wittner BS, Spencer JA, Yu M, Pely A, Engstrom A, Zhu H, Brannigan BW, Kapur R, Stott SL, Shioda T, Ramaswamy S, Ting DT, Lin CP, Toner M, Haber DA, Maheswaran S (2014) Circulating tumor cell clusters are oligoclonal precursors of breast cancer metastasis. Cell 158(5):1110–1122. https://doi.org/10.1016/j.cell.2014.07.013

60. Pyrgaki C, Liu A, Niswander L (2011) Grainyhead-like 2 regulates neural tube closure and adhesion molecule expression during neural fold fusion. Dev Biol 353(1):38–49. https://doi.org/10.1016/j.ydbio.2011.02.027

61. Werth M, Walentin K, Aue A, Schonheit J, Wuebken A, Pode-Shakked N, Vilianovitch L, Erdmann B, Dekel B, Bader M, Barasch J, Rosenbauer F, Luft FC, Schmidt-Ott KM (2010) The transcription factor grainyhead-like 2 regulates the molecular composition of the epithelial apical junctional complex. Development 137 (22):3835–3845. https://doi.org/10.1242/dev.055483

62. Aue A, Hinze C, Walentin K, Ruffert J, Yurtdas Y, Werth M, Chen W, Rabien A, Kilic E, Schulzke JD, Schumann M, Schmidt-Ott KM (2015) A Grainyhead-like 2/Ovo-like 2 pathway regulates renal epithelial barrier function and lumen expansion. J Am Soc Nephrol 26(11):2704–2715. https://doi.org/10.1681/ASN.2014080759

63. Gao X, Vockley CM, Pauli F, Newberry KM, Xue Y, Randell SH, Reddy TE, Hogan BL (2013) Evidence for multiple roles for grainyhead-like 2 in the establishment and maintenance of human mucociliary airway epithelium.[corrected]. Proc Natl Acad Sci U S A 110(23):9356–9361. https://doi.org/10.1073/pnas.1307589110

64. Senga K, Mostov KE, Mitaka T, Miyajima A, Tanimizu N (2012) Grainyhead-like 2 regulates epithelial morphogenesis by establishing functional tight junctions through the organization of a molecular network among claudin3, claudin4, and Rab25. Mol Biol Cell 23 (15):2845–2855. https://doi.org/10.1091/mbc.E12-02-0097

65. Mlacki M, Kikulska A, Krzywinska E, Pawlak M, Wilanowski T (2015) Recent discoveries concerning the involvement of transcription factors from the Grainyhead-like family in cancer. Exp Biol Med 240 (11):1396–1401. https://doi.org/10.1177/1535370215588924

66. Cieply B, Riley P, Pifer PM, Widmeyer J, Addison JB, Ivanov AV, Denvir J, Frisch SM (2012) Suppression of the epithelial-mesenchymal transition by Grainyhead-like-2. Cancer Res 72(9):2440–2453. https://doi.org/10.1158/0008-5472.CAN-11-4038

67. Chung VY, Tan TZ, Tan M, Wong MK, Kuay KT, Yang Z, Ye J, Muller J, Koh CM, Guccione E, Thiery JP, Huang RY (2016) GRHL2-miR-200-ZEB1 maintains the epithelial status of ovarian cancer through transcriptional regulation and histone modification. Sci Rep 6:19943. https://doi.org/10.1038/srep19943

68. Nevil M, Bondra ER, Schulz KN, Kaplan T, Harrison MM (2017) Stable binding of the conserved transcription factor grainy head to its target genes throughout *Drosophila melanogaster* development. Genetics 205 (2):605–620. https://doi.org/10.1534/genetics.116.195685

69. Boglev Y, Wilanowski T, Caddy J, Parekh V, Auden A, Darido C, Hislop NR, Cangkrama M, Ting SB, Jane SM (2011) The unique and cooperative roles of the grainy head-like transcription factors in epidermal development reflect unexpected target gene specificity. Dev Biol 349(2):512–522. https://doi.org/10.1016/j.ydbio.2010.11.011

70. Watanabe K, Villarreal-Ponce A, Sun P, Salmans ML, Fallahi M, Andersen B, Dai X (2014) Mammary morphogenesis and regeneration require the inhibition of EMT at terminal end buds by Ovol2 transcriptional repressor. Dev Cell 29(1):59–74. https://doi.org/10.1016/j.devcel.2014.03.006
71. Lee B, Villarreal-Ponce A, Fallahi M, Ovadia J, Sun P, Yu QC, Ito S, Sinha S, Nie Q, Dai X (2014) Transcriptional mechanisms link epithelial plasticity to adhesion and differentiation of epidermal progenitor cells. Dev Cell 29(1):47–58. https://doi.org/10.1016/j.devcel.2014.03.005
72. Kitazawa K, Hikichi T, Nakamura T, Mitsunaga K, Tanaka A, Nakamura M, Yamakawa T, Furukawa S, Takasaka M, Goshima N, Watanabe A, Okita K, Kawasaki S, Ueno M, Kinoshita S, Masui S (2016) OVOL2 maintains the transcriptional program of human corneal epithelium by suppressing epithelial-to-mesenchymal transition. Cell Rep 15(6):1359–1368. https://doi.org/10.1016/j.celrep.2016.04.020
73. Dai X, Schonbaum C, Degenstein L, Bai W, Mahowald A, Fuchs E (1998) The ovo gene required for cuticle formation and oogenesis in flies is involved in hair formation and spermatogenesis in mice. Genes Dev 12 (21):3452–3463. https://doi.org/10.1101/gad.12.21.3452
74. Nair M, Bilanchone V, Ortt K, Sinha S, Dai X (2007) Ovol1 represses its own transcription by competing with transcription activator c-Myb and by recruiting histone deacetylase activity. Nucleic Acids Res 35(5):1687–1697. https://doi.org/10.1093/nar/gkl1141
75. Hong T, Watanabe K, Ta CH, Villarreal-Ponce A, Nie Q, Dai X (2015) An Ovol2-Zeb1 mutual inhibitory circuit governs bidirectional and multi-step transition between epithelial and mesenchymal states. PLoS Comput Biol 11(11):e1004569. https://doi.org/10.1371/journal.pcbi.1004569
76. Watanabe K, Liu Y, Noguchi S, Murray M, Chang JC, Kishima M, Nishimura H, Hashimoto K, Minoda A, Suzuki H (2019) OVOL2 induces mesenchymal-to-epithelial transition in fibroblasts and enhances cell-state reprogramming towards epithelial lineages. Sci Rep 9(1):6490. https://doi.org/10.1038/s41598-019-43021-z
77. Chakrabarti R, Hwang J, Andres Blanco M, Wei Y, Lukacisin M, Romano RA, Smalley K, Liu S, Yang Q, Ibrahim T, Mercatali L, Amadori D, Haffty BG, Sinha S, Kang Y (2012) Elf5 inhibits the epithelial-mesenchymal transition in mammary gland development and breast cancer metastasis by transcriptionally repressing Snail2. Nat Cell Biol 14(11):1212–1222. https://doi.org/10.1038/ncb2607
78. Yeung TL, Leung CS, Wong KK, Gutierrez-Hartmann A, Kwong J, Gershenson DM, Mok SC (2017) ELF3 is a negative regulator of epithelial-mesenchymal transition in ovarian cancer cells. Oncotarget 8 (10):16951–16963. https://doi.org/10.18632/oncotarget.15208
79. Gondkar K, Patel K, Krishnappa S, Patil A, Nair B, Sundaram GM, Zea TT, Kumar P (2019) E74 like ETS transcription factor 3 (ELF3) is a negative regulator of epithelial-mesenchymal transition in bladder carcinoma. Cancer Biomark 25(2):223–232. https://doi.org/10.3233/CBM-190013
80. Ng AY, Waring P, Ristevski S, Wang C, Wilson T, Pritchard M, Hertzog P, Kola I (2002) Inactivation of the transcription factor Elf3 in mice results in dysmorphogenesis and altered differentiation of intestinal epithelium. Gastroenterology 122(5):1455–1466
81. Oliver JR, Kushwah R, Wu J, Pan J, Cutz E, Yeger H, Waddell TK, Hu J (2011) Elf3 plays a role in regulating bronchiolar epithelial repair kinetics following Clara cell-specific injury. Lab Invest 91(10):1514–1529. https://doi.org/10.1038/labinvest.2011.100
82. Albino D, Longoni N, Curti L, Mello-Grand-M, Pinton S, Civenni G, Thalmann G, D'Ambrosio G, Sarti M, Sessa F, Chiorino G, Catapano CV, Carbone GM (2012) ESE3/EHF controls epithelial cell differentiation and its loss leads to prostate tumors with mesenchymal and stem-like features. Cancer Res 72(11):2889–2900. https://doi.org/10.1158/0008-5472.CAN-12-0212
83. Tymms MJ, Ng AY, Thomas RS, Schutte BC, Zhou J, Eyre HJ, Sutherland GR, Seth A, Rosenberg M, Papas T, Debouck C, Kola I (1997) A novel epithelial-expressed ETS gene, ELF3: human and murine cDNA sequences, murine genomic organization, human mapping to 1q32.2 and expression in tissues and cancer. Oncogene 15 (20):2449–2462. https://doi.org/10.1038/sj.onc.1201427
84. Oliver JR, Kushwah R, Hu J (2012) Multiple roles of the epithelium-specific ETS transcription factor, ESE-1, in development and disease. Lab Invest 92(3):320–330. https://doi.org/10.1038/labinvest.2011.186
85. Laffin B, Wellberg E, Kwak HI, Burghardt RC, Metz RP, Gustafson T, Schedin P, Porter

WW (2008) Loss of singleminded-2s in the mouse mammary gland induces an epithelial-mesenchymal transition associated with up-regulation of slug and matrix metalloprotease 2. Mol Cell Biol 28(6):1936–1946. https://doi.org/10.1128/MCB.01701-07

86. Acloque H, Ocana OH, Matheu A, Rizzoti K, Wise C, Lovell-Badge R, Nieto MA (2011) Reciprocal repression between Sox3 and snail transcription factors defines embryonic territories at gastrulation. Dev Cell 21 (3):546–558. https://doi.org/10.1016/j.devcel.2011.07.005
87. Barrallo-Gimeno A, Nieto MA (2005) The snail genes as inducers of cell movement and survival: implications in development and cancer. Development 132(14):3151–3161. https://doi.org/10.1242/dev.01907
88. Traylor-Knowles N, Hansen U, Dubuc TQ, Martindale MQ, Kaufman L, Finnerty JR (2010) The evolutionary diversification of LSF and Grainyhead transcription factors preceded the radiation of basal animal lineages. BMC Evol Biol 10:101. https://doi.org/10.1186/1471-2148-10-101
89. Barrallo-Gimeno A, Nieto MA (2009) Evolutionary history of the snail/scratch superfamily. Trends Genet 25(6):248–252. https://doi.org/10.1016/j.tig.2009.04.001
90. Pattabiraman DR, Bierie B, Kober KI, Thiru P, Krall JA, Zill C, Reinhardt F, Tam WL, Weinberg RA (2016) Activation of PKA leads to mesenchymal-to-epithelial transition and loss of tumor-initiating ability. Science 351(6277):aad3680. https://doi.org/10.1126/science.aad3680
91. Park SM, Gaur AB, Lengyel E, Peter ME (2008) The miR-200 family determines the epithelial phenotype of cancer cells by targeting the E-cadherin repressors ZEB1 and ZEB2. Genes Dev 22(7):894–907. https://doi.org/10.1101/gad.1640608
92. Gregory PA, Bert AG, Paterson EL, Barry SC, Tsykin A, Farshid G, Vadas MA, Khew-Goodall Y, Goodall GJ (2008) The miR-200 family and miR-205 regulate epithelial to mesenchymal transition by targeting ZEB1 and SIP1. Nat Cell Biol 10(5):593–601. https://doi.org/10.1038/ncb1722
93. Bracken CP, Gregory PA, Kolesnikoff N, Bert AG, Wang J, Shannon MF, Goodall GJ (2008) A double-negative feedback loop between ZEB1-SIP1 and the microRNA-200 family regulates epithelial-mesenchymal transition. Cancer Res 68(19):7846–7854. https://doi.org/10.1158/0008-5472.CAN-08-1942
94. Burk U, Schubert J, Wellner U, Schmalhofer O, Vincan E, Spaderna S, Brabletz T (2008) A reciprocal repression between ZEB1 and members of the miR-200 family promotes EMT and invasion in cancer cells. EMBO Rep 9(6):582–589. https://doi.org/10.1038/embor.2008.74
95. Abba ML, Patil N, Leupold JH, Allgayer H (2016) MicroRNA regulation of epithelial to mesenchymal transition. J Clin Med 5(1). https://doi.org/10.3390/jcm5010008
96. Zhou BP, Deng J, Xia W, Xu J, Li YM, Gunduz M, Hung MC (2004) Dual regulation of snail by GSK-3beta-mediated phosphorylation in control of epithelial-mesenchymal transition. Nat Cell Biol 6 (10):931–940. https://doi.org/10.1038/ncb1173
97. Peinado H, Del Carmen Iglesias-de la Cruz M, Olmeda D, Csiszar K, Fong KS, Vega S, Nieto MA, Cano A, Portillo F (2005) A molecular role for lysyl oxidase-like 2 enzyme in snail regulation and tumor progression. EMBO J 24(19):3446–3458. https://doi.org/10.1038/sj.emboj.7600781
98. Jung HY, Fattet L, Tsai JH, Kajimoto T, Chang Q, Newton AC, Yang J (2019) Apical-basal polarity inhibits epithelial-mesenchymal transition and tumour metastasis by PAR-complex-mediated SNAI1 degradation. Nat Cell Biol 21(3):359–371. https://doi.org/10.1038/s41556-019-0291-8
99. Weng M, Wieschaus E (2016) Myosin-dependent remodeling of adherens junctions protects junctions from snail-dependent disassembly. J Cell Biol 212(2):219–229. https://doi.org/10.1083/jcb.201508056
100. Campbell K, Casanova J, Skaer H (2010) Mesenchymal-to-epithelial transition of intercalating cells in drosophila renal tubules depends on polarity cues from epithelial neighbours. Mech Dev 127(7-8):345–357. https://doi.org/10.1016/j.mod.2010.04.002
101. Nagafuchi A, Shirayoshi Y, Okazaki K, Yasuda K, Takeichi M (1987) Transformation of cell adhesion properties by exogenously introduced E-cadherin cDNA. Nature 329 (6137):341–343. https://doi.org/10.1038/329341a0
102. Mege RM, Matsuzaki F, Gallin WJ, Goldberg JI, Cunningham BA, Edelman GM (1988) Construction of epithelioid sheets by transfection of mouse sarcoma cells with cDNAs for chicken cell adhesion molecules. Proc Natl

Acad Sci U S A 85(19):7274–7278. https://doi.org/10.1073/pnas.85.19.7274
103. Auersperg N, Pan J, Grove BD, Peterson T, Fisher J, Maines-Bandiera S, Somasiri A, Roskelley CD (1999) E-cadherin induces mesenchymal-to-epithelial transition in human ovarian surface epithelium. Proc Natl Acad Sci U S A 96(11):6249–6254. https://doi.org/10.1073/pnas.96.11.6249
104. McCrea PD, Gottardi CJ (2016) Beyond beta-catenin: prospects for a larger catenin network in the nucleus. Nat Rev Mol Cell Biol 17(1):55–64. https://doi.org/10.1038/nrm.2015.3
105. Conacci-Sorrell M, Simcha I, Ben-Yedidia T, Blechman J, Savagner P, Ben-Ze'ev A (2003) Autoregulation of E-cadherin expression by cadherin-cadherin interactions: the roles of beta-catenin signaling, slug, and MAPK. J Cell Biol 163(4):847–857. https://doi.org/10.1083/jcb.200308162
106. Pece S, Gutkind JS (2000) Signaling from E-cadherins to the MAPK pathway by the recruitment and activation of epidermal growth factor receptors upon cell-cell contact formation. J Biol Chem 275 (52):41227–41233. https://doi.org/10.1074/jbc.M006578200
107. Rubsam M, Mertz AF, Kubo A, Marg S, Jungst C, Goranci-Buzhala G, Schauss AC, Horsley V, Dufresne ER, Moser M, Ziegler W, Amagai M, Wickstrom SA, Niessen CM (2017) E-cadherin integrates mechanotransduction and EGFR signaling to control junctional tissue polarization and tight junction positioning. Nat Commun 8(1):1250. https://doi.org/10.1038/s41467-017-01170-7
108. Li R, Liang J, Ni S, Zhou T, Qing X, Li H, He W, Chen J, Li F, Zhuang Q, Qin B, Xu J, Li W, Yang J, Gan Y, Qin D, Feng S, Song H, Yang D, Zhang B, Zeng L, Lai L, Esteban MA, Pei D (2010) A mesenchymal-to-epithelial transition initiates and is required for the nuclear reprogramming of mouse fibroblasts. Cell Stem Cell 7(1):51–63. https://doi.org/10.1016/j.stem.2010.04.014
109. Samavarchi-Tehrani P, Golipour A, David L, Sung HK, Beyer TA, Datti A, Woltjen K, Nagy A, Wrana JL (2010) Functional genomics reveals a BMP-driven mesenchymal-to-epithelial transition in the initiation of somatic cell reprogramming. Cell Stem Cell 7(1):64–77. https://doi.org/10.1016/j.stem.2010.04.015
110. Li Q, Hutchins AP, Chen Y, Li S, Shan Y, Liao B, Zheng D, Shi X, Li Y, Chan WY, Pan G, Wei S, Shu X, Pei D (2017) A sequential EMT-MET mechanism drives the differentiation of human embryonic stem cells towards hepatocytes. Nat Commun 8:15166. https://doi.org/10.1038/ncomms15166
111. Ward C, Volpe G, Cauchy P, Ptasinska A, Almaghrabi R, Blakemore D, Nafria M, Kestner D, Frampton J, Murphy G, Buganim Y, Kaji K, Garcia P (2018) Fine-tuning Mybl2 is required for proper mesenchymal-to-epithelial transition during somatic reprogramming. Cell Rep 24 (6):1496–1511. e1498. https://doi.org/10.1016/j.celrep.2018.07.026
112. Downing TL, Soto J, Morez C, Houssin T, Fritz A, Yuan F, Chu J, Patel S, Schaffer DV, Li S (2013) Biophysical regulation of epigenetic state and cell reprogramming. Nat Mater 12(12):1154–1162. https://doi.org/10.1038/nmat3777
113. Fatehullah A, Tan SH, Barker N (2016) Organoids as an in vitro model of human development and disease. Nat Cell Biol 18 (3):246–254. https://doi.org/10.1038/ncb3312
114. Shaw TJ, Martin P (2016) Wound repair: a showcase for cell plasticity and migration. Curr Opin Cell Biol 42:29–37. https://doi.org/10.1016/j.ceb.2016.04.001

Part III

Methods to Study EMT in In Vivo Systems

Chapter 8

Live Imaging of Epithelial-Mesenchymal Transition in Mesoderm Cells of Gastrulating *Drosophila* Embryos

Lingkun Gu and Mo Weng

Abstract

Epithelial-mesenchymal transitions (EMTs) drive the generation of cell diversity during both evolution and development. More and more evidence has pointed to a model where EMT is not a binary switch but a reversible process that can be stabilized at intermediate states. Despite our vast knowledge on the signaling pathways that trigger EMT, we know very little about how EMT happens in a step-wise manner. Live imaging of cells that are undergoing EMT in intact, living, animals will provide us valuable insights into how EMT is executed at both the cellular and molecular levels and help us identify and understand the intermediate states. Here, we describe how to image early stages of EMT in the mesoderm cells of live *Drosophila melanogaster* embryos and how to image contractile myosin that suspends EMT progression.

Key words Live imaging, Drosophila embryo, Epithelial-mesenchymal transition, Adherens junctions, Myosin

1 Introduction

The epithelial-mesenchymal transition (EMT) is a process where highly organized epithelial cells lose cell polarity and junctions and leave epithelial sheets to become migratory mesenchymal cells. The emergence of EMT allowed the evolution from animals with epithelial cells only, to animals with diverse tissue types [1]. Accordingly, during embryogenesis, the first EMT occurs upon gastrulation to generate three germ layers, which lay out the foundations of different tissues [1]. Despite EMT being first discovered in development, it has also been found to play a central role during tumorigenesis [2, 3]. Recent advances in EMT studies suggest that cells that activate EMT programs often only progress partially along the epithelial-mesenchymal axis, expressing a mixture of epithelial and mesenchymal markers. In addition, EMT can be followed by its reverse process MET, such as in the case of endoderm development, epithelial wound healing, and tumor metastasis. The concept of

Kyra Campbell and Eric Theveneau (eds.), *The Epithelial-to Mesenchymal Transition: Methods and Protocols*, Methods in Molecular Biology, vol. 2179, https://doi.org/10.1007/978-1-0716-0779-4_8,

discrete cell states along the epithelial-mesenchymal spectrum of phenotypes has been raised [4, 5]. Therefore, observing EMT progression of living animals is a powerful tool to understand what the intermediate cell states are, and how cells reach these intermediate states.

One of the central steps during EMT is the loss of cell-cell junctions, so that the cells restricted within the epithelial sheet can dissociate from each other [6]. As the basic structure to connect cells and support cell shapes, cell-cell junctions not only respond to EMT signaling, but also are regulated by many other intrinsic and extrinsic cues, including other signals or physical tension. Therefore, cell-cell junctions such as adherens junctions can act as a platform where other inputs modulate EMT progression. Indeed, EMT during development often happens simultaneously with other morphogenic events. During gastrulation in *Drosophila melanogaster* (*Drosophila*), the mesoderm primordium undergoes both EMT and mechanical tension-driven epithelial folding. While the EMT program aims to disassemble junctions, epithelial folding requires strong junctions to maintain tissue integrity and effect tissue-wide shape changes. It was shown that mechanical tension not only drives epithelial folding but also overrides the EMT program to strengthen adherens junctions [7, 8]. This provides an excellent system to study EMT in the context of epithelial morphogenesis.

Drosophila embryos have several characteristics that facilitate high-quality live imaging and in order to understand early embryo morphogenesis, it is important to do live imaging properly. The adherens junctions in early *Drosophila* embryos are very close to the embryo surface and therefore can be steadily imaged with laser scanning confocal microscopy. Before gastrulation, *Drosophila* embryos consist of a single layer of epithelial cells surrounding the yolk with apical surfaces facing outside (Fig. 1a). Cells are about 30 μm tall and 6–7 μm wide, with spot adherens junctions localized in a relatively narrow zone within 6–7 μm below the apical surface (Fig. 1b). This single layer of epithelial cells is generated during a process called cellularization, where cell membranes of the syncytial embryo gradually ingress between nuclei and grow deeper to eventually enclose the nuclei and cytoplasm surrounding the yolk. Cellularization happens 2 h after egg laying and it takes 50 min to 1 h for the cellularization membrane front to reach the end of the cytoplasm. Therefore, the depth of cellularization front from the embryo surfaces can be used to estimate the embryo's age for live imaging (Fig. 2). Spot adherens junctions are formed in all the cells during early cellularization and appear as clusters of cadherin-catenin complexes around the subapical region and only become more belt-like much later in the development. No other junctions, such as septate junctions and focal adhesions, have formed during this early stage of development [9]. Immediately

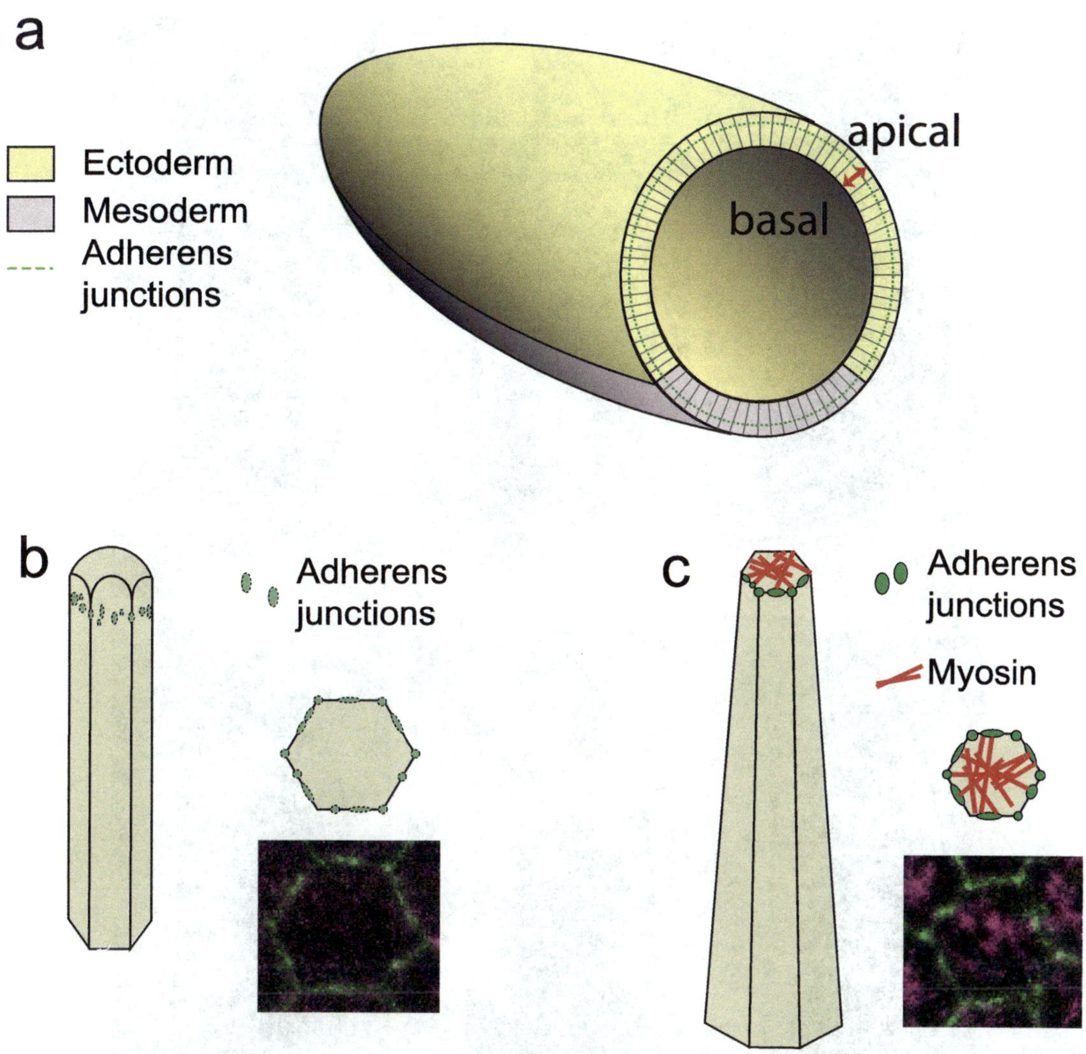

Fig. 1 Schematics of an early *Drosophila* embryo. (**a**) Before gastrulation, *Drosophila* embryos have a single layer of epithelial cells with the apical surface facing outside. (**b**) A mesodermal epithelial cell before apical constriction with spot adherens junctions concentrated at the subapical position. Also shown is the schematic of the *en face* view of cells from their apical surfaces/embryo surface. Below the schematic is a confocal image of such a view, E-cadherin::GFP in green, and myosin::mCherry in magenta. (**c**) A mesodermal epithelial cell during apical constriction with reinforced adherens junctions at the very apical edges of the cell and contractile actomyosin network on the apical cortex. The *en face* views in a schematic and a confocal image

following cellularization, gastrulation starts with the epithelial folding of the mesoderm to form the ventral furrow. This ventral furrow formation is driven by mechanical force generated from myosin contraction, on the apical surface of the mesoderm. This apical myosin contraction leads to apical constriction (Fig. 1b, c). The bulk of contractile or activated myosin is on the very apical surface of the cells and the fibers extend to the adherens junctions. In response to the myosin contraction, adherens junctions in the

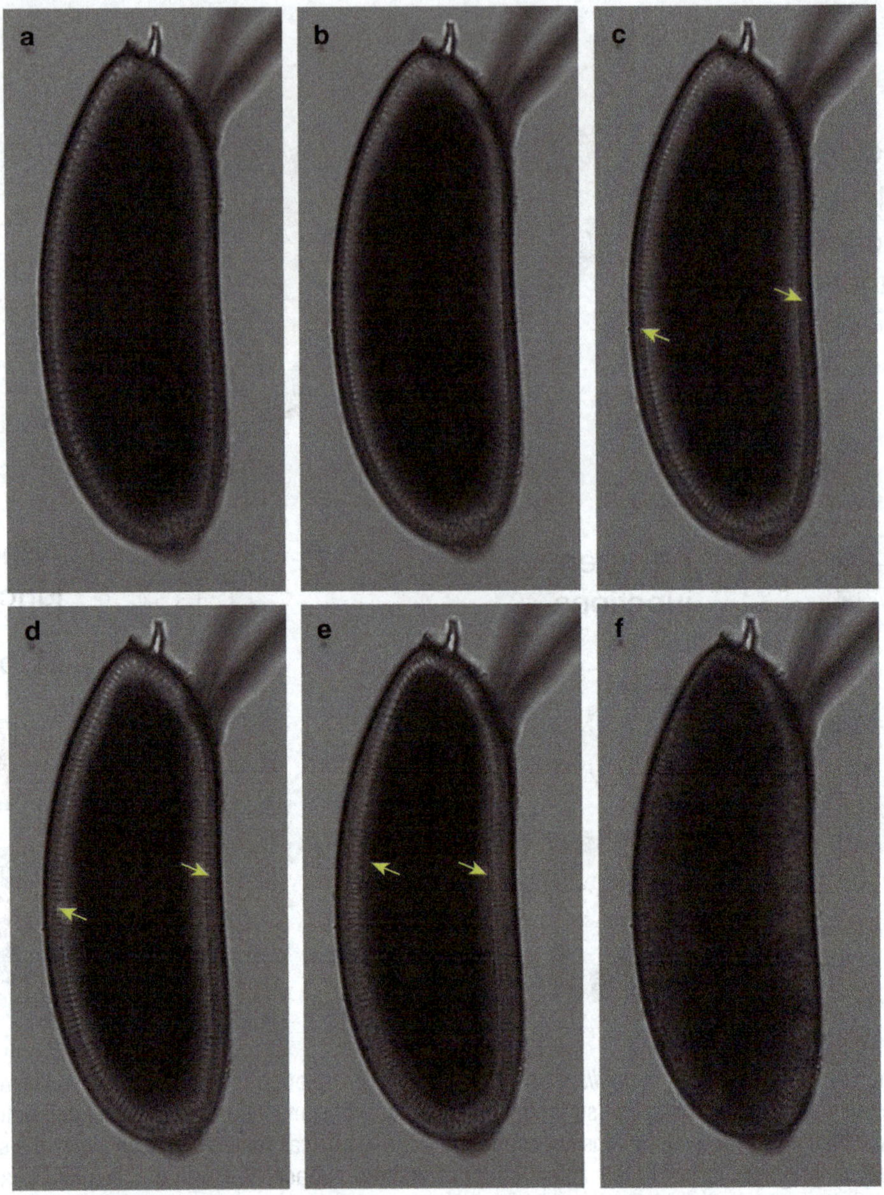

Fig. 2 Bright-field views of a *Drosophila* embryo from cellularization to gastrulation. In all images, ventral side is left and the anterior side points up. Yolk appears dark while the cytoplasm appears as a transparent circle. (**a**) The end of syncytial nuclear cycle 12. The bump-like objects at the periphery of the egg are nuclei. Yolk and cytoplasm are separating, but the boundary is unclear. Embryos of this age or younger are more fragile and should be handled with more care. (**b**) The end of the last syncytial nuclear division (cycle 13). Nuclei appear to be smaller due to smaller spacing and slight elongation. Yolk and cytoplasm have separated further. The cellularization front is hardly visible. (**c**) Early cellularizing embryo. The cellularization front becomes visible as a straight, sharp, and thin line, largely in parallel to the embryo surface (yellow arrows). Nuclei are further elongated. The yolk/cytoplasm boundary becomes sharper. (**d**) Cellularization proceeds further and the cellularization front is readily visible (yellow arrows). Pick embryos of this age or younger for imaging junctions loss and reversal during EMT and epithelial folding. (**e**) Cellularization on the ventral side is almost finished.

mesoderm migrate toward apical direction and are strengthened—eventually become highly concentrated at the apical edge of the cells [7].

Adherens junction levels during the EMT of *Drosophila* mesoderm primordium do not change unidirectionally. Instead, there is a transient upregulation following the initial downregulation [7]. EMT in those cells is driven by the transcription factor Snail, a conserved master regulator of EMT in many other systems [10]. In the early *Drosophila* embryo, Snail is zygotically expressed specifically in the mesoderm and gradually accumulates to high levels from before cellularization to gastrulation [7, 11]. This high level of Snail eventually drives the mesoderm to undergo EMT [12]. Although the disassociation of cells only occurs after the folding of the mesoderm, it was shown that the adherens junctions start to be broken down before the epithelial folding event [7]. This is consistent with many of Snail's transcriptional targets being activated or repressed before ventral furrow formation [13–15]. However, during the epithelial folding, the contractile myosin, which drives the apical constriction and eventually the folding of the mesodermal epithelium, relocates and strengthens adherens junctions despite the continued expression of Snail (Figs. 2 and 3). Such a reverse in junction levels is possible because junctions are downregulated by Snail at the step of assembly and disassembly [8]. Only after the mesoderm is completely internalized and myosin is inactivated, does Snail-dependent junction disassembly resume, and mesodermal cells undergo a full EMT.

The approach described here relies on mounting the embryos directly on the cover glass taking advantage of the fact that both the embryos and the glass are hydrophobic and therefore stick to each other. This method for live imaging embryos avoids the use of glue and allows continued readjustment of the orientation of the embryo. While it creates a small footprint of the embryo on the glass, within this area light goes through only the vitelline membrane, glass, and emersion oil, thus greatly improving image quality. We will discuss how to identify healthy embryos of the right stages and orientation, as this is also critical for the efficient imaging. Finally, how fast imaging can be achieved through the reduction of photobleaching will also be discussed.

Fig. 2 (continued) Nuclei on the ventral side appear to be less aligned, indicating the initiation of apical constriction. Gastrulation begins. (**f**) Gastrulation proceeds with the ventral furrow already formed and the posterior midgut reaching the dorsal side

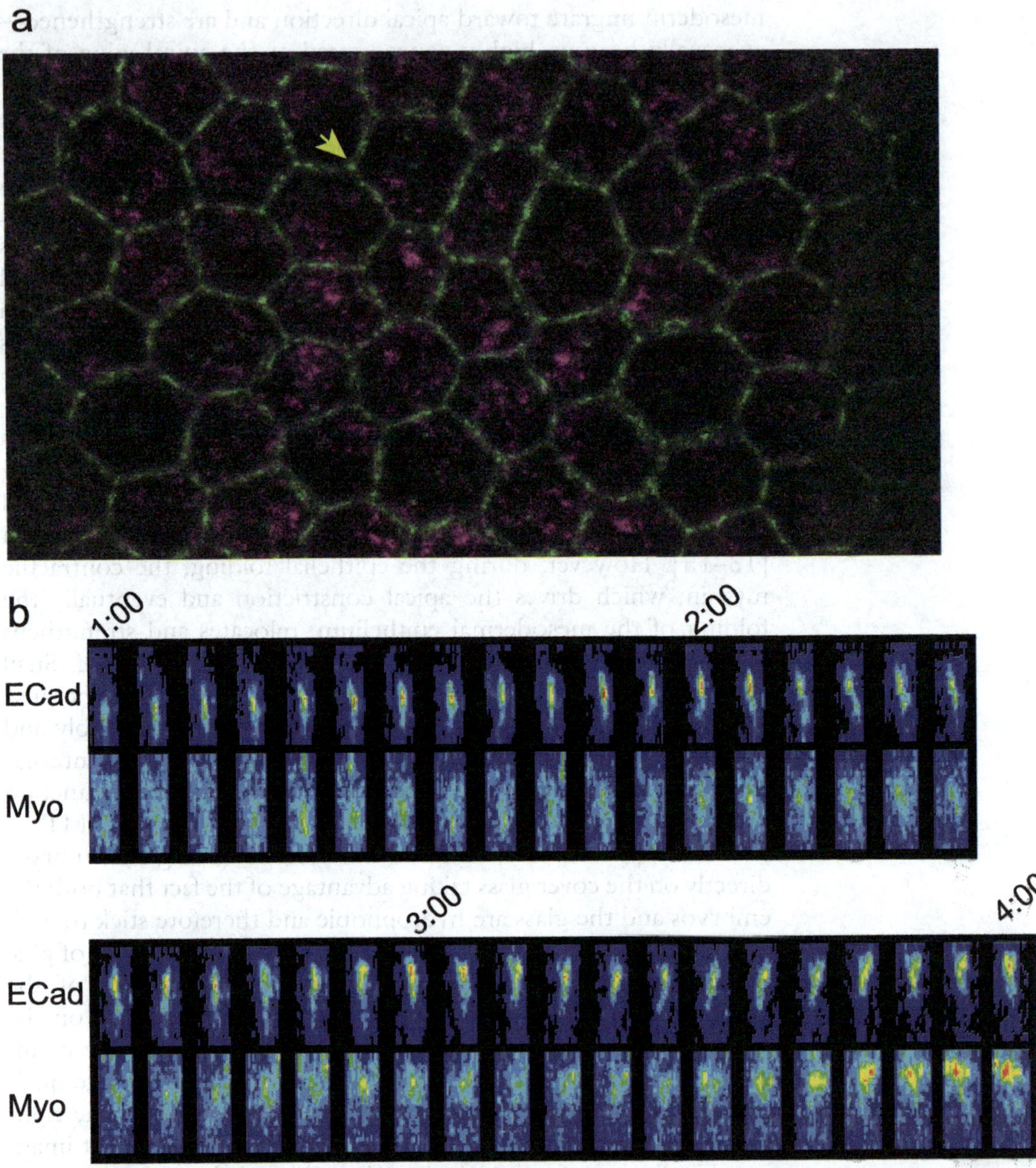

Fig. 3 Examples of confocal images of live embryos expressing E-Cad::GFP and myosin::mCherry (myosin regulatory light chain). (**a**) A *z*-projection of ventral cells undergoing apical constriction. E-**C**adherin is in green and **M**yosin in magenta. The arrow indicates the tracked junction cluster shown in (**b**). (**b**) An example of a junction cluster tracked for 3 min. The junction cluster is 3D reconstructed from the confocal stacks. Images are pseudocolored to reflect the fluorescence intensities. The upper stripe is E-cadherin::GFP and lower one is myosin::mCherry within the same volume. The temporal resolution is 5 s

2 Materials

2.1 Embryo Collection

1. Plastic cups with punctured small holes.
2. 60 mm apple juice agar plates (to make 500 ml: Drosophila Agar type II 9 g, sucrose 12 g, apple juice 250 ml, H_2O 250 ml).
3. Yeast paste.
4. 60 mm petri dish lid.

2.2 Embryo Preparation and Mounting

1. Stereomicroscope with a transmitted light source (e.g., Zeiss Stemi 508).
2. 200 μm pipettor or glass Pasteur pipette.
3. Fine forceps.
4. Paint brushes: one with most bristles removed and another one with only one bristle.
5. 35 mm glass bottom Petri dish.
6. Absorbent C-fold paper towels: regular size and cut into about 2 cm × 2 cm pieces.
7. Air permeable membrane.
8. Water in a wash bottle.
9. Halocarbon oil 27.
10. 4% sodium hypochlorite/bleach: dilute household bleach with water (household bleach from grocery stores usually contains 6% or 8% sodium hypochlorite).

2.3 Live Imaging

1. Laser scanning confocal microscope.
2. Plan Apochromatic 63× oil immersion objective.

3 Methods

3.1 Embryo Collection

1. Place 40 to 60 young flies expressing desired fluorescent proteins in the egg collection cup and cover the opening with an apple juice plate with a drop of yeast paste. Keep the cup upside down with a rubber band securing the plate to the cup. Change the plate every day to feed the flies for about 2 days so the flies start to lay abundant amounts of eggs.
2. On the day of imaging, change to a room-temperature apple juice plate with yeast paste on it and collect the eggs for desired length of time. Keep the cup in a dark and undisturbed place to increase egg laying. To image embryos from early cellularization to gastrulation, collect eggs for at least 2 h. As flies lay more eggs at this temperature, keeping the egg collection cup

in 25 °C is preferred, unless the experiment requires a different temperature.

3. After at least 2 h of collection, change the egg collection cup to a new apple juice plate. Pour Halocarbon oil 27 on the plate with collected embryos to identify embryos with the desired stages.

3.2 Embryo Staging and Dechorionation

1. Submerged in Halocarbon oil, the interior of the embryos become visible for recognizing developmental stages. Embryo preparation alone routinely takes 20 min. Therefore, picking embryos of sufficiently early stages is necessary. To image mesoderm-EMT before apical constriction, identify the embryos just starting cellularization or earlier (Fig. 2). Go through the plate, pick up around 10–15 embryos of the desired stages with forceps. If there are too many embryos to handle while searching around the plate, collect them to a marked place on the agar plate.
2. Sort through the embryos collected to find the 7–10 best ones and pick them up together with forceps. Slowly open the forceps so the embryos are clustered on one leg of the forceps. Dry the oil on a small piece of paper towel. Leave the cluster of embryos on the edge of the paper towel in order to clean the oil on the forceps. Dry the oil as much as possible as it prevents the bleach from reaching the chorion.
3. Put several drops of 4% sodium hypochlorite/bleach on a small piece of paper towel. Just enough to cover the texture of the paper towel. This allows the embryos to completely submerge in the liquid without running off the paper due to liquid overflowing.
4. Place the cluster of embryos in the bleach on the paper towel. Embryos usually get loosened and fall into the bleach once the cluster touches the bleach.
5. Use the paint brush with a few bristles to separate embryos by gently sweeping them around and against the paper towel. This also removes the residual oil on the embryos. From this point one, the embryos will be kept on this small paper towel until they are mounted on the glass (*see* **Note 1**).
6. While the embryos are being dechorionated, prepare the water drops to wash embryos. Evenly distribute three large drops of water around the inside rim of a 60 mm petri dish lid. Make three such lids with a total of 9 drops of water. Do not leave the embryos in bleach for more than 1 min.
7. Get embryos out of the bleach by gently picking up the small piece of paper towel from a corner with forceps. Dry this small piece of paper towel on regular paper towels. Do not over dry. Wash the small paper by touching a drop of water, previously

prepared, with the corner of the paper grabbed by the forceps. Then dry on the regular paper towel. Repeat this with all 9 drops of water so the embryos are washed thoroughly. Keep the damp paper towel with embryos in a lid of a 35 mm dish (*see* **Notes 2** and **3**).

3.3 Embryo Mounting

1. Clean the glass of the 35 mm glass bottom dish with Kimwipes soaked in 70% ethanol. Do not touch the rim of the well to prevent rubbing off debris.
2. Pick up each embryo one by one from the paper towel using the paint brush with a single bristle. Land the embryos on the glass in the dish. Avoid placing embryos on any remaining debris on the glass (*see* **Note 4**).
3. After placing all embryos on the glass, put a few drops of water on top of the embryos. Rock the dish so that the whole well is covered by the water. Due to both the glass and vitelline membrane being hydrophobic, embryos stick to the glass underwater without floating. An air bubble will often form attaching to each embryo.
4. Carefully pick up each embryo using the one-bristle paint brush, within the water, leaving the air bubble on the glass. Do not lift the embryo outside the water, otherwise the embryo will be lost.
5. Choose a clean area, adjust the angles of the paint brush so that the side of the embryo to be imaged is landed on the glass. To image the mesoderm, have the ventral posterior side of the embryo landing first on the glass. Gently lay the embryo down to the middle of its ventral side on the glass. Try to arrange the embryos in one or two rows to facilitate switching between embryos with high magnification objective (*see* **Notes 5** and **6**).
6. Once all embryos are orientated correctly, cover the well of water with a small piece of air permeable membrane. Place the hydrophilic side of the membrane down and let the water spread to the entire membrane. Let the excessive water run away from the membrane so that the membrane sticks to the petri dish (*see* **Note 7**).

3.4 Confocal Imaging

1. Put a drop of immersion oil on the glass bottom of the dish where the embryos are located on the other side of the glass. Place the dish on the motorized stage of the confocal microscope (*see* **Note 8**).
2. Use the 63×/1.4 oil objective (Plan Apochromatic). It can cover about 10 × 20 cell on the ventral side of the embryo. Watch the embryos from above the stage and move them to the center of the objective. Embryos can be seen as white spots by

bare eyes when the bright-field light is on. There is no need for epifluorescence mode (*see* **Note 9**).

3. Under bright field, adjust the focus and move the stage slightly to find the shadow of the unfocused embryo and then focus properly. Go through the rows of the embryos to find one of appropriate stages by examining the depth of cellularization. Further inspect the orientation of the embryo: the embryo should appear very symmetric including the pole cells at the posterior end. Damaged embryos can show visible leakage of yolk droplets. Finally move the focus to the surface of the embryo to facilitate focusing under confocal mode.
4. Set the laser power to low levels and visualize the embryo in live mode. Do a final check on the age, orientation, and healthiness of the embryo. Red fluorescent protein is almost always weaker and faster bleached than GFP so use 488 nm laser alone for this purpose if possible (*see* **Note 10**).
5. Once the embryo to be imaged is identified, adjust the laser power, gain, zoom factor, and temporal resolution. Find a balance between them to minimize photobleaching and maximize the image quality (*see* **Note 11**).

4 Notes

1. Bleach can lose potency over time and the dechorionation becomes slow. It is tempting to push and roll the embryo to facilitate dechorionation. However, doing so often results in embryo damage or embryos rolling inside the vitelline membrane. This will lead to difficulties in properly orienting the embryos when trying to precisely image the mesoderm. Furthermore, embryos rolled inside the vitelline membrane always roll back to the original position exactly during ventral furrow formation. Therefore, the cells being imaged will be moving in or out of the field of view.
2. Usually preparing water drops and paper towels for the washing step is sufficient for dechorionation. While not all embryos will appear to be dechorionated before washing, they will by the first two to three rounds of drying and washing cycle. As the water rushes through the small paper towel from its corner, all embryos should be completely dechorionated, if they are not already. When the bleach is not potent enough, drying the small paper towel and reapplying the bleach can significantly speed up the dechorionation process.

3. Do the washing and picking-up under the stereomicroscope to prevent losing embryos. Pick up different corners of the small paper towel when embryos are washed from one corner to another.
4. Plastic bristles are hydrophobic like the vitelline membrane so usually the embryo sticks to the bristle without problems. For the one-bristle brush, the thicker it is, the easier it is to pick up embryos as the contact area between the bristle and embryos is larger. Obviously, bristles cannot be too thick as it becomes difficult to manipulate. One way to make such a brush is to cut one bristle off from the root and stick the tip of this bristle back to the other end of the pen brush. This way the root, which is thicker, becomes the tip of the one-bristle brush. Flattening the bristle also increases its area.
5. During this step, once the embryos are on the glass they are exposed to dry air. It is preferable to do this quickly as possible to avoid drying the embryos too much. As the orientation of the embryos will be adjusted later, landing the embryos on its dorsal side is often the safest and easiest way. Due to its geometry, the dorsal side is physically sturdier and requires less care. It is possible to place many embryos down on the glass within a short period of time without damaging them or pushing the embryos to roll inside the vitelline membrane.
6. It is crucial to orient the embryo precisely so that between the desired tissues and objective there is only immersion oil, glass, and vitelline membrane. Again, to decrease the chance of the embryos rolling inside the vitelline membrane, avoid rolling the embryos from side to side. Rocking the embryos from anterior to posterior is more acceptable. If the embryo does not land precisely in the middle of the ventral side, lift the embryo up from anterior and before the posterior leaves the glass adjust the paint brush angle to place the embryo down from posterior to anterior.
7. Placing an air permeable membrane on the well has several functions. It maintains the embryos in a very thin layer of water with sufficient air exchange. Meanwhile it prevents this thin layer of water from evaporation too fast while imaging. Furthermore, as this membrane sticks to the well due to atmosphere pressure, there is no worry that the water may spill out when transported to the microscope and placed on the stage. It also allows imaging through upright microscope, as long as there is sufficient working distance, since the petri dish can be placed upside down with the membrane holding up the water in the well.

8. Placing the oil on the dish where the embryos are rather than on the objective has several advantages. First it prevents the oil from running down the objective before being held between the objective and the cover glass on the bottom of the dish. It also restricts the oil to the area where the embryos are, so that the oil does not spread over large area, resulting in too little oil between the objective and area to be imaged.
9. A 60× or 63× oil objective is necessary to get sufficient quality images of detailed morphology of adherens junctions and contractile myosin.
10. Determine the exact age of the embryo by the depth of the cellularization, as shown by any proteins localizing around the cellularization front such as E-cadherin or myosin. DIC can be helpful to determine the age if no fluorescent proteins in the embryo associate with cellularization structures.

    The pattern of E-cadherin or myosin around cellularization front in the precisely oriented embryo should appear very symmetric. Since cellularization proceeds faster in the mesoderm, the pattern will appear asymmetric in a slightly tilted embryo.

    Normally the footprint of the ventral side of the embryo on the glass is a smooth oval. When embryos leak content due to damage, the footprint will appear abnormally big and show wrinkles due to loss of internal pressure. The outline of the footprint can also appear wavy. Additionally, avoid embryos that have landed on debris. Damaged embryos often also show defects in cellularization. The cellularization front in a healthy embryo will appear very even on both the *XY* plane or in the *z* direction, both in terms of cell sizes and cellularization speeds. Cells in damaged embryos may have shrinking cells surrounded by stretched cells. If the significantly damaged cell is outside field of view, cells in the field of view often appear stretched to a certain direction. Do not image such an embryo.
11. To image embryos from cellularization to gastrulation, a good starting point of imaging format is the following: physical size of each pixel: 0.12 μm; digital dimension of each image: 1024 × 512; *Z* stack depth: 10–15 μm with 1 μm interval; temporal interval: 1 min. Scanning goes much faster along *X* than *Y*, to save time, always use more pixels along *X* rather than the other way around. For most confocal microscopes, each stack will take much less than 1 min. Therefore, there is no need to use bidirectional imaging as less frequent and spread-out exposures to lasers appears to reduce photobleaching compared to concentrating all the laser dosage in a short period of time. At the end of an imaging session, zoom out to examine if the imaged area is significantly darker than the rest of the embryo.

To image adherens junction dynamics and track junction clusters during apical constriction, the following format can be tested first: physical size of each pixel: 0.12 μm; digital dimension of each image: 512 × 256; *Z* stack depth: 7 μm with 0.5 μm interval; temporal interval: 5 s. *Z* Piezo driver is helpful to image at such a high temporal resolution. Bidirectional mode may be used depending on the speed of the microscope.

Acknowledgments

We would like to thank Rolin Sauceda for comments on the manuscript. The research in Weng lab is supported by UNLV startup and the National Institute of Health (R00 HD088764).

References

1. Thiery JP, Sleeman JP (2006) Complex networks orchestrate epithelial-mesenchymal transitions. Nat Rev Mol Cell Biol 7:131–142
2. Lambert AW, Pattabiraman DR, Weinberg RA (2017) Emerging biological principles of metastasis. Cell 168:670–691
3. de Craene B, Berx G (2013) Regulatory networks defining EMT during cancer initiation and progression. Nat Rev Cancer 13:97–110
4. Li W, Kang Y (2016) Probing the fifty shades of EMT in metastasis. Trends Cancer 2:65–67
5. Dongre A, Weinberg RA (2019) New insights into the mechanisms of epithelial–mesenchymal transition and implications for cancer. Nat Rev Mol Cell Biol 20:69–84
6. Nieto MA, Huang RYYJ, Jackson RAA, Thiery JPP (2016) Emt: 2016. Cell 166:21–45
7. Weng M, Wieschaus E (2016) Myosin-dependent remodeling of adherens junctions protects junctions from snail-dependent disassembly. J Cell Biol 212:219–229
8. Weng M, Wieschaus E (2017) Polarity protein Par3/bazooka follows myosin-dependent junction repositioning. Dev Biol 422:125–134
9. Tepass U, Hartenstein V (1994) The development of cellular junctions in the drosophila embryo. Dev Biol 161(2):563–596. https://doi.org/10.1006/dbio.1994.1054
10. Lamouille S, Xu J, Derynck R (2014) Molecular mechanisms of epithelial-mesenchymal transition. Nat Rev Mol Cell Biol 15:178–196
11. Bothma JP, Norstad MR, Alamos S, Garcia HG (2018) LlamaTags: a versatile tool to image transcription factor dynamics in live embryos. Cell 173(7):1810–1822.e16. https://doi.org/10.1016/j.cell.2018.03.069
12. Lim J, Thiery JP (2012) Epithelial-mesenchymal transitions: insights from development. Development 139:3471–3486
13. Shishido E, Higashijima S, Emori Y, Saigo K (1993) Two FGF-receptor homologues of drosophila: one is expressed in mesodermal primordium in early embryos. Development 117:751–761
14. Vincent S, Wilson R, Coelho C, Affolter M, Leptin M (1998) The drosophila protein Dof is specifically required for FGF signaling. Mol Cell 2:515–525
15. Manning AJ, Peters KA, Peifer M, Rogers SL (2013) Regulation of epithelial morphogenesis by the gprotein-coupled receptor mist and its ligand fog. Sci Signal. https://doi.org/10.1126/scisignal.2004427

Chapter 9

Zebrafish Neural Crest: Lessons and Tools to Study In Vivo Cell Migration

Zain Alhashem, Macarena Alvarez-Garcillan Portillo, Mint Ravinand Htun, Anton Gauert, Luis Briones Montecinos, Steffen Härtel, and Claudia Linker

Abstract

The study of cell migration has been greatly enhanced by the development of new model systems and analysis protocols to study this process in vivo. Zebrafish embryos have been a principal protagonist because they are easily accessible, genetically tractable, and optically transparent. Neural crest cells, on the other hand, are the ideal system to study cell migration. These cells migrate extensively, using different modalities of movement and sharing many traits with metastatic cancer cells. In this chapter, we present new tools and protocols that allow the study of NC development and migration in vivo.

Key words Zebrafish, Cell migration, Cell tracking, Photoconversion, Cell dissociation, Gal4/Kalt4, Tamoxifen, UAS, In vivo imaging, Clonal analysis

1 Introduction

The live observation of biological processes is fascinating, which is why cinematography was adopted by biologists as soon as it was developed. Indeed, filming was first used as an experimental tool rather than an entertainment device! In the late nineteenth century, a horse trot was documented, and the exquisite choreography of its legs was analysed in detail [1]. Soon after, in 1903, the first movies of microscopic specimens were made depicting cheese mites [2]. Since then, light microscopy, image acquisition, and analysis techniques have made steady advances, letting us appreciate new levels of detail and complexity. Initially progress was made in cultured cells, but recently the advances of animal models have allowed imaging of live cells in their physiological context at high spatial and temporal resolution. In this sense, the teleost *Danio*

Electronic supplementary material: The online version of this chapter (https://doi.org/10.1007/978-1-0716-0779-4_9) contains supplementary material, which is available to authorized users.

Kyra Campbell and Eric Theveneau (eds.), *The Epithelial-to Mesenchymal Transition: Methods and Protocols*, Methods in Molecular Biology, vol. 2179, https://doi.org/10.1007/978-1-0716-0779-4_9,

rerio has emerged as the model of choice for such studies as it presents many advantages [3]: zebrafish are easily raised in the laboratory, genetically tractable and produce large cohorts of externally developing embryos, which are optically transparent. Imaging live zebrafish embryos has allowed the study of dynamic cell behaviour in great detail and highlighted the importance of cell and tissue interactions.

The study of cell migration is one of the areas that have particularly benefited from live imaging in zebrafish. The germ primordium, lateral line, and immune cells are examples of migratory model systems that have been studied. Our laboratory has concentrated on the analysis of neural crest cells (NC). NC are an ideal model to study cell migration as they arise early in development, being readily accessible to manipulation and imaging. Furthermore, they present a variety of migratory modalities using distinct substrates for movement. They also share many characteristics with metastatic cells, including genetic regulatory networks and cellular behavior [4, 5].

NC arise early in development and migrate extensively invading most of the embryonic tissues and differentiating into a plethora of derivatives from neurons and pigment, to chromaffin cells [6]. NC are induced at the end of gastrulation between the open neural plate and prospective epidermis. After neural tube closure, NC reside at its dorsal-most region from where they migrate ventrally in a rostro-caudal sequence. Cranial NC (CNC) initiate movement first, forming streams with large numbers of collectively migrating mesenchymal cells. These groups move superficially between the epidermis and the neural tissue (Fig. 1). In the trunk, NC start migrating later in development, colonizing the medial part of each body segment. Trunk NC (TNC) take two different routes around the somites: the medial or the lateral pathway [7]. In the medial pathway, cells wedge between the neural tube and the notochord covering the medial region of the somite in the anteroposterior axis. These form unicellular chains of collectively migrating mesenchymal cells. Alternatively, TNC can take the lateral pathway moving superficially between the somite and the epidermis as single mesenchymal cells.

The study of NC migration requires labelling, imaging, tracking, and quantitative analysis of cell movement at the individual and collective level within a group; and each step presents specific challenges. The first is obtaining suitable fluorescent reporter lines that specifically label NC cells. Several different lines have been generated using the regulatory elements of the NC genes FoxD3, Crestin, and Sox10 (Table 1). Most of these lines bear cytoplasmic or membrane fluorescent proteins, which hinder the identification and tracking of single cells within tightly packed groups. To circumvent this problem, we have developed the Sox10:mG^{kg312Tg} line [8] in which chromatin is marked with red fluorescence

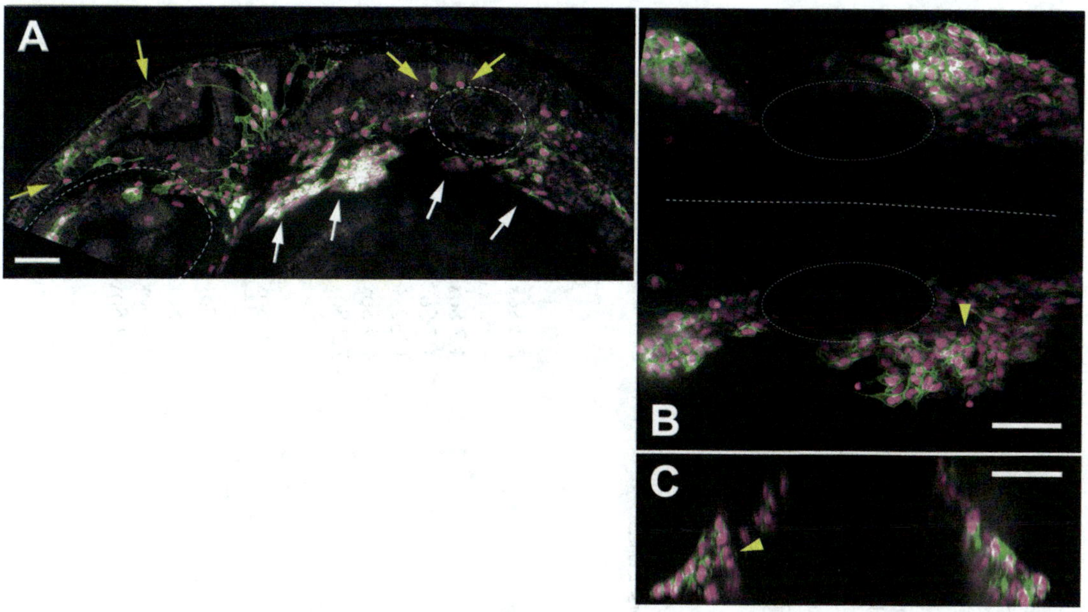

Fig. 1 Cranial Neural Crest. (**a**) Nuclear and membrane fluorescent NC cells in the cranial region of a 24hpf Sox10:mG^{kg312Tg} embryo. Yellow arrows point to superficial cells migrating in isolation, white arrows point to large groups migrating collectively into the branchial arches. Z projection of a lateral view. (**b**) Coronal view of the same embryo, showing migratory streams anterior and posterior to the otic vesicles (dotted oval). Dorsal midline shown by the dotted white line. Yellow arrow heads indicating a dividing cell, shown in the transversal plane in (**c**) Scale bars are 50 μm. Anterior to the left

(H2B-monomeric Cherry) and plasma membrane with green fluorescence (GPI-GFP). This combination allows to concomitantly follow both the morphology and movement of individual cells. It has the additional advantage of retaining nuclear labelling upon nuclear membrane breakage, permitting the study of mitotic events in detail (Fig. 2) and the tracing of cell lineage unambiguously (Fig. 3). We regularly image these embryos overnight (16–20 h) and have recorded movies lasting several days. These generate large amounts of data which commercial analysis packages handle with difficulty. To solve this problem, we have assembled a pipeline of tools that can handle large data files. The first step, often disregarded, is to stabilize the images by correcting all movement that does not correspond to cell migration (e.g., drift, rotation, deformation, or growth; Supplementary Movie 1). This is best performed by obtaining a correction matrix using anatomical landmarks on bright field images. This correction matrix is then applied to the fluorescent images, and nuclear tracking is performed in three dimensions (3D). It is important to track cells in 3D, as it permits the separation of nuclei in densely packed groups (e.g., CNC streams; Fig. 3a–c), or of cells that describe parallel trajectories in different z planes (e.g., TNC chains on each side of the embryo; Fig. 3d, e). Most importantly, it allows to accurately

Table 1
List of available zebrafish transgenic lines to monitor NC development

Line name	Allele	Description	Regulatory element	References
FoxD3 promotor				
foxd3:GFP	zf15Tg	Cytoplasmic GFP, not strongly expressed in NC before 24hpf.	14 zFoxD3	[15]
foxd3-citrine foxd3-mCherry	ct110aGt ct110aRGt	Cytoplasmic fluorescent protein under the control of endogenous promotor. Gene/trap screen. Earliest reporter of NC, expressed from neurula stages. Downregulated by 24hpf and expressed in muscle cells from ≈20hpf. Homozygous embryos are null mutants. Heterozygous do not show phenotype and can be used as reporter.	Endogenous	[16]
Crestin promoto				
Crestin:EGFP	cz3337Tg	Cytoplasmic EGFP.	4.5 Crestin	[17]
Crestin:Gal4-VP16,14x UAS-E1b:EGFP	t32231Tg	Cytoplasmic GFP driven through Gal4 activation. Not strongly expressed in NC before 24hpf. Expressed in dorsal EVL	1 Crestin	
Sox10 promotor				
Cytoplasmic				
7.2sox10:EGFP	ir937Tg	Cytoplasmic EGFP	7.2 Sox10	[18]
4.9Sox10:GFP	ba2Tg	Cytoplasmic EGFP	4.9Sox10	
4.725Sox10:GFP	ba3Tg	Cytoplasmic EGFP	4.7 Sox10	
4.725Sox10:GFP	ba4Tg	Cytoplasmic EGFP	4.7 Sox11	[19]
1252Sox10:GFP	ba5Tg	Cytoplasmic EGFP	1252 Sox10	
sox10:DsRed	el10Tg	Cytoplasmic RFP	4.9 Sox10	[20]
sox10:TagRFP	co26Tg	Cytoplasmic RFP	7.2 Sox10	[21]
Membrane				
sox10:EGFP CAAX	el375Tg	Membrane EGFP	4.9 Sox10	[22]
sox10:mCherryCAAX	el361Tg	Membrane RFP	4.9 Sox10	
4.9sox10:LY-tomato	ir1040Tg	Membrane RFP	4.9 Sox10	[23]
4.9sox10:LY-GFP	ir866Tg	Membrane GFP	4.9 Sox10	
sox10:EGFP-Hsa.HRAS	sl3Tg	Membrane EGFP	7.2 Sox10	[24]

sox10(7.2):Mrfp	vu234Tg	Membrane RFP	7.2 Sox10	[25]
Nuclear and membrane				
sox10:mG	kg312Tg	Chromatin/nuclear RFP and membrane GFP	4.9 Sox10	[8]
Photoconvertible				
sox10:Kaede	zf393Tg	Cytoplasmic photoconvertible green to red fluorescent protein	7.2 Sox10	[26]
Sox10:H2B-Dendra2	kg329Tg	Chromatin/nuclear photoconvertible green to red fluorescent protein	4.9 Sox10	This paper
sox10:EOS	w9Tg	Cytoplasmic photoconvertible green to red	4.9 Sox10	[27]
sox10:NLS-Eos	w18Tg	Nuclear photoconvertible green to red	4.9 Sox10	
sox10:KikGR	el2Tg	Cytoplasmic photoconvertible green to red	4.9 Sox10	[28]
Gal4				
sox10:GAL4-VP16	km6Tg	Gal4 transcriptional activator	4.9 Sox10	[29]
sox10:GAL4-VP16	sq9Tg	Gal4 transcriptional activator	7.2 Sox10	
Sox10:H2BmCherry-2A-Kalt4ER	kg328Tg	Chromatin/nuclear RFP and Gal4-ER^{T2} fusion transcriptional activator	4.9 Sox10	This paper
sox10:GAL4-VP16-IRES-EGFP	el159Tg	Gal4 transcriptional activator. Does not express a fluorescent marker as the IRES-GFP is not functional.	4.9 Sox10	[20]
Cre				
sox10:Cre	ba74Tg	Cre recombinase	4.7 sox10	[30]
sox10:Cre-ERT2,myl7:GFP	t007tg	ERT2-Cre fusion recombinase	4.9 sox10	[31]

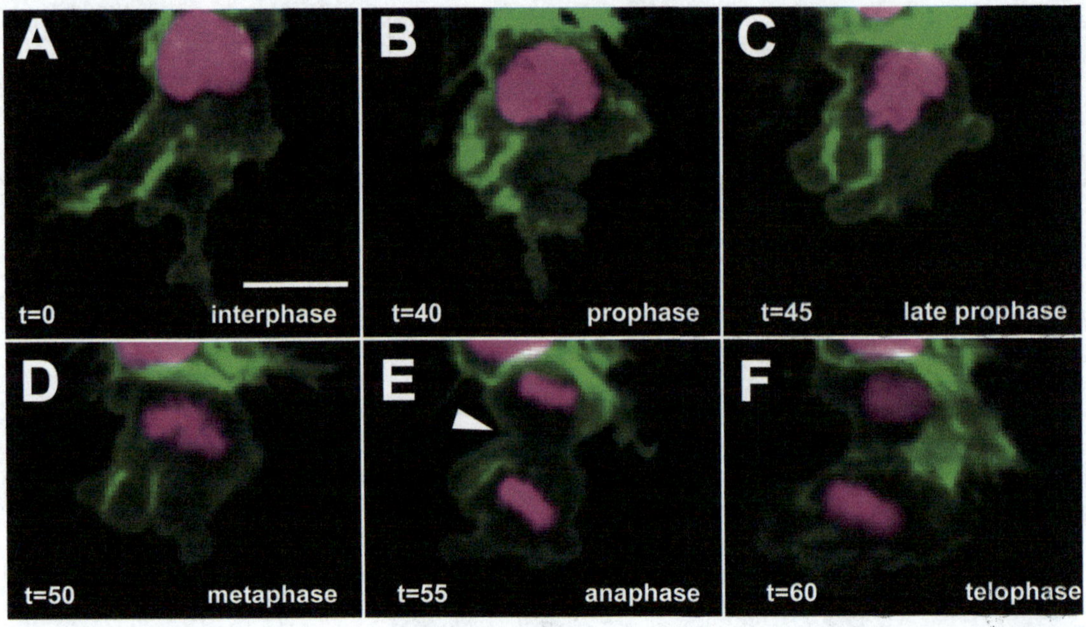

Fig. 2 Phases of division in a migrating trunk neural crest cell. (**a**) Sox10:mG^{kg312Tg} transgenic during interphase. Chromatin condensation can be observed during prophase (**b**), and membrane labels the plane of division during cytokineis, indicated by the white arrowhead in (**e**). Scale bar 10 μm. *t* time in minutes

calculate path distances, which are largely underestimated when tracking in 2D as the movement in z is disregarded. Together, this imaging setup and analysis tools permit the study of NC migration in wild type conditions. The next step is to define the molecular pathways regulating this process. General signalling perturbations such as mRNA injections or analysis of mutant embryos can be misleading, as these manipulations may generate alterations of the NC population early in development and/or induce defects in other tissues. Hence, any observed NC migratory phenotypes under such conditions may only be the result of indirect perturbations. Appropriate assessment of the signals regulating NC migration requires the analysis of gain- and loss-of-function conditions in a tissue-specific and time-controlled manner. To attain this goal, we have taken advantage of the UAS/Gal4 system and generated a new Sox10:Kalt4ER^{kg328Tg} line (Fig. 4). The UAS system has been extensively used in drosophila [9] and shown to work in zebrafish [10, 11]. In the Sox10:Kalt4ER^{kg328Tg} transgenic, all NC have their chromatin labelled with RFP and express the transcriptional activator Gal4 fused to the hormone-binding domain of the oestrogen receptor (ER^{T2}). In the absence of oestrogen, Gal4ER^{T2} is retained inactive in the cytoplasm. Upon addition of the estrogen analogue tamoxifen, Gal4ER^{T2} is translocated to the nucleus where it binds and activates expression of any UAS-driven transgene. This system can induce protein overexpression within 30 min of

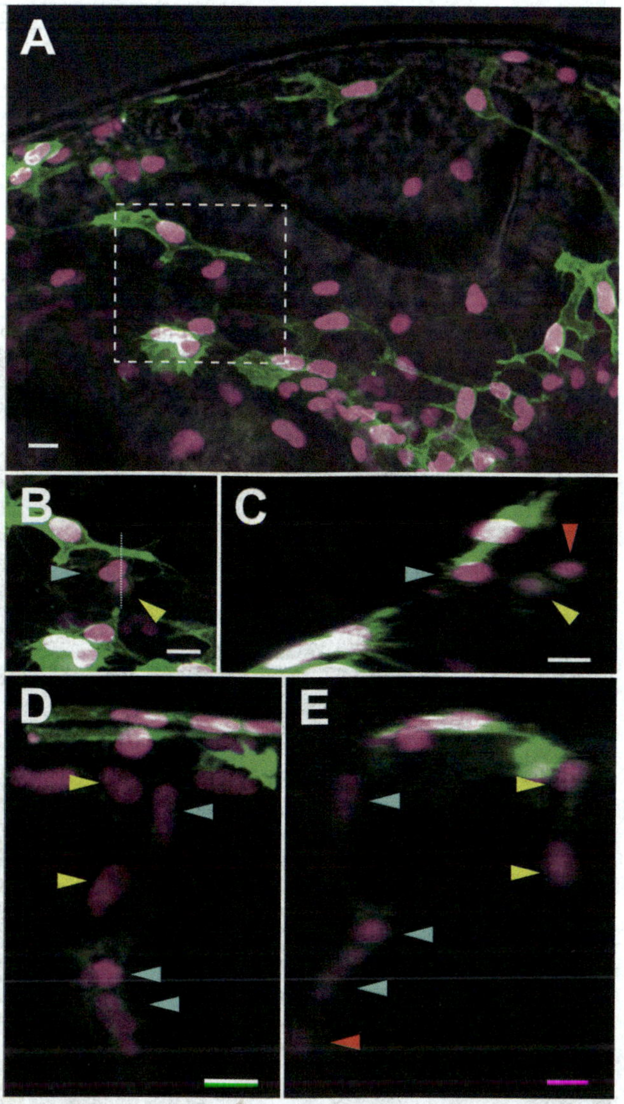

Fig. 3 Nuclei separation in 3D. (**a**) Z projection of a dorsal view of cranial neural crest from a 24hpf Sox10:mG^{kg312Tg} embryo. (**b**) Enlargement of the region boxed in (**a**) Cyan and yellow arrowheads point to adjacent nuclei. (**c**) Transversal plane at the level indicated by the dotted line in (**b**). Cyan and yellow arrowheads point to the nuclei shown in (**b**), which are at different z planes clearly separated from each other. Moreover, an obscured third nucleus that is yet in a different z plane, is indicated by a red arrowhead. (**d**) Z projection of a lateral view of trunk neural crest from a 24hpf Sox10:mG^{kg312Tg} embryo. Migrating cells align in a single chain. (**e**) Transversal view of the same embryo showing that some of the aligned nuclei in the z projection, actually form part of the contralateral chain at a different z plane. Cyan arrowheads indicating cells in one of the chains, yellow arrowheads pointing to cells in the contralateral chain. Red arrow pointing to a nucleus that cannot be differentiated from its neighbours in the z projection. Scale bar 10 μm. Anterior to the left

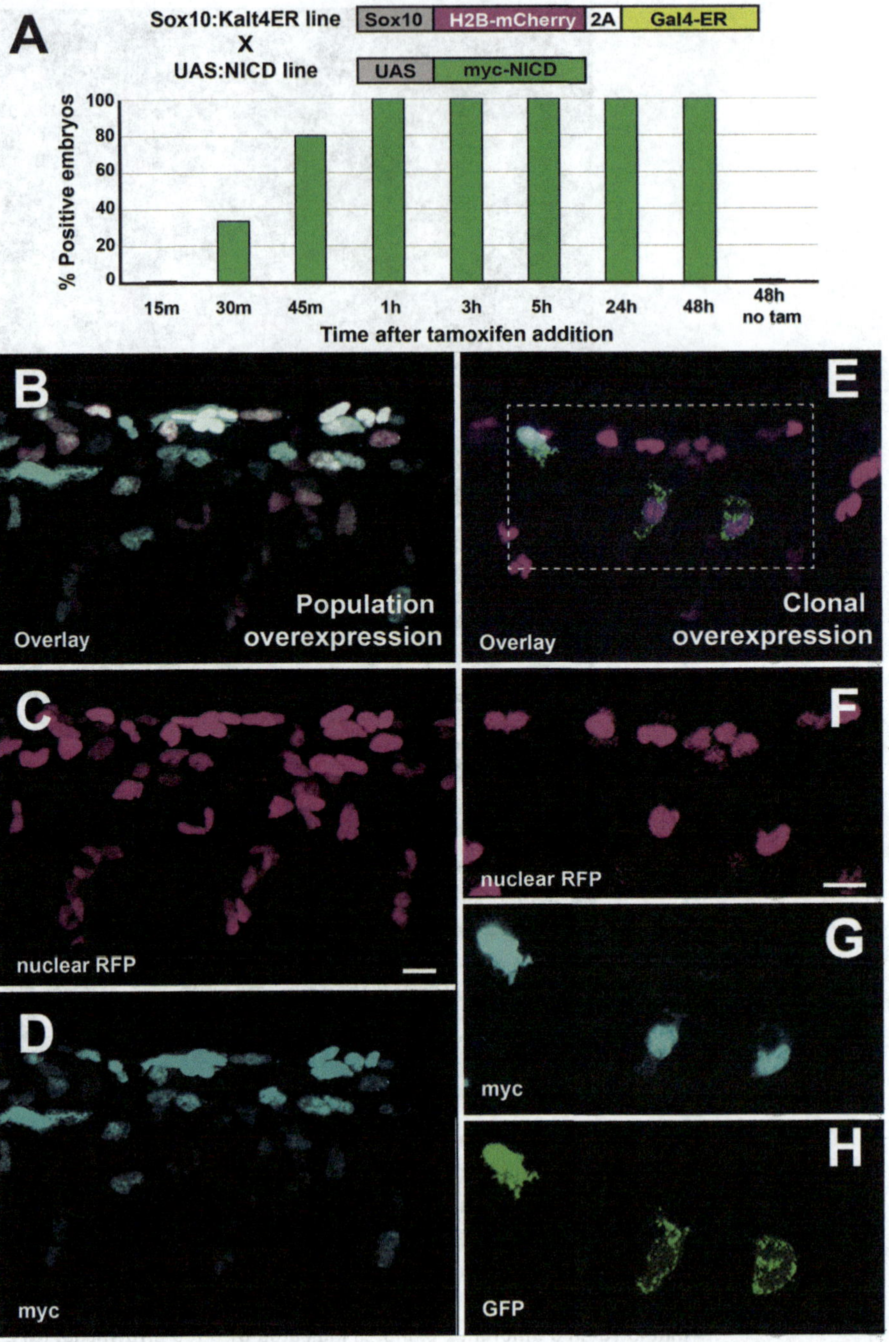

Fig. 4 Induction of protein expression in neural crest only with time control A-D. Induction of expression in the entire neural crest population. A. Diagram of the lines crossed, Sox10:Kalt4ERkg328Tg and UAS controlling the NICD-myc fusion protein 32. Time-course of myc detection after tamoxifen addition at 11hpf. B. Overlay of nuclear RFP (C) and antibody staining for myc detection after 14h of tamoxifen addition (D). E-H. All neural crest cells are labelled by nuclear RFP (E and F), while only clones that inherited the injected plasmid, express myc (G) and GFP (H) induced by tamoxifen after 13h incubation. Dotted square in E, enlarged in F-H. Anterior to the right, dorsal top. Scale bar 10μm

tamoxifen addition, allowing a tight time control (Fig. 4a–d). Loss-of-function conditions can be obtained by the use of UAS-driven dominant negative proteins. This setup allows the overexpression of protein in the entire NC population. Moreover, clonal induction in a restricted number of NC cells is also possible. To do this, one-cell stage Sox10:Kalt4ER^{kg328Tg} embryos are injected with plasmid DNA of any UAS-driven construct of interest (Fig. 4e–h). Plasmid DNA is unevenly distributed in the early embryo, leading to a small number of random cells inheriting it. Only cells fated to the NC lineage that have inherited the plasmid will express the protein of interest upon tamoxifen addition, allowing for mosaic expression. Hence, the Sox10:Kalt4ER^{kg328Tg} transgenic is a versatile tool that will allow the fine molecular dissection of NC migration. Finally, we can label single NC at defined position within the group in vivo using the Sox10:Dendra2^{kg329Tg} line. In this transgenic, all NC cells express Dendra2, a photoconvertible fluorescent protein fused to H2B, labelling chromatin with GFP; upon UV illumination, Dendra2 irreversibly changes emission to the RFP spectrum, allowing to label single or groups of nuclei in vivo within a moving population (Fig. 5). Cells labelled in this way can then be isolated by fluorescence-activated cell sorting (FACS) and used for a number of purposes, including the generation of cDNA libraries of cells at specific locations. The combination of these tools allowing live imaging, single cell labeling, and temporal and spatial control of protein expression will bring a new understanding of the molecular control underlying NC migration in vivo. Moreover, these tools can be readily adapted to other model systems.

In this chapter, we present a number of imaging tools and quantitative analysis protocols our laboratory has developed to study NC migration. These include new zebrafish transgenic lines that allow labelling, imaging and over expression of proteins of interest in NC cells specifically, as well as imaging protocols and the use of quantitative analysis techniques.

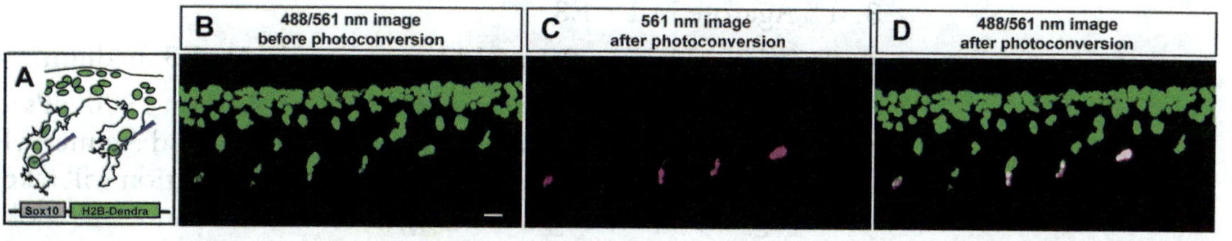

Fig. 5 Labelling of single neural crest cells by photoconversion. (**a**) Diagram of the Sox10:Dendra2^{kg329Tg} transgene and embryo in which a single cell is illuminated with a 405 nm laser (blue beam). (**b**) Dendra2 fluoresce green in all neural crest nuclei before photoconversion. (**c**) Dendra2 switches to RFP fluorescence once illuminated with the 405 nm laser. (**d**) Overlay of the channels

2 Materials

For all solutions, ultrapure sterile water is used (prepared by purifying deionized water, 18 MΩ-cm at 25 °C, and autoclaving). Store all solutions at room temperature unless otherwise indicated.

2.1 Microscopy Setup

1. Upright compound microscope with a 20× or 40× high numerical aperture dipping lens (*see* **Note 1**).
2. Having a motorized stage will allow imaging several embryos in every session greatly enhancing experimentation efficiency. This is especially important, when working with genetic condition in which not all specimens are carriers. Having said that, a motorized stage is not strictly required.
3. Temperature regulation. This can be obtained by having a temperature chamber enclosing the microscope head, or by having a tight regulation of the temperature of the room. In either case, make sure to stabilize temperature of all the components for at least 1 h before initiating imaging, and that the temperature is the same at every imaging session.
4. Confocal or spinning disk heads. The use of a spinning disk head will greatly reduce the time of imaging, allowing to image a greater number of embryos per session and reducing phototoxicity.

2.2 Mounting Embryos and Imaging

1. E3 fish medium. Prepare a 60× stock solution (1 L): 17.2 g of NaCl, 0.76 g of KCl, 2.9 g of $CaCl_2 \cdot 2H_2O$, 4.9 g of $MgSO_4 \cdot 7H_2O$. Adjust volume to 1 L of water. Dilute to a 1× working solution as needed.
2. Tricaine (MS-222, Sigma A5040-100G). Prepare a 15 mM stock solution: Add 0.2 g of tricaine powder and 1 mL of Tris–HCl 1 M pH 9.5 to 49 mL dH_2O. Adjust pH to 7 and store 1 mL aliquots at −20 °C. After thawing, tricaine aliquots can be stored at 4 °C for a few days (no more than 1 week).
3. 1% Agarose in 1× E3.
4. 0.6% Low Melting Point (LMP) agarose in 1× E3 medium.
5. 3% Methyl cellulose (Sigma M-0387, Viscosity 1500 cPs). Heat 100 mL 1× E3 in a closed bottle to 60 °C, add 3 g methyl cellulose and mix with a magnetic stirrer. The solution will take a long time (overnight or longer) to dissolve and will look slimy. To completely dissolve the powder, put the solution at −20 °C and stir every 30 min until it almost freezes, then put it rocking at 4 °C for 2 days. To remove any undissolved powder, centrifuge the solution at 13,200 RPM for 30 min, this should make the solution clear. Store at −20 °C.
6. Custom-made casting mould (Fig. 6).

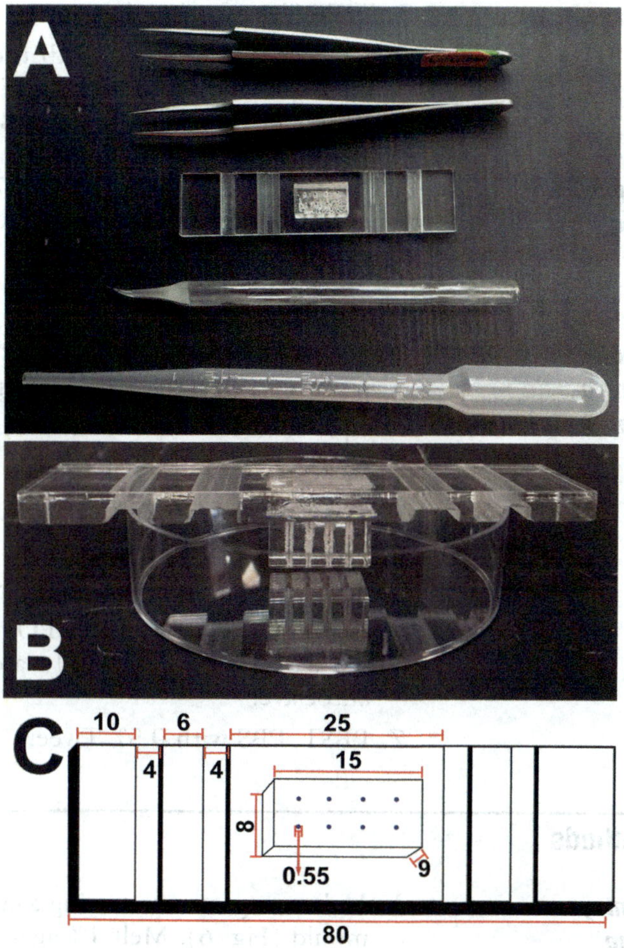

Fig. 6 Tools and mould used to mount embryos for imaging. (**a**) Forceps, mould, hair knife, and plastic transfer pipet. (**b**) mould mounted on a 60 mm petri dish before addition of agarose. (**c**) Mould diagram. Dimensions in mm

7. Commercially available heating block/plate set at 40 °C.
8. Plastic dishes (*see* **Note 2**).
9. Plastic transfer pipettes.
10. Hair knife: Make a thin end glass Pasteur pipet by pulling it on a flame until it splits (it should look like a sharp pencil). Cut the end and add some melted wax to glue a small piece of dark hair (about 1 cm). Approach the hair to the flame very slightly to make a ball shape. Be careful, it will burn very quickly.
11. High-precision grade tweezers numbers 5 and 3.
12. Stereoscope.
13. Upright confocal microscope with a 20× or 40× lens and a motorized stage if multiple embryos are to be recorded.

2.3 Image Analysis

1. TIFF files of a time-lapse stack.
2. High specification PC with the following software: Java 6 or later (preferentially Java 8), Fiji (Image J), Interactive Data Language (IDL), MATrix LABoratory (MatLab).

2.4 Tamoxifen Induction

4-HydroxyTamoxifen (Sigma H6278-10MG, or an equivalent reference from other providers). Prepare a 5 mM stock solution: 10 mg of tamoxifen powder in 5.1 mL ethanol. Aliquot and store at −20 °C. Dilute to a working concentration of 0.5 μM in E3.

2.5 Photoconversion and Embryo Dissociation

1. Glass bottom petri dishes (35 mm).
2. FACSMax Cell dissociation solution (Amsbio T200100 or an equivalent reference from other providers).
3. Heat-inactivated bovine serum.
4. Trypsin-EDTA 0.25% cell culture standard.
5. Cooled tabletop centrifuge.
6. Sterile PBS buffer. Add 8 g of NaCl, 0.2 g of KCl, 1.44 g of Na_2HPO_4, 0.24 g of KH_2PO_4 to 800 mL of water. Adjust the pH to 7.4 with HCl. Add water to a total volume of 1 L and autoclave.
7. PBST, PBS with 0.5% Tween.

3 Methods

3.1 Embryo Mounting

1. Make an agarose cast using a custom-made 8- or 16-pin plastic mould (Fig. 6). Melt 1% agarose in E3, wait for it to cool to approximately 60 °C. Add 0.6 mL of fresh tricaine to 5.4 mL of agarose. Place the plastic mold onto a 60 mm petri dish, pour agarose making sure no bubbles are trapped around the pins (*see* **Note 3**).
2. Let the agarose set and remove the plastic mould. It should leave 8 or 16 rounded, and evenly spaced holes of the size of the embryo yolk.
3. Pre-warm the agarose mould on top of a 40 °C heating block.
4. Mix 0.9 mL of melted 0.6% LMP and 0.1 mL of tricaine stock solution in a 1.5 mL tube and leave in the heating block set at 40 °C. Make sure that the temperature does not exceed 42 °C as higher temperatures can kill the embryos.
5. Dechorionate the embryos to be mounted in E3/tricaine (1 mL tricaine stock solution for 25 mL of E3) using n5 tweezers.
6. As soon as you take the LMP agarose from the heating block, it will start to set. Hence, you have to be as quick as possible in

the following transfer and positioning of the embryos. Position the LMP agarose tube and the plate with embryos under the stereomicroscope. Using a plastic pipette, transfer 8 embryos with as little medium as possible (no more than 0.1 mL) into the LMP agarose tube.

7. Quickly take back the embryos into the transfer pipette with 0.3 mL of LMP agarose and place on top of the agarose mould. Using the hair knife, position the embryos on the agarose holes so that the yolk sits into the holes. Orient the embryos either in a lateral or dorsal position. Let the LMP solidify and the embryos set for about 10 min.

Alternatively, for short time-lapse movies, it is possible to mount the embryos in 3% methyl cellulose. Make a shallow agarose mould, transfer the embryos to an Eppendorf tube containing 0.9 mL 3% methyl cellulose and 0.1 mL of tricaine, and then transfer the embryos to the agarose mould with little methyl cellulose. Use a hair knife to orient the embryos.

3.2 Time-Lapse Imaging

1. Place the embryo dish on the motorized stage and add dipping medium (E3/tricaine) up to the rim of the dish.
2. Select the objective to be used (*see* **Note 3**) and focus on one of the embryos at the edge of the plate. Find the cells of interest.
3. Set up the acquisition protocol in the software, recording the required fluorescent channels. Always take a similar stack of the bright field channel that will be required to correct 3D drift of the movie (*see* Subheading 3.3). For an overnight time-lapse, set the lasers power and exposure time to the minimum levels to generate a dim but visible signal (*see* **Note 4**). In our setup, a 15–20% lasers power, 400ms exposure and 180 gain, with 2 × 2 binning of the camera.
4. The z step and range of the stack will depend on your specific experiment. For good resolution movies of migrating trunk neural crest on lateral view, we use a 2 μm z step and a total stack of 70 μm (or 140 μm to acquire both sides of the embryo), leaving some space above and below the region of interest in case of z drift. For high resolution images, that would allow volume reconstruction, a 1 μm z step is required, in which case long-term (>10 h) imaging is not possible due to photo-bleaching.

   The time step and total length of imaging will vary depending on the specific question (*see* **Note 5**). If a motorized stage is in place, the time interval between images will determine how many embryos can be imaged simultaneously.
5. Check that all the embryos are in focus and start the acquisition. After 1 h of acquisition, check that all embryos are still in focus and have not drifted; refocus if at all necessary.

6. Make sure that the dipping medium does not completely evaporate by filling the dish to the rim before leaving for the night. Position an open bottle of water near the stage and within the microscope heat chamber if one is installed, to provide ambient humidity.
7. Naming files. A logic and standardized naming system will save a lot of time and effort. For files to appear in chronological order, use first the date year_month_day (2020_01_10), followed by the name of the person in charge of the experiment, and indication of the type of sample (e.g., WT for wild type). In parallel, an Excel database with all recorded movies is kept including information of the experiment, identity of each of the embryos in the file, and imaging parameters. Always maintain the name as you process the images as it will be easier to trace back.

3.3 Image Processing: Drift Correction

During imaging, samples often move or drift within the field of view. To obtain accurate tracking data, it is essential that these movements are corrected. This will ensure that any movement recorded is due to cell displacement and not to changes in the overall position of the sample (Supplementary Movie 1 shows an example of drift correction). The idea is to use in focus anatomical landmarks of the bright field image to generate a transformation matrix of deformation and drift correction, to then apply this matrix to each fluorescent channel plane.

1. Open the imaged file including all channels in Fiji (*see* **Note 6**).
2. Display it as a hyperstack including all channels, time points, and z planes. Trim any frame at the beginning and/or end of the movie that is not required.
3. Separate all channels by clicking on "Image > Colour > Split channels." Save the fluorescent channels in separate folders as individual frames by selecting "Save As > Image Sequence"; use TIFF format.
4. Open the bright field stack and select one z plane in which anatomical landmarks are clear (e.g., otic vesicle or somite borders). Make a stack with one z plane for each time point in which the chosen landmarks are in focus. "Save As > Image Sequence" use TIFF format and create separate folder to save them.

   To correct the drift using Fiji, follow **steps 5–11**.
5. Click on "Plugins > Registration > Register Virtual Stack Slices."
6. For the "Source Directory," select the folder where the in focus bright field stack was saved.
7. For the "Output Directory," create a new folder.

8. Maintain the feature extraction model as "Rigid," with the Registration model as "Rigid–translate + rotate."
9. Select the "Save transforms" option and choose a directory to store them. This location will be important in **steps 5** and **6**.
10. Select one frame of reference to which all other frames will be adjusted and apply correction. Once the correction is calculated, a window with the corrected ("Registered") stack will open and the transformation matrices saved in the folder selected in **step 5-e**, as XML files.
11. Run through the registered movie and make sure the correction is adequate. If the correction does not work, *see* **Note 7**.

    In order to apply the transformation matrix to the whole stack of fluorescent images, we must generate as many matrices as number of images. To do so, follow **steps 12–19**.
12. Open IDL.
13. Copy the script in the file "Matrix.txt" (Supplementary Script 1) onto a new workbook. To avoid mistakes, select all the script in the text file by pressing Ctrl+Shift+a, then copy by pressing Ctrl+c and paste onto the MATLab workbook with Ctrl+v.
14. Write the number of time points and the total number of Z-slices of the movie in the script, where it says:

    ;Total Time
    Times=____
    ;Total Z-slices
    Z=____
15. Add the path to the matrices that were obtained in **step 4-e** in the script, where it says:

    ;PATH
    Path=____
16. This path can be found by right-clicking on one of the XML matrix files, selecting "Properties" and copying (Ctrl+C) the information on the "Location" section. After pasting (Ctrl+V) the "Location" on the IDL script, add one backslash (\) (e.g., Path = 'C:\User\Experiments\Movies\Movie_1\Matrices\').
17. Save the file with the new information (Ctrl+S).
18. Click on the "Compile" button, shaped as two yellow cogs, located at the top-left corner of the IDL screen.
19. Run the script by typing "copyMatrix" in the command line of IDL and pressing Enter. This will create a folder named "matrix" containing the correct number of XML files.

    Apply the transformation to all other channels of your movie by following **steps 20–25**.

20. In ImageJ, select the plugin "Transform Virtual Stack Slices" by clicking "Plugins," "Transform," "Transform Virtual Stack Slices."
21. In the source directory, select the folder containing the images for one of the channels.
22. In the output directory, create a new folder where the corrected images will be saved.
23. In the transform directory, select the folder containing the correct number of matrices created in **step 6.g**.
24. Keep the "interpolate" option selected and press "OK."
25. Repeat **step 6-b–f** for every channel in the movie.

    Adjust all movies to the same frame size, and equivalent orientation by following **steps 26–28**.
26. Orientation: All samples must be oriented in the same way to homogenize the angle of migration. For in vivo imaging, the position of the anterior-posterior (AP) and the dorsal-ventral (DV) axes must remain constant in all movies (e.g., *X* axis: anterior to posterior; *Y* axis: dorsal to ventral; *Z* axis: left to right). To do this, open the file in Fiji select "Image > Transform > Flip Horizontal or Flip Vertical or Rotate" until all axis are positioned as required. It is advisable to take note of sequence applied (flips and/or rotation angles) and include these in the database. When rotating, click the option "Enlarge Image To Fit Result" in order to not lose data. Apply the same transformations to all channels of the movie.
27. Framing: Ensure that the samples are centred within the frame of the movie. If the sample is off-centre, select the "Rectangle" tool in Fiji (the left-most button on the Fiji menu) and draw a rectangle around the sample so that it is located in the middle. Then select "Image > Crop" to make this the new frame.
28. Size: Once all the corrections have been done, the file can be returned to its original size by selecting "Image > Adjust > Canvas Size." Maintain this to a standard size for all stacks.

3.4 Image Processing: 3D Tracking Using View5D

With your corrected image you can proceed to semi-automatically track cells/nuclei/objects in 3D. After opening the image and making sure its size is correct you will indicate the objects to follow in the first frame they appear. Thereafter the program will try to automatically follow them over time. You can define the parameters to find the objects (e.g., intensity levels, maximal radius in which to look in the next time point, etc.). Once the automatic tracking is done, you can manually correct it.

1. Open the corrected stack in Fiji and ensure the properties of the image correspond to real units. Go to "Image > Properties" and correct number of slices and time frames, as well as the

Unit of Length (e.g., from "Pixels" to "μm"), Width, Height, and Depth that correspond to the size of voxel. If your instrument is properly calibrated this information should be found, and correctly read by Fiji, from the metadata (*see* **Note 8**).

2. Start View5D through "Image > Stack > View5D > start viewer."
3. In the window that opens, make sure that the dimension Order is the one your data has. Usually CZT (channel, z, time). Make sure that SliceCount, Size Z; TimeCount, Number of Time point and Colours are correct. Tick the option "Aspects with data scaling" and press "OK." This will open the 5D viewer, which divides the screen in four: left-top is the *XY* plane; right-top is the *XZ* plane; left-bottom is the *YZ* plane; and right bottom is the line of commands. The actions taken will only apply to the part of the screen the cursor is on, which is highlighted by a colour frame on the image.
4. Adjust the size and colours for optimal visualization. To adjust the size, hover over the *XY* top-left screen and press Shift+A ("**A**") to zoom in or "**a**" to zoom out. To adjust the colours, hover over the different coloured squares representing each channel on the bottom-right corner and press "**c**" until the desired colour is reached (list of View5D commands Supplementary Table 1).
5. Allow for the full movie to load onto the viewer by scrolling through it once. You can scroll backwards or forward in time by pressing "**,**" and "**.**" or using the slider at the righthand side of the image.
6. Define the parameters to display, the directory where to save the tracking and the parameter tracking by opening the "Marker positioning" window. Position the cursor in any of the three image windows (*XY*, *XZ*, or *YZ*) and press "**n**."
7. Choose the directory where to save the markers (tracking) by pressing "MarkerFileOut."
8. Choose the display mode of the tracking in the image by ticking the boxes you find convenient in the list on the top-left corner. Here, "line" applies to the line connecting each consecutive marker in the track; "list" is the number associated with each track in the movie (1, 2, 3, etc.); "trees" are the individual tracks, referred to as trees for the branches representing daughter cells (*see* **step 21** for an explanation on tracking with cell divisions). All the "Display" or "Show" options on this list do not affect the tracking, only how the tracks are seen on the screen; tick or untick them as convenient throughout the process. *Make sure the option* "Annotate with List Nr." is selected. 5DView will randomly assign colours to the tracks; hence it is possible that two tracks have the same

colour. Number assignation will allow to easily distinguish tracks in this situation.

9. Set up parameters for automatic maximum finding and sub-pixel position by center of mass. These parameters determine how the program localizes the nucleus after it has moved from the previous time point. A very low radius will minimize jumping to a different nucleus but will lose the tracked nucleus in big movements. A high radius will allow for following big leaps in migration, but it will make it easier for the track to jump to a different nucleus. What constitutes a "low" or "high" value will depend on the relative size of the nuclei and the distance between them and will need to be decided by trial and error. In the case of TNC, appropriate values for *X*, *Y*, and *Z* range between 3 and 8. These can be adjusted throughout the tracking to accommodate for more or less clustered migration.
10. Press "OK" to start tracking.
11. Set the cursor on top of the nucleus to track in the first time point it appears and press "**m**." This nucleus will now have track number 1. At this point take note of which track correspond to which cell, so later on you can compile data from different cell populations (e.g., cell1 at the front of the migratory group, Leader; cell 2 in the middle of the group, Follower 2; or cell 1a Leader's front daughter; etc.)
12. Press "**W**" (Shift+w) for automatic tracking of the nucleus. A track line will appear. The colour of the track line can be changed by pressing "**w**" while working on that track.
13. Use the keys "**9**" and "**0**" to circulate through the track and ensure that it follows the selected nucleus in all planes *XY*, *XZ*, and *YZ* (look at the different windows, all should display the corresponding track).

    To correct an erroneous location along the track, drag the corresponding mark to the appropriate location follow **steps 14–18**.
14. If the track returns to the wrong nucleus after dragging it to the right one, open the Marker positioning menu and remove the ticks from "Use automatic maximum finding" and "Sub-pixel position by of Mass." This will make the tracking manual until the boxes are ticked again, allowing for the correct positioning of the marker.
15. Once the marker is positioned correctly, advance to the following time point by pressing "**0**." The track will still follow the trajectory traced automatically by the computer beforehand. You can manually correct it as before or delete the rest of the track and re-track the path for the remaining time points. To delete the track in this and all following time points press "**Q**" (Shift+q).

16. Open the Marker positioning menu by clicking "**n**" and tick the "Use automatic maximum finding" and "Subpixel position by Centre of Mass" boxes again.
17. Return to the last marker available in the track by pressing "**9**" and "**0**."
18. Press "**W**" (Shift+w) again to restart the tracking.
19. Once track 1 is correct and complete, select a different nucleus at the beginning of its migration and press "**k**." This will create marker number 2. Repeat **steps 11–13** for each nucleus you need to track.
20. To save all tracks, hover the cursor over command window (bottom right) and press "**m**." Importantly, do this periodically as the program does not save automatically and you may lose all your work if the program or the computer freezes. Pressing "**m**" shows the raw data on the command window and creates a text file in the selected "MarkerFileOut" folder with your tracking data.

    In case of *cell division* during a track, refer to **steps 21** or **22** to either follow only one daughter cell as a continuation of the on-going track (**step 21**) or to follow both daughter cells after division (**step 22**).
21. To follow only one daughter cell as part of the same track (a single track showing cell1 plus one of its daughters, but ignoring the second daughter), place the marker on that daughter cell once the division is completed and press "**W**" (Shift+w) to continue the track (if the program has not done so automatically).
22. To record the division in the data file and follow both daughter cells as related tracks, go to the first time point after the division and press "\" (backslash). This will create two new related tracks labeled "1a" and "1b." Move the "1a" marker to one of the daughter cells; move to the next time point (second time point after division) and delete all tracking after this point by pressing "**Q**"; press "**W**" (Shift+w) and follow **steps 8–11** to re-track the first daughter path. Do the same for the second daughter.

3.5 Data Analysis: View5D Output

The output from View5D will be saved in the selected folder as a text file. By default, it will be named "markers," or have any name assigned in **step 6**. This can be opened in Microsoft Excel to easily access the data. This file will include 24 columns with all the tracking data. The ones that are relevant for this work are:

3.5.1 Position in Pixels

The position of each nucleus in every axis at any given time point measured in pixels are in columns C, D, and E called "PosX [pixels]," "PosY [pixels]," and "PosZ [pixels]," respectively. This information should **not** be used for analysis since the units will be arbitrary.

3.5.2 Position in μm (and Other Specified Units)

This is the same information as described above but corrected for the dimensions as detailed in the properties of the original file. That information was adjusted in **step 1** of Subheading 3.2. It is detailed in columns J, K, and L, named "X [μm]," "Y [μm]," and "X [μm]," respectively. If the units are different to μm, the names will display those.

3.5.3 Time

There are two columns named "Time [time]," column G and N. They both display the same information: the time point corresponding to the values for *X*, *Y*, and *Z* in that row. For cells tracked from the beginning of the movie, this list will start at 0. Otherwise it will start at the time point where the track was initiated.

3.5.4 List Name

The last column on the file, column X, is titled "ListName" and it tells us which cell the track corresponds to. If you have taken note of their identities during the tracking (*see* Subheading 3.4, **step 7**), you can identify each cell by the number in this column.

Calculation of speed, velocity, and persistence is performed using simple Euclidian formulas or as per Gorelik and Gautreau [12].

3.5.5 Visualization of 3D Tracks in 2D

It is usually easier to generate 2D representations of View5D tracks to visualize and qualitatively compare several tracks at a time. We perform this step by uploading View5D tracks onto the "Manual Tracking" plugin of Fiji. This can be performed manually, if you wish to select precisely which tracks to visualize, for example only cells of a certain type; or automatically if you wish to have a representation of all the tracks.

This step requires to generate an Excel sheet that contains tracking (*X*, *Y* position and time) and cell identity information in a manner that Manual Tracking can read.

For manual conversion, follow **steps 1–9**:

1. Create a new workbook on Microsoft Excel and copy the information in the following columns for each of the tracks to be visualized.
2. Name the columns as follows: "Track n" on cell B1, "Slice n" on C1, "X" on D1, and "Y" on E1 (leave column A free, see below).
3. Under "Track n," on column B, add the number of the track. Start with the track with the lowest number; it is easier to maintain the numbers View5D assigned to each track.

4. "Slice n," on column C, is time information. View5D assigns 0 (t_0) to the first tracking time point, while Manual Tracking assigns 1 (t_1). Hence, copy column "Time [time]" from the View5D file (either column G or N) plus one (1) into column C.
5. Under "X" and "Y," on columns D and E, copy the information corresponding to the X and Y coordinates in pixels ("PosX [pixels]" and "PosY [pixels]," in the View5D file).
6. To add a second cell, go to the row below the last row corresponding to this cell, and perform **steps 3–5**, adding a two (2) in column B (or the next number track in View5D).
7. Repeat this process for all cells of interest, adding 3, 4, 5, etc. as their corresponding value on column B.
8. Once you have added all the tracks that you wish to visualise, add a correlative number list (1, 2, 3...) on column A, starting on cell A2 and continuing until the last time point of the last cell.
9. Save this file as file type "Text (Tab delimited)."

   For automatic conversion, follow **steps 10–16**:
10. Open the tracking file from the View5D output on Microsoft Excel.
11. Save this file as file type "Excel 97-2003 Workbook."
12. Open MatLab.
13. Copy the script in the file "transform_to_manual_track.txt" (Supplementary Script 2) onto a new workbook.
14. Run the script by pressing the green arrow button on the top ribbon.
15. Find the Excel file that you saved on **step 2**.
16. A window called "Data Specification" will pop up that requires information on the Excel file. The first space is the name of the Tab containing the tracking data. The second space is the total number of time points. The third one is the total number of rows containing information (it will be the number of the row immediately above the row with "#Statistics Summary," at the bottom of the file). Press OK. This will create a file on the same directory that has your Excel file, called "name.xls_markers_-ManualTrack.txt."

   Once you have the text file, upload it to Manual Tracking.
17. Open the 3D movie in Fiji.
18. Create a 2D maximal projection ("Image > Stack > Z project..." a window will open to select the type of projection. Select Max Intensity, or the preferable mode of projection and click "OK").

19. Open "Manual Tracking" plugin "Plugins > Tracking > Manual Tracking."
20. On the "Tracking" window, click "Load Previous Track File." Fiji will ask you if you are sure that you want to load an existing file; say "Yes."
21. Select the text file you created previously, that will be uploaded in a new data window.
22. At this point you can select how to overlay your tracks onto the movie in the "*Drawing*" section of the Tracking window. The top row ("Dots" / "Progressive Lines" / "Dots & Lines") will create a separate movie only with the tracks over a black background. The bottom row ("Overlay Dots" / "Overlay Lines" / "Overlay Dots & Lines") will create a separate movie with the tracks over the projection of your movie. "Dots" overlays a dot on top of each tracked cell; "Lines" overlays the track, progressing through time; "Dots & Lines" does both at the same time.
23. You can now save this new file as a separate movie.

3.6 Time-Controlled Induction of Protein Expression in Neural Crest

Transgenic embryos from the Sox10:Kalt4ER^{kg328Tg} line express the Gal4ER fusion protein in all NC, which upon addition of tamoxifen will move to the nucleus and initiate transcription of any UAS-driven transgenes. This system allows protein overexpression only in NC at any required time of development.

1. Cross Sox10:Kalt4ER^{kg328Tg} transgenic fish to the desired UAS regulated line for expression in the entire neural crest population (*see* **Note 9**). Alternatively, for clonal induction, inject UAS-driven plasmid DNA into one-cell stage embryos (*see* **Note 10**).
2. Incubate the embryos at 28.5 °C until the required stage. Alternatively, to obtain stages from 11hpf onwards in the morning, maintain embryos at 28.5 °C until 50% epiboly, then transfer them to 19 °C overnight (*see* **Note 11**).
3. Embryos will be about 10hpf at 9:00 next morning. Move them to 28.5 °C until the desired stage.
4. Add tamoxifen to the E3 medium to a final concentration of 0.5 μM. Measure the necessary amount of E3 in a 50 mL Falcon tube; add the embryos to be treated in their chorions and set the volume correctly; dilute the 5 mM stock tamoxifen 10,000 times. Pour in to a desired Petri dish and incubate at 28.5 °C until the desired stage. Expression of UAS-driven proteins is observed after 30 min of tamoxifen addition and is maximal after 1 h (Fig. 4).

5. If time-lapse movies are required, add 0.5 μM tamoxifen to the agarose mould, LMP and medium indicated in the imaging protocol described.

3.7 Photo-Labelling of Individual Neural Crest Cells

1. Obtain Sox10:Dendra2^{kg329Tg} embryos and incubate at 28.5 °C or 19 °C (see above) until the desired stage. Hetero- and homozygous embryos are bright enough to carry out the experiment, but homozygous embryos are easier to work with as the fluorescence intensity is higher.
2. At the desired developmental stage, dechorionate and transfer to E3/tricaine medium to mount for photoconversion.
3. Add 0.1 mL of tricaine to 0.9 mL of 3% methyl cellulose in a 1.5 mL tube, vortex and let stand for 10 min for bubbles to come up. With a plastic pipette transfer the embryos with the absolute minimum amount of medium into the tube.
   (a) If using an upright microscope, embryos can be placed and oriented in an agarose mould as explained before (*see* Subheading 3.1), or simply mounted in methyl cellulose: add a thin layer (0.3–0.5 mL) of methyl cellulose to a plastic Petri dish where the embryos will be located. Transfer the embryos from the 1.5 mL tube into the drop and orient in a line next to each other using a hair knife.
   (b) If using an inverted microscope mount in a drop of methyl cellulose as explained but use a glass bottom dish; in this case, make sure gently push in the embryos so they touch the glass.
4. Put the embryo dish on the microscope stage and focus on the embryos on the desired area. Take a quick image with 488 nm illumination as reference image. Limit the exposure to the 488 nm laser to the minimum as it can photoconvert Dendra2 at high power.
5. Run a trial experiment to determine exposure time and power to efficiently photoconvert Dendra2 using the 405 nm laser. Start with minimum laser power and exposure times. Dendra2 is very sensitive and will totally bleach easily (green and red emission).
6. Once the photoconversion conditions have been established, select regions of interest around single or group of nuclei required (Fig. 4). Make sure the region of interest selected is as small as possible to avoid photoconversion of adjacent cells (*see* **Note 12**).
7. After photoconversion, take a quick image using the 561 nm laser to check that that desired cells have been photoconverted successfully.

3.8 Embryo Dissociation

It is important that once the embryos have been dissected cells are dissociated, sorted and frozen on dry ice in less than 1 h.

1. Defrost all solutions (FACsMax solution, FBS, trypsin, etc.) at 4 °C overnight and keep on ice when starting the experiment. Pre-cool a table top centrifuge to 4 °C.
2. Take the photoconverted embryos out of the methyl cellulose by delicately flowing E3 with a plastic pipette over the embryos. Transfer the embryos into fresh E3 + tricaine medium.
3. Using n3 tweezers, manually dissect the head and yolk from the tail (to be dissociated). Using one pair of forceps, pinch right under the yolk at the beginning of the yolk stalk, and with the second forceps cleave the tail away.
4. Immediately transfer the tail to a cooled 1.5 mL tube on ice using a long glass Pasteur pipette carrying over as little E3 as possible. Aim to a minimum of 10–20 embryos per tube.
5. Spin down the embryos using the pre-cooled 4 °C centrifuge at 400 × *g* for 1 min.
6. Remove the supernatant under a dissecting microscope, be careful not to disturb the pellet.
7. Add 200 μL of trypsin-EDTA. Wash a P200 tip in sterile PBST, by suctioning and discarding the solution. Proceed to dissociate the pellet by vigorously pipetting up and down for 10 min or longer until no clumps of tissue remain.
8. Add 5% of HS-FBS to stop the reaction.
9. Spin down in pre-cooled 4 °C centrifuge at 400 × *g* for 5 min.
10. Remove supernatant under dissection microscope, be careful not to disturb the pellet.
11. Slowly add 200 μL of FACsMax solution without disturbing the pellet. Spin down in pre-cooled 4 °C centrifuge at 400 × *g* for 5 min.
12. Remove supernatant under dissection scope, be very careful not to disturb the pellet and repeat **steps 10–12**.
13. Resuspend the dissociated cell pellet in 300 μL of FACsMax solution using PBST washed tips. Transfer into 35 μm filtered FACS collection tube and keep on ice.
14. Proceed immediately to the FACS machine. Gating conditions depend on the sorter used and should be experimentally determined. At the end of sorting, samples can be immediately processed or frozen at −80 °C.

4 Notes

1. Water immersion and long distance lenses are required. The best compromise between magnification and numerical aperture needs to be found. Numerical apertures below 1.0 are not ideal as they will require longer exposure and higher lens power, which will produce photo damage. If using a 40× lens, we advise to take 2–3 adjacent fields of view and stitch them together, ensuring that the cells of interest do not go out of plane during the experiment. Also, place an open container with water next to the microscope stage to prevent evaporation.
2. Regular commercial plastic petri non-cell culture dishes of any size can be used. We routinely use 60 mm and 35 mm, but the size of the dish will depend on the microscope inset and the lens size. If using a motorized stage to make several movies in one session, the lens has to be able to freely travel from one embryo to the next without touching the walls of the dish. Small dishes are more suitable when treating embryos with drugs as smaller amounts are required. In which case it is important to add drug to the agarose mould.
3. Bubbles can form underneath the plastic cast. To avoid that, spray the cast with 70% ethanol before positioning it on top of the agarose. After the agarose hardens, leave it to air dry for 20 min to evaporate any alcohol.
4. It is important that the fluorescent protein to be imaged is expressed at good levels before starting to image. We have observed that imaging cells with low fluorescent levels by increasing laser power quickly bleaches the initial protein present and cells are not able to accumulate it thereafter, generating a black movie.
5. The time step has to be adapted to the question at hand. For example, to study actin flow in neural crest, the faster the acquisition time the better it is, although a 2 min interval is generally good. To study neural crest differentiation, the acquisition time can be set to 30 min or 1 h intervals, which allows for a longer total time of the movie.
6. The type of files obtained will depend on the software of the system used. Fiji will read most of the common microscopy software directly. If yours is not directly read, export your movies as TIFF series with maximal resolution. Open them in Fiji clicking "File > Import > Image sequence," a Sequence Option window will open. Indicate the total number of images, starting image and increment, and whether to sort the images numerically. For example, if your channels are one after the other and you have 30 z steps and ten time points, then for the

first channel you will need to say number of images 300, starting image 1, increment 30 and sort numerically. This will generate a stack of the first channel. For the second channel would be number of images 300, starting image 301, increment 30 and sort numerically; and so on.

7. There are a number of variables that can cause the Registration to fail. If the new stack has not been corrected properly (e.g., has black frames or incorrect rotations), try one or both of these steps:
   (a) Check the focused stack created in **step 4** and make sure that all frames display the anatomical features in focus, without drastic changes between consecutive frames.
   (b) Adjust the brightness of the stack "Image > Adjust > Brightness/Contrast" to make the landmarks more evident.
8. If you are not sure about the calibration of your instrument, you can calibrate it by taking an image of a graded stage reticule and calculate the *XY* length of a pixel. Use the smallest reticule you can find, a micrometre scale 10 mm in 0.1 mm division is suitable, take an image. Using the measuring tool from the microscope software, measure the number of pixels each reticule unit spans. Or open your reticule image in Fiji, draw horizontal straight line spanning one reticule unit using the line tool, and pressing shift as you draw. Press "m" the number of pixels will be under length in the Results table. Calculate the pixel length in μm.
9. Gal4ER fusion was generated using the Kalt4 form of the Gal4 activator [10] and the extracellular domain of the human oestrogen receptor.
10. Plasmid DNA of UAS-driven constructs with or without transposable elements (Tol2 [13] or ACDS [14]) has been successfully expressed after injection. We have generated double UAS vectors driving the expression of super folded GFP in one strand and any protein of interest in the other. In our hands, super folded GFP and Scarlett are the brightest green and red fluorescent proteins, respectively. The use of the zebrafish codon sequence has a major impact in the expression levels in both cases. In all cases, 20 ng of plasmid DNA is injected. Transposase mRNA is injected when required; the concentration of RNA to inject needs to be experimentally defined as is batch dependent. It is very important to inject into the cell of early one-cell stage embryos, when the cell is still flat surrounding the yolk. The number of clones obtained is drastically reduced if embryos are injected once the first cell has plumped up.

11. Embryos transferred to 19 °C before 10% epiboly will not survive. Embryos transferred to temperatures below 19 °C at any point will die.
12. In our hands, maximal intensity under 561 nm laser is attained about 1 h after photoconversion. For this reason, embryos are illuminated with 405 nm laser and then incubated for 1 h before further processing either to dissociation or imaging. Also, photoconverted Dendra2 is stable for about 36 h after photoconversion.

Acknowledgments

We are in debt to Nicolas Daudet for his help and support. The Linker lab has been supported by grants from the Wellcome Trust, Royal Society, and the Medical Research council. SH and LB are funded by ICM P09-015-F, DAAD 57220037 & 57168868, CORFO 16CTTS-66390, Fondecyt 1181823, FONDEF 19I10334.

References

1. Stramer BM, Dunn GA (2015) Cells on film—the past and future of cinemicroscopy. J Cell Sci 128:9–13. https://doi.org/10.1242/jcs.165019
2. BFI. Cheese Mites (1903) | BFI National Archive. n.d.
3. Meyers JR (2018) Zebrafish: development of a vertebrate model organism. Curr Protoc Essent Lab Technol 16:e19. https://doi.org/10.1002/cpet.19
4. Brabletz T, Kalluri R, Nieto MA, Weinberg RA (2018) EMT in cancer. Nat Rev Cancer
5. Delloye-Bourgeois C, Castellani V (2019) Hijacking of embryonic programs by neural crest-derived neuroblastoma: from physiological migration to metastatic dissemination. Front Mol Neurosci 12. https://doi.org/10.3389/fnmol.2019.00052
6. Le Douarin NM, Kalcheim C (1999) The neural crest, 2nd edn. Cambridge University Press, Cambridge
7. Raible DW, Wood A, Hodsdon W, Henion PD, Weston JA, Eisen JS (1992) Segregation and early dispersal of neural crest cells in the embryonic zebrafish. Dev Dyn 195:29–42. https://doi.org/10.1002/aja.1001950104
8. Richardson J, Gauert A, Briones Montecinos L, Fanlo L, Alhashem ZM, Assar R et al (2016) Leader cells define directionality of trunk, but not cranial, neural crest cell migration. Cell Rep 15:2076–2088. https://doi.org/10.1016/j.celrep.2016.04.067
9. Duffy JB (2002) GAL4 system indrosophila: a fly geneticist's swiss army knife. Genesis 34:1–15. https://doi.org/10.1002/gene.10150
10. Distel M, Wullimann MF, Köster RW (2009) Optimized Gal4 genetics for permanent gene expression mapping in zebrafish. Proc Natl Acad Sci U S A 106:13365–13370. https://doi.org/10.1073/pnas.0903060106
11. Collins RT, Linker C, Lewis J (2010) MAZe: a tool for mosaic analysis of gene function in zebrafish. Nat Methods 7:219–223. https://doi.org/10.1038/nmeth.1423
12. Gorelik R, Gautreau A (2014) Quantitative and unbiased analysis of directional persistence in cell migration. Nat Protoc 9:1931–1943. https://doi.org/10.1038/nprot.2014.131
13. Kwan KM, Fujimoto E, Grabher C, Mangum BD, Hardy ME, Campbell DS et al (2007) The Tol2kit: a multisite gateway-based construction kit forTol2 transposon transgenesis constructs. Dev Dyn 236:3088–3099. https://doi.org/10.1002/dvdy.21343
14. Chong-Morrison V, Simoes FC, Senanayake U, Carroll DS, Riley PR, Sauka-Spengler T (2018) Re-purposing Ac/Ds transgenic system for CRISPR/dCas9 modulation of enhancers and non-coding RNAs in zebrafish. BioRxiv

2018:450684. https://doi.org/10.1101/450684
15. Gilmour DT, Maischein H-M, Nüsslein-Volhard C (2002) Migration and function of a glial subtype in the vertebrate peripheral nervous system. Neuron 34:577–588. https://doi.org/10.1016/S0896-6273(02)00683-9
16. Hochgreb-Hägele T, Bronner ME (2013) A novel FoxD3 gene trap line reveals neural crest precursor movement and a role for FoxD3 in their specification. Dev Biol 374:1–11. https://doi.org/10.1016/j.ydbio.2012.11.035
17. Kaufman CK, Mosimann C, Fan ZP, Yang S, Thomas AJ, Ablain J et al (2016) A zebrafish melanoma model reveals emergence of neural crest identity during melanoma initiation. Science 351:aad2197. https://doi.org/10.1126/science.aad2197
18. Hoffman TL, Javier AL, Campeau SA, Knight RD, Schilling TF (2007) Tfap2 transcription factors in zebrafish neural crest development and ectodermal evolution. J Exp Zoolog B Mol Dev Evol 308B:679–691. https://doi.org/10.1002/jez.b.21189
19. Dutton JR, Antonellis A, Carney TJ, Rodrigues FS, Pavan WJ, Ward A et al (2008) An evolutionarily conserved intronic region controls the spatiotemporal expression of the transcription factor Sox10. BMC Dev Biol 8:105. https://doi.org/10.1186/1471-213X-8-105
20. Das A, Crump JG (2012) Bmps and Id2a act upstream of Twist1 to restrict Ectomesenchyme potential of the cranial neural crest. PLoS Genet 8:e1002710. https://doi.org/10.1371/journal.pgen.1002710
21. Blasky AJ, Pan L, Moens CB, Appel B (2014) Pard3 regulates contact between neural crest cells and the timing of Schwann cell differentiation but is not essential for neural crest migration or myelination: PARD3 IN NC MIGRATION AND SCHWANN CELL MYELINATION. Dev Dyn 243:1511–1523. https://doi.org/10.1002/dvdy.24172
22. Askary A, Mork L, Paul S, He X, Izuhara AK, Gopalakrishnan S et al (2015) Iroquois proteins promote skeletal joint formation by maintaining chondrocytes in an immature state. Dev Cell 35:358–365. https://doi.org/10.1016/j.devcel.2015.10.004
23. Schilling TF, Pabic PL, Hoffman TL (2010) Using transgenic zebrafish (Danio rerio) to study development of the craniofacial skeleton. J Appl Ichthyol 26:183–186. https://doi.org/10.1111/j.1439-0426.2010.01401.x
24. Smith CJ, Morris AD, Welsh TG, Kucenas S (2014) Contact-mediated inhibition between oligodendrocyte progenitor cells and motor exit point glia establishes the spinal cord transition zone. PLoS Biol 12:e1001961. https://doi.org/10.1371/journal.pbio.1001961
25. Kucenas S, Takada N, Park H-C, Woodruff E, Broadie K, Appel B (2008) CNS-derived glia ensheath peripheral nerves and mediate motor root development. Nat Neurosci 11:143–151. https://doi.org/10.1038/nn2025
26. Dougherty M, Kamel G, Grimaldi M, Gfrerer L, Shubinets V, Ethier R et al (2013) Distinct requirements for wnt9a and irf6 in extension and integration mechanisms during zebrafish palate morphogenesis. Development 140:76–81. https://doi.org/10.1242/dev.080473
27. Prendergast A, Linbo TH, Swarts T, Ungos JM, McGraw HF, Krispin S et al (2012) The metalloproteinase inhibitor Reck is essential for zebrafish DRG development. Development 139:1141–1152. https://doi.org/10.1242/dev.072439
28. Balczerski B, Matsutani M, Castillo P, Osborne N, Stainier DYR, Crump JG (2012) Analysis of Sphingosine-1-phosphate signaling mutants reveals endodermal requirements for the growth but not dorsoventral patterning of jaw skeletal precursors. Dev Biol 362:230–241. https://doi.org/10.1016/j.ydbio.2011.12.010
29. Chung A-Y, Kim P-S, Kim S, Kim E, Kim D, Jeong I et al (2013) Generation of demyelination models by targeted ablation of oligodendrocytes in the zebrafish CNS. Mol Cells 36:82–87. https://doi.org/10.1007/s10059-013-0087-9
30. Rodrigues FSLM, Doughton G, Yang B, Kelsh RN (2012) A novel transgenic line using the Cre-lox system to allow permanent lineage-labeling of the zebrafish neural crest. Genesis 50:750–757. https://doi.org/10.1002/dvg.22033
31. Mongera A, Singh AP, Levesque MP, Chen Y-Y, Konstantinidis P, Nusslein-Volhard C (2013) Genetic lineage labeling in zebrafish uncovers novel neural crest contributions to the head, including gill pillar cells. Development 140:916–925. https://doi.org/10.1242/dev.091066
32. Scheer N, Riedl I, Warren JT, Kuwada JY, Campos-Ortega JA (2002) A quantitative analysis of the kinetics of Gal4 activator and effector gene expression in the zebrafish. Mech Dev 112:9–14

Chapter 10

Live Imaging of the Neural Crest Cell Epithelial-to-Mesenchymal Transition in the Chick Embryo

Mary Cathleen McKinney and Paul M. Kulesa

Abstract

Live embryo imaging may provide a wealth of information on intact cell and tissue dynamics, but can be technically challenging to sustain embryo orientation and health for long periods under a microscope. In this protocol, we describe an in vivo method to mount and image cell movements during the epithelial-to-mesenchymal transition (EMT) of neural crest cells within the chick dorsal neural tube. We focus on describing the collection of images and data preparation for image analysis throughout the developmental stages HH15-21 in the chick trunk. Trunk neural crest cell EMT is crucial to development of the peripheral nervous system and pigment cell patterning. The methods we describe may also be applied to other cell and tissue phenomena at various chick developmental stages with some modifications.

Key words Chick, In vivo, Whole embryo culture, Time-lapse imaging, Neural crest, EMT

1 Introduction

Static imaging limits our understanding of the mechanisms underlying in vivo cell and tissue dynamics. However, to resolve individual cell behaviors within intact embryos present several technical challenges. Early-stage embryos are often fragile and difficult to fluorescently mark and manipulate. Later-stage embryos are often difficult to keep oriented and in the field of view of the microscope objective due to vibrations of the heart, tissue growth of normal morphogenesis, thermal drift due to a heated chamber, or shrinkage due to dehydration. One of the crucial phenomena of early development is the epithelial-to-mesenchymal transition of neural crest cells enabling them to migrate and form peripheral structures. Neural crest cells exit from the dorsal neural tube in a rostral-to-caudal manner starting near the midbrain [1]. Following single cell behaviors in the dorsal neural tube and neural crest cell emigration has proved difficult due to the challenges described above. Following this protocol, we were able to observe behaviors within the

Kyra Campbell and Eric Theveneau (eds.), *The Epithelial-to Mesenchymal Transition: Methods and Protocols*, Methods in Molecular Biology, vol. 2179, https://doi.org/10.1007/978-1-0716-0779-4_10,

chick trunk neural tube and examining sites in the periphery to which their progeny migrate, we discovered a more complex, unordered emigration of trunk neural crest cells than previously thought [2]. Our extensive experience with in vitro slice culture and whole embryo imaging in chick and quail embryos allowed us to define the conditions required for single cell resolution of fluorescently labeled cells during the epithelial-to-mesenchymal transition of the trunk neural crest cells. Here, we present a method to culture whole chick embryo explants and maintain embryo orientation and health over a 24 h period. Although our approach described here is an exciting step forward in allowing direct observation and measurement of complex cell behaviors during trunk neural crest cell epithelial-to-mesenchymal transition, our methods may be applicable for a wider range of phenomena and in other research organisms.

2 Materials

2.1 Embryology and Labeling

1. Fertilized chicken eggs.
2. DiI labeling solution: 50 μg DiI powder (C7000, Thermo Fisher, or equivalent), 10 μL ethanol, 90 μL 0.3 M sucrose.
3. Microinjector (Picospritzer, Parker Hannifin, or equivalent).
4. Glass needles pulled to a shape you prefer (BF100-50-10, Sutter Instrument, or equivalent).
5. Whatman filter paper.
6. Embryology tools (forceps, tungsten needle with holder, small scissors).
7. Ink Solution: Sterilized 10% black pen ink in PBS.
8. 10 mL and 1 mL syringes.
9. 18- and 25-gauge needles.
10. Ringer's Solution: 7.2 g Sodium Chloride, 0.37 g Potassium Chloride, 0.17 g Calcium Chloride per liter, pH 7.4.
11. Phosphate Buffered Saline pH 7.4.
12. Optional: fluorescent protein vector(s) and a square wave electroporator (ECM 830, BTX; or equivalent).

2.2 Mounting

1. Approximately 2.5 cm diameter coin and pencil.
2. 60 mm plastic petri dishes.
3. EC Culture Medium: egg albumen, agar solidifying agent (Bacto Agar, 214050, BD Diagnostics, or equivalent), Penicillin–Streptomycin, PBS.
4. 49° water bath or similar.

2.3 Microscopy and Image Analysis

1. Upright confocal microscope with environmental chamber and smaller stage chamber to limit dehydration.
2. Fiji software with plugin Correct 3d Drift [3, 4].
3. Optional: Imaris (Bitplane AG).

3 Methods

3.1 Prepare Culture Dishes

1. The preparation of culture dishes may be performed up to 1 week in advance.

   The critical steps are outlined here but can also be found in [5]. Crack an egg into a large dish and use a 10 mL syringe with 18 g needle to extract thin albumen. Repeat until you have collected 30 mL of thin albumen. Place in 49 °C water bath or similar heated incubator.
2. Dissolve 180 mg of Agar into 30 mL PBS, then reduce temperature to 49 °C by placing in water bath. If this solution is too hot, it will cook the egg albumen so allow enough time to cool to 49 °C.
3. Mix Agar solution and egg albumen, then add 600 μL penicillin/streptomycin and immediately pour into 60 mm petri dishes and cover with lid. About 7 mL per dish (*see* **Note 1**).
4. After the medium has solidified, the culture dishes can be stored at 4 °C in a container with a wet paper towel for about 7 days before drying out or finding bacterial growth on medium.

3.2 Label Embryos

1. Incubate embryos in humidified egg incubator to desired developmental stage.
2. Add Penicillin–Streptomycin to ink and Ringer's solution in a small petri dish. Use a 1 mL syringe and 25G needle to hold this solution. This will be injected underneath the embryo to provide contrast to visualize.
3. Cut a window in the eggshell large enough to allow for visualization and microinjection of the embryo.
4. Inject the ink solution underneath embryo by inserting the needle into the yolk outside of the area of the embryo.
5. Remove the vitelline membrane in the area above the embryo that is to be fluorescently labeled using a tungsten needle or fine forceps. Hydrate the embryo frequently by gently dripping Ringers solution onto the vitelline membrane.
6. Inject DiI solution into neural tubes of embryos using picospritzer and glass needles. Take care to keep the DiI contained within the neural tube since it will label all tissue it touches.
7. Reseal eggs with tape.

8. Optional: Depending on the experiment, embryos can be injected then electroporated with DNA vectors in combination with or instead of DiI. For electroporation protocol, *see* [6].
9. Incubate embryos until desired developmental stage.

3.3 Create a Few Tools

To assist with mounting embryos on EC culture, create paper washers following **steps 1** and **2**:

1. Trace coin on Whatman filter paper with pencil and cut circle with fine scissors.
2. Cut out center hole in paper leaving an approximately 0.5 cm ring of paper. Keep paper clean and dry.

   A large environmental chamber around the entire stage, objective and condenser unit is useful to eliminate thermal drift of the sample. However, without a small chamber inside to keep the sample hydrated, the EC culture medium will dehydrate and constantly change the focal plane of the embryo or its health will fail. To construct this small chamber, follow **steps 3** and **4**.
3. Cut sheets of plastic to make a box with a bottom the same size as the microscope stage insert approximately 13 × 11 × 2.5 cm. The lid of the box should be removable with a single hole roughly the same size as the objective to be used. Lids can be created for various objectives but we do not recommend using an immersion objective for this protocol (*see* **Note 2**).
4. Seal the lid to the bottom with vacuum grease carefully expelled from a 10 mL syringe. In addition, humidity can be added to the chamber by incorporating a slow flow of humidified air through a bubbler or wet paper towels within the chamber.

3.4 Mounting Embryos and Imaging

1. Turn on the microscope and environmental chamber to evenly heat components to 37 °C at least 2 h before the samples are ready. Pre-warm EC culture dishes to 37 °C, 30 min before working with embryos.
2. Screen embryos for health, stage, and proper labeling. Neural crest EMT events happen over several hours in the trunk, but to stage match experiments we recommend choosing an embryo where the neural tube is well labeled, anterior neural crest can be observed to be exiting the neural tube but focus the imaging on a somite level where the neural crest has yet to emerge. This way, all embryos from different days of experiments can be time-aligned to the first neural crest cell to exit.
3. Open egg and create a large window for manipulation. Place one paper ring centered over embryo (Fig. 1a). While the ring is resting on the embryo, peel back more vitelline membrane

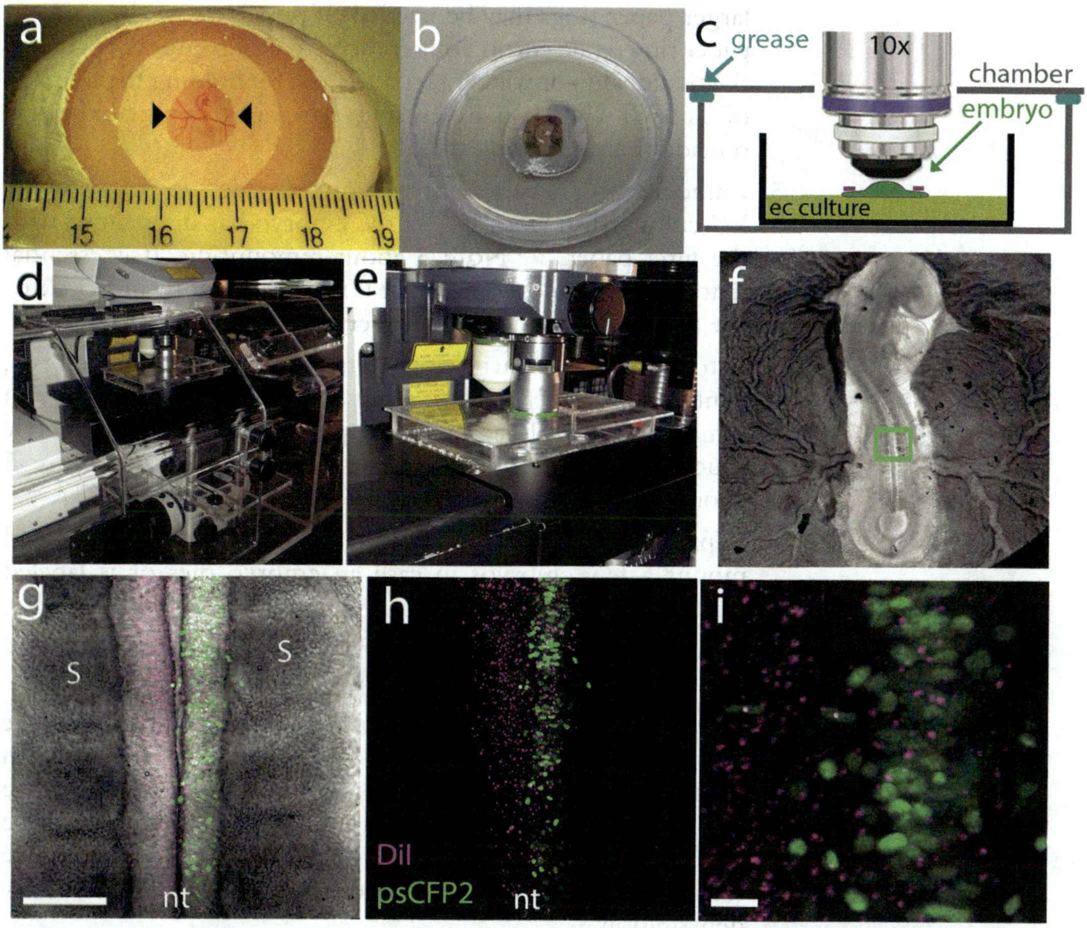

Fig. 1 Confocal imaging a 3-day chick embryo on an upright microscope. (**a**) Embryo inside windowed egg with scale bar (centimeters) and paper ring. (**b**) Embryo in EC culture dish dorsal side up with paper ring. (**c**) Schematic of chamber on microscope stage. Gray small chamber is sealed with vacuum grease and can have parafilm around the objective to reduce evaporation from the sample. Petri dish with embryo is placed inside this chamber with the lid off and objective is brought down through a hole in the top of the chamber that is custom sized for the objective. Paper ring (purple) is left on embryo. A dish of water, damp towel, or humidified air can be inserted inside chamber. (**d**) Microscope with large, heated environmental chamber surrounding entire stage of microscope and (**e**) the small chamber sitting on stage. (**f**) Example bright-field image of embryo inside microscope with region of interest highlighted in green box. (**g–h**) Example images of neural tube (nt) labeled with DiI (purple) and psCFP2 (green) and surrounding tissues, including somites (s) in (**g**). Scale bar is 100 μm. (**i**) Magnified image of similar embryo showing migrating neural crest cells and the neural tube. Scale bar is 20 μm

over the area to be imaged with a tungsten needle or forceps and tack down by sticking it to the filter paper. Let albumen run over paper while in the egg and moisten any dry parts with Ringer's solution.

4. Cut around outside of paper ring to remove embryo. Do not cut any vasculature. Depending on the developmental stage,

larger paper rings may be needed. If possible, cut alongside of paper ring while cutting out embryo to crimp edges of paper to the tissue. Dull or wide forceps are best for holding the free side of embryo while you are cutting. Attempt to only hold tissue outside the vascular plexus.

5. Immediately transfer embryo dorsal side up to a warm EC culture dish (Fig. 1b). If the embryo has disengaged with the filter paper, *see* **Note 3** about remounting. It is essential that the extra-embryonic tissue is stretched tightly in a circle or the tissue will not develop properly.
6. If too many yolk particles have transferred with the embryo, gently drop Ringer's solution just to the side of the embryo and remove yolk with a narrow-tipped pipette. Avoid directly touching the embryo. *See* **Note 4** about yolk and **Note 5** about excess water in the dish.
7. Replace lid to EC culture dish and place in incubator and mount a few embryos in case an embryo dies or is labeled poorly.
8. Optional: Multiple embryos can be placed in one EC culture dish for multi-positional imaging *see* **Note 6**.
9. Place EC culture dish without lid into the small box on the stage of the microscope (Fig. 1c–e). Seal the lid of the box with vacuum grease and place parafilm around objective to reduce the gap between the objective and lid. It will probably not be airtight since some flexibility is needed to move the objective for a *z*-stack but air flow should be reduced enough to slow dehydration.
10. Set up imaging parameters for your label and region of interest on the microscope (Fig. 1f–i). Take great care to not allow contact between the objective and the embryo. If the objective pulls on the tissue, the embryo will either tear or not lay evenly taut and thus not develop properly. Clean the objective and start with a new embryo.
11. The embryo in this environment is fragile and should be exposed to a minimum of laser light. EMT events occur at the dorsal most part of the neural tube but you may wish to image the neural crest as they migrate deeper into the tissue but the more deeply you image, the longer the image will take to acquire. Less than 1 min of continuous excitation is ideal. Frequency of images should be more than double the amount of time exposed to the laser, i.e., 1 min of imaging, 2 min or more of rest. For the best automated tracking, images should be taken frequently enough that a cell at least partially overlaps its position in the previous image. Longer time intervals are fine but will require more hand-corrections to the track. Depending on what type of information you require, you

may be interested in images 30 s apart or maybe 5–10 min for a global view of the cells in motion.

12. To ease image processing, do not change the size of the imaging region, laser intensity, or detector gain throughout the experiment.
13. Check the microscope once an hour for the first 3 h, then check the image every few hours to re-adjust the focus and potentially *XY* position to keep the region of interest in focus as the embryo grows and shifts position.

3.5 Image Processing

1. Open Fiji software and concatenate all time-lapses that were saved and restarted through the experiment.
2. Drift, growth, and the heartbeat of the embryo all contribute to mis-alignment of the images over many hours. After concatenation, align the image using the descriptor-based series registration plugin [3]. If the embryo has moved a lot during imaging, first try adjusting the parameters in the plugin. If this fails, partial registration over different blocks of time can be aligned separately then concatenated back together.
3. If the embryo drifted or grew by a large amount in the *Z* direction, sometimes registration may work in the *XY* direction but not in Z. Imaris (Bitplane Inc) has a registration algorithm that may suffice, or a manual registration can be done if the XT module is available.
4. Tracking of cells or other analysis can be performed at this point to answer your specific scientific questions.

4 Notes

1. Volume of medium. 7 mL of EC culture medium is an estimate. More culture medium will provide a buffer if the sample begins to dehydrate on the microscope, but more medium will raise the sample and may interfere with the working distance on the microscope.
2. Objectives. A water dipping objective is not recommended for this method because the embryonic tissue needs access to oxygen. The area covered by even a small droplet of water will become necrotic or pull on the rest of the tissue. A long working distance objective is ideal; however, a 20× 0.8NA or 10× 0.5NA are possibilities.
3. Remounting. If the embryo has fallen off the paper or become folded, it may be corrected by flooding the EC culture dish with Ringer's solution and gently reorienting the embryo. Remove most of the water, then lay down a new paper ring.

Remove the extra water with a p200 pipettor or a narrow-tipped transfer pipette.

4. Yolk particles. Yolk is widely autofluorescent and will obscure a bright-field image. If a considerable amount of yolk is above the embryo, imaging will be more difficult. If gently washing the embryo with fresh Ringer's solution in the imaging dish cannot be accomplished, place the embryo into a fresh dish of Ringer's solution and gently blow the yolk off with a transfer pipette. The embryo can then be placed back on an EC culture dish. Transferring the embryo should be avoided so the tissue does not fold and the embryo stays warm. Small amounts of yolk beneath the embryo may not be in the region of interest and can be left alone.
5. Excess water. If there is too much water underneath the embryo, it will slide in the culture dish during transfer to the microscope and during imaging. If there is enough water around the embryo to collect with a pipette, do so using either a p200 pipettor or a narrow-tipped transfer pipette. Do not pull on tissue, instead pull from outside the paper ring even tipping the dish slightly while collecting the water.
6. Mounting multiple embryos. 35 mm dishes can also be used for smaller embryos (<HH13) but the 60 mm provides more room for manipulation of the embryo. Two embryos may be mounted in one 60 mm petri dish. It is easier to accomplish this with younger embryos since they are smaller and thinner so the possibility of touching the microscope objective to the embryo is lessened. A small amount of vacuum grease can be used to stabilize the dish to the stage for multi-position imaging but may not be necessary.

References

1. Kulesa PM, Gammill LS (2010) Neural crest migration: patterns, phases and signals. Dev Biol 344(2):566–568. https://doi.org/10.1016/j.ydbio.2010.05.005
2. McKinney MC, Fukatsu K, Morrison J, McLennan R, Bronner ME, Kulesa PM (2013) Evidence for dynamic rearrangements but lack of fate or position restrictions in premigratory avian trunk neural crest. Development 140 (4):820–830. https://doi.org/10.1242/dev.083725
3. Preibisch S, Saalfeld S, Schindelin J, Tomancak P (2010) Software for bead-based registration of selective plane illumination microscopy data. Nat Methods 7(6):418–419. https://doi.org/10.1038/nmeth0610-418
4. Schindelin J, Arganda-Carreras I, Frise E, Kaynig V, Longair M, Pietzsch T, Preibisch S, Rueden C, Saalfeld S, Schmid B, Tinevez JY, White DJ, Hartenstein V, Eliceiri K, Tomancak P, Cardona A (2012) Fiji: an open-source platform for biological-image analysis. Nat Methods 9(7):676–682. https://doi.org/10.1038/nmeth.2019
5. Chapman SC, Collignon J, Schoenwolf GC, Lumsden A (2001) Improved method for chick whole-embryo culture using a filter paper carrier. Dev Dyn 220(3):284–289. https://doi.org/10.1002/1097-0177(20010301)220:3
6. McLennan R, Kulesa PM (2007) In vivo analysis reveals a critical role for neuropilin-1 in cranial neural crest cell migration in chick. Dev Biol 301 (1):227–239. https://doi.org/10.1016/j.ydbio.2006.08.019

Chapter 11

Exploiting *Drosophila melanogaster* Wing Imaginal Disc Eversion to Screen for New EMT Effectors

Sofia Golenkina, Rosemary Manhire-Heath, and Michael J. Murray

Abstract

In the early stages of *Drosophila melanogaster* (*Drosophila*) metamorphosis, a partial epithelial-mesenchymal transition (pEMT) takes place in the peripodial epithelium of wing imaginal discs. Blocking this pEMT results in adults with internalized wings and missing thoracic tissue. Using peripodial GAL4 drivers, GAL80^{ts} temporal control, and UAS RNAi transgenes, one can use these phenotypes to screen for genes involved in the pEMT. Dominant modifier tests can then be employed to identify genetic enhancers and suppressors. To analyze a gene's role in the pEMT, one can then visualize peripodial cells in vivo at the time of eversion within the pupal case using live markers, and by dissecting, fixing, and immunostaining the prepupae. Alternatively, one can analyze the pEMT ex vivo by dissecting out wing discs and culturing them in the presence of ecdysone to induce eversion. This can provide a clearer view of the cellular processes involved and permit drug treatments to be easily applied.

Key words Epithelial-mesenchymal transition, Wing, Imaginal disc, Eversion, RNAi, Culture

1 Introduction

Genetic screens in *Drosophila melanogaster* (*Drosophila*) have provided enormous insight into a wide range of cellular and developmental processes, and molecular pathways. These include both forward screens employing chemical mutagenesis and transposon insertions [1] and reverse genetic screens using sequence-specific gene disruption methods such as RNAi [2]. However, few screens have specifically targeted epithelial-mesenchymal transition (EMT) events, despite the enormous importance of EMTs in development, cancer metastasis, and wound healing [3]. Part of the reason for this is that the two best documented EMTs in *Drosophila*, the formation of the mesoderm and endoderm [4–6], occur early in embryogenesis, which creates challenges for loss-of-function

Sofia Golenkina and Rosemary Manhire-Heath contributed equally with all other contributors.

Kyra Campbell and Eric Theveneau (eds.), *The Epithelial-to Mesenchymal Transition: Methods and Protocols*, Methods in Molecular Biology, vol. 2179, https://doi.org/10.1007/978-1-0716-0779-4_11,

screens. Firstly, since most genes in *Drosophila* are not haplo-insufficient, only homozygous mutant embryos will reveal a phenotype. This can involve significant extra work since one must isolate and maintain mutant alleles over wild-type balancer chromosomes and then identify homozygotes using transgenic markers on the balancers such as lacZ transgenes. Secondly, due to maternal gene product, only a few genes will exhibit early zygotic phenotypes [7]. For example, despite known roles for RhoA in cellularization [8] and ventral furrow formation [9], zygotic RhoA mutant embryos exhibit only late embryonic phenotypes such as defects in dorsal closure [10]. In addition, although RNAi through the female germline can generate early embryonic phenotypes [11], strong tissue-specific RNAi at such early stages is not effective.

A later EMT event that is better suited to genetic screens is wing disc eversion. This occurs during the early stages of metamorphosis when imaginal discs undergo major morphogenetic movements to create the adult body. Topologically, imaginal discs are epithelial sacs with their apical surfaces oriented inwards. Hence, during eversion, discs must somehow turn themselves inside out. Traditionally this was thought to involve a "turning-out-your-pocket" mechanism [12]. This model was overturned in 2004 by Pastor-Pareja et al. [13], however, when it was shown that one side of the disc, the squamous peripodial epithelium (PE), undergoes a partial EMT (pEMT) event that creates perforations through which the rest of the wing disc can emerge. The morphogenetic movements of the PE have been subsequently well documented using live imaging ex vivo [14, 15]. Recently, a similar mechanism has been demonstrated for leg discs [16], suggesting that all imaginal discs may use this method.

Wing eversion presents as a useful system for identifying genes involved in EMT in other systems, as it exhibits many of the classic hallmarks of EMT. PE cells express matrix metalloproteases (MMPs) to break down the basement membrane, lose apico-basal polarity, delocalize cell-cell junction proteins such as E-Cadherin (E-Cad), and extend F-Actin rich protrusions as they invade the overlying larval epidermal cells. The wing discs from either side of the larva subsequently undergo an epithelial migration, eventually meeting up at the dorsal midline where they fuse to form the final pupal/adult epithelium [17]. PE cells also undergo apoptosis as the retraction of the epithelium proceeds but our group and others [13] have found that eversion will still take place when one inhibits apoptosis with p35, so it appears separable from the pEMT itself. Mutations in genes required for eversion such as the Jun N-terminal Kinase (JNK) pathway and MMPs result in easily scored adult phenotypes such as missing wings and thoracic clefts [18–20]. Also, since eversion occurs late in development, tissue-specific RNAi is very effective. Disc eversion, therefore, provides an accessible system to screen for, and analyze, genes involved in EMT.

In this chapter we describe how one can use peripodial GAL4 drivers, UAS RNAi libraries, and the temperature-sensitive repressor, GAL80^{ts}, to screen for genes involved in the peripodial pEMT using adult eversion phenotypes (Fig. 1). For genes with partially penetrant phenotypes, screens for enhancer/suppressor genes can

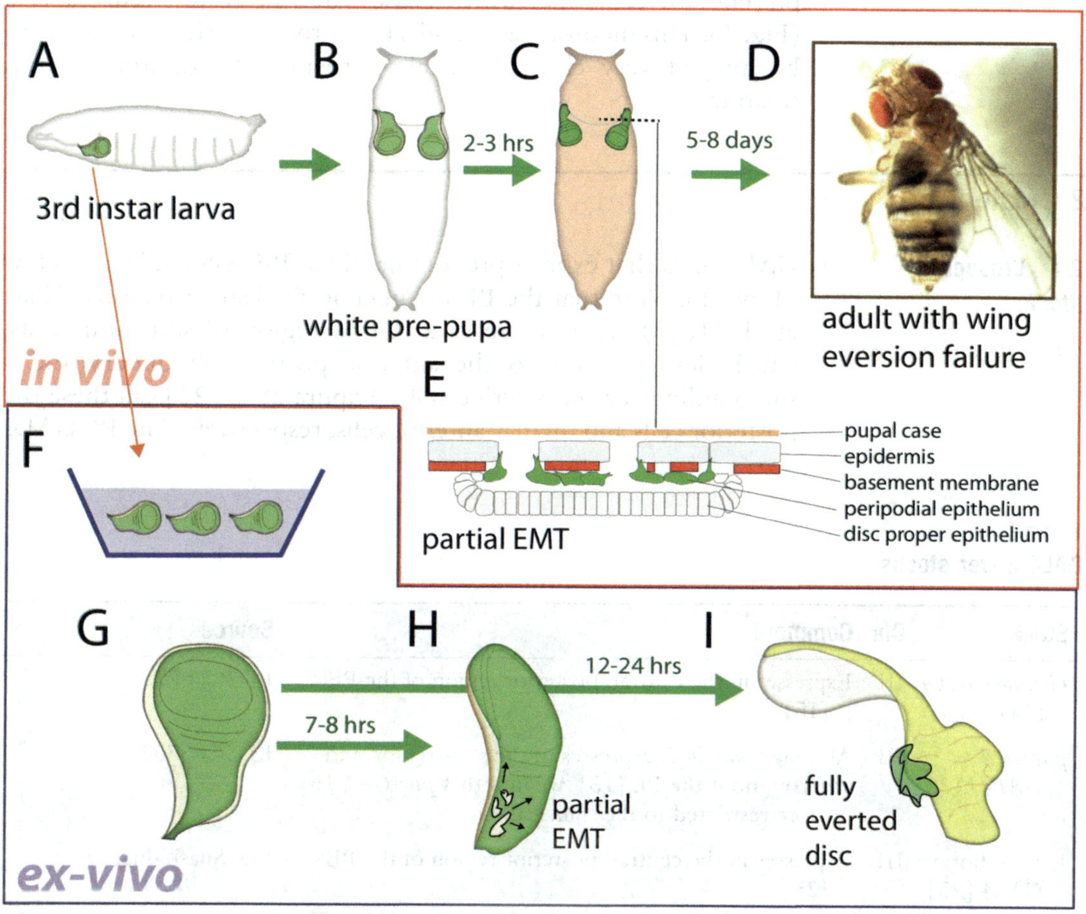

Fig. 1 Overview. (**a**) Wandering third-instar larva with one wing disc highlighted. (**b**) A white prepupa which has everted spiracles and ceased movement but not yet begun to tan. (**c**) Approximately 2–3 h later, the attachment point at the stalk of the wing discs has moved laterally within the pupal case and the pEMT is underway. (**d**) Approximately 5 days later an eclosed adult showing a single eversion failure, whereby one wing disc is still internalized and half the thorax is missing. (**e**) Schematic of the stage when the PE cells of the wing disc (green) undergo a pEMT, whereby they become motile and dismantle their epithelial junctions, secrete MMPs to degrade the basement membrane (red), and invade through the overlying epidermis (gray). (**f**) Wing discs can be isolated from larvae and cultured in the presence of ecdysone. (**g**) Wing disc at 0 h. (**h**) After 7–8 h, the pEMT is underway and perforations in the peripodial epithelium (green) are forming and expanding outwards (arrows). The DP cells are undergoing morphogenetic movements that bring the dorsal and ventral halves of the wing blade together. Contractions in the cuboidal stripe cells flanking the squamous central peripodial cells (not shown) cause the wing disc to bend [15]. (**i**) Leaving the disc for longer periods results in a fully everted disc in which the PE has completely retracted over the DP cells and forms a clump, while the wing blade has become more flattened and "wing-like"

be achieved either by co-expression of UAS RNAi constructs, or by reducing gene copy with mutants or deficiencies [21]. To examine the cellular basis for adult eversion phenotypes one can then use fluorescent transgenes and immunostaining to highlight key drivers and markers of the EMT process and examine PE cells in vivo, at the time of the EMT, or dissect out wing discs and culture them in the presence of ecdysone, which causes the pEMT to occur ex vivo (Fig. 1). This method has the advantage that the PE cells are able to be imaged very clearly, and also provides the option of drug treatments.

2 Materials

2.1 Drosophila Stocks

2.1.1 GAL4 Driver Stocks

GAL4 lines that express predominantly in PE cells are listed below (Table 1). Note that the PE has genetically distinct regions. Aldaz et al. [15] describe these as a central region of squamous cells, which lies posterior to the anterior/posterior boundary and a surrounding region of stripe cells. Tripura et al. [22] call these the posterior cells and medial-anterior cells, respectively. The PE GAL4

Table 1
GAL4 driver stocks

Stock	Chr	Comment	Source
Grunge-GAL4 [24]	III	Expresses in the central/posterior region of the PE [15]	BDSC#6773
puckered-GAL4 [13]	III	Although *puc-lacZ* expresses in the stalk region and throughout the PE [13], we find that *puc-GAL4* is more restricted to the stalk region	BDSC#6762
Ultrabithorax-GAL4 [25]	III	Expresses in the central/posterior region of the PE [25]	L.S. Shashidhara
odd-GAL4 [26]	II	Expresses in the medial-anterior/stripe cells [15, 22, 25]	L.S. Shashidhara
pannier-GAL4 [27]	III	Expresses in the stalk region in both PE cells and DP cells that will form the central notum	BDSC#3039 and other stocks
tubP:GAL80^{ts} [28]	X, II, III	A temperature-sensitive GAL4 repressor, GAL80^{ts}, under the control of the alphaTub84B promoter	BDSC#7017, #7019 and other stocks
UAS-Dicer-2 [23]	X, II, III	Co-expression of Dicer-2 can enhance the RNAi process	BDSC#24648, #24650, #24651 and other stocks

drivers listed below drive predominantly in one or other of these domains. Also listed is *pnr-GAL4* which express strongly in both the PE cells, near the stalk region, and the underlying disc proper (DP) epithelial cells that will give rise to the central notum region. Although each of the peripodial GAL4 drivers can elicit eversion phenotypes, we find *Ubx-GAL4* is the most effective at producing clear eversion-failure phenotypes such as internalized wings. The *pnr-GAL4* driver can also elicit eversion failure, though more commonly it results in disruption of the later epithelial migration and fusion events, and produces thoracic clefts [23]. Also listed are stocks carrying GAL80^{ts} and UAS-Dicer-2 constructs, which can be combined with the GAL4 constructs to achieve more effective driver stocks.

2.1.2 RNAi Libraries

Three public stock collections are available for GAL4-UAS RNAi knockdown (Table 2). These have been thoroughly described in a recent methods paper [2] and so will be only briefly covered here. We have achieved strong eversion phenotypes with UAS lines from each of these libraries.

Table 2
UAS RNAi libraries

Library	Source	Comment	URL
VDRC [23]	The Vienna Drosophila Resource Centre, Vienna, Austria	12,671 stocks covering 91% of protein-coding genes. Three libraries are available: GD library: Long hairpin dsRNA constructs, random P-element insertions KK library: Long hairpin dsRNA constructs, attP insertions on chromosome II [insertions are potentially at two sites: 40D3 and 30B3] [29, 30] shRNA library: Short hairpin RNA constructs inserted at {attP}40 at 25C6	http://stockcenter.vdrc.at
NIG-FLY	The National Institute for Genetics, Mishima, Japan	11,910 stocks covering 45% of protein-coding genes. Long hairpin dsRNA constructs, random P-element insertions	http://www.shigen.nig.ac.jp/fly/nigfly/
TRiP [11, 31, 32]	The Transgenic RNAi Project at the Harvard Medical School, Boston, USA	11,454 stocks covering 70% of protein-coding genes. Long hairpin dsRNA constructs (VALIUM1 and VALIUM10) OR short hairpin constructs (VALIUM 20, VALIUM21, and VALIUM22). Inserts on chromosomes II ({attP}40 at 25C6) and III ({attP}2 at 68A4)	http://www.flyrnai.org/TRiP-HOME.html

2.1.3 gRNA Libraries

An alternative way of achieving sequence/tissue-specific gene disruption is to use the CrispR/Cas9 system. This approach uses a guide RNA (gRNA) construct for a gene of interest, which is ubiquitously expressed under the control of an RNApol III promoter. When this is combined with a GAL4 driver and UAS-Cas9, one can achieve tissue-specific somatic cutting of the gene of interest [33]. Genome-wide gRNA libraries are currently being created as part of the Harvard Medical School Transgenic RNAi Project (TRiP), available from the BDSC, and by NIG-FLY. UAS-Cas9 stocks are available at both the BDSC and NIG-FLY stock centers.

2.1.4 Fluorescent Reporter Stocks for EMT Factors

Analysis of the changes undergone by PE cells undergoing pEMT is facilitated by fluorescent markers for cellular and extracellular molecules that play important roles in EMT. Listed below (Table 3) are a useful set of such fly stocks which label molecules associated with the basement membrane (BM), adherens junctions (AJs), F-Actin, and myosin. While analysis of cellular and extracellular changes is straightforward with ex vivo cultured discs, obtaining clear visualization of the pEMT event as it occurs in vivo is challenging, due to the complex environment. For in vivo analysis, transgenic fluorescent reporters are useful as their expression can be restricted to peripodial cells using genetic means (i.e., GAL4 drivers or direct promoter fusions) to make them stand out from the surrounding cells [13] and they can remove the need for immunostaining. We have listed some fluorescent UAS lines here that are useful, such as the F-Actin reporters, but many other UAS fluorophore stocks are available at the stock centers. We have also included reporters for the JNK pathway which plays a central role in wing disc eversion, and the netrin receptor Frazzled which our lab has identified as an inhibitor of the EMT event [21]. In addition, there are several other genes and pathways that are not listed here, which have been shown to play a role in eversion such as Pvr [18] and Dpp [20]. Some care should be taken when using fluorescently tagged genes to visualize key mediators of EMT such as E-Cad and F-Actin as the use of non-endogenous regulation (e.g., ubiquitous promoters, GAL4/UAS) and the tag itself can alter the normal levels of expression, sub-cellular localization, and stability/dynamics (*see* **Notes 1** and **2**).

2.2 Solutions

1. Shields and Sang M3 insect medium.
2. Penicillin-Streptomycin.
3. Fetal calf serum.
4. 20-Hydroxyecdysone.
5. Triton X-100.
6. Glycerol.
7. Oil for adhering pupa to glass (e.g., Liquid paraffin or voltalef 10S).

Table 3
Fluorescent reporter stocks for EMT factors

EMT feature	Stock	Comment	Source
Basement membrane	Viking-GFP [34]	GFP exon trap inserted in Collagen IV	DGRC#110692
Adherens junction	UAS-alpha-cat-GFP [35]	GFP fusion to the AJ protein alpha Catenin	DGRC#109002
Adherens junction	Ubi-p63E-DE-Cadherin-GFP [36] *See* **Note 1**	GFP fusion to Epithelial Cadherin under the control of the ubiquitin promoter	BDSC#58471
Myosin	sqh[AX3];P{sqh-GFP.RLC}-42 [37]	GFP-tagged myosin regulatory light chain under endogenous regulatory region, in null background	BDSC#57144
F-actin	UAS-LifeAct-GFP, UAS-Lifeact-Ruby [38] *See* **Note 2**	Actin-binding peptide fused to GFP and Ruby (red fluorescence)	BDSC#35544, #35545 and others
JNK pathway	*puc*^{E69}*-lacZ* [39]	Enhancer trap in the JNK target gene *puckered*	DGRC#109029
JNK pathway	puc$^{A251.1F3}$-lacZ	Enhancer trap in the JNK target gene *puckered*	BDSC#11173
JNK pathway	msn^{06946}-lacZ [40]	Enhancer trap in the Jun kinase kinase kinase kinase, misshapen [41]	BDSC#11707
JNK pathway	*Tre-eGFP* *Tre-DsRedT4* [42]	Synthetic JNK reporters with four binding sites (Tre elements) for c-Jun, placed upstream of a basal promoter	BDSC#59010, BDSC#59011, BDSC#59012
JNK pathway	*mRE-DsRedT4* [42]	Negative control line mRE-DsRedT4 in which the consensus site is scrambled	BDSC#59013
Frazzled	Venus-frazzled	Venus exon trap in *frazzled*, between the signal sequence and the first extracellular IgG domain [21]	DGRC#115148

8. Culture medium: This is as described in Aldaz et al. [14]. Supplement 5 ml of Shields and Sang M3 insect medium, with 25 μl of penicillin-streptomycin, 100 μl fetal calf serum, and 1 μl 1 mg/ml 20-hydroxyecdysone. Make fresh each time it is required.
9. Phosphate buffered saline (PBS) 10× Stock solution: Add 3.33 g $NaH_2PO_4{\cdot}2H_2O$ + 22.4 g Na_2HPO_4 to ~800 ml dH_2O, adjust pH to 7.4, add 102.2 g NaCl and adjust volume to 1 L and store at room temperature.

10. PBST: 1× PBS, 0.1% Triton X.
11. PBST3: 1× PBS + 0.3% Triton X.
12. 10× PB = 2 M Na_2HPO_4, 2 M NaH_2PO_4, pH 7.4.

2.3 Antibodies and Stains for EMT Factors

2.3.1 Adherens Junctions

• Rat-anti-DE-Cadherin	DCAD2, Developmental Studies Hybridoma Bank (DSHB)
• Rat-anti-alpha-Catenin	DCAT-1, DSHB
• Mouse-anti-armadillo	N27A1, DSHB

2.3.2 F-Actin

• Rhodamine-Phalloidin	PHDR1, Cytoskeleton, Inc.
• Phalloidin-iFluor 488 (and spectral variants)	ab176753, Abcam

2.3.3 JNK-Pathway Activation

• Phospho-JNK	9251, Cell Signaling Technology
• Mouse-anti-MMP1	3A6B4, DSHB

2.3.4 Frazzled Localization

• Rabbit-anti-Frazzled	Antibody to the C-terminal domain [41] (Available from M.J.M. and others)

2.3.5 Basement Membrane

Although we and others have used previously a commercial antibody that recognized *Drosophila* LanB1 (Abcam Rabbit-anti-laminin β1 ab47650), it is no longer available, and we are unaware of other commercial antibodies that recognize *Drosophila* ECM components (i.e., Collagen IV, Laminin, Nidogen, Perlecan). Antibodies for Perlecan [43], Nidogen [44, 45], and laminins, e.g., [46] should be requested from the appropriate researchers. Our group has confirmed that, like Coll IV and Lan B1, Nidogen is present in the BM of the wing discs.

2.4 Equipment for Pupal Dissections and Tissue Culture

1. Molded glass cavity/staining blocks with hemispherical cavity.
2. Two pairs of sharp forceps.
3. Vannas spring micro-scissors.
4. Paintbrush.
5. 90 mm petri dishes.
6. iBiDi culture dishes.
7. A medium sized (e.g., ~15 cm × 15 cm × 10 cm) plastic box to act as a humidified chamber.

3 Methods

3.1 RNAi Screen for Genes Involved in Eversion

1. Establish a GAL4 driver stock that incorporates GAL80^{ts} (*see* **Note 3**). One can also include *UAS-Dicer-2* to enhance the strength of the RNAi effect [23] (*see* **Note 4**).
2. In a vial, cross five female virgins of the driver stock, between 2 and 5 days old, to 5 males (2–5 days old) and leave at the restrictive temperature (e.g., ~22–23 °C) for 3 days (*see* **Note 5**).
3. Turn adult parent flies into a new vial each day for 3 days, i.e., Day 1—vial#1; Day 2—vial#2; Day 3—vial#3. On Day 4, remove adults from vial#3. Shift vials to 29 °C 2 days after the adults have been removed (i.e., Day 4—move vial#1 to 29 °C, Day 5—vial#2, Day 6—vial#3) (*see* **Note 6**).
4. From Day 11 onwards, phenotypically score adult flies. This includes both eclosed adults and pharate adults that are still within the pupal case (Fig. 2 and *see* **Note 7**). To score pharate adults, one can remove prepupae from the vial by carefully dislodging them by taking hold of the anterior spiracles with forceps or pushing laterally with a paintbrush. Place prepupae in 80% ethanol for scoring. If air bubbles impair optical clarity (Fig. 2d), pierce the case to allow ethanol to displace the air. Although there is a wide spectrum of phenotypes, we find it useful to categorize them as follows (in increasing level of severity):
   - Mild defects of one or two wings (e.g., smaller, crumpled, mispositioned) but thorax normal (Fig. 2g).
   - Both wings everted but thorax disrupted, usually in the form of a midline cleft (Fig. 2a, d, e).
   - Single eversion failure resulting in an internalized wing and missing half of the thorax (Fig. 2b, f)
   - Double eversion failure, where both wings are internalized and completely lacking a thorax.
   - Early pupal lethal—when a pupa has failed to develop to adulthood (*see* **Note 8**).
5. When it is clear that no further pupae will develop to adulthood (typically ~Day 15), one can fill the vial with 80% ethanol, dislodge pupae from walls with a paintbrush, and tip out the contents into a 90 mm petri dish to complete the scoring (Fig. 2c and *see* **Note 9**).

3.2 Screening for Enhancers/Suppressors

Having identified a gene whose knockdown causes partially penetrant eversion failure, one can then use this genotype as a convenient vehicle for screening for suppressors and enhancers. We find that genotypes whose eversion phenotypes are partially penetrant

Fig. 2 Adult eversion phenotypes. (**a**) Adult with a thoracic cleft and missing left wing. (**b**) Single wing eversion failure with an internalized wing (darkened patch, arrowhead) and missing half-thorax. The right-hand thorax (arrow) and wing are normal. (**c**) Sorting of prepupae in ethanol. (**d**) Pharate adult with thoracic cleft inside pupal case. Bubbles can obscure the phenotype (arrowhead). (**e**) Pharate adult with thoracic cleft removed from pupal case. (**f**) Pharate adult with single eversion failure shown from the dorsal and ventral (**f**′) side. (**g**) A crumpled wing phenotype. The right-hand wing is smaller and mispositioned (**g**), whereas the left wing is normal (**g**′, **g**″, arrowhead)

(e.g., ~30–40% for *Ubx > netA.IR* [21]) provide a very sensitive background to detect mild suppression and enhancement effects. We speculate that it is the nature of the pEMT event, that once a perforation has formed there is a tendency for the disruption and retraction of the PE to continue to completion, possibly because the PE is under tension at the time of the pEMT [15]. Thus, genetic perturbations that have a small effect on the formation of perforation may have an amplified effect in terms of adult phenotypes.

1. Create a new driver stock that incorporates the UAS RNAi construct, the GAL4 driver, and the *GAL80*^{ts} construct, and keep it at restrictive temperatures.
2. Cross this new driver stock to UAS RNAi stocks for other genes and quantify the proportions of eversion defect categories to see if phenotypic severity is increased or reduced.

3. Since adding an extra UAS line could potentially suppress phenotypes by titrating GAL4 away from the target UAS RNAi construct, it is worthwhile including some negative controls such as UAS-GFP, in your tests.

3.3 Analyzing the pEMT

Having identified genes required for eversion, one can then investigate how they affect the pEMT. To analyze the cellular events at the time of the pEMT, there are several approaches. Most straightforward is to allow a prepupa to develop until the time of pEMT (~2–3 h APF) and then either image it live, through the pupal case (*see* Subheading 3.4), or dissect and fix it and apply immunostaining (*see* Subheading 3.5). An alternative is to induce the pEMT ex vivo by dissecting larvae/prepupae and culturing them in the presence of ecdysone (*see* Subheading 3.6). This method was established by Milner over 40 years ago [47] and has since been substantially updated and improved by Aldaz et al. [14]. This method can be applied to wing discs from wandering third-instar larvae, or prepupae that have been allowed to develop until just before the pEMT.

3.4 Live Visualization of the Peripodial pEMT In Vivo

To visualize peripodial cells in vivo as they undergo the pEMT and break through the epidermis, one must carefully stage prepupae from the time of pupariation, i.e., the white prepupa stage when larvae have ceased movement and everted their spiracles, but not yet begun to tan [48]. Analysis can be carried out either on living prepupae, using fluorescent transgene reporters, or by dissecting and fixing prepupae at the time of eversion. Both approaches are detailed below. The simplest approach is to leave the prepupa intact and image through the pupal case with a laser confocal. This approach has been used to visualize the moment when the pEMT occurs and at the later stages during the epithelial migration of thorax closure [13, 21]. Although this is the least intrusive approach, and technically the simplest, image quality is significantly impacted by the pupal case.

1. Remove prepupae from the vial by dislodging them gently with a paintbrush or using forceps to take hold of the anterior spiracles. Take care not to puncture or deform them. Clean any food off with water and a paintbrush.
2. Adhere the prepupa to the base of the iBidi dish with a small drop of liquid paraffin (or other oil such as voltalef 10S) orienting the prepupa such that the events one is examining are closest to the glass base. For example, to visualize the pEMT events, one must have the lateral anterior region downwards. To visualize the epithelial migration during thorax closure, the operculum should be downwards (*see* **Notes 10** and **11**).

3. Periodically capture *z*-stacks of the prepupa on an inverted confocal in a room with temperature kept at 25 °C, or with a stage heater. For us, a typical imaging setup is a Nikon A1R inverted confocal, 20× PL APO Objective, 512 × 512 pixels, 5–10 slices of 3 μm, taken every 60–90 s (*see* **Note 12**).

3.5 Immunohistochemistry of the Peripodial pEMT In Vivo

1. Remove prepupae from the vial by dislodging them gently with a paintbrush or using forceps to take hold of the anterior spiracles and place them in the fixative solution of 4% formaldehyde in PBST. Take care not to puncture or deform them. Dislodge any food by holding them with forceps and using a paintbrush. Remove floating debris with a P1000 tip and replenish the fixative solution.
2. Using scissors, cut the tips off the posterior spiracles to allow the fixative to enter and stabilize the internal tissues. Then use two sharp forceps to pull apart the pupal case, anterior to the midpoint (0.3–0.4 of the pupal length, from the anterior), separating it into anterior and posterior halves, taking care not to deform the pupal case. Fix for 20 min (*see* **Note 13**).
3. While fixation is proceeding, using forceps, remove as much of the internal tissue (i.e., fat body, gut, salivary glands) as possible, without disturbing the dorso-lateral wing disc region (*see* **Note 14**). One can further increase the accessibility of the wing disc area by cutting the anterior half with micro-scissors, into dorsal (containing the wing discs) and ventral parts. This will also further facilitate removal of unwanted tissue.
4. Wash off the fixative 3–4 times with PBST3.
5. Immunostaining of the pupal tissue is then carried out using standard methods—but with base solution of PBST3, unless using the DCAD2 antibody in which case PBS is replaced with PB.

3.6 Ex Vivo Eversion

The protocol for ex vivo eversion is as described in Aldaz et al. [14]. We use ex vivo eversion in two assays (Fig. 1). The first is a simple overnight assay, whereby discs are given enough time to fully evert if they are able, and then scored as having everted successfully, partially or not at all (Fig. 3). The other assay is to culture the discs up to the time of the onset of the pEMT (~7–8 h) and then fix and stain them (Fig. 4).

1. Select wandering third-instar larvae and transfer them to a cavity block with PBS. Wash off food and excess debris and then transfer larvae to 70% ethanol for 5 min to disinfect them. Wash again in PBS, and transfer to a cavity block containing culture medium.

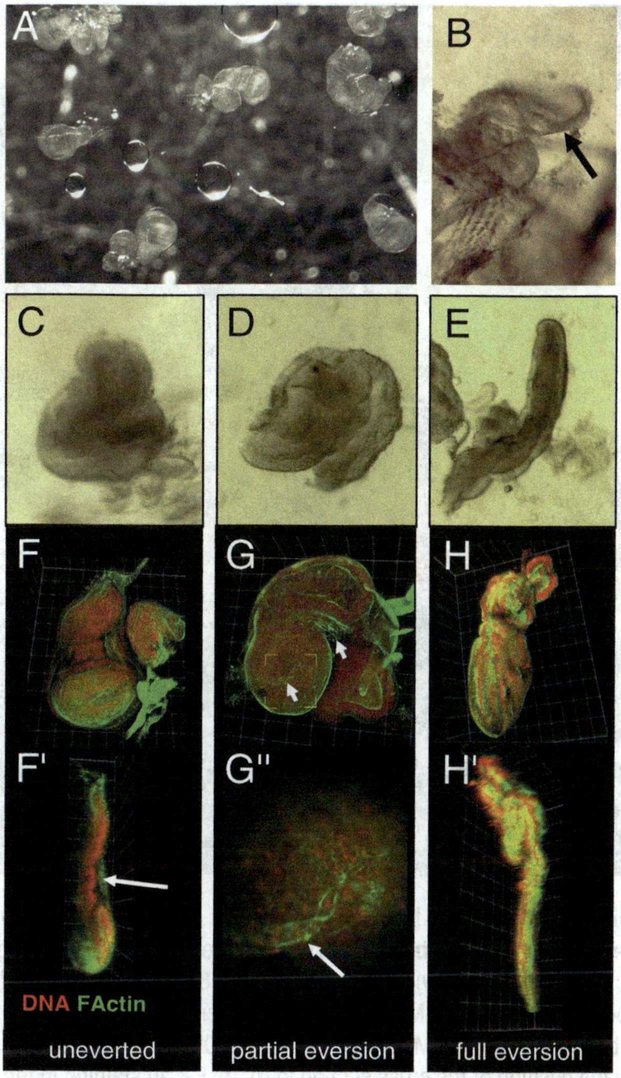

Fig. 3 Overnight eversion assay. (**a**) Third-instar wing discs that have been separated from the body wall, immersed in culture medium. Some still have the haltere and T3 leg disc attached. (**b**) A disc, which remains attached to the body wall, has undergone eversion. (**c–e**) Categories of eversion success. Wing discs can remain uneverted, such that they maintain the original shape (**c**), be partly everted (**d**), or fully everted (**e**). (**f**) Three-dimensional view of an uneverted disc stained with DAPI and Rhodamine Phalloidin, showing the PE stretched over the DP epithelium (**F′**, arrow). (**g**) A partially everted disc in which the PE can be seen to have only partly retracted over the disc (**G′**, arrow). (**h**) Fully everted wing disc

2. Dissect wing imaginal discs from the larva (*see* **Note 15**). One can either isolate the discs (Fig. 3a) or leave them attached to the internal surface of the body wall (Fig. 3b). We find eversion can proceed equally well in either case, though determining whether eversion has been successful is easier when the discs are separated.

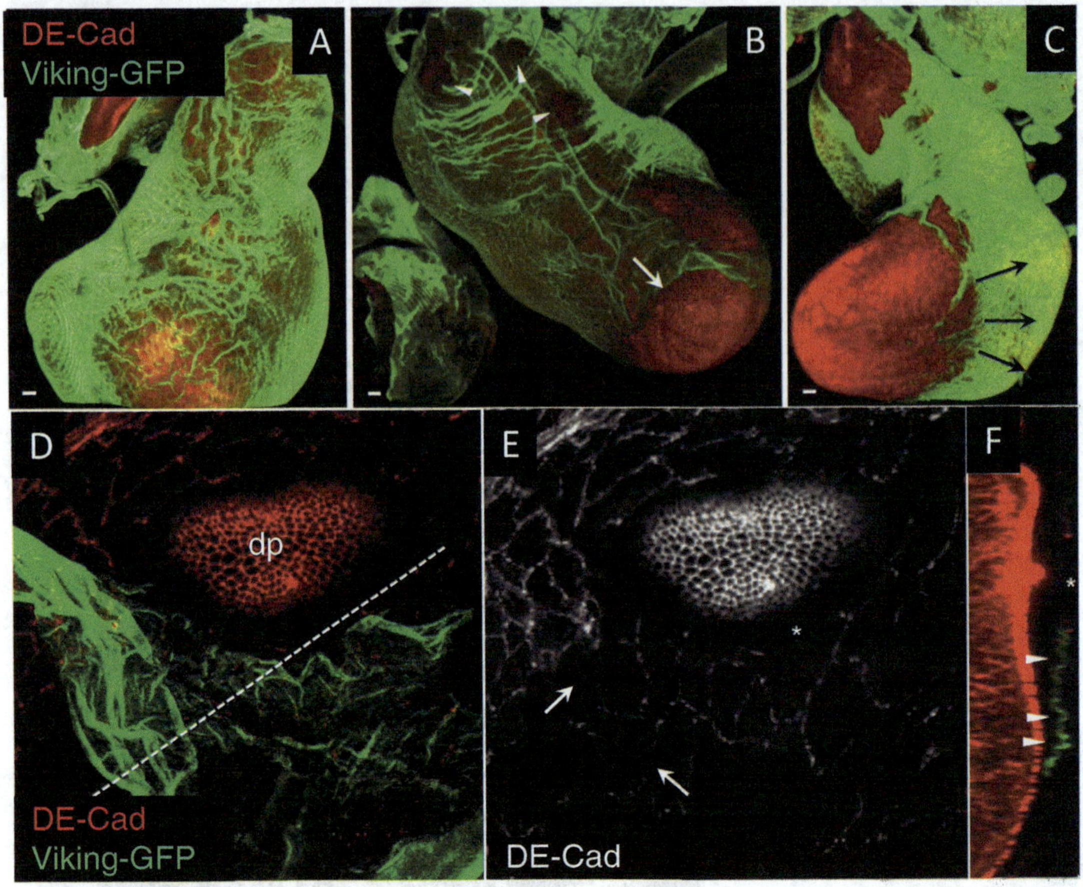

Fig. 4 Discs cultured for 8 h, undergoing pEMT. (**a**) A third-instar wing disc before culturing. Viking-GFP completely covers the disc. (**b**) After 8 h culture, the BM is becoming degraded near the stalk (arrowheads) and at the tip of the blade (arrow). (**c**) Another disc at 8 h culture. The BM is degrading and pulling back over the PE. (**d**) Degradation of the BM (green) and adherens junctions (red) in the region of the stalk. A hole can be seen in both the PE and BM, with cells of the DP epithelium (dp) showing from underneath. Dashed line indicates the position of the cross section in (**f**). (**e**) E-Cad staining in the AJ surrounding the hole is fragmented and weak (arrows). (**f**) Patches of BM sit over the PE but the apposition is not tight (arrowheads). Discs are shown as maximum projections

3. Transfer wing discs to a new cavity block with ~250–500 μl culture medium.
4. Place a 60 mm petri dish lid over the cavity block, place in a humidified chamber, and move chamber to an incubator at the desired temperature (29 °C if using *GAL80*^{ts}). Our chamber is a simple plastic box with wet paper towel lining the base which we enclose in saran wrap.
5. To analyze the onset of pEMT, culture discs for 7–8 h, and then fix them in 4% formaldehyde in PBS for 15 min. Wash several times in PBST and then continue with immunostaining using standard methods.

6. To mount wing discs for imaging, it is important to control the degree to which the coverslip touches and compresses the wing disc by suspending it at an appropriate height using supports on either side. For cultured discs, which will have become more three-dimensional following the morphogenetic movements of eversion, we find one coverslip (held in place with small drops of glycerol) on either side works well. This allows the coverslip to lightly touch the discs, which helps prevent them from moving around, without squashing them. To completely preserve the 3D shape something thicker is needed (e.g., we sometimes use two coverslips on either side). For discs that have not been cultured, one often wishes to flatten the disc to more easily visualize the entire PE in one focal plane, and for this something thinner than a coverslip is needed (e.g., we cut square pieces of plastic from thin autoclave bags, ~0.05 mm thick). Glycerol is then used to fill the space under the coverslip, and nail-polish used to hold everything in place (*see* **Note 16**).
7. Analyze the immunostained discs on a confocal microscope (*see* **Note 17**). Key events of the pEMT that are expected to occur include the degradation and retraction of the basement membrane (Fig. 4), the reduction, dismantling and fragmentation of the adherens junctions (Fig. 4e), the activation of the JNK pathway, and a reduction of Frazzled expression levels [21] (*see* **Note 16**).
8. Alternatively, one can leave the discs overnight and score them the next day. We find discs usually fall into three easily scored categories. Successful eversion whereby the wing blade becomes quite flattened and "wing-like" (Fig. 3b, e, h), unsuccessful eversion, where the wing disc remains relatively flat and the peripodial epithelium is intact (Fig. 3c, f), and partially successful eversion where the disc has clearly undergone some morphogenetic movements and shows some signs of having produced a wing blade, but it is still balled up and constrained inside a partially retracted peripodial epithelium (Fig. 3d, g).

4 Notes

1. The degradation and downregulation of E-Cad during the pEMT that one can see with immunostaining may not be as apparent with this GFP fusion since it is under the control of a strong, constitutive promoter.
2. Although there are direct *UAS-Actin-GFP* fusions available, these may cause phenotypes. Reporters like *UAS-Lifeact-GFP* and *UAS-GFP-Moesin*^{ABD} [38] (i.e., GFP fused to the actin-binding domain of Moesin) are preferable as they decorate the filaments but do not incorporate into the filaments.

3. The protocol assumes a GAL4 driver stock in which $GAL80^{ts}$ has been included to allow for a temperature shift regime that restricts knockdown to the 2 days of third-instar development. This avoids embryonic/larval lethality for genes with essential roles in early development. The alternative is to carry out the entire cross at 29 °C, which will achieve maximum knockdown and may be more appropriate for some RNAi stocks. In past screens, we have found it useful to carry out crosses using both temperature regimes, in parallel.
4. Test a new driver stock with UAS positive control stocks that should block eversion. For example, stocks that inhibit the JNK pathway, such as the RNAi line *UAS-bsk.IR*, or the dominant negative *UAS-bsk*^{DN} (BDSC#1386), *UAS-Timp*, to block the action of MMPs, or the *UAS-netA.IR* line (VDRC#108577). Test the GAL4 driver with and without *UAS-Dicer-2* and with and without the *GAL80*^{ts}. Dicer-2 can enhance the strength of the RNAi effect [23]—though this is only of value when using long hairpin RNAi libraries.
5. The number of pairs in the cross is dependent upon the food. With the rich molasses-based media we use, five females provide enough larvae to strongly establish the vial (i.e., churn up the top layer of food) after a day of egg-laying. We recommend using females for the GAL4 driver, both for the convenience of collecting virgins of only one genotype, but also because we tend to find stronger phenotypes when using female drivers.
6. With this regime a vial will be shifted to 29 °C when the progeny is between 2 and 3 days old, which at 25 °C corresponds to larvae just beginning second instar to larvae just beginning third instar. Thus, development will proceed at 29 °C for at least the entirety of the third instar. Holding the crossed parents at the slightly lower temperature of room temperature will mean the period at 29 °C is slightly longer. We restrict the egg lay to 1 day, so that the number of days over which flies reach adulthood is minimized to decrease the likelihood that one will stop scoring too soon. It is important that all progeny that are capable of reaching adult stages are counted since the more slowly developing progeny often have the stronger phenotypes. We turn adults three times so as to maximize the *n*-value for one cross.
7. It is common for a cross to produce only pharate adults. Since the effects of pEMT failure can be just as easily scored in pharate adults as eclosed adults, we do not concern ourselves with the difference. The wings of pharate adults are uninflated but it is still easy to determine if they are normal or affected (e.g., Fig. 2g).

8. Although it is not necessarily the case that phenotypes such as crumpled wings and thoracic clefts would be caused by failure in the pEMT, we believe that this is often the case. RNAi lines that we know block the pEMT and cause wing eversion failure (e.g., *UAS-netA.IR*), also produce these milder phenotypes. We speculate that partial failure of the pEMT may impede subsequent epithelial migration resulting in clefts, or subsequent development of the wing, resulting in weak wing phenotypes.
9. If one is not ready to score the phenotypes, the ethanol-filled vial can be sealed with saran wrap to prevent evaporation and kept for later analysis.
10. Although the suction effect of the oil is often sufficient to hold the case securely to the coverslip, one can also use physical objects (e.g., small pieces of aluminum or glass) to support the pupal case at a suitable angle.
11. An alternative approach has been developed by Aldaz et al. [14] whereby one partially dissects the prepupa into "quarters" in culture medium, orients the pupal fragment with the open side towards the glass and images from the "inside out" to avoid imaging through the pupal case. This method produces high quality results but requires significant skill in dissection techniques.
12. Capturing movements of a three-dimensional in vivo event involves trade-offs between the conflicting demands of obtaining clear images at each time point (i.e., high x–y pixel resolution, small z-step size, strong laser, long-dwell times), achieving good temporal resolution, and avoiding photodamage to the tissue. The choice of objective may be constrained by the three-dimensional structure of the pupal case. For example, when imaging thorax closure through the operculum, the anterior spiracles stick out, preventing the surface of the operculum from coming closely in contact with the glass, necessitating a longer working distance objective, typically of lower magnification.
13. Others have used overnight fixation in 1× PBS + 0.3 Triton X [20].
14. Removing tissue is critical as it will absorb antibodies and staining reagents and inhibit them from penetrating into the region where the wing disc attaches to the body wall.
15. A common method used to dissect out wing discs is to hold the middle of the larva with one pair of forceps and the mouthparts with another pair, and then pull out the mouthparts, bringing with it the CNS, associated imaginal discs, and wing discs. Alternatively, one can hold the larva in the middle with both forceps side-by-side and pull the larva apart into two halves.

One can then grab the mouthparts on-end with one pair of forceps, stabilize the body wall with the other pair, and push the larva inside out. We find this method is less likely to damage the delicate peripodial epithelium. Isolated wing discs can be conveniently transferred in a P1000 tip. If the disc is still attached to the body wall, we transfer using a P1000 tip which has had some of the tip removed with scissors to create a larger bore. If discs are sticking to the inside of the tip, it is worth trying other brands as some use rougher plastic than others.

16. It is imperative that partly everted wing discs are handled very gently to avoid damaging the very delicate PE. Since one is screening for the formation of holes, it is important not to introduce artifactual holes. Normally, one can distinguish these from real holes as they will have occurred after fixation and have sharply demarcated edges.
17. Since the PE can become highly convoluted and three-dimensional by the time of eversion analysis of the PE breakdown requires a *z*-stack with high resolution in z (typically 0.5–1 μm *z*-step) so that one can have continuity in the *z*-direction to detect changes to the AJs (i.e., diminution, fragmentation, and delocalization of DE-Cadherin) even when the tissue is oriented orthogonally to the image plane.

Acknowledgments

We thank members of the Murray lab for advice and support, Vishal Chaturvedi for critical reading of the manuscript, and Isabelle Lohrey for experimental tests. This work was supported by a National Collaborative Grant from the National Breast Cancer Foundation, and a University of Melbourne RGSS grant to M.J.M.

References

1. St Johnston D (2002) The art and design of genetic screens: *Drosophila melanogaster*. Nat Rev Genet 3:176–188. https://doi.org/10.1038/nrg751
2. Kaya-Copur A, Schnorrer F (2016) A guide to genome-wide in vivo RNAi applications in *Drosophila*. In: Drosophila: methods and protocols, Methods in molecular biology, vol 1478. Springer, New York, pp 117–143. https://doi.org/10.1007/978-1-4939-6371-3_6
3. Thiery JP, Acloque H, Huang RYJ, Nieto MA (2009) Epithelial-mesenchymal transitions in development and disease. Cell 139:871–890. https://doi.org/10.1016/j.cell.2009.11.007
4. Leptin M, Grunewald B (1990) Cell shape changes during gastrulation in Drosophila. Development 110:73–84
5. Tepass U, Hartenstein V (1994) Epithelium formation in the Drosophila midgut depends on the interaction of endoderm and mesoderm. Development 120:579–590
6. Campbell K, Whissell G, Franch-Marro X et al (2011) Specific GATA factors act as conserved inducers of an endodermal-EMT. Dev Cell 21:1051–1061. https://doi.org/10.1016/j.devcel.2011.10.005
7. Gryzik T, Müller H-AJ (2004) FGF8-like1 and FGF8-like2 encode putative ligands of the FGF receptor Htl and are required for mesoderm

migration in the Drosophila gastrula. Curr Biol 14:659–667. https://doi.org/10.1016/j.cub.2004.03.058

8. Crawford JM, Harden N, Leung T et al (1998) Cellularization in *Drosophila melanogaster* is disrupted by the inhibition of rho activity and the activation of Cdc42 function. Dev Biol 204:151–164. https://doi.org/10.1006/dbio.1998.9061
9. Barrett K, Leptin M, Settleman J (1997) The Rho GTPase and a putative RhoGEF mediate a signaling pathway for the cell shape changes in Drosophila gastrulation. Cell 91:905–915
10. Magie CR, Meyer MR, Gorsuch MS, Parkhurst SM (1999) Mutations in the Rho1 small GTPase disrupt morphogenesis and segmentation during early Drosophila development. Development 126:5353–5364
11. Ni J-Q, Zhou R, Czech B et al (2011) A genome-scale shRNA resource for transgenic RNAi in Drosophila. Nat Methods 8:405–407. https://doi.org/10.1038/nmeth.1592
12. Fristrom D, Fristrom JW (1993) The metamorphic development of the adult epidermis. In: Bate M, Alfonso Martinez Arias A (eds) The development of *Drosophila melanogaster*. Cold Spring Harbor Laboratory Press, Plainview, pp 843–897
13. Pastor-Pareja JC, Grawe F, Martín-Blanco E, García-Bellido A (2004) Invasive cell behavior during Drosophila imaginal disc eversion is mediated by the JNK signaling cascade. Dev Cell 7:387–399. https://doi.org/10.1016/j.devcel.2004.07.022
14. Aldaz S, Escudero LM, Freeman M (2010) Live imaging of Drosophila imaginal disc development. Proc Natl Acad Sci U S A 107:14217–14222. https://doi.org/10.1073/pnas.1008623107
15. Aldaz S, Escudero LM, Freeman M (2013) Dual role of myosin II during Drosophila imaginal disc metamorphosis. Nat Commun 4:1761–1710. https://doi.org/10.1038/ncomms2763
16. Proag A, Monier B, Suzanne M (2019) Physical and functional cell-matrix uncoupling in a developing tissue under tension. Development 146:dev172577–9. https://doi.org/10.1242/dev.172577
17. Murray MJ (2015) Drosophila models of metastasis. AIMS Genet 2(1):25–53. https://doi.org/10.3934/genet.2015.1.25
18. Ishimaru S, Ueda R, Hinohara Y et al (2004) PVR plays a critical role via JNK activation in thorax closure during Drosophila metamorphosis. EMBO J 23:3984–3994. https://doi.org/10.1038/sj.emboj.7600417
19. Srivastava A, Pastor-Pareja JC, Igaki T et al (2007) Basement membrane remodeling is essential for Drosophila disc eversion and tumor invasion. Proc Natl Acad Sci U S A 104:2721–2726. https://doi.org/10.1073/pnas.0611666104
20. Martín-Blanco E, Pastor-Pareja JC, Garcia-Bellido A (2000) JNK and decapentaplegic signaling control adhesiveness and cytoskeleton dynamics during thorax closure in Drosophila. Proc Natl Acad Sci U S A 97:7888–7893
21. Manhire-Heath R, Golenkina S, Saint R, Murray MJ (2013) Netrin-dependent downregulation of Frazzled/DCC is required for the dissociation of the peripodial epithelium in Drosophila. Nat Commun 4:1–10. https://doi.org/10.1038/ncomms3790
22. Tripura C, Chandrika N-P, Susmitha V-N et al (2011) Regulation and activity of JNK signaling in the wing disc peripodial membrane during adult morphogenesis in Drosophila. Int J Dev Biol 55:583–590. https://doi.org/10.1387/ijdb.103275ct
23. Dietzl G, Chen D, Schnorrer F et al (2007) A genome-wide transgenic RNAi library for conditional gene inactivation in Drosophila. Nature 448:151–156. https://doi.org/10.1038/nature05954
24. Gibson MC, Lehman DA, Schubiger G (2002) Lumenal transmission of decapentaplegic in Drosophila imaginal discs. Dev Cell 3:451–460
25. Pallavi SK, Shashidhara LS (2003) Egfr/Ras pathway mediates interactions between peripodial and disc proper cells in Drosophila wing discs. Development 130:4931–4941. https://doi.org/10.1242/dev.00719
26. Larsen C, Franch-Marro X, Hartenstein V et al (2006) An efficient promoter trap for detection of patterned gene expression and subsequent functional analysis in Drosophila. Proc Natl Acad Sci U S A 103:17813–17817. https://doi.org/10.1073/pnas.0607652103
27. Calleja M, Moreno E, Pelaz S, Morata G (1996) Visualization of gene expression in living adult Drosophila. Science 274:252–255
28. McGuire SE, Mao Z, Davis RL (2004) Spatiotemporal gene expression targeting with the TARGET and gene-switch systems in Drosophila. Sci STKE 2004:pl6. https://doi.org/10.1126/stke.2202004pl6
29. Green EW, Fedele G, Giorgini F, Kyriacou CP (2014) A Drosophila RNAi collection is subject to dominant phenotypic effects. Nat Methods 11:222–223. https://doi.org/10.1038/nmeth.2856
30. Vissers JHA, Manning SA, Kulkarni A, Harvey KF (2016) A Drosophila RNAi library modulates Hippo pathway-dependent tissue growth.

Nat Commun 7:10368. https://doi.org/10.1038/ncomms10368

31. Ni J-Q, Liu L-P, Binari R et al (2009) A Drosophila resource of transgenic RNAi lines for neurogenetics. Genetics 182:1089–1100. https://doi.org/10.1534/genetics.109.103630
32. Ni J-Q, Markstein M, Binari R et al (2008) Vector and parameters for targeted transgenic RNA interference in *Drosophila melanogaster*. Nat Methods 5:49–51. https://doi.org/10.1038/nmeth1146
33. Port F, Strein C, Stricker M et al (2019) A large-scale resource for tissue-specific CRISPR mutagenesis in Drosophila. Elife 9:e53865. https://doi.org/10.7554/eLife.53865. PMC7062466. PMID: 32053108
34. Morin X, Daneman R, Zavortink M, Chia W (2001) A protein trap strategy to detect GFP-tagged proteins expressed from their endogenous loci in Drosophila. Proc Natl Acad Sci U S A 98:15050–15055. https://doi.org/10.1073/pnas.261408198
35. Oda H, Tsukita S (1999) Dynamic features of adherens junctions during Drosophila embryonic epithelial morphogenesis revealed by a Dalpha-catenin-GFP fusion protein. Dev Genes Evol 209:218–225
36. Oda H, Tsukita S (2001) Real-time imaging of cell-cell adherens junctions reveals that Drosophila mesoderm invagination begins with two phases of apical constriction of cells. J Cell Sci 114:493–501
37. Royou A, Field C, Sisson JC et al (2004) Reassessing the role and dynamics of nonmuscle myosin II during furrow formation in early Drosophila embryos. Mol Biol Cell 15:838–850. https://doi.org/10.1091/mbc.E03-06-0440
38. Hatan M, Shinder V, Israeli D et al (2011) The Drosophila blood brain barrier is maintained by GPCR-dependent dynamic actin structures. J Cell Biol 192:307–319. https://doi.org/10.1083/jcb.201007095
39. Ring JM, Martinez-Arias A (1993) Puckered, a gene involved in position-specific cell differentiation in the dorsal epidermis of the Drosophila larva. Dev Suppl:251–259
40. Mattila J, Omelyanchuk L, Kyttälä S et al (2005) Role of Jun N-terminal Kinase (JNK) signaling in the wound healing and regeneration of a *Drosophila melanogaster* wing imaginal disc. Int J Dev Biol 49:391–399. https://doi.org/10.1387/ijdb.052006jm
41. Su YC, Treisman JE, Skolnik EY (1998) The Drosophila Ste20-related kinase misshapen is required for embryonic dorsal closure and acts through a JNK MAPK module on an evolutionarily conserved signaling pathway. Genes Dev 12:2371–2380
42. Chatterjee N, Bohmann D (2012) A versatile ΦC31 based reporter system for measuring AP-1 and Nrf2 signaling in Drosophila and in tissue culture. PLoS One 7:e34063. https://doi.org/10.1371/journal.pone.0034063
43. Friedrich MV, Schneider M, Timpl R, Baumgartner S (2000) Perlecan domain V of *Drosophila melanogaster*. Sequence, recombinant analysis and tissue expression. Eur J Biochem 267:3149–3159. https://doi.org/10.1046/j.1432-1327.2000.01337.x
44. Wolfstetter G, Shirinian M, Stute C et al (2009) Fusion of circular and longitudinal muscles in Drosophila is independent of the endoderm but further visceral muscle differentiation requires a close contact between mesoderm and endoderm. Mech Dev 126:721–736. https://doi.org/10.1016/j.mod.2009.05.001
45. Dai J, Estrada B, Jacobs S et al (2018) Dissection of Nidogen function in Drosophila reveals tissue-specific mechanisms of basement membrane assembly. PLoS Genet 14:e1007483–e1007431. https://doi.org/10.1371/journal.pgen.1007483
46. Kumagai T, Yokoyama H, Goto A et al (2000) Screening for Drosophila proteins with distinct expression patterns during development by use of monoclonal antibodies. Biosci Biotechnol Biochem 64:24–28. https://doi.org/10.1271/bbb.64.24
47. Milner MJ (1977) The eversion and differentiation of *Drosophila melanogaster* leg and wing imaginal discs cultured in vitro with an optimal concentration of beta-ecdysone. J Embryol Exp Morphol 37:105–117
48. Bainbridge SP, Bownes M (1981) Staging the metamorphosis of *Drosophila melanogaster*. J Embryol Exp Morphol 66:57–80

Chapter 12

Visualizing Mouse Embryo Gastrulation Epithelial-Mesenchymal Transition Through Single Cell Labeling Followed by Ex Vivo Whole Embryo Live Imaging

Wallis Nahaboo, Bechara Saykali, Navrita Mathiah, and Isabelle Migeotte

Abstract

Epithelial-mesenchymal transition (EMT) is often studied in pathological contexts, such as cancer or fibrosis. This chapter focuses on physiological EMT that allows the separation of germ layers during mouse embryo gastrulation. In order to record individual cells behavior with high spatial and temporal resolution live imaging as they undergo EMT, it is very helpful to label the cells of interest in a mosaic fashion so as to facilitate cell segmentation and quantitative image analysis. This protocol describes the isolation, culture, and live imaging of E6.5–E7.5 mouse embryos mosaically labeled in the epiblast, the epithelium from which mesoderm and endoderm layers arise through EMT at gastrulation.

Key words EMT, Mouse embryo, Live imaging

1 Introduction

The passage from an epithelial to a mesenchymal cell organization (epithelial-to-mesenchymal transition or EMT) is essential for gastrulation, a key morphogenetic event that establishes the three layers of the animal body plan (ectoderm, mesoderm, and endoderm). In this context, EMT renders cells competent for motility, but is also coupled to a switch of cell fate. In amniotes, it occurs at a transient organizing center called the primitive streak [1–3]. In mouse, the overlap of multiple signaling gradients coming from embryonic epiblast, as well as extraembryonic ectoderm and visceral endoderm, creates a permissive zone in the posterior epiblast for primitive streak establishment at around embryonic day (E) 6.

Prior to the onset of gastrulation, the epiblast consists of a columnar pseudostratified epithelium sitting on a basal lamina. EMT involves degradation of basal membrane proteins, downregulation of cell–cell junction proteins, loss of polarity, and acquisition of motility. Live imaging showed that epiblast cells in the

Kyra Campbell and Eric Theveneau (eds.), *The Epithelial-to Mesenchymal Transition: Methods and Protocols*, Methods in Molecular Biology, vol. 2179, https://doi.org/10.1007/978-1-0716-0779-4_12,

primitive streak region delaminate through apical constriction (leading to the so-called "bottle shape") followed by retraction of the apical process and extrusion on the basal side [4, 5]. Cells subsequently acquire a mesenchymal shape with an array of filopodia and migrate away from the streak [6].

Technically, live imaging analysis of whole mouse embryos during gastrulation remains challenging. Embryos have already implanted, so culture conditions must compensate for the removal from the maternal environment. Mouse embryos cannot be fully constrained and are very photosensitive. Rapid embryo growth, complex optical properties, and the requirement of optically scattering, auto-fluorescent serum for proper development are further obstacles for high-resolution imaging.

Recent advances in fluorescent reporters, embryo culture, and microscopy have nevertheless allowed visualization and tracking of entire embryonic populations in real time thanks to improvements in spatial and temporal resolution, physical coverage of the embryo, and long-term imaging capability through reduction of phototoxicity [7, 8]. Laser scanning confocal and multiphoton microscopy are the most common tools for imaging embryo development. Both provide excellent spatial resolution; multiphoton microscopy allows higher penetration depth, minimizes exposure as it relies on low energy photons of high wavelength, and reduces the background signal due to multiphoton absorption. Light sheet imaging further diminishes phototoxicity; in particular, an adaptive light sheet imaging approach has recently allowed tracking single cell behavior for 48 h between E6.5 and E8.5 [9].

Important insight has been obtained through qualitative analysis of live imaging data, but quantitative analysis remains a bottleneck. Nuclear reporters are best suited for cell tracking, and several algorithms have been successfully developed to segment nuclei [9]. Cell shape information requires the use of membrane reporters, and it is practically extremely complex to properly segment cell contours when all cells are fluorescent. One way to circumvent this issue is to simplify the model through single cell labeling. This can be achieved through either injection or electroporation of nucleic acids into individual or groups of cells [10], use of photomodulatable proteins that can be activated or converted in a region of interest, or regulation of the expression of fluorescent proteins through lineage-specific inducible promoters. Here we describe a protocol allowing labeling a portion of epiblast cells at a chosen ratio in order to observe cells' shape changes as they undergo EMT during mouse embryo gastrulation.

We use hydroxy (OH)-tamoxifen-induced recombination of Rosa$^{mT/mG}$ [11] through Sox2Cre-ER^{T2} [12] (*see* below) activation to achieve mosaic labeling of the epiblast. Cells in the PS acquiring either a round or a bottle shape can be segmented and tracked to identify their destiny (within the epiblast or mesoderm layers) as well as that of their progeny when division occurs. A similar approach could be used to record the dynamics of cell organelles or specific proteins during EMT. The portfolio of conditional fluorescent reporter mouse lines is indeed constantly increasing [13].

2 Materials

2.1 Large Equipment

1. Stereomicroscope with transmitted light and magnification ranging from approximately 10× to 40× for embryo dissection.
2. Humidified CO_2 incubator.
3. Laser scanning confocal or 2-photon inverted microscope (Fig. 1) with 5×, 10×, 20 or 25×, and 40× objectives (for example, a LD LCI Plan Apochromat 25×/0.8 Imm Korr DIC M27 (WD 570 um) objective or LD C Apochromat 40×/ 1.1 W Korr M27 (WD 620 μm) for detailed acquisitions). 5× and 10× dry objectives are used to identify and position samples, and 20 or 25× immersion objectives are used to record a global view of the embryo. To visualize individual cells, 40× immersion objective is usually most appropriate. The microscope must be placed on an anti-vibration air table. The on-stage environmental culture chamber allows temperature control (37 °C), proper humidification, and gas control ($CO_2/O_2/N_2$) (*see* **Note 1**).
4. Computer workstation with sufficient RAM, adequate graphic card for the image data acquisition and processing software, and consequent storage capacity (as an example, we use a computer equipped with Dual Intel Xeon Processor (Eight Core HT, 3.4 GHz Turbo, 25 MB), 64 GB 1866MHz DDR3 ECC RDIMM RAM, NVIDIA 980 GTX graphics card, and running Windows 7, 64 bit). Acquisition software available with commercial microscopes (such as Zeiss' ZEN, Leica's LASX or MetaMorph) and open-source software such as ImageJ or Fiji allow basic image analysis. Commercially available software packages provide additional features for 3D/4D analysis and representation. However, there is no generic software for detailed analysis of cell contours that will be efficient in any 3D context. Depending on the volume of data, one can either develop dedicated software or rely on manual segmentation, which is greatly facilitated by the use of a graphic tablet (such as Wacom CINTIQ 13HD or equivalent).

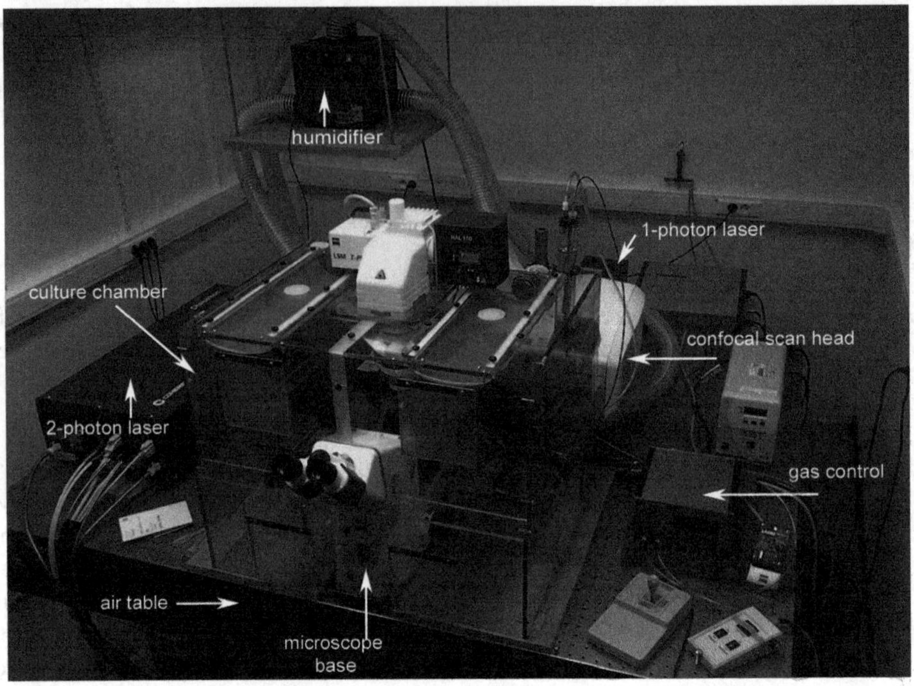

Fig. 1 Microscope setup

2.2 Small Equipment

1. Dumont #5 forceps (Inox or Dumostar).
2. Small surgical scissors.
3. Tungsten needles and needle holder.
4. 35 mm plastic Petri dishes.
5. 35 mm Glass-bottom No. 1.5 dishes (MatTek) or micro-slides of 15 conical wells with #1.5 tissue treated optically compatible polymer bottom (for example, micro-angiogenesis slides Ibidi Cat. No. 81506 or equivalent).
6. Embryo-tested lightweight mineral oil.
7. Glass pipettes or plastic transfer pipettes or P20 pipette for embryo transfer (all have to be previously soaked in culture medium to avoid embryos sticking to the pipette tip).
8. Hydroxytamoxifen suspended at 100 mg/mL in ethanol 100%, and diluted in sesame oil to a final concentration of 10 mg/mL.
9. 1 mL Insulin syringe 25 g × 16 mm.

2.3 Media

1. Dissection medium: Dulbecco's modified Eagle medium (DMEM)/F-12 supplemented with 10% fetal bovine serum and 1% Penicillin-Streptomycin and L-glutamine and 15 mM HEPES.
2. Culture Medium: Embryos in the peri-gastrulation (E5.5–E7.75) period are cultured in 50% DMEM/F12 with L-glutamine and without phenol red for optimal imaging, 50% rat

serum, 1% Penicillin-Streptomycin. Rat serum is commercially available (for example Janvier) or can be homemade [8]. It is best collected from males older than 12 weeks. If the commercial serum is delivered frozen, it should be gently thawed on ice, aliquoted in small amounts (0.5–2 mL) and stored at −80 °C. Rat serum aliquots are heat-inactivated at 56 °C for 30 min prior to preparation of the culture medium.

2.4 Mouse Lines

Mice colonies must be maintained in a certified animal facility in accordance with local guidelines. The local ethics committee must approve all experiments. The mouse lines used in this protocol are the mT/mG [11] and the Sox2-Cre ER^{T2} [12] lines available at the Jax Laboratory (*see* **Note 2**). $Rosa^{mT/mG}$ is a cell membrane-targeted, two color fluorescent Cre-reporter allele. Cell membrane tdTomato (mT) expression is universal, and Cre recombinase expressing cells (and their descendants) have membrane-localized EGFP (mG) expression replacing the red fluorescence. The Sox2-Cre ER^{T2} knock-in mice have the *Sox2* open reading frame replaced with a Cre ER^{T2} fusion gene. *Sox2* is expressed in the inner cell mass at E3.5 then in the epiblast. Cre ER^{T2} is a fusion of the Cre recombinase with a mutant form of the human estrogen receptor that does not bind its natural ligand at physiological concentration but binds 4-hydroxytamoxifen. Cre ER^{T2} can only gain access to the nuclear compartment after exposure to tamoxifen. A portion of epiblast cells proportional to the amount of tamoxifen will be labeled with GFP, and those located in the primitive streak will undergo EMT, allowing recording cell behavior as they exit the primitive streak to become mesenchymal.

3 Methods

3.1 Mice

2–3 female mT/mG homozygous mice (at least 6 weeks old) are placed in a cage with a single Sox2-CreER male (referred to as a breeding triangle). Females are checked for the presence of a vaginal plug the following morning. The day of plug detection is counted as embryonic day 0.5 (E0.5), since mating is assumed to have occurred at the midpoint of the dark period. Pregnant females are injected intraperitoneally with preheated (37 °C) 0.1 mg/g OH-tamoxifen (roughly 2 mg in 200 μL sesame oil) at E6.25 and E6.75 (*see* **Note 3**) and euthanized by cervical dislocation at E7.25 for collection of embryos. The dosage of tamoxifen can be adjusted to the degree of mosaicism one is trying to achieve.

3.2 Collection of Embryos (Fig. 2)

1. Before starting the dissection, equilibrate and pre-warm the culture medium by placing it in a tube slightly opened in a humidified incubator at 37 °C and 5% CO_2 in atmosphere for at least 1 h. The dissection medium should be at room temperature.

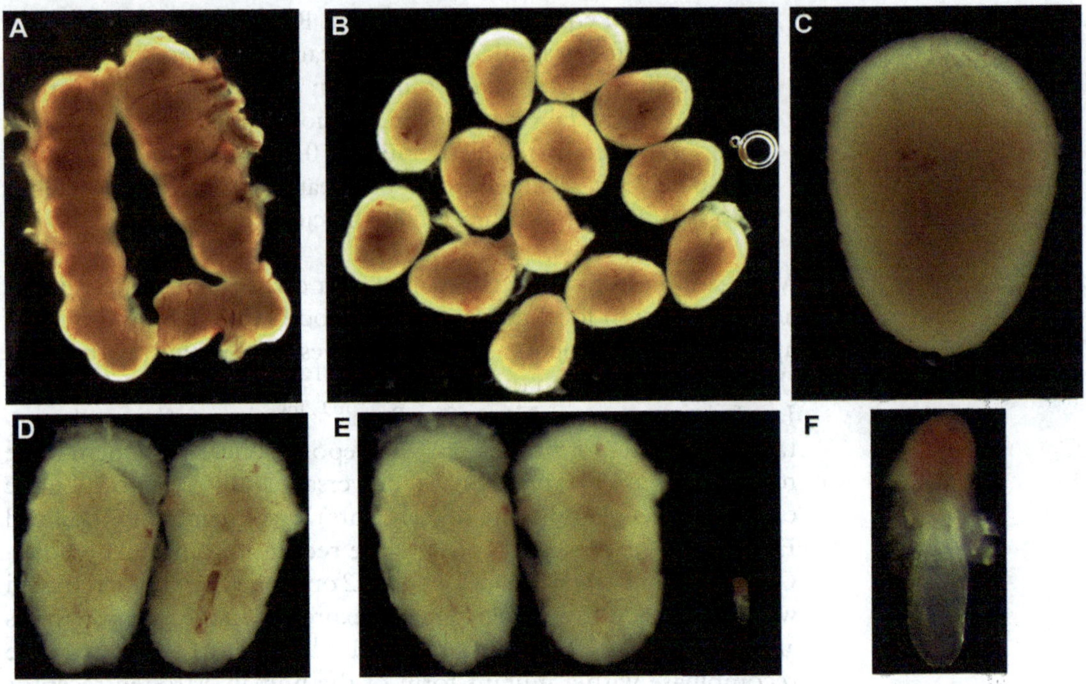

Fig. 2 Mouse embryo dissection. (**a**) The uterus is dissected. (**b**) Deciduae are extracted from the uterus. (**c**, **d**) Deciduae are cut open, to reveal the embryo. (**e**) The embryo is removed from the decidua and (**f**) liberated from the Reichert's membrane

2. After sacrificing the female, dissect out the uterus (remove as much fat as possible) and place it in a dish of dissection medium.
3. Under a stereomicroscope, dissect deciduae out of the uterus. Make a slit on the decidua side. When the slit is almost as important as the height of the decidua, make a slit on the side of one decidua and pull out the decidua by the opening. It is best to use coarse forceps at this stage.
4. Place deciduae in a dish with clean medium to improve visibility. Deciduae are pear shaped, and embryos are within their broader section. Hold the broad extremity with one hand, and slice the thinner portion in 2 using not too sharp forceps, in order to be able to open the decidua as an open book. Carefully remove embryos from the deciduae with closed forceps by digging under the embryo and gently pushing it.
5. Place embryos in yet another dish with clean medium. Remove Reichert's membrane with either fine pristine forceps or sharpened tungsten needles. It is essential that embryo and ectoplacental cone are left intact, and damaged embryos should not be used for imaging. Transfer the embryo into culture medium with a pipette or tip so as to transfer a minimal amount of dissection medium and place in the incubator (*see* **Notes 4** and **5**).

6. Place individual embryos in conical wells containing 30 μL of culture medium. Carefully add sterile water into the dedicated ridges of the microslide. It must not overflow so it does not dilute the culture medium, and so the lid of the microslide does not get sealed to its bottom by the capillarity of the water. It is advised to let embryos recover from dissection for 1 h before proceeding to imaging.

3.3 Static Culture and Imaging of Gastrulating Embryos

Embryos can develop normally under the microscope for up to 24 h.

1. Pre-warm the incubation chamber to 37 °C and equilibrate CO_2/O_2 levels.
2. Transfer the microslide from the incubator to the microscope incubation chamber.
3. Several embryos can be imaged in a single experiment. As they grow and are not physically attached, it is not practical to use the automatic positioning program. Rather, position in *X*, *Y*, and determination of optimal *Z*-stack have to be set up for each embryo at each time-point. Before starting the time-lapse, one should ensure optimal positioning of the embryos (Fig. 3). At that stage of development, anterior-posterior axis can be determined through morphology alone. To visualize gastrulation EMT, they should lie either on their posterior or lateral side, which are natural positions due to the ellipsoid embryo shape. Embryos can be gently flipped in the conical well using coarse forceps in a closed position, or a plastic tip.
4. It is essential to minimize embryo exposure to light (*see* **Note 6**).

3.4 Image Analysis

Imaging data should be stored on two separate servers as backup. Images are viewed and analyzed using an image analysis program installed on a powerful workstation. Confocal and two photon generated images are 3D reconstructed, projected on the *Z* axis, and their brightness and contrast enhanced (Fig. 4). For time lapses acquisitions, embryos must be registered using a drift correction tool. Embryos are highlighted on each *Z* slice to generate a 3D virtual embryo. The same embryo is registered across time by matching its center of gravity. Individual cells are then segmented through manually highlighting cell membranes along the *Z* slices using a graphic tablet. The cells are then tracked over time using the imaging program tracking algorithm. Tracking and cell shape data can then be funneled through a dedicated python script [14] for statistical analysis.

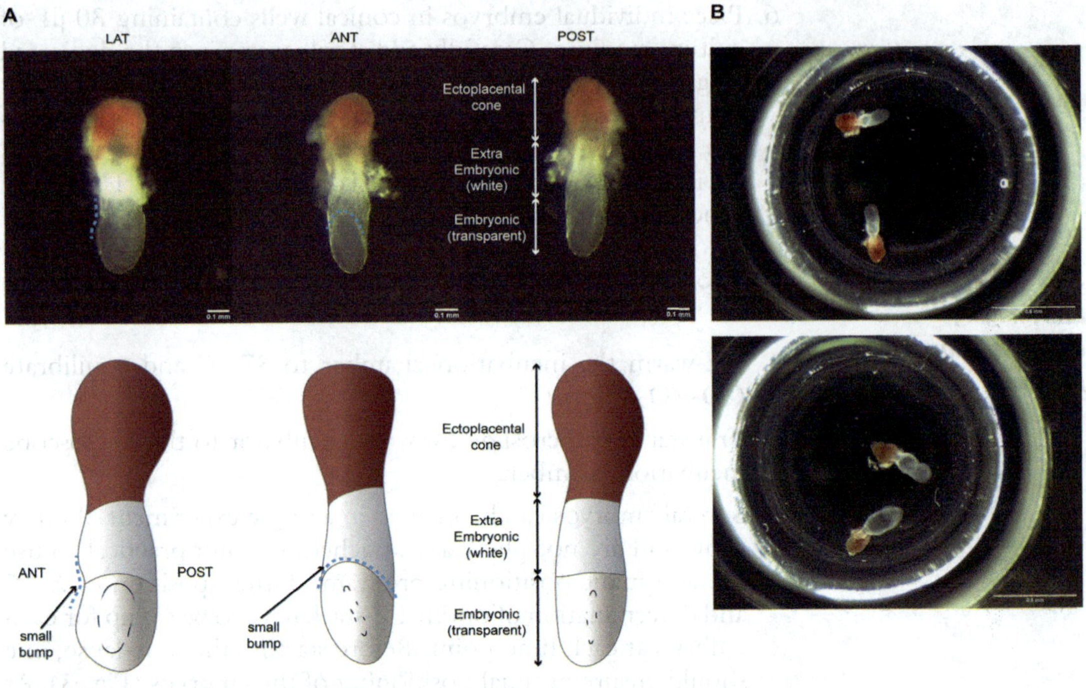

Fig. 3 Embryo culture and orientation for live imaging. (**a**) Images of the same embryos (top) oriented in lateral (left), anterior (middle), and posterior (right) positions. Extraembryonic (white) region is between the ectoplacental cone (red) and the embryonic region (transparent). Schematics and annotations (bottom) described the embryo morphology. Blue lines follow the bump formed on the anterior side of the embryo. Scale bar represents 0.1 mm. (**b**) Images of embryos in a well with culture media before imaging (top) and after imaging (bottom) of 6 h. Scale bar represents 0.5 mm

4 Notes

1. It is essential to culture embryos in a CO_2 controlled environment (5%) as HEPES-buffered media are toxic for the embryos when used for long periods of time. It is ideal to culture E6.5–E8 embryos at 5% O_2 [8]. However, embryos grow adequately in 5% CO_2 in atmosphere (20% O_2) for the usual imaging time frame (8–12 h) [7].
2. Mouse lines are backcrossed into an outbred CD1 background as it results in bigger litter sizes, and bigger embryos for a similar morphological stage that display better resistance to phototoxicity.
3. Introduction of tamoxifen is toxic even at low concentration before E4.5, and embryonic toxicity is dose-dependent between E5.5 and E7.5. In our system, hydroxytamoxifen quantities (0.1 mg/g) sufficient to trigger efficient recombination could only be injected from E6.25 onwards while preserving embryo health.

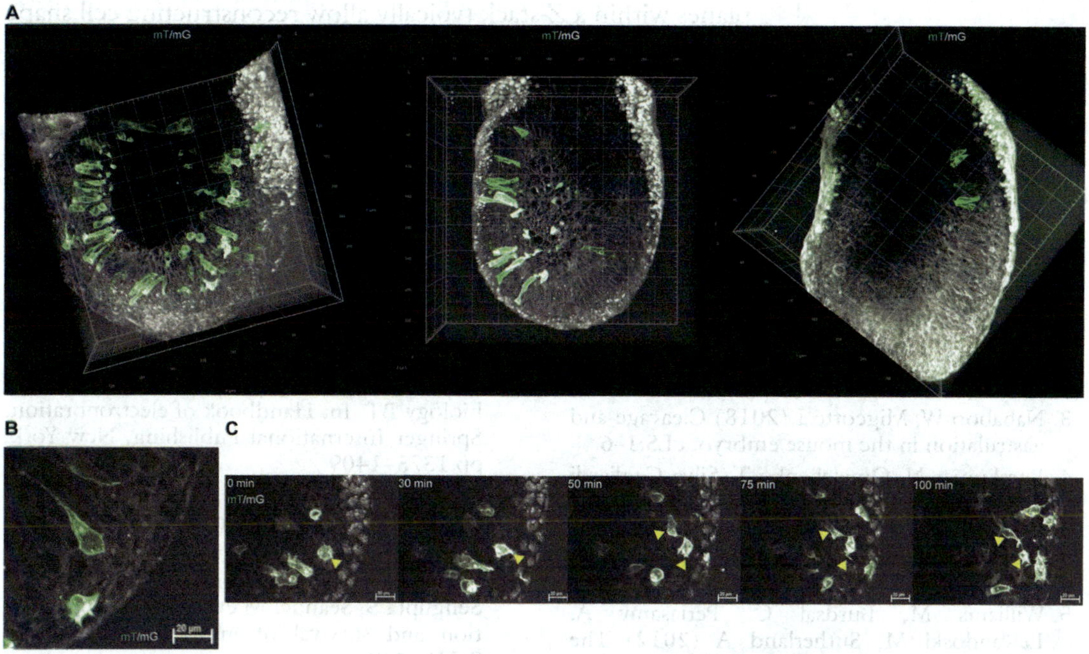

Fig. 4 Confocal imaging of embryos expressing mT/mG under Sox2Cre-ER recombination. (**a**) 3D reconstructions of live embryos. For the same dose of hydroxytamoxifen, the amount of converted cells (green) is variable: high (left), asymmetric (middle), and low (right) expressions. All embryos are oriented in a lateral position with the posterior side to the right. (**b**) 2D section of a converted cell inside the epiblast displaying a bottle shape. Scale bar represents 20 μm. (**c**) Time-lapse of 2D sections of an embryo where we can follow mitosis and daughter cells (yellow arrows). Scale bar represents 10 μm. Green represents membrane GFP and gray membrane Tomato

4. Embryos should be staged according to anatomical landmarks [15], as there is great shape and size variability between and even within litters dissected at a defined time.
5. Speed is essential during dissection, so a learning curve is to be expected before embryo survival is optimal. It is best to dissect two or three deciduae, transfer embryos in the incubator, and then proceed to the next batch.
6. Options include reducing laser power (use minimum laser power to be able to distinguish cell shape, not to obtain a perfect image. Signal-noise ratio can be optimized though post-imaging processing); reducing the time embryos are exposed to the laser (by decreasing the frequency of scans, increasing the size of optical sections and/or scan speed). For the latter, imaging two fluorescent labels is done on one channel track, 1024 × 1024 resolution at maximal scanning speed and 2× signal averaging. Since only 1/3 to 1/4 of the cells are labeled, cells can be easily tracked even when imaged every 20 min. Cell shape changes are progressive, so we rarely miss essential events at this rate. 3 μm intervals between individual

planes within a *Z*-stack typically allow reconstructing cell shape in 3D. Embryos are ideally imaged for around 8–12 h between E7 and E7.75. Embryos should then be cultured for an additional 6–12 h after imaging to check for fitness.

References

1. Arnold SJ, Robertson EJ (2009) Making a commitment: cell lineage allocation and axis patterning in the early mouse embryo. Nat Rev Mol Cell Biol 10:91–103
2. Ramkumar N, Anderson KV (2011) SnapShot: mouse primitive streak. Cell 146:488–488.e2
3. Nahaboo W, Migeotte I (2018) Cleavage and gastrulation in the mouse embryo. eLS:1–6
4. Ramkumar N, Omelchenko T, Silva-Gagliardi NF, McGlade CJ, Wijnholds J, Anderson KV (2016) Crumbs2 promotes cell ingression during the epithelial-to-mesenchymal transition at gastrulation. Nat Cell Biol 18(12):1281–1291
5. Williams M, Burdsal C, Periasamy A, Lewandoski M, Sutherland A (2012) The mouse primitive streak forms in situ by initiation of epithelial to mesenchymal transition without migration of a cell population. Dev Dyn 241:270–283
6. Ferrer-Vaquer A, Viotti M, Hadjantonakis A-K (2010) Transitions between epithelial and mesenchymal states and the morphogenesis of the early mouse embryo. Cell Adhes Migr 4:447–457
7. Srinivas S (2010) Imaging cell movements in egg-cylinder stage mouse embryos. Cold Spring Harb Protoc 5:1386–1394
8. Nowotschin S, Garg V, Piliszek A, Hadjantonakis A-K (2019) Ex utero culture and imaging of mouse embryos BT. In: Vertebrate embryogenesis: embryological, cellular, and genetic methods. Springer, New York, pp 163–182
9. McDole K, Guignard L, Amat F, Berger A, Malandain G, Royer LA et al (2018) In Toto imaging and reconstruction of post-implantation mouse development at the single-cell level. Cell 175:1–18
10. Gosse C, Zhao X, Migeotte I, Suárez-Boomgaard D, Hue I, Degrelle S et al (2017) The use of electroporation in developmental biology BT. In: Handbook of electroporation. Springer International Publishing, New York, pp 1375–1409
11. Muzumdar MD, Tasic B, Miyamichi K, Li L, Luo L (2007) A global double-fluorescent Cre reporter mouse. Genesis 45:593–605
12. Arnold K, Sarkar A, Yram MA, Polo JM, Sengupta S, Seandel M et al (2013) Regeneration and survival of mice. Cell Stem Cell 9:317–329
13. Kiyonari H, Kaneko M, Abe T, Shioi G, Aizawa S, Furuta Y et al (2019) Dynamic organelle localization and cytoskeletal reorganization during preimplantation mouse embryo development revealed by live imaging of genetically encoded fluorescent fusion proteins. Genesis 57:1–9
14. Saykali B, Mathiah N, Nahaboo W, Racu M-L, Hammou L, Defrance M et al (2019) Distinct mesoderm migration phenotypes in extra-embryonic and embryonic regions of the early mouse embryo. Elife 8:e42434
15. Downs KM, Davies T (1993) Staging of gastrulating mouse embryos by morphological landmarks in the dissecting microscope. Development 118:1255–1266

Chapter 13

Fluorescence Recovery After Photobleaching to Study the Dynamics of Membrane-Bound Proteins In Vivo Using the *Drosophila* Embryo

Joshua Greig and Natalia A. Bulgakova

Abstract

The epithelial-to-mesenchymal transition is a highly dynamic cell process and tools such as fluorescence recovery after photobleaching (FRAP), which allow the study of rapid protein dynamics, enable the following of this process in vivo. This technique uses a short intense pulse of photons to disrupt the fluorescence of a tagged protein in a region of a sample. The fluorescent signal intensity after this bleaching is then recorded and the signal recovery used to provide an indicator of the dynamics of the protein of interest. This technique can be applied to any fluorescently tagged protein, but membrane-bound proteins present an interesting challenge as they are spatially confined and subject to specialized cellular trafficking. Several methods of analysis can be applied which can disentangle these various processes and enable the extraction of information from the recovery curves. Here we describe this technique when applied to the quantification of the plasma membrane-bound E-cadherin protein in vivo using the epidermis of the late embryo of *Drosophila melanogaster* (*Drosophila*) as an example of this technique.

Key words Protein dynamics, Membrane proteins, E-cadherin, Cell-cell adhesion, FRAP, Diffusion, Endocytosis

1 Introduction

The maintenance of epithelial tissue, epithelial-to-mesenchymal transitions (EMTs), and the migration of mesenchymal cells relies on the dynamic turnover of transmembrane proteins, such as cadherins and integrins [1–5]. These membrane proteins are confined in space, with diffusion along the plane of the lipid bilayer, but they can move in and out of the plasma membrane using the specialized cellular mechanism of endocytic internalization and recycling [6–9]. Alterations to this dynamic turnover, both diffusional and endocytic trafficking, result in profound changes in cell behaviour, and in some cases can induce or prevent EMT [10, 11]. Therefore, the total amount of a transmembrane protein in a specimen

Kyra Campbell and Eric Theveneau (eds.), *The Epithelial-to Mesenchymal Transition: Methods and Protocols*, Methods in Molecular Biology, vol. 2179, https://doi.org/10.1007/978-1-0716-0779-4_13,

provides only part of the information about cell–cell interactions and cellular behaviour. To fully understand these processes, one must measure the dynamics of the protein of interest in vivo in living cells and tissues.

Fluorescence recovery after photobleaching (FRAP) is a fluorescent microscopy technique which allows one to do this precisely and has made a significant impact on the understanding of the functions of proteins and the regulation of cell adhesion and migration [12–16]. In essence, the technique relies on the disruption of the signal from a fluorescently tagged protein in a portion of a sample, which is achieved by bleaching a region with a short, intense pulse of photons. One then measures the fluorescence in the bleached area for a period following the bleaching to record the recovery of the fluorescence signal. Importantly, the proteins in the bleached area are still present but are merely "dark" due to the loss of fluorescence. Therefore, what is measured is the exchange of proteins between the bleached area and the rest of the cell. Carefully designed FRAP experiments not only enable the determination of the overall dynamics of a transmembrane protein, but also allow one to distinguish the contributions of both the diffusional and endocytic trafficking processes and to determine the specific changes in each.

The performance of a FRAP experiment is reliant upon the following steps: preparation of the sample, calibration of the microscope settings, performing and acquiring the data on the microscope, analysing the data, fitting the recovery curve, and interpreting the results. Here, we use an example of FRAP performed on the membrane-bound epithelial cadherin (E-cad) molecule which has been tagged with the fluorophore EGFP [17]. The theoretical aspects of FRAP are extensively described in other publications [18–23]. We therefore focus on the practical aspects of performing FRAP, which will allow anyone with access to a confocal microscope to do the whole procedure from sample preparation to the final result.

We and other research groups have shown that the E-cad signal recovers in *Drosophila* and mammalian cells by both diffusional and endocytic recycling mechanisms, which are kinetically distinct [1, 15, 24, 25]. Here, we specifically use the example of the epidermis of the stage 15 *Drosophila* embryo. However, aside from sample preparation, the same protocol, calibration, and analysis are applicable for other proteins and cell types. Finally, we outline the main considerations for designing and analysing FRAP experiments.

2 Materials

2.1 Flies

ubi::E-cad-GFP (Bloomington, 58471 or 58742) flies can be used as the source of E-cad-GFP (*see* **Note 1**). The copy number of E-cad-GFP and endogenous E-cad loci must be equal in the control and experimental animals, as diffusional coefficients calculated from recovery curves depend on the total concentration of the tagged protein [26]. Alternatively, flies with endogenously tagged *shg*:: E-cad-GFP (Bloomington 60584) can be used. In our experience, ubiquitously expressed and endogenously tagged E-cad behave identically in FRAP experiments [27].

2.2 Reagents and Equipment

1. Apple juice plates for the collection of embryos: 60 mm plates filled with 10 ml of apple juice agar media (for 50 plates: 15 g agar, 400 ml distilled water, 100 ml apple juice, 7.5 ml 10% nipagin in absolute ethanol) and baking yeast.
2. Collection chambers, which tightly fit the apple juice plates and allow oxygen access to the flies.
3. Embryo preparation: bleach and deionized H_2O.
4. Filtration nets: using a razor blade, cut a 15 ml Falcon tube at about the 11.5 ml mark and make a hole of about 1 cm diameter in its lid. Then, cut a nylon net out of a 100 μm pore cell strainer, and assemble the embryo filtration net by inserting the net between the lid and tube.
5. Dissection needles.
6. Microscope slides.
7. Heptane glue: incubate a 5 cm length of adhesive tape (Sellotape 1447052 is used due to no toxic effects) with 2 ml heptane for at least 24 h prior to conducting experiments in a sealed glass vial.
8. Halocarbon oil 27.
9. A confocal microscope with a 488 laser and sufficiently sensitive detectors.
10. Software packages: Fiji (https://fiji.sc), Microsoft Excel, and appropriate statistical packages, e.g., Matlab (https://www.mathworks.com/products/matlab.html), R (https://www.r-project. org), and GraphPad Prism (https://www.graphpad.com/scientific-software/prism/).

3 Methods

3.1 Embryo Collection

1. For synchronized egg collections, set at least 50 virgin female and 20 male flies to mate for 1 day in a collection chamber with an apple juice plate attached to the bottom of the chamber at

25 °C. A measure of yeast paste should be applied to apple juice plates with a spatula before placing with the flies in the embryo collection chambers.

2. Collect embryos at 25 °C for a 1.5-h time interval, the plates used for these collections should have only a small measure of yeast and be prewarmed at 25 °C before use. Allow them to develop at 18 °C for 21 h to reach the desired developmental stage, at the end of dorsal closure corresponding to late stage 15.
3. Dislodge the embryos from the surface of the apple juice agar by applying a small measure of deionized water and brushing surface with a paint brush.
4. Dechorionate the embryos by immersing in a 1:1 sodium hypochlorite (bleach) and deionized water solution for 5 min followed by filtration through an in-house made embryo filtration net and extensive washing with deionized water.

3.2 Mounting of Samples on Microscope Slides

1. Prepare the imaging microscope slide (Fig. 1a): attach two 22 × 22 coverslips at either end of a microscope slide using a spot of nail varnish to create a bridge, then either add a strip of adhesive tape across the central channel on the slide between the two coverslips (orthogonal to the long axis of the slide) or add a few drops of heptane glue, spread it thinly with another coverslip, and allow heptane to evaporate (Fig. 1b). Keep the slide covered with a plate to avoid dust.
2. Transfer dechorionated embryos to apple juice agar segments on a microscope slide using a paint brush (Fig. 1c, d).
3. Select the correct genotype using a fluorescent stereomicroscope and specific fluorescent markers, e.g., the presence of the fluorophore of interest or specific fluorescently tagged balancer chromosomes. For stage 15 embryos, we routinely use balancers with GFP driven in the mesoderm with *twist* promoter (Bloomington 6662 and 6663) or YFP expressed in mandibular and maxillary segments with *Deformed* promoter (Bloomington 8578 and 8704). Then transfer the desired embryos to an adjacent segment of apple juice agar, while orientating them relative to one another and according to their anterior-posterior and dorso-ventral axis (*see* **Note 2**, Fig. 1e, f).
4. Transfer embryos to prepared slide (*see* **step 1**) by pressing this slide delicately against the apple juice segment containing the aligned embryos (*see* **Note 3**, Fig. 1g).
5. Add 50 μl of halocarbon oil over the embryos, so that embryos are fully covered (add dropwise) and leave for 10 min (Fig. 1h).

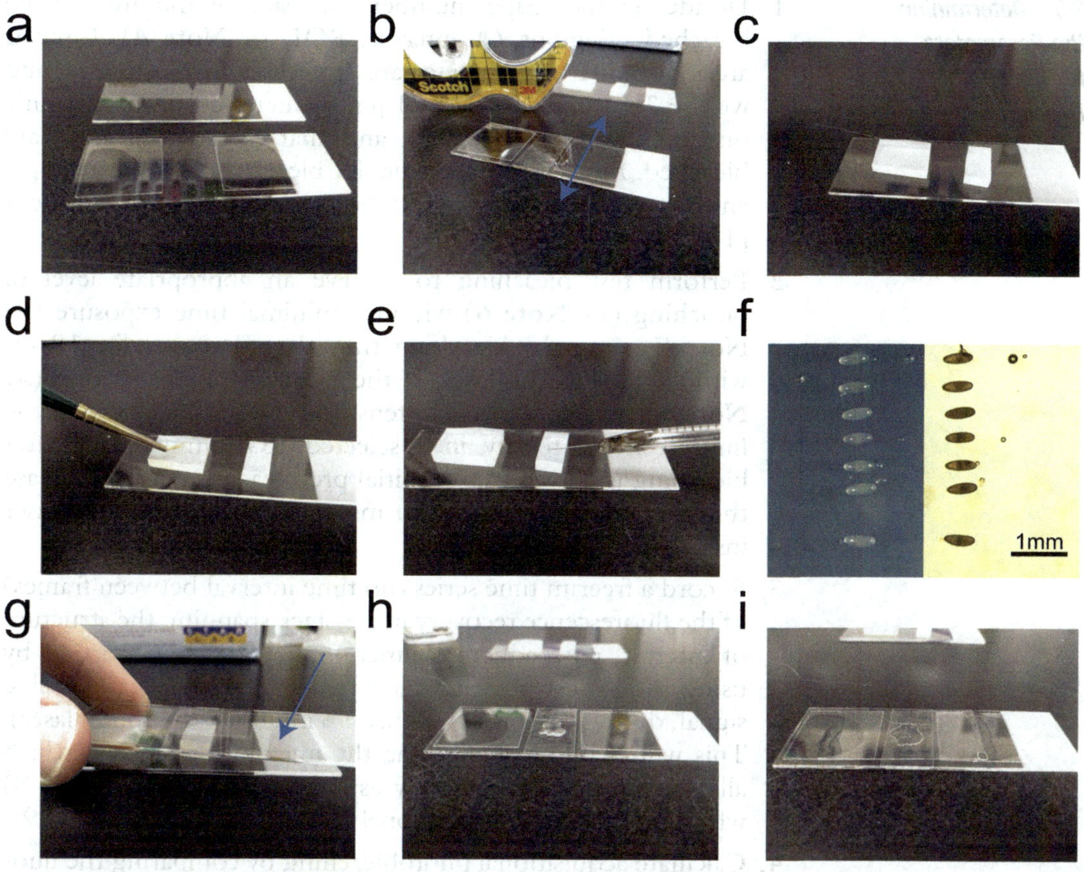

Fig. 1 Preparation of the embryo samples for live imaging. (**a–i**) The complete steps of the process of preparing the microscope slides. (**a**) Bridge created by affixing two 22 × 22 coverslips by a droplet of nail varnish at either edge of the microscope slide (slide farther away is before the coverslips are added). (**b**) On the slide with the bridge, a strip of scotch tape is placed between coverslips occupying the canal area (arrowheads show direction). (**c**) Two segments of apple juice agar are placed on a separate microscope slide. (**d**) After dechorionation, the embryos are transferred onto one of the apple juice segments by using a paint brush. (**e**) Use a needle or forceps to transfer the desired embryos to the adjacent apple juice segment and align them in correct orientation. (**f**) Positioning of the embryos on the apple juice segments. The embryos are orientated with anterior left. Most are orientated with ventral side up, the bottom embryo is positioned laterally to show the bean shape and curvature which can aid in distinguishing the dorsal and ventral sides. The image shows the same embryos under different illuminations. (**g**) Transfer of the embryos from the apple juice segment to the imaging slide. (**h**) Incubation with 50 μl of halocarbon oil added dropwise to the embryos. (**i**) Finished slide ready for imaging. After incubation with halocarbon oil, a larger 22 × 40 coverslip is overlaid and the short ends sealed with nail varnish

6. Apply a 22 × 40 coverslip over the embryos; it is best to gradually lower the coverslip to minimise bubble entrapment. Seal the ends of the coverslip by using nail varnish and allow to dry for few minutes before taking to the microscope; this prevents any slippage of the coverslip or wet nail varnish coming into contact with the lens of the microscope (Fig. 1i).

3.3 Determining the Parameters for the FRAP Experiments

1. Decide on the shape, number, and size of the areas to be bleached (*R*egions *O*f *I*nterest, ROI, *see* **Note 4**). Circular areas of 0.5–1 μm diameter are applicable in most cases, and we use 2–3 circular areas of 1 μm diameter ensuring that only one event occurs for a cell and that no adjacent cells are bleached. *Important*: The same size bleach spot within a sample and between genotypes must be used during an experiment [18] (*see* **Note 5**).
2. Perform test bleaching to achieve an appropriate level of bleaching (*see* **Note 6**) within a minimal time exposure (*see* **Note 7**). Start by bleaching the selected region for 10 ms with 100% 488 nm laser. If the bleaching is too strong (*see* **Note 6**), reduce the laser intensity and bleaching time. If the fluorescence intensity in a selected ROI immediately after bleaching is above 40% of initial pre-bleach intensity, increase the bleaching time up to 20 ms or/and use a 405 nm laser instead.
3. Record a freerun time series (no time interval between frames) of the fluorescence recovery in a *z*-stack spanning the structure of interest, e.g., adherens junction, for 2–5 min starting by using the lowest power, which yields sufficient clearly visible signal, during this phase (we use 1% intensity of 488 nm laser). This will enable to determine the imaging parameters which allow recording the recovery as fast as possible (*see* **Note 8**) while minimizing any additional photobleaching (*see* **Note 9**).
4. Calculate acquisitional photobleaching by comparing the fluorescence intensity of the control (unbleached) ROI at the end of the time series to its pre-bleach value; and the initial recovery (between the first and second time points after bleaching, *see* Subheading 3.5). Use these values to adjust the imaging parameters. If the acquisitional bleaching is more than 15%, reduce the laser power or increase the scanning speed (*see* **Note 9**). If the initial recovery is greater than 5%, reduce intervals between time points by reducing image resolution, applying digital zoom, or/and increasing scanning speed (*see* **Note 8**). If the initial recovery is below 1%, increase the interval between time points.
5. Repeat **steps 3** and **4** until desired levels of acquisitional bleaching and initial recovery are achieved, while maintaining sufficient spatial resolution (ROI below 10 pixel in diameter are likely to introduce large noise due to even mild movements in *XY* plane). For E-cad-GFP, we use 6× 0.38 μm sections which span the entire depth of adherens junction. Each section is 320 × 320 pixel, with a spatial resolution of 0.093 μm/pixel, and is taken every 20 s. We use a 63× magnification lens with a numerical aperture (NA) of 1.4, 1–2% laser power, 2 μs/pixel dwell with amplification of the Hv/gain of the PMT for optimal image acquisition.

6. Calculate the recovery at the five last time points to ensure that the duration of recording the recovery is sufficient to enable measurement and identification of several components (*see* **Note 10**). If the change in the fluorescence intensity of the ROI during these time points is greater than 1% of initial intensity, increase the time of recording. In this case, one must ensure that acquisitional photobleaching remains low. If the change determined is lower than 1%, calculate the recovery at the previous five time points. Use the last time point, at which the recovery of the final five time points remains below 1%, to determine the duration required to record the recovery.

3.4 Bleaching and Acquisition

1. Select a region of the epidermis for the experiment, set up a *z*-stack, and select the ROI (Fig. 2a–d).
2. Set up the following experiment sequence (according to specific software and microscope): a 4D *z*-stack and time series at selected resolution, time interval, and duration with bleaching activated just before the start of the second or third *z*-section.
3. For measuring the dynamics of E-cad-GFP in *Drosophila* embryos, use 8–10 embryos with 2–3 measurements per embryo. Although the recovery of individual bleach events is likely to be noisy and subjected to fluctuations in intensities, this size of the dataset enables one to obtain a good and stable averaged recovery curve (Fig. 3).

3.5 Signal Intensity Measurements and Data Processing

1. Open the raw ".oib/.czi" files and use the grouped *z*-projector plugin to compile each time point as an average intensity projection (*see* **Note 11**): for E-cad-GFP average intensity complied by 6 for each stack taken, yields 45 time points. We use Fiji (https://fiji.sc) for this purpose.
2. Select a region in the center of a cell without bleached junctions and measure intensity using a ROI of the same size as used for the bleached spots across the time series (*see* **Note 12**, Fig. 2e).
3. Perform the same for a control region, which is a junction between two cells whose E-cad was not bleached, and for the bleached regions (*see* **Note 13**, Fig. 2e).
4. Subtract background and normalize fluorescence intensity of the bleached region as following: $I_n = (F_n - BG_n)/(FC_n - BG_n)$, where F_n is intensity of the bleached ROI at the time point n, FC_n is intensity of the control unbleached ROI of the same size at the plasma membrane at the time point n, and BG_n is background intensity, measured with the same size ROI in cytoplasm at the time point n.
5. Calculate the relative recovery at each time point using the following formula: $R_n = (I_n - I_1)/(I_0 - I_1)$, where I_n, I_1, and I_0 are the normalized intensities of bleached ROI at time

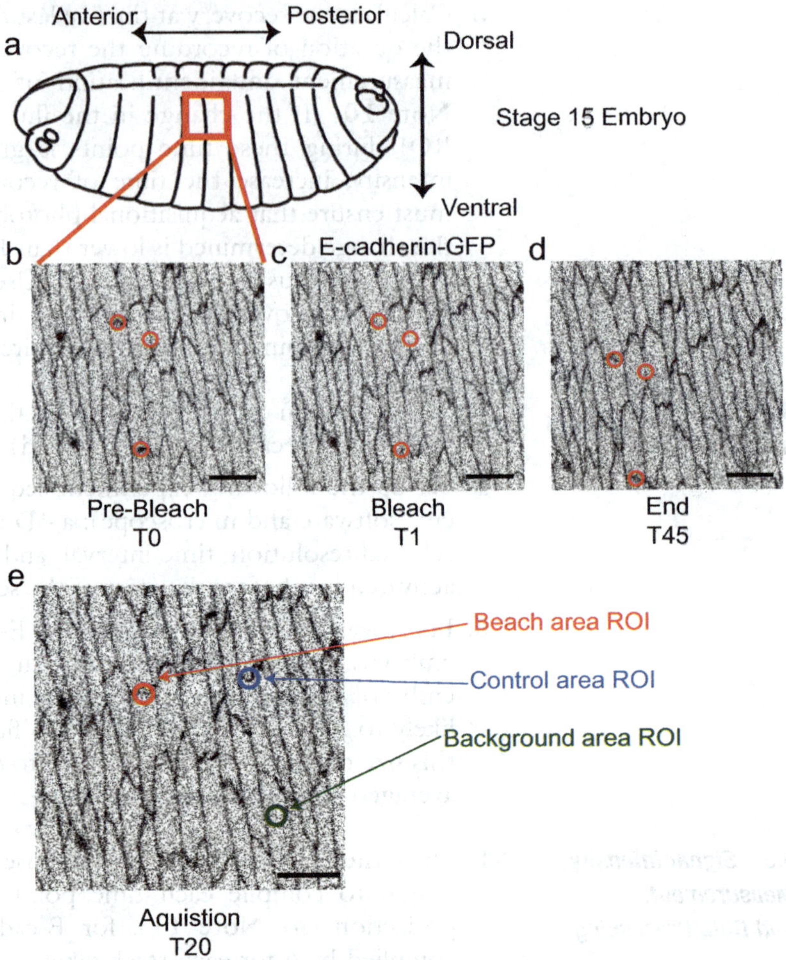

Fig. 2 Performing FRAP and data acquisition. (**a**) Cartoon representation of a late stage *Drosophila* embryo. The area representing the dorsolateral epidermis which is used for the FRAP experiment is highlighted by the square. (**b**) The cells of the epidermis of the stage 15 embryo have a distinct rectangular morphology: with long anterior-posterior (AP) and short dorsal-ventral (DV) cell borders. The regions of interest (ROI) which are to be bleached are shown (red circles), in this case, two DV and one AP borders are indicated. The ROI have the same shape and diameter to maintain a consistent diffusion coefficient. A *z*-stack of six slices is positioned to bleach the signal in the ROI in the *z*-axis (Time = −20 s). (**c**) Signal in the ROI immediately after bleaching (Time = 0 s). Some signal is still visible in the ROI, an initial indicator that bleaching was not too strong. (**d**) The area and bleached ROI at the end of the experiment (total 15 min for E-cad-GFP in the embryonic epidermis). More signal is now evident in the ROI and the tissue has shifted position during the course of the experiment. Each of the *z*-stacks is projected for each time point giving a total number of 45 time points for this experiment (time = 900 s). (**e**) The three measurements which are taken for each time point for analysis: the bleached ROI in which the signal recovery after bleaching is recorded (red); the control area (blue) to account for bleaching during acquisition; and the background signal (green). Scale bar = 10 μm

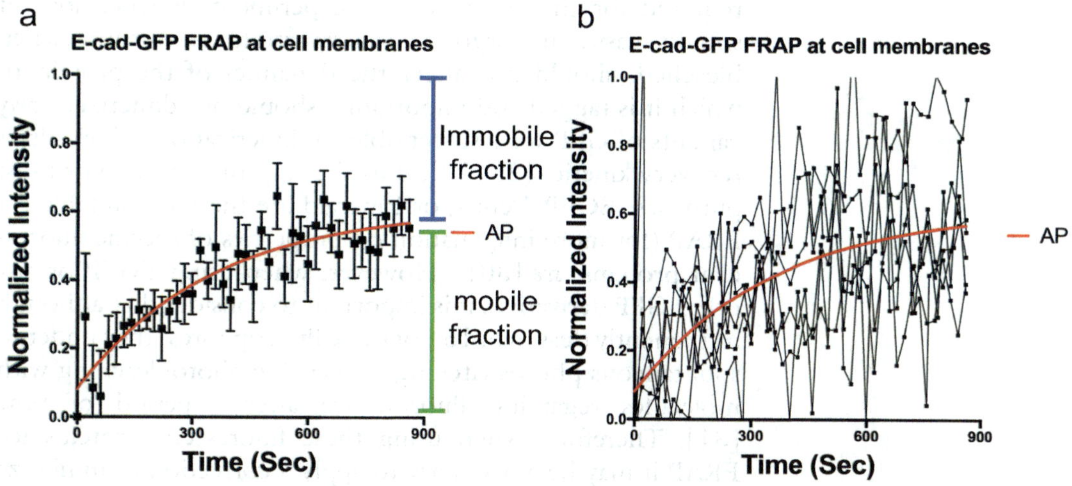

Fig. 3 Plotting recovery curves. (**a**, **b**) Representation of the recovery of E-cad-GFP in the embryonic epidermis. (**a**) Recovery of the E-cad-GFP signal at the long (AP) cell borders. The signal recovers to approximately 50% of the fraction which was bleached. The signal which recovers corresponds to the protein which is exchanged at the time-scale of the experiment, and therefore is termed the mobile fraction or the unstable fraction. The proportion of the bleached signal which does not recover, meaning the protein does not exchange at this time-scale, is called the immobile or stable fraction. The line represents the nonlinear fit curve and the individual points are means with the bars representing the SEM. (**b**) The same data but showing the individual traces for each of the replicates with the line of best fit from the nonlinear analysis shown in red. This graph indicates the variability inherent in FRAP experiments and the necessity of having a sufficient sample size to mitigate this effect and derive valid and accurate conclusions from any comparative analysis ($n = 6$ embryos, each embryo had two borders bleached and the average of these was used as an embryo average)

point n, immediately after photobleaching, and before photobleaching, respectively (Fig. 3).

6. Perform nonlinear regression analysis to test for the best fit model using suitable statistical package, e.g., GraphPad Prism (https://www.graphpad.com/scientific-software/prism). Fit the recovery to a single exponential model in a form of $f(t) = 1 - F_{\text{im}} - A_1 e^{-t/T_{\text{fast}}}$, and to bi-exponential model in a form of $f(t) = 1 - F_{\text{im}} - A_1 e^{-t/T_{\text{fast}}} - A_2 e^{-t/T_{\text{slow}}}$, where F_{im} is a size of the immobile fraction, T_{fast} and T_{slow} are the half-times, and A_1 and A_2 are the amplitudes of the fast and slow components of the recovery. Use an F-test to choose the model and compare datasets (*see* **Notes 14** and **15** and Fig. 3).

4 Notes

1. Several theoretical aspects should be taken into consideration when selecting the fluorophore. The ideal fluorophore for FRAP purposes has the optical properties of being bright, yet sufficiently photostable for the long low-level excitation

required for the course of the experiment. Furthermore, it should easily undergo a non-reversible conversion when bleached, should not affect the dynamics of the protein to which it is tagged and importantly should not dimerize. Early variants of GFP were susceptible to dimerization, which alters recovery kinetics [28, 29], thus skewing the results. For most purposes, EGFP, Venus, or Emerald are the most suitable for FRAP (for more information on properties of specific fluorescent proteins, *see* [30]). However, when using EGFP or any other GFP derivatives it is important to consider that although infrequently (less than 15% of these fluorophores) they undergo spontaneous photoswitching—reversible photobleaching with molecules regaining fluorescence after a period of time [31]. Therefore, when using these fluorescent proteins for FRAP it may be a necessary to apply a correction to minimize the contribution of photoswitching in the signal recovery [32]. Finally, tagging proteins with EGFP can have an effect on its behaviour [33, 34]. This can be mitigated if one is careful in the selection of the site of tagging, in particular avoiding locations which might affect protein functions or interactions. However, whenever possible we would recommend confirming that the tagged protein retains normal behaviour and is able to substitute for the endogenous protein. Additionally, it is crucial to use the same tagged proteins in control and experimental animals or cells.

2. *Drosophila* embryos are bean-shaped with convex side corresponding to embryo ventral side and concave to its dorsal side. The anterior of the embryos is marked by the presence of micropyle. Therefore, to orient embryos correctly for imaging using an upright microscope they should be aligned with dorsal side down on the segment of apple juice agar. We also recommend aligning all embryos in the same anterior-posterior orientation as this simplifies both the transfer to the slide and imaging.
3. When transferring embryos to a slide, it is important to bear in mind that the embryos are comparatively fragile. Therefore, when pressing the surface of the adhesive strip microscope slide to the embryos positioned on the apple juice segments it is recommended that a delicate amount of pressure is applied and only for a brief moment. If possible, use a dark bench to better visualise the embryos on the apple juice segments which appear similar to small rice grains to the unaided eye.
4. Circular areas of a Gaussian intensity profile provide the most straightforward system for the following analysis, although an approach for calculating diffusion coefficients when using arbitrary bleach area geometries has been proposed [35]. The size

of the bleached area must be small enough so that the unbleached protein is in excess, so that the contribution of bleached protein in the exchange can be neglected. Alternatively, a whole-cell FRAP can be used to determine the rate of new protein production [36].

5. While fixing the size of a bleach spot is important during each series of experiments, due to the change in the diffusional coefficient with bleach spot size and diameter [15, 23], a comparison of the recovery when applying different bleach spot sizes can be used to confirm and test the diffusional component of the recovery. This is due to the half-time of the recovery being affected by the spot size in the case of recovery primarily through diffusion, but not when binding reactions predominate over diffusion [23].
6. It is important to achieve a sufficient photobleaching depth (the amount of signal lost relative to pre-bleach) for a meaningful recovery while avoiding overbleaching. If over 80% of the initial signal is bleached, a Gaussian approximation of the bleached region is no longer valid, meaning that most common mathematical models cannot be applied [22, 28]. Additionally, excessive bleaching might cause photodamage due to localized heating [18, 37]. One approach which one can use to test for any changes in protein dynamics due to photodamage is sequential bleaching. Namely, to bleach an area, allow it to recover, then bleach it again with the same parameters. In the absence of photodamage, the recovery of the second bleach event will have the same dynamics (number and half-times of components) but will be close to complete (proportionally to the extent of the first bleaching round) due to the immobile fraction having already been bleached in the first round. This control for photodamage has been applied by our group and others [15, 38] and is an important consideration when starting a new FRAP experiment on a newly tagged protein or in a new system. In our experience, the best means of producing consistent results without photodamage is achieved when the fluorescent protein is bleached to 20–40% of the initial fluorescent intensity. For E-cad-GFP, we use eight scans with 50–70% laser power using a 488 nm wavelength laser.
7. The time to achieve sufficient bleaching must be minimized for two reasons. First, one needs to consider that protein dynamics do not cease while photobleaching is being performed, which causes misinterpretation of the kinetics due to the corona effect [39]. Longer bleaching times will lead to bigger bleached areas due to the bleached protein moving out of the spot where laser is applied. Secondly, localized heating mentioned above increases with bleaching time, which could intensify potential damage to cells and proteins [37]. Photobleaching times below

20 ms, and ideally within 6–12 ms, were suggested to be appropriate for determining diffusion coefficients of monomeric GFP [29, 37], and approaches to measure actual bleached area have also been suggested [29].

8. To achieve sufficient temporal resolution during the acquisition of fluorescence recovery, it is necessary to record the recovery at intervals which are smaller than the half-time of the fastest recovery component. Although as fast an acquisition as 1/10 of the half-time is recommended in some publications [28], in our experience, for recording E-cad-GFP, 20 s intervals are sufficient to detect a diffusional component with the half-time of about 25 s. For calibration, the aim is to achieve sufficient temporal resolution while maintaining maximal possible spatial resolution.

9. The photobleaching during the acquisition of the fluorescence recovery should be minimized as much as possible to obtain the most accurate information about the dynamics [40]: deviation by more than 10–15% is usually suggestive of excessive acquisitional bleaching [28].

10. It has been suggested that the duration of recovery acquisition should be 7–10 times longer than the characteristic half-time of the slowest detected component [28]. In our experience, three times longer acquisition is sufficient for a reliable estimation of both half-times and maximum recovery [15]. Furthermore, theoretically the signal ought to completely recover with a given time. For E-cad in the *Drosophila* this has been calculated to be a period of 2 h, through recovery of the "immobile" fraction by slow E-cad degradation and the addition of newly synthesized protein [41].

11. We use average intensity projections instead of maximum projections, as it considers not only protein concentration (intensity), but also distribution along the *z*-axis, i.e., junction width, and is reflective of the total protein amount. For example, a junction with the same E-cad density, but shorter in the *z*-axis will result in the same intensity when a maximum projection is used as a control and might skew the interpretation of the results.

12. There are several plugins which allow one to perform registration of a time series, i.e., compensation for tissue movement over time, such as "Register Virtual Stack Slices" in ImageJ. However, even when using such scripts, we recommend manually tracking or checking to control for ROI position when measuring intensities. For example, changes in the membrane curvature are observed in epithelial cells, which might shift the ROI position even if the cell position is correctly registered.

13. The best are control regions which are located in as close z position to the bleached region as possible and have a relatively similar intensity, as the side-effects of image acquisition such as z-drift and photobleaching will be comparable, enabling more accurate normalization. Therefore, while one control region might be possible to use for all bleached ROIs, it might be advisable to select a different appropriate control region for each bleached ROI in a time series.
14. Although more accurate equations exist for the various hypothetical methods of how protein exchanges and the signal recovers, e.g., diffusion only, diffusion-coupled, diffusion-uncoupled [23], we find that a general fit by using a sum of exponential components provides a valid linearization of a set of nonlinear first-order differential equations according to Lyapunov's first method [42, 43].
15. If the recovery is fit by a single exponential model, and appropriate tests are performed to determine whether the recovery occurs via diffusion or reaction-dominant mechanisms, it is possible to obtain additional information about protein dynamics. In the case of a pure-diffusion dominant recovery, the exact solution exists for a circular bleach area in form: $f(t) = e^{-T_D/2t}(I_0(T_D/2t) + I_1(T_D/2t))$, where I_0 and I_1 are modified Bessel functions of the first kind. In this case, $T_D = w^2/D_f$, where w is the radius of the circular beam, and D_f is the diffusion coefficient [20, 21]. If the recovery is reaction dominant, the following solution describes the recovery: $f(t) = 1 - C_{eq}e^{-k_{off}t}$, where C_{eq} is a constant which depends only on the dissociation (off-rate, k_{off}) and association (pseudo-on-rate, k_{on}) of the reaction [20, 44]. We would like to highlight that in this case the rate of the reaction depends only on the off-rate and does not reflect the on-rate, which can be a common misapprehension when interpreting the results of FRAP experiments [4].

References

1. Erami Z, Timpson P, Yao W, Zaidel-Bar R, Anderson KI (2015) There are four dynamically and functionally distinct populations of E-cadherin in cell junctions. Biol Open 4:1481–1489
2. Foote HP, Sumigray KD, Lechler T (2013) FRAP analysis reveals stabilization of adhesion structures in the epidermis compared to cultured keratinocytes. PLoS One 8:e71491
3. Indra I, Choi J, Chen C-S, Troyanovsky RB, Shapiro L, Honig B, Troyanovsky SM (2018) Spatial and temporal organization of cadherin in punctate adherens junctions. Proc Natl Acad Sci U S A 115:E4406–E4415
4. Wehrle-Haller B (2007) Analysis of integrin dynamics by fluorescence recovery after photobleaching. In: Coutts AS (ed) Adhesion protein protocols. Humana Press, Totowa, NJ, pp 173–201
5. Cluzel C, Saltel F, Lussi J, Paulhe F, Imhof BA, Wehrle-Haller B (2005) The mechanisms and dynamics of αvβ3 integrin clustering in living cells. J Cell Biol 171:383–392
6. Changede R, Sheetz M (2017) Integrin and cadherin clusters: a robust way to organize adhesions for cell mechanics. Bioessays 39:e201600123

7. Nanes BA, Kowalczyk AP (2012) Adherens junction turnover: regulating adhesion through cadherin endocytosis, degradation, and recycling. Subcell Biochem 60:197–222
8. Woichansky I, Beretta CA, Berns N, Riechmann V (2016) Three mechanisms control E-cadherin localization to the zonula adherens. Nat Commun 7:10834
9. Nader GPF, Ezratty EJ, Gundersen GG (2016) FAK, talin and PIPKIγ regulate endocytosed integrin activation to polarize focal adhesion assembly. Nat Cell Biol 18:491–503
10. Lu Z, Ghosh S, Wang Z, Hunter T (2003) Downregulation of caveolin-1 function by EGF leads to the loss of E-cadherin, increased transcriptional activity of beta-catenin, and enhanced tumor cell invasion. Cancer Cell 4:499–515
11. Wheelock MJ, Shintani Y, Maeda M, Fukumoto Y, Johnson KR (2008) Cadherin switching. J Cell Sci 121:727–735
12. Pasapera AM, Schneider IC, Rericha E, Schlaepfer DD, Waterman CM (2010) Myosin II activity regulates vinculin recruitment to focal adhesions through FAK-mediated paxillin phosphorylation. J Cell Biol 188:877–890
13. Stutchbury B, Atherton P, Tsang R, Wang D-Y, Ballestrem C (2017) Distinct focal adhesion protein modules control different aspects of mechanotransduction. J Cell Sci 130:1612–1624
14. Nanes BA, Chiasson-MacKenzie C, Lowery AM, Ishiyama N, Faundez V, Ikura M, Vincent PA, Kowalczyk AP (2012) p120-catenin binding masks an endocytic signal conserved in classical cadherins. J Cell Biol 199:365–380
15. Bulgakova NA, Grigoriev I, Yap AS, Akhmanova A, Brown NH (2013) Dynamic microtubules produce an asymmetric E-cadherin–Bazooka complex to maintain segment boundaries. J Cell Biol 201:887–901
16. Ratheesh A, Gomez GA, Priya R, Verma S, Kovacs EM, Jiang K, Brown NH, Akhmanova A, Stehbens SJ, Yap AS (2012) Centralspindlin and α-catenin regulate Rho signalling at the epithelial zonula adherens. Nat Cell Biol 14:818–828
17. Oda H, Tsukita S (2001) Real-time imaging of cell-cell adherens junctions reveals that Drosophila mesoderm invagination begins with two phases of apical constriction of cells. J Cell Sci 114:493–501
18. De Los SC, Chang C-W, Mycek M-A, Cardullo RA (2015) FRAP, FLIM, and FRET: detection and analysis of cellular dynamics on a molecular scale using fluorescence microscopy. Mol Reprod Dev 82:587–604
19. Day CA, Kraft LJ, Kang M, Kenworthy AK (2012) Analysis of protein and lipid dynamics using confocal fluorescence recovery after photobleaching (FRAP). Curr Protoc Cytom. Chapter 2:Unit2.19
20. Sprague BL, Pego RL, Stavreva DA, McNally JG (2004) Analysis of binding reactions by fluorescence recovery after photobleaching. Biophys J 86:3473–3495
21. Soumpasis DM (1983) Theoretical analysis of fluorescence photobleaching recovery experiments. Biophys J 41:95–97
22. Axelrod D, Koppel DE, Schlessinger J, Elson E, Webb WW (1976) Mobility measurement by analysis of fluorescence photobleaching recovery kinetics. Biophys J 16:1055–1069
23. Sprague BL, McNally JG (2005) FRAP analysis of binding: proper and fitting. Trends Cell Biol 15:84–91
24. Bulgakova NA, Brown NH (2016) Drosophila p120-catenin is crucial for endocytosis of the dynamic E-cadherin-Bazooka complex. J Cell Sci 129:477–482
25. de Beco S, Gueudry C, Amblard F, Coscoy S (2009) Endocytosis is required for E-cadherin redistribution at mature adherens junctions. Proc Natl Acad Sci U S A 106:7010–7015
26. Safranyos RG, Caveney S, Miller JG, Petersen NO (1987) Relative roles of gap junction channels and cytoplasm in cell-to-cell diffusion of fluorescent tracers. Proc Natl Acad Sci U S A 84:2272–2276
27. Greig J, Bulgakova NA (2018) E-cadherin endocytosis is modulated by p120-catenin through the opposing actions of RhoA and Arf1. bioRxiv 477760
28. Klein C, Waharte F (2010) Analysis of molecular mobility by fluorescence recovery after photobleaching in living cells. Chapter 12 in, Microscopy: Science, Technology, Applications and Education. Editors: Mendez-Vilas A, Diaz J. Publisher: Formatex Research Center
29. Kitamura A, Kinjo M (2018) Determination of diffusion coefficients in live cells using fluorescence recovery after photobleaching with wide-field fluorescence microscopy. Biophys Physicobiol 15:1–7
30. Kremers G-J, Gilbert SG, Cranfill PJ, Davidson MW, Piston DW (2011) Fluorescent proteins at a glance. J Cell Sci 124:157–160
31. Sinnecker D, Voigt P, Hellwig N, Schaefer M (2005) Reversible photobleaching of enhanced green fluorescent proteins. Biochemistry 44:7085–7094

32. Mueller F, Morisaki T, Mazza D, McNally JG (2012) Minimizing the impact of photoswitching of fluorescent proteins on FRAP analysis. Biophys J 102:1656–1665
33. Jian X, Cavenagh M, Gruschus JM, Randazzo PA, Kahn RA (2010) Modifications to the C-terminus of Arf1 alter cell functions and protein interactions. Traffic 11:732–742
34. Hoffmann A, Dannhauser PN, Groos S, Hinrichsen L, Curth U, Ungewickell EJ (2010) A comparison of GFP-tagged clathrin light chains with fluorochromated light chains in vivo and in vitro. Traffic 11:1129–1140
35. Blumenthal D, Goldstien L, Edidin M, Gheber LA (2015) Universal approach to FRAP analysis of arbitrary bleaching patterns. Sci Rep 5:11655
36. Huang J, Huang L, Chen Y-J, Austin E, Devor CE, Roegiers F, Hong Y (2011) Differential regulation of adherens junction dynamics during apical–basal polarization. J Cell Sci 124:4001–4013
37. Axelrod D (1977) Cell surface heating during fluorescence photobleaching recovery experiments. Biophys J 18:129–131
38. Lalo U, Allsopp RC, Mahaut-Smith MP, Evans RJ (2010) P2X1 receptor mobility and trafficking; regulation by receptor insertion and activation. J Neurochem 113:1177–1187
39. Weiss M (2004) Challenges and artifacts in quantitative photobleaching experiments. Traffic 5:662–671
40. Wu J, Shekhar N, Lele PP, Lele TP (2012) FRAP analysis: accounting for bleaching during image capture. PLoS One 7:e42854
41. Iyer KV, Piscitello-Gómez R, Paijmans J, Jülicher F, Eaton S (2019) Epithelial viscoelasticity is regulated by mechanosensitive E-cadherin turnover. Curr Biol 29:578–591.e5
42. Parks PC (1992) A. M. Lyapunov's stability theory—100 years on. IMA J Math Control Inf 9:275–303
43. Lyapunov AM (1992) The general problem of the stability of motion. Int J Control 55:531–534
44. Bulinski JC, Odde DJ, Howell BJ, Salmon TD, Waterman-Storer CM (2001) Rapid dynamics of the microtubule binding of ensconsin in vivo. J Cell Sci 114:3885–3897

Chapter 14

Methods to Generate and Assay for Distinct Stages of Cancer Metastasis in Adult *Drosophila melanogaster*

Jamie Adams, Andreu Casali, and Kyra Campbell

Abstract

Metastasis underlies the majority of cancer-related deaths. Until recently, research on this complex multi-step process has been hindered by a lack of genetically tractable experimental models amenable to high-throughput analyses. This was recently overcome with the development of a model of metastatic colorectal cancer (CRC) in adult flies, which relies on the activation of a partial-epithelial-to-mesenchymal transition (EMT) in intestinal tumors. In this model, tumor cells are labeled with both GFP and luciferase reporters, enabling high-throughput analyses. We report here the detailed protocol for generating the model, and assaying for primary tumor burden and distinct stages of metastasis, including the number of circulating tumor cells and secondary metastases.

Key words Cancer metastasis, *Drosophila*, Colorectal cancer

1 Introduction

Metastasis is the leading cause of cancer-related deaths globally. It is a complex multistep process which involves the invasion of cells from the primary tumor into surrounding tissues, migration and dissemination through the body, and colonization of distant sites. Invertebrate models such as *Drosophila melanogaster* (*Drosophila*) are emerging as powerful tools to investigate the mechanisms underlying malignancy and identify novel therapeutics [1]. While there are many anatomical and physiological differences between humans and flies, the number of cellular and organismic aspects of tumorigenesis that they share is remarkable. Additionally, of particular relevance to modeling colorectal cancer (CRC), there are striking similarities between the intestinal tracts of mammals and of *Drosophila* [2]. Experimentally, *Drosophila* are highly amenable to genetic manipulation, with a wealth of sophisticated genetic tools and publically available reagents. Combined with their short life-cycle, this allows for the analysis of large number of individuals

Kyra Campbell and Eric Theveneau (eds.), *The Epithelial-to Mesenchymal Transition: Methods and Protocols*, Methods in Molecular Biology, vol. 2179, https://doi.org/10.1007/978-1-0716-0779-4_14,

at a level of detail and rapidity that far surpasses current mammalian models.

Two of the most common initiating events in human CRC is activation of WNT and RAS signaling pathways [3]. This can be modeled in intestinal clones in adult *Drosophila* by combining mutations in the negative regulators of the Wnt pathway, *Apc* and *Apc2*, with overexpression of the oncogenic form of Ras, UAS-Ras [V12] using the midgut stem cell driver, Escargot (Esg)-Gal4—generating the so-called “ApcRas” model [4]. This genetic manipulation leads to the formation of tumor-like overgrowths with many similarities to human CRC tumors, including increased proliferation, a block in cell differentiation and cell polarity, and disrupted organ architecture [5]. However, there are no secondary tumors found in these flies, with ApcRas clones confined to the gut and surrounded by a thick layer of basement membrane. Therefore, these tumors can be considered benign CRC tumors.

Key steps of metastasis include invasion/dissemination of cells from primary tumors, migration through the body, colonization of distant sites, and growth of secondary metastases. Research on the metastatic process has been hindered by a lack of genetically tractable experimental models reproducing these key steps, which are also amenable to high-throughput analyses. Recently, when investigating the requirement of an epithelial-to-mesenchymal transition (EMT) for metastatic dissemination, we found that over-expression of the EMT master regulator Snail (Sna) in ApcRas intestinal tumors induces tumor metastasis in adult *Drosophila* [6]. In the “ApcRasSna” model, Sna drives a partial-EMT in intestinal tumor cells, which, although they retain some epithelial markers, subsequently break through the basal lamina of the midgut, undergo a collective migration, and seed polyclonal metastases. In this model, secondary growths can be found in multiple distant locations outside the gut, including the abdomen, thorax, and head [6]. Tumor cells generated in this metastatic CRC model are labeled with both GFP and luciferase reporters. Here we report a detailed protocol for firstly generating the ApcRasSna model, and then for assaying for each step of the metastatic process, including notes on how it can be made high-throughput. We expect that this can be exploited to examine the effects of many different risk factors such as diet, alcohol, toxins, and social behavior on tumor metastasis, as well as to perform rapid large-scale genetic and drug screens.

2 Materials

2.1 Flies

Lines for making controls (clones with just the reporters: GFP and luciferase)

- *yw hsp-flipase; esg Gal4, UAS-GFP/CyO; UAS-Luciferase FRT82B Gal80/ TM6b.*

- *yw hsp-flipase; Sp/CyO; FRT82B Gal80/TM6b.*

Lines for making ApcRas clones (benign CRC)

- *yw hsp-flipase; esg Gal4, UAS-GFP, UAS-RasV12/CyO; UAS-Luciferase FRT82B Gal80/TM6b.*
- *yw hsp-flipase; Sp/CyO; FRT82B Apc2N175KApcQ8/TM6b.*

Lines for making ApcRasSna clones (metastatic CRC)

- *yw hsp-flipase; esg Gal4, UAS-GFP, UAS-RasV12/CyO; UAS-Lucifersase FRT82B Gal80/TM6b.*
- *yw hsp-flipase; UAS-Sna/CyO; FRT82B Apc2N175KApcQ8/TM6b.*

$Apc2^{N175K}$ is a loss-of-function allele, Apc^{Q8} is a null allele, UAS-Ras^{V12} is a gain-of-function transgene, and UAS-Sna is a wild-type transgene. Stocks were obtained from Bloomington Stock Center, except UAS-Sna (kind gift of Kumar lab).

2.2 Equipment and Reagents

1. Parafilm.
2. A fluorescence stereo microscope.
3. Two pairs of sharp forceps.
4. Microscope slides.
5. Phosphate-Buffered Saline (PBS).
6. Oil 10 S, VOLTALEF.
7. Electron microscope grade paraformaldehyde.
8. Bovine serum albumin (BSA).
9. Mounting medium such as Vectashield or Fluoromount.
10. Triton X-100.
11. Anti-GFP antibody.
12. Phalloidin-TRITC.
13. Confocal microscope.
14. Tungsten needle.
15. Luciferase reporter assay kit.
16. A microplate reader that can detect luminescence.
17. 96-Well microplates.

2.3 Image Analysis

For image analysis, we recommend using ImageJ Fiji distribution (https://imagej.net/Fiji/Downloads). In principle Fiji should run on all major computer platforms, including Microsoft Windows, Mac OS, and Linux.

3 Methods

In this section, we describe the workflow to assess flies for distinct steps of metastatic cancer; it is subdivided into four sections. Firstly, the flies with clones of mutant intestinal cells are generated, control lines are generated in parallel. Next, the flies are screened for secondary tumors. Third, the primary tumor burden is analyzed by either detecting clones visually by GFP or by enzymatic reaction to detect for luciferase. Finally, the hemolymph of the flies is screened for circulating tumor cells. In the notes, we will discuss the advantages and disadvantages of detecting GFP vs luciferase as we go through each detection assay.

3.1 Generation of the Flies for Analysis

1. Cross 30 females and 10 males in each vial according to the mating schemes depicted in Materials. Collect females that lack CyO and TMBb balancers.
2. Collect females over a period of 6 days. On the seventh day, add 20 female and 3 male (*see* **Note 1**) flies to a vial. Wrap the tops with parafilm and heat shock for 1 h in a 37 °C water bath. The vials should be totally submerged to ensure that all flies are subjected to the heat, and spread as equally as possible for a uniform heat shock (Fig. 1).
3. Flies are placed in fresh vials of food after the heat shock and flipped every 4 days. This is to ensure that flies are not stuck to the food.

3.2 Screening Flies for Secondary Metastases

1. Screen flies every 3–4 days for metastases on the fluorescence stereo microscope. Primary tumors in ApcRasSna flies grow in the midgut and Malpighian tubules (MpTs, the kidneys of the fly). These appear as small dots or lumps of GFP (Fig. 2a). Generally, secondary tumors can be distinguished from primary tumors as they tend to be larger, brighter, and often closer to the surface than tumors in the intestinal tract (Fig. 2b).
2. Confirm metastasis by dissecting out the gastrointestinal tract. To do this, first anaesthetize the fly with CO_2, and place in a drop of PBS on a microscope slide. Cut the head cleanly off the fly with a pair of forceps. Next pinch the anus of the fly with one forceps, and the thorax of the fly between stripes 3 and 4 with the other (Fig. 3a). Gently pull until the proventriculus, and the 4 MpTs pop out (Fig. 3b) (*see* **Note 2**).
3. Check and see if masses of GFP positive cells separated from the gastrointestinal tract come out in the hemolymph (Fig. 4a) and to see if any GFP positive masses are left in the body (Fig. 4b). These tumors can then be imaged live by mounting in 10 S oil, or fixed and stained, using standard methods for fixing and staining adult midguts (*see* **Note 3**).

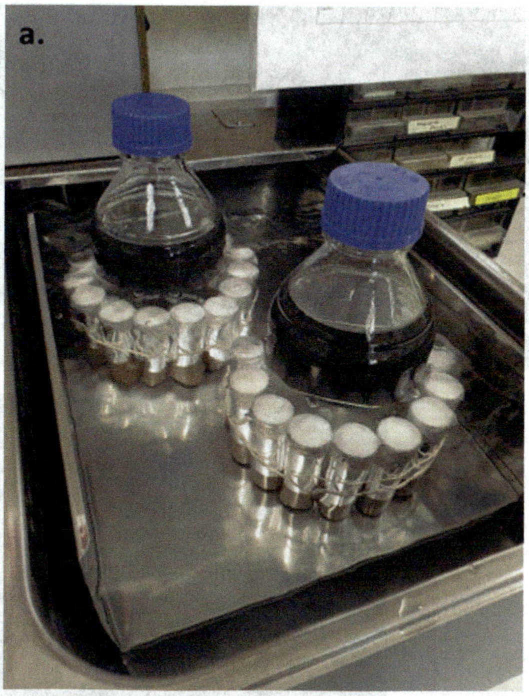

Fig. 1 Generating the flies with intestinal clones. The setup for ensuring that all flies are equally heat shocked and tubes made air-tight before submersion

3.3 Detailed Characterization of Primary Tumor Burden in Flies

1. For a detailed characterization of primary tumor burden, dissect the gastrointestinal tract from ten female mated flies at a precise time point, e.g., 3 weeks after heat shock. Dissect off the MpTs (*see* **Note 4**).
2. Fix the midguts (*see* **Note 3**) and immunostain for GFP to mark the clones, and Phalloidin to demark the midgut.
3. Image a single *z*-plane of the entire midgut on a confocal microscope with a 10× lens (*see* **Note 5**). On average a female midgut is around 0.4 cm long, therefore imaging in tile-scan mode is the most efficient way of acquiring this data. If this is not available, then image the entire midgut and stitch the multiple images together on Fiji.
4. Import a single image of an entire midgut into Fiji and use the freehand selection to draw around tumors. Use the measure tool to measure the area of each clone. Add up the total area of the midgut covered by GFP, divide by the total area of the gut and multiply by 100. This will give the total GFP% coverage of the midgut. In a second measurement which we find is a more robust measure for tumors is to calculate the average size of

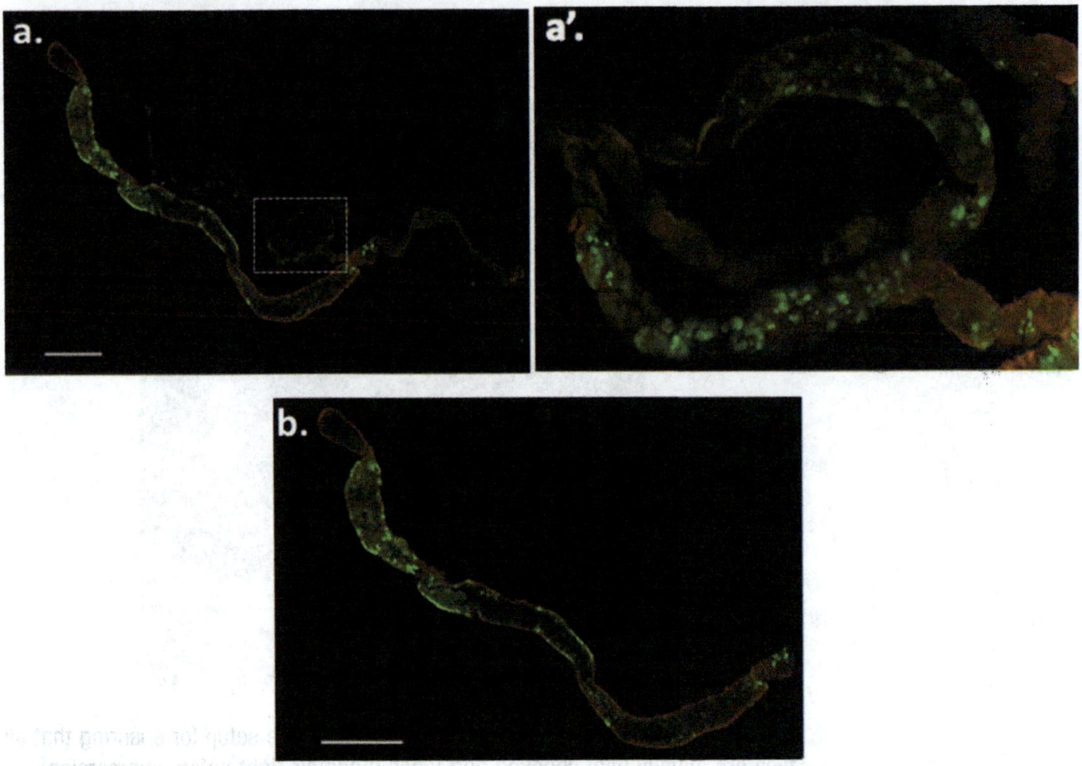

Fig. 2 Primary tumors are present in the midgut and MpTs. (**a**) The MpTs are prone to primary tumors and if not dissected off during staining, should be eliminated prior to analysis in Fiji. (**a′**) shows an enlargement of the boxed region in (**a**). (**b**) Other structures non-specific to the analysis (hindgut, ovaries, crop) should also be removed prior to analysis

clones per gut. This increases significantly in tumorigenic situations.

5. This analysis can be extended in Fiji to calculate a number of additional measures that taken together can be indicative of the tumor burden—such as the GFP% coverage in a particular region and the average cell number per clone.

3.4 High-Throughput Analysis of Total Tumor Burden in Flies

1. For high-throughput analyses, the presence of the luciferase reporter can be taken advantage of. To understand how luciferase activity is related to cell number, we previously sorted the GFP^+ cells from midgut-induced ApcRasSna clones by fluorescence-activated cell sorting. A linear correlation ($r = 0.9994$, $p = 0.0006$) was observed between the number of cells isolated and the amount of luciferase activity detected, allowing us to accurately detect down to 10 cells [6].
2. To measure the tumor burden in whole flies, first squash flies with a 200 μl pipette tip in lysis buffer. This can be done either with individual flies or in small batches.

Fig. 3 Screening for secondary metastases. (**a**, **b**) Flies visualized using a fluorescence stereo microscope. The fly chosen for further dissection and analysis is indicated via an arrow due to clear presence of GFP in the thoracic region and shown enlarged in (**b**). (**a**) Primary tumors appear as small dots or lumps of GFP inside the body cavity highlighted by dashed white boxes. (**b**) Secondary metastases are larger, bright, and often closer to the surface than tumors in the intestinal tract. Here we show a thoracic tumor, outlined by a dashed white box, clearly visible via GFP

3. Remove a fraction of the liquid for analysis and pipette into a 96-well microplate.
4. Add substrate to the lysate (as directed by manufacturer of Luciferase reporter kit).
5. Within 5 min (*see* **Note 6**) read the bioluminescence levels on a microplate reader.
6. Raw luciferase data is then compared with the control conditions. Controls should be read in parallel in each individual experiment.

3.5 High-Throughput Analysis of Circulating Tumor Cell Number

1. Hemolymph is extracted from whole adult flies according to the instructional video published by Laura Musselman (https://www.youtube.com/watch?v=im78OIBKlPA). The fly abdomen is pierced where it meets the thorax with a tungsten needle.
2. A 0.5 ml Eppendorf is pierced with an *x* needle, the lid is cut off and placed in a 1.5 ml Eppendorf containing lysis buffer.

Fig. 4 Dissecting the gastrointestinal tract using a stereo microscope. (**a**) First the head of an anaesthetized fly is removed using forceps. Next, pinch the anus of the fly with one set of forceps (solid arrowhead) and the connection between the thorax and the abdomen of the fly with the other (dotted arrowhead). (**b**) Gently pull until the entire intestinal tract (including the ovaries, the crop, and Malpighian tubules—outlined via dotted lines) has come out of the abdomen. Finally, remove organs non-essential for analysis of the midgut (ovaries, crop, Malpighian tubules). If organs are not removed, they can be eliminated during image analysis

3. The pierced fly is placed inside the 0.5 ml Eppendorf, and the tubes are spun at 2348 × *g* for 5 min at 4 °C. This can be done either with individual flies or in small batches.
4. Pipette the solution containing the hemolymph and lysis buffer into a 96-well microplate.
5. Add substrate to the lysate (as directed by manufacturer of Luciferase reporter kit).
6. Within 5 min (*see* **Note 6**) read the bioluminescence levels on a microplate reader.
7. Raw luciferase data is then compared with the control conditions. Controls should be read in parallel in each individual experiment.

4 Notes

1. The guts of males and females as well as virgin females and mated females are physiologically and morphologically distinct [7], therefore to keep data constant we normally analyze mated females. For this reason, a small number of males are added to the vials, but only the females are analyzed.

2. The MpTs become very fragile when they have tumors, and prone to breaking. They should be dissected with care and checked to make sure they are intact. This is to ensure that MpTs tumors are not mistaken for secondary tumors.
3. To fix adult midguts (or secondary metastases), incubate in PBS and 4% electron microscope grade formaldehyde for 30 min. Rinse the samples three times with PBT-BSA (PBS, 4% BSA, 0.1% Triton X-100). Next, incubate with the primary antibody overnight at 4 °C followed by agitation with the secondary antibody for 2 h at room temperature. Finally, rinse the samples three times with PBT-BSA and mount in an appropriate mounting medium such as Vectashield or Fluoromount.
4. The ureter and base of the MpTs are prone to large bloated tumors (Fig. 5a). To focus analyses on the primary tumor burden in the midgut, either dissect off the MpTs during sample preparation, or they can be eliminated during image analysis on Fiji.
5. While capturing images on multiple *z*-planes leads to a more precise analysis, given the large size of midguts, this is very time consuming to both acquire and analyze. We have found that selecting a representative *z*-plane by eye, and averaging over a sample size of 10 guts, has so far proved robust in our analyses of a number of different genotypes over a range of time points.
6. The timing between adding the substrate and reading the plate does not necessarily have to be 5 min, but as the bioluminescence resulting from the reaction decreases with time, it needs to be constant across experiments.

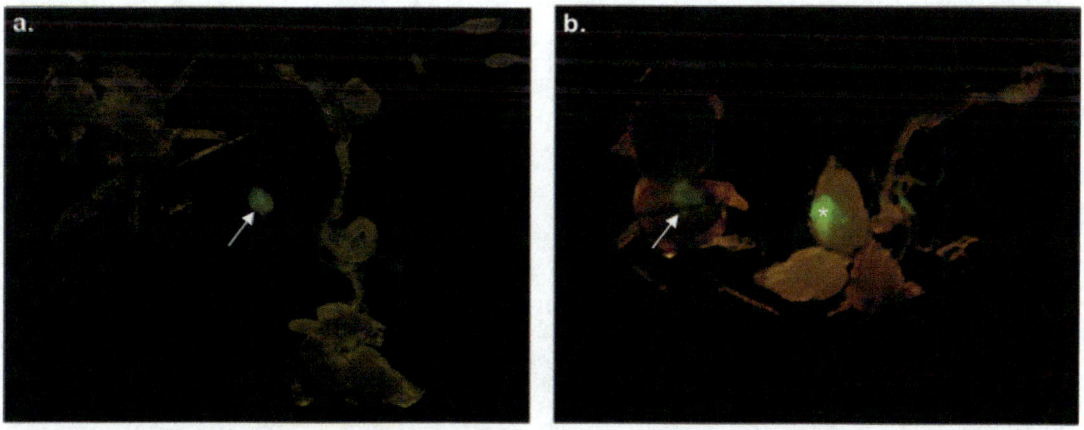

Fig. 5 Examples of secondary tumors under the fluorescence stereo microscope. (**a**) A secondary tumor that came out with the hemolymph (arrow) during dissection. (**b**) Secondary tumors left in the body cavity (arrow) and invading into the ovaries (asterisk)

Acknowledgements

We are thankful to the rest of the Campbell lab for helpful discussions. This work was supported by a Wellcome Trust/Royal Society Sir Henry Dale Fellowship (Grant number 204615/Z/16/Z).

References

1. Gonzalez C (2013) *Drosophila melanogaster*: a model and a tool to investigate malignancy and identify new therapeutics. Nat Rev Cancer 13 (3):172–183
2. Casali A, Batlle E (2009) Intestinal stem cells in mammals and *Drosophila*. Cell Stem Cell 4 (2):124–127
3. Jackstadt R, Sansom OJ (2016) Mouse models of intestinal cancer. J Pathol 238(2):141–151
4. Martorell O, Merlos-Suarez A, Campbell K et al (2014) Conserved mechanisms of tumorigenesis in the *Drosophila* adult midgut. PLoS One 9(2): e88413
5. Matorell O, Merlos-Suárez A, Campbell K, Barriga F, Christov C, Miguel-Aliaga I, Batlle E, Casanova J, Andreu Casali A (2014) Conserved mechanisms of tumorigenesis in the Drosopihla adult midgut. PLos One 9(2): e88413
6. Campbell K, Rossi F, Adams J et al (2019) Collective cell migration and metastases induced by an epithelial-to-mesenchymal transition in *Drosophila* intestinal tumors. Nat Commun 10 (1):2311
7. Miguel-Aliaga I, Jasper H, Lemaitre B (2018) Anatomy and physiology of the digestive tract of *Drosophila melanogaster*. Genetics 210 (2):357–396

Part IV

In Vivo Imaging of MET

Chapter 15

4D Live Imaging and Analysis of Chick Embryo Somites

Gi Fay Mok, James McColl, and Andrea Münsterberg

Abstract

Avian (chick) embryos are an established and accessible model organism making them ideal for studying developmental processes. Chick embryos can be harvested from the egg and cultured allowing real-time observations and imaging. Here, we describe ex vivo culture and preparation of somite tissue followed by time-lapse multi-photon microscopy, image capture and processing. We applied this approach to perform live imaging of somites, the paired segments in vertebrate embryos that form in a regular sequence on either side of the neural tube, posteriorly from presomitic mesoderm (psm). Somites give rise to cell lineages of the musculoskeletal system in the trunk such as skeletal muscle, cartilage and tendon, as well as endothelial cells. Until recently it was not possible to observe the cellular dynamics underlying morphological transitions in live tissue, including in somites which undergo epithelial-to-mesenchymal transitions (EMT) during their differentiation. In addition to the experimental setup, we describe the analytical tools used for image processing.

Key words Live imaging, Somites, Chick embryo, Multi-photon microscopy, Image processing

1 Introduction

The chick embryo is a classic vertebrate model for studying developmental processes in amniote species [1]. Embryos are easily accessible as they develop outside the mother and this enables in vivo manipulation experiments. They can be removed from the egg and cultured for a few days or, as described in this chapter, tissue slices can be dissected and cultured for live imaging [2]. Recent advances in genetic modification and new transgenic lines with fluorescently labeled cells will further improve cell imaging approaches [3]. Embryo morphogenesis is complex, and a better understanding of the underlying processes is of fundamental importance. Morphogenesis is closely linked to and driven by cellular dynamics such as proliferation, cell growth, cell movement, and also cell migration, which often occur rapidly over short timescales. However, it has been surprisingly challenging, to date, to image, capture, and analyze these processes in live tissues.

Kyra Campbell and Eric Theveneau (eds.), *The Epithelial-to Mesenchymal Transition: Methods and Protocols*, Methods in Molecular Biology, vol. 2179, https://doi.org/10.1007/978-1-0716-0779-4_15,

Somites are paired segments that form during vertebrate development on either side of the midline structures, the neural tube and notochord. They are transient embryonic structures which give rise to progenitors that form the musculoskeletal system—such as cartilage, tendon, muscle, dermis, and endothelial cells. They are generated in a regular sequence—every 90 min in chick embryos, from the pre-segmented mesoderm (PSM) at the posterior end of the embryo [4]. The newest formed somites comprise a sphere of epithelial cells encapsulating a central cavity, the somitocoel, containing mesenchymal cells. As somites mature and begin to differentiate, the ventral domain dissociates, via epithelial-to-mesenchymal transition (EMT), to form the sclerotome, which gives rise to the axial skeleton. The dorsal domain remains epithelial to form the dermomyotome and myotome, which give rise to all trunk and appendicular skeletal muscles [5].

Although much has been learned about the signaling pathways and key regulatory genes that are important for coordinating the fate decisions of these cells [6], very little is known about the cellular dynamics that underlie the morphological transitions during somite differentiation. To observe cellular behaviors and rearrangements that contribute to the morphogenesis of differentiating somites, we used live multi-photon imaging of transgenic chick embryos, expressing a membrane-bound GFP. Here, we describe a method for dissecting somites together with their neighboring tissues—the neural tube and notochord, lateral plate mesoderm, surface ectoderm and endoderm, which provide signaling cues essential for specification of somite derivatives [6]. The tissue slice can be maintained in culture in order to image somite differentiation at cellular resolution (*see* Fig. 1). In optimized conditions, somites are still exposed to their native signaling environment. They grow over time and individual cells are very active, they increase in size, divide, change shape, move and extend filopodia (*see* examples of observed cell behavior in Figs. 2 and 3) [2]. This approach can potentially be adapted to study the morphogenesis and individual cell behavior across a range of tissues within an embryo.

2 Materials

2.1 Harvesting Avian Embryos

1. Penicillin/Streptomycin, 100,000 U/mL.
2. Fetal calf serum, FCS.
3. Buffered culture medium such as F12 supplemented with nutrients. We routinely use Glutamax Ham's F12 medium (Invitrogen cat. no. 31765).
4. 30 mm petri dish.

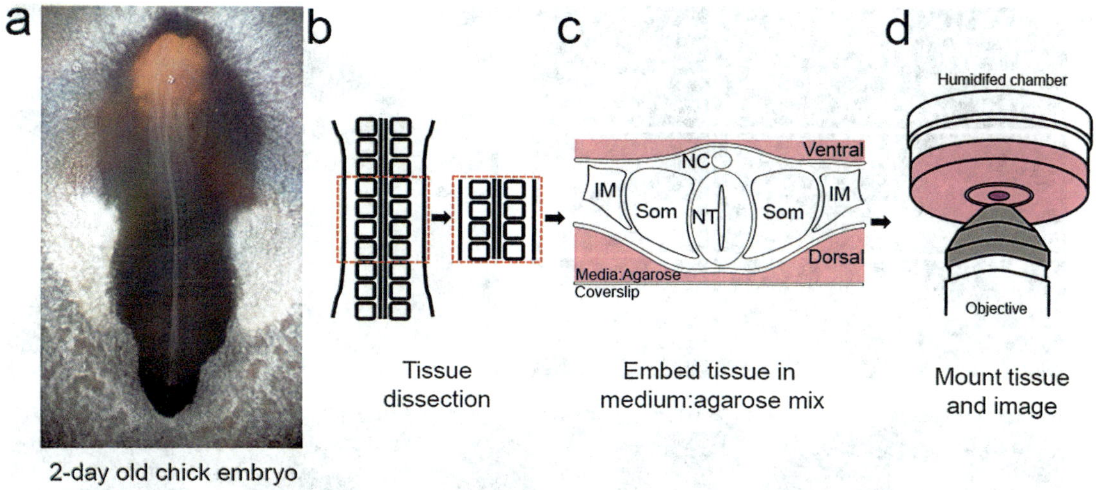

Fig. 1 Experimental setup. (**a**) The bright-field image shows a typical embryo, the white rectangle indicates the region shown in the schematic in (**b**), with paired somites flanking the neural tube and notochord in the axial midline. (**b**) Red stippled rectangle represents the tissue slice that is dissected and flipped around so that the dorsal side is facing the coverslip. (**c**) Tissue section schematic shows how the tissue is embedded in the agarose/medium mixture, indicated in pink (*IM* intermediate mesoderm, *NT* neural tube, *Nc* notochord, *Som* somite). (**d**) The tissue sits in a 35 mm dish with a N°1 glass coverslip base. Samples are imaged using a multi-photon microscope with an inverted stage and 20× objective

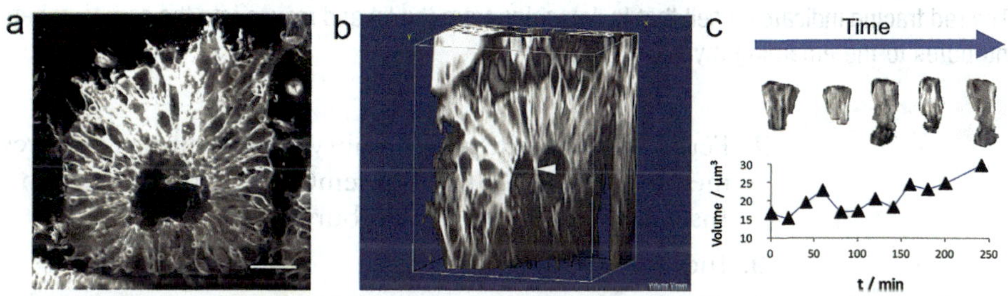

Fig. 2 Example of volume measurements and 3D rendering of cells over time. (**a**) Optical section of an epithelial somite with the central somitocoel viewed from the top of the somite (dorsal), white arrow indicates a cell that is descending into the somitocoel. (**b**) Section through the somite, white arrow depicts the same cell. (**c**) 3D volume view using ImageJ 3D viewer shows that this cell undergoes changes in shape and volume over time. Volume measurements are shown

5. Insect pins.
6. Fine incision scalpels (pfm Medical ophthalmic blades cat. no. 200300715 or equivalent).
7. Low melting point agarose.
8. Micro-pipette with fine tips and 20 μL volume.

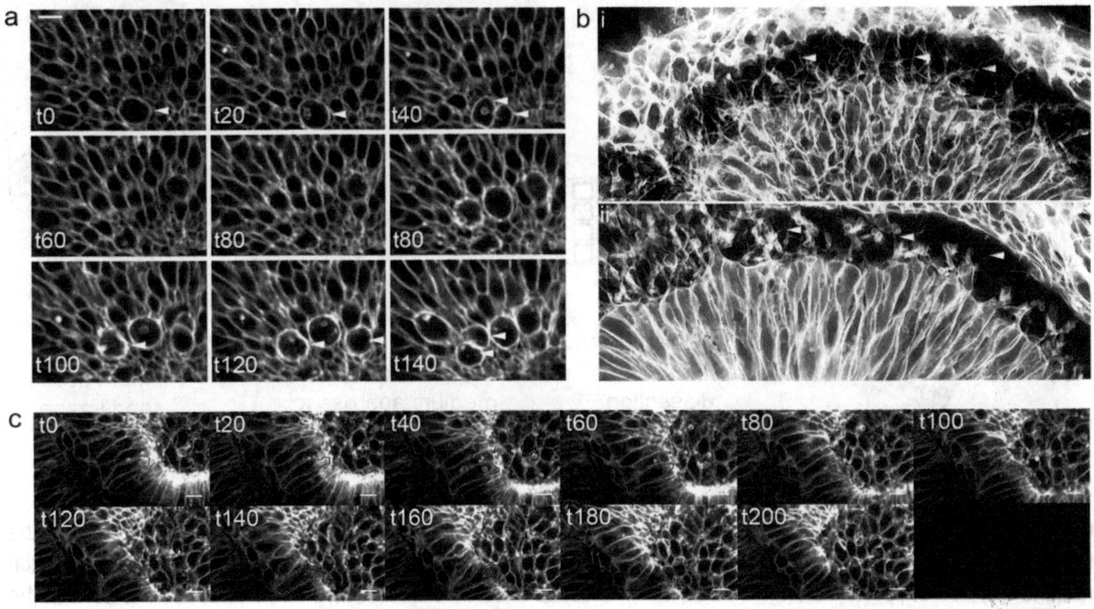

Fig. 3 Examples of cellular dynamics and behaviors that can be observed. (**a**) The panels show a series of images in the same optical plane from $t = 0$ to $t = 140$ min. Cells that round up and then divide can be seen and are indicated by white arrow heads, scale bar = 10 μm. (**b**) The top panel (1) shows an optical section with somite cells extending filopodia toward the overlying ectoderm, (2) is a deeper optical section where cells extending filopodia toward the intermediate mesoderm laterally can be observed. (**c**) The panels show a series of images focusing on the medial lip of the somite, adjacent to the neural tube visible in the bottom corner on the left. The red tracing indicates a cell that is detaching from the lip and moves into the somite, where it most likely contributes to the emerging myotome ($t =$ min)

9. Fertile transgenic eggs; embryos ubiquitously expressing a membrane-bound GFP (memGFP), obtained from Roslin Institute, University of Edinburgh, Scotland, UK.
10. Incubator (38 °C).
11. Silicone elastomer (Sylgard 184 silicone, Dow Corning, USA or equivalent).
12. Non-coated 35 mm glass-bottom dishes with 7 mm glass diameter.

2.2 Time-Lapse Video-Microscopy

1. Inverted two-photon microscope equipped with 10 and 20× objectives and a motorized stage for multiple *XYZ* acquisition (*see* **Note 1**).
2. Environmental chamber to control temperature (38 °C) and CO_2 (5%) (*see* **Notes 2** and **3**).
3. A software with time-lapse and multiple *XYZ* positions modules to control the microscope (*see* **Note 4**).

2.3 Image Processing

1. Computer workstation with 8 Gb RAM and capability to run MatLab and ImageJ.
2. Between 1 and 2 Tb storage for archiving raw, denoised, and analyzed data (*see* **Note 5**).

3 Methods

3.1 Tissue Dissection, Culture, and Mounting

1. Incubate fertile eggs (*see* **Note 6**), transgenic for memGFP at 38 °C in a humidified incubator until HH14/15 or desired stage (*see* **Note 7**).
2. Harvest embryos (*see* **Note 8**) in culture medium, pre-warmed to 38 °C, in petri dishes coated with silicone lining the base of the dishes (*see* **Note 9**).
3. Pin down embryos on the silicone dish using insect pins, stretching the extraembryonic tissue (*see* **Note 10**).
4. Use fine scalpels to dissect the explants containing five to six somite pairs; for imaging of the most posterior epithelial somites, ensure that the pre-segmented mesoderm is included. Ensure that the somites are flanked by intermediate and lateral plate mesoderm, with surface ectoderm above and endoderm below, as well as notochord and neural tube in the axial midline (*see* **Note 11**).
5. Prepare low melting point agarose in culture medium (2% w/v) (no phenol red) with 10% serum and 1% penicillin/streptomycin.
6. Place a small amount (200 μL) of the agarose/medium on the base of the glass slide, on the glass-bottom dishes.
7. While setting, transfer the dissected tissue (*see* **Note 12**) and position it with the dorsal side toward the base (*see* **Note 13**), so that the imaging starts from dorsal through to the ventral side on an inverted microscope as shown in Fig. 1.
8. Add warm agarose/medium on the somite explants until it is fully covered.
9. Place the dishes on ice for 5 min to fully set.
10. Then place inside a humidified and heated chamber (38 °C) for microscopy (*see* **Notes 14** and **15**).

3.2 Long-Term Two-Photon Microscopy

1. Acquire images at 10× magnification to view a string of 3–4 somites (depending on stage/size) or 20× magnification to view an individual somite, starting at the interface and imaging up through the tissue at 510 nm *Z* slices. Return to the interface and repeat every 20 min. Record images as Tiff files (*see* **Notes 16** and **17**).

2. To de-noise the Tiff files, download the nd-safir de-noising code. The code is available at http://serpico.rennes.inria.fr/doku.php?id=software:nd-safir:index.
3. Open a command prompt and type: cd C:\ndsafir-3.0-win64.
4. Type: ndsafir—i 'input\input.tif'—o 'output\output.tif'. Replace input with your input image and replace output with your output image.
5. Repeat for all files and once denoised the images can be analyzed (*see* **Notes 18** and **19**).

3.3 Tracking and Image Analysis

1. First create a time-lapse stack using a single slice in either MatLab or ImageJ.
2. Invert the stack in ImageJ, go to "edit/invert."
3. Bandpass filter the stack to remove high and low frequency noise, go to "Process/FFT/Bandpass filter."
4. Subtract background using a rolling ball radius of 20 pixels, go to "Process/Subtract background."
5. Save the processed stack, go to "file/Save as."
6. Track the processed stack using ImageJ trackmate plugin, go to "plugins/Tracking/Trackmate."
7. Determine the parameters for cell detection in trackmate and apply to all further data sets.
8. Run the analysis and save the results file.
9. To determine the directional concerted movement, isolate the angle results from the results file. Plot the sum of all the values and S.D.
10. For angle plots take the angle data, import into MatLab, and apply angle plot in MatLab.
11. Plot track speed using analysis packages such as excel, MatLab or (for this work) we used Origin.

3.4 Somite Size Quantification

1. Open file in ImageJ.
2. Find the center of the somite and use the threshold tool ("image\adjust\threshold") to determine the area of the somite center slice.
3. Set measurements to record perimeter ("Analyse/Set measurements," then select perimeter). Measure perimeter, go to "Analyse/measure."
4. To model the volume of the somite, we assume the perimeter to be equal to the circumference. From the circumference, we calculate the radius using $C = 2\pi r$.
5. Next we use the radius to model the volume of a sphere using $V = 4/3\pi r^3$.

6. Calculate the volume of all other somites and plot.
7. For the somitocoel, use the draw tool to determine the perimeter by hand and then calculate as a sphere.

3.5 Morphology Analysis

1. Take a single slice from the top of the somite at time zero and the same slice at a later time.
2. Download imageJ correlator plugin at https://imagej.nih.gov/ij/plugins/image_correlator.html.
3. Run the image correlator plugin in ImageJ to determine the correlation factor.
4. Repeat for all samples and plot the mean and S.D. for the top of all somites.
5. Repeat for mid and bottom sections.

3.6 Real-Time Accurate Cell-Shape Extractor (RACE) Analysis

1. Take the image stack of interest and crop out a region at the lip and the region opposite the lip of the somite in *X*, *Y*, *Z*. Save these regions.
2. Download source code from *https://bitbucket.org/jstegmaier/race*.
3. Run one set of data and adjust threshold and watershed levels so that all cells are detected. By eye, determine that no artifacts are being detected. Log threshold and watershed values and apply these to all data sets.
4. Run all data and plot results.

4 Notes

1. Time-lapse images were captured using a TriM Scope II (multiphoton) microscope from LaVision Biotec GmbH, Bielefeld, Germany equipped with a Chameleon Vision II pulsed laser from Coherent Inc., Daventry, UK. Imaging used a 20× Plan-Apochromat objective (0.8 NA) on a Zeiss AxioObserver D1, and fluorescence (525 ± 25 nm) captured using a high-sensitivity GaAsP detector.
2. Temperature (38 °C) was maintained using a Heating Insert P S1 controlled by TempModule S and AxioVision software (Carl Zeiss UK, Cambridge).
3. Set chamber to 38 °C and the lid to the chamber to 38.5 °C to avoid condensation.
4. We used ImSpectorPro software.
5. Each image file will be 30–100 Gb in size.
6. Use freshly delivered eggs that have been stored for no more than 1 week at 16–17 °C.

7. Use the Hamburger-Hamilton stage series [7] as a guide for incubation times, but establish empirically the number of days/hours required as this will vary with age of flock and season.
8. Only choose well-developed embryos for harvesting.
9. For silicone plates, ensure that there is a coating of at least 3 mm thick on the bottom of the dish. Alternative methods could be used such as agarose.
10. Ensure that embryos are pinned down under tension using the extraembryonic tissues. This aids the dissection and enhances tissue integrity and survival.
11. It is important to perform the dissections of the desired tissue "chunk" rapidly and effectively.
12. Use a fine pipette tip to transfer the tissue. Depending on tissue size, it may be necessary to cut the tip to create a wider opening. Part-fill the tip with agarose/medium mixture so as not to dilute the agarose/medium mixture on the dish during the tissue transfer.
13. Use fine plastic tip to position the tissue.
14. Clean glass on chamber with alcohol and lens wipe tissue to ensure no debris or scratches.
15. We describe the techniques to prepare dissected somites for imaging; however, the method can also be modified for imaging whole embryos at younger stages.
16. When setting up time-lapse imaging using multi-photon microscopes, a trial run will help determine orientation, depth, and duration of images needed to be captured to achieve optimal image quality.
17. When imaging deeper into the tissue, the light will scatter. Hence it is important to increase the laser power exponentially when moving deeper into the tissue.
18. For the analysis, make sure images are well aligned and the system does not drift. If drift does occur, use the registration plugin (ImageJ) to correct or discard the drifting data.
19. When tracking the inverted images, first confirm by eye on a small data set that the program is picking out individual cells and that they are tracked well. If not, you may need to adjust the threshold and tracking parameters.

References

1. Stern CD (2005) The chick; a great model system becomes even greater. Dev Cell 8(1):9–17. https://doi.org/10.1016/j.devcel.2004.11.018
2. McColl J, Mok GF, Lippert AH, Ponjavic A, Muresan L, Munsterberg A (2018) 4D imaging reveals stage dependent random and directed cell motion during somite morphogenesis. Sci

Rep 8(1):12644. https://doi.org/10.1038/s41598-018-31014-3

3. Davey MG, Balic A, Rainger J, Sang HM, McGrew MJ (2018) Illuminating the chicken model through genetic modification. Int J Dev Biol 62(1–2–3):257–264. https://doi.org/10.1387/ijdb.170323mm
4. Benazeraf B, Pourquie O (2013) Formation and segmentation of the vertebrate body axis. Annu Rev Cell Dev Biol 29:1–26. https://doi.org/10.1146/annurev-cellbio-101011-155703
5. Christ B, Scaal M (2008) Formation and differentiation of avian somite derivatives. Adv Exp Med Biol 638:1–41
6. Brent AE, Tabin CJ (2002) Developmental regulation of somite derivatives: muscle, cartilage and tendon. Curr Opin Genet Dev 12(5):548–557
7. Hamburger V, Hamilton HL (1951) A series of normal stages in the development of the chick embryo. J Morphol 88(1):49–92

Chapter 16

In Vivo Analysis of the Mesenchymal-to-Epithelial Transition During Chick Secondary Neurulation

Elena Gonzalez-Gobartt, Guillaume Allio, Bertrand Bénazéraf, and Elisa Martí

Abstract

The neural tube in amniotic embryos forms as a result of two consecutive events along the anteroposterior axis, referred to as primary and secondary neurulation (PN and SN). While PN involves the invagination of a sheet of epithelial cells, SN shapes the caudal neural tube through the mesenchymal-to-epithelial transition (MET) of neuromesodermal progenitors, followed by cavitation of the medullary cord. The technical difficulties in studying SN mainly involve the challenge of labeling and manipulating SN cells in vivo. Here we describe a new method to follow MET during SN in the chick embryo, combining early in ovo chick electroporation with in vivo time-lapse imaging. This procedure allows the cells undergoing SN to be manipulated in order to investigate the MET process, permitting their cell dynamics to be followed in vivo.

Key words Mesenchymal-to-epithelial transition, Secondary neurulation, Neural tube formation, Chick embryo, In ovo electroporation, In vivo time-lapse imaging, Cell dynamics

1 Introduction

During development, the entire vertebrate peripheral nervous system (PNS) is formed through the epithelial-to-mesenchymal transition (EMT) of neuroepithelial cells, which generates migratory neural crest cells, one of the best studied examples of physiological EMT [1–3]. By contrast, the mesenchymal-to-epithelial transition (MET) is the reverse process and plays an important role during organogenesis, as well as in the elongation of the caudal nervous system, a process known as secondary neurulation (SN). Indeed, the formation of the vertebrate neural tube (NT) involves two different morphogenetic events, primary (PN) and secondary

Electronic supplementary material: The online version of this chapter (https://doi.org/10.1007/978-1-0716-0779-4_16) contains supplementary material, which is available to authorized users.

Kyra Campbell and Eric Theveneau (eds.), *The Epithelial-to Mesenchymal Transition: Methods and Protocols*, Methods in Molecular Biology, vol. 2179, https://doi.org/10.1007/978-1-0716-0779-4_16,

neurulation (SN) [4, 5]. During PN, the lateral ends of the anterior neural plate elevate and the bilateral neural folds fuse with each other to form the anterior NT [6–8]. By contrast, in SN mesenchymal neuromesodermal progenitors are recruited to elongate the caudal body axis and drive the caudal elongation of the NT. The MET of neuromesodermal progenitors (NMPs) is the central event of SN, along with the formation of a compact nerve cord and its subsequent cavitation to form the caudal NT. In human embryos, the transition from the primary to the secondary NT occurs at the lumbosacral level; therefore the development of the lumbar, sacral, coccygeal and equinal cord largely involves SN [9–12].

Here, we propose that the chick embryo is an ideal model to understand the MET events that occur during amniotic SN. While manipulating and imaging mammalian embryos is still problematic, the chick embryo provides an easily accessible model in which the stages of development can be readily identified [13]. At early stages of development the avian body plan is very similar to that of mammals, and these embryos have excellent optical properties as they are thin and planar [14]. Furthermore, the SN region in the chick embryo extends up to the lumbar region [15, 16], closely resembling human development, whereas SN only occurs at the tail level in mice [17–19].

In the chick, SN starts when undifferentiated NMPs converge onto the dorsal midline, adopt a neural cell identity and undergo MET. The complete MET process can be followed in stage HH15 chick embryos, as different degrees of polarization exist along the anteroposterior axis (Fig. 1). The first cells to undergo MET are located in the periphery of the medullary cord, while the central cells remain mesenchymal until the very end of the process. It is between these two cell populations that small cavities of varied size and shape form, later coalescing to form a single central lumen (Fig. 1c–g) [20, 21].

However, the analysis of SN in vivo has always been technically difficult. Cell tracing studies have identified the epiblast region occupied by cells that undergo SN in the future (the preSN region), located caudo-medially to Hensen's Node at stage HH9 chick embryos [22–26]. However, the NT is still open in the posterior domains of stage HH9 chick embryos, so for electroporation the DNA must be injected on top of the epiblast and the electrodes positioned above and below the embryo (Fig. 2a, b). Earlier embryonic stages have always been electroporated ex ovo in this way [27, 28], which facilitates the electroporation of the flat epiblast, although cultured embryos do not grow to stage HH15 with normal body elongation [29].

Here we propose a new method for the in vivo analysis of MET during SN that overcomes all these technical difficulties. The method involves combining early in ovo chick embryo electroporation with time-lapse imaging in a culture setup specifically adapted

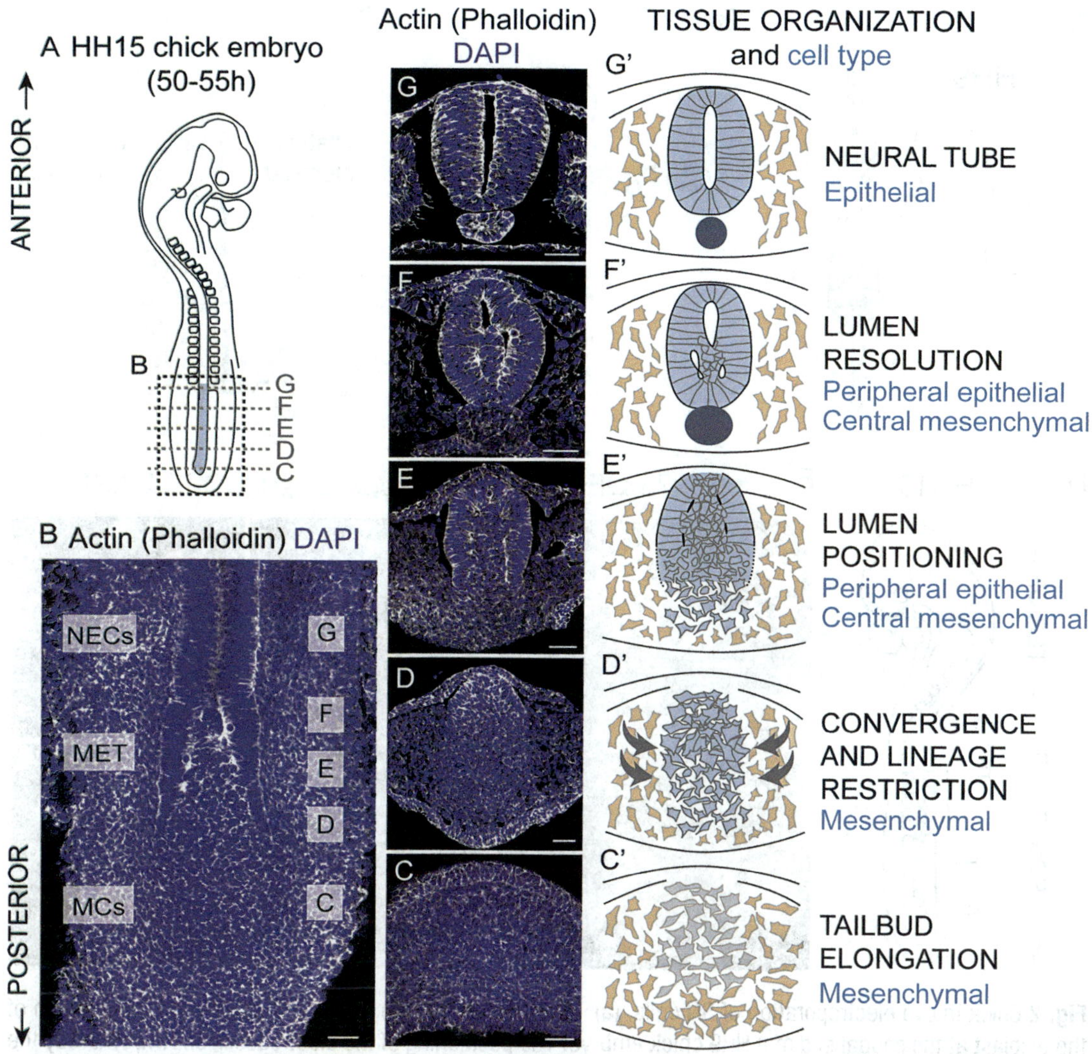

Fig. 1 Chick secondary neurulation (SN). (**a**) Drawing of a HH15 chick embryo showing the caudal region where SN is taking place. The progression of SN can be followed along the anteroposterior axis of the same embryo. (**b**) Dorsal view of the boxed region in (**a**) showing the distribution of actin (Phalloidin) in white. The cells in posterior regions are mesenchymal cells (MCs), those in the intermediate regions are undergoing the mesenchymal-to-epithelial transition (MET) and the anterior cells are neuroepithelial cells (NECs). The transverse sections in (**c**)–(**g**) correspond to different levels along the anteroposterior axis. Scale bar = 40 μm. (**c–g**) Transverse sections at different anteroposterior levels showing the distribution of actin. Scale bar = 40 μm. (**c′–g′**) Schematic representation of chick SN showing major cell and tissue rearrangements. Neural progenitors are shown in light blue, the surrounding mesoderm is in brown and the notochord appears in dark blue. (**c′–d′**) Chick SN starts with the convergence of neuromesodermal progenitors in the centre of the tissue and the formation of a solid medullary cord (gray arrows). (**e′**) Cells located dorsally and at the periphery of the medullary cord are the first to undergo the MET. Epithelialization propagates ventrally, although the cells in the center of the tissue remain mesenchymal and small lumens open up between the peripheral epithelial and central mesenchymal cell populations. (**f′**) The small cavities formed coalesce in a dorsoventral gradient to form a single central lumen and the mesenchymal cells that remain in the center are finally cleared from the lumen. (**g′**) The result of this MET and central clearing process is that a hollow neural tube is formed that is surrounded by neuroepithelial cells

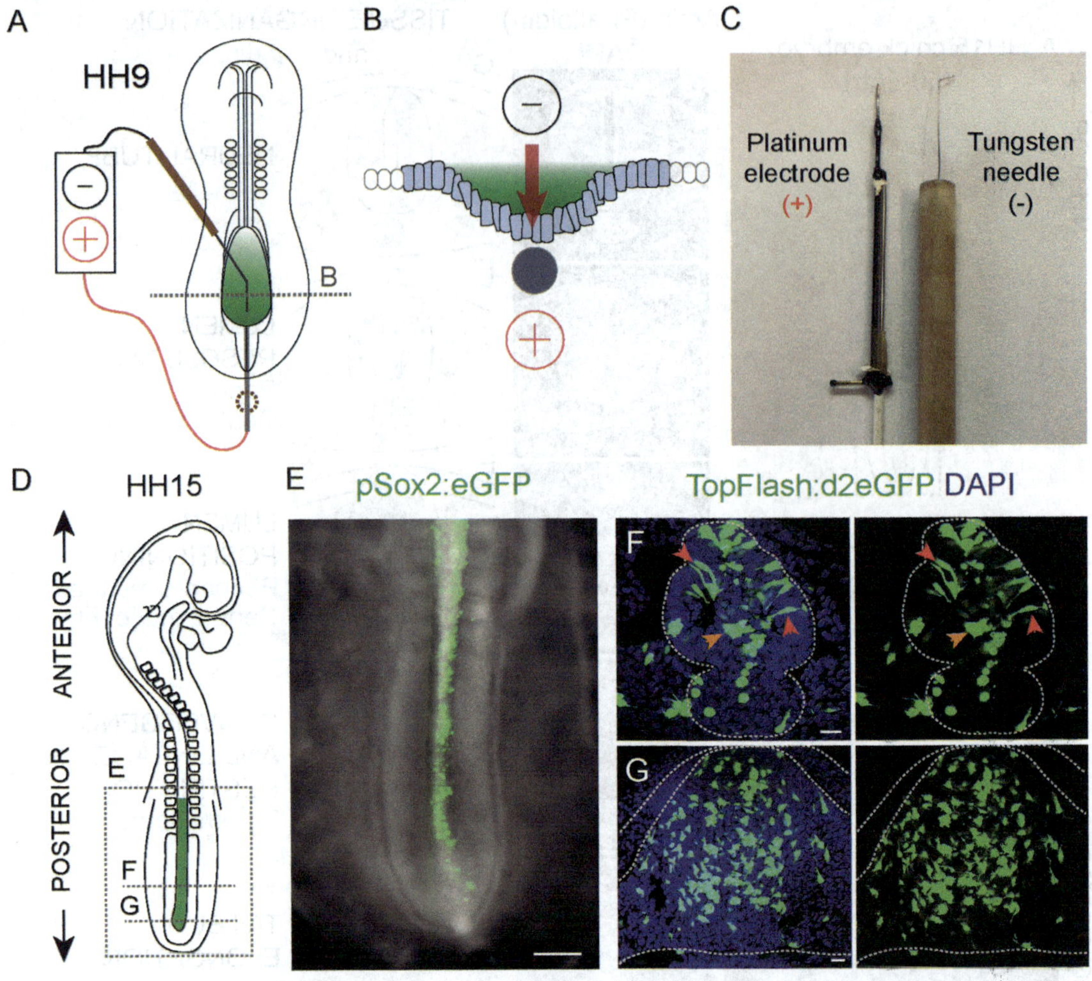

Fig. 2 Chick in ovo electroporation of SN cells. (**a**) The DNA mix (green) is injected into the concave region of the epiblast at the caudal end of a HH9 chick embryo. The positioning of the electrodes is shown, whereby the positive platinum electrode is placed below the embryo through a hole made in the vitelline membrane, while the tungsten needle connected to the negative electrode is positioned on top of the embryo. Both electrodes are positioned parallel to the anteroposterior axis of the embryo. (**b**) Scheme of a transverse section of a chick embryo at the level of the dotted line in (**a**). The epiblast is electroporated by applying the current from top to bottom (red arrow). (**c**) Image of the two electrodes used. (**d**) Schematic representation of a HH15 chick embryo showing the region electroporated using our method 24 h-post-electroporation (hpe). (**e**) Dorsal view of the boxed region in (**d**) following the electroporation of pSox2:eGFP. Both the caudal neural tube and the SN region are efficiently electroporated. Scale bar = 200 μm. (**f**, **g**) Transverse sections at the two levels indicated in (**d**) showing the electroporation of TopFlash:d2eGFP. Panel (**f**) is at the level where the lumen is forming, and where both peripheral epithelial (red arrows) and central mesenchymal cells (orange arrow) are labeled. Panel (**g**) shows the efficient electroporation of the posterior progenitor cells. Scale bar = 40 μm

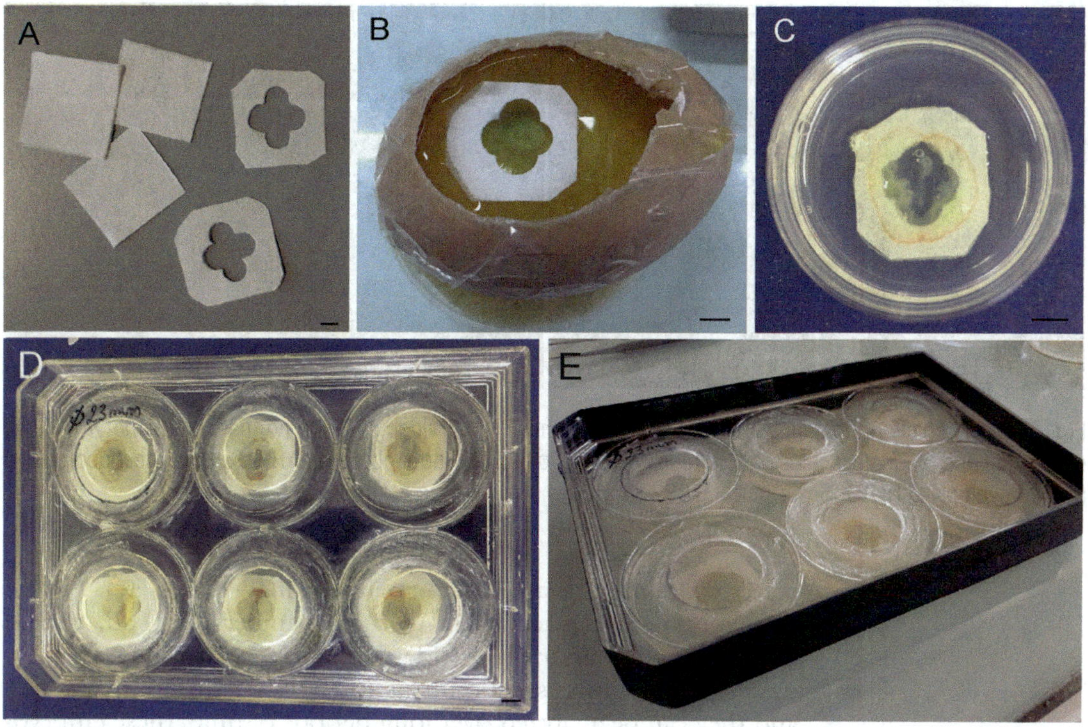

Fig. 3 Embryo culture for in vivo time-lapse imaging. (**a**) Paper rings are prepared by making a cloverleaf hole in the center of 2 × 2 cm squares of Whatman grade 1 filter paper with a paper punch and cutting off the corners. (**b**) The paper ring is laid over the embryo inside the egg, so that the embryo is positioned in the center. (**c**) After carefully cutting out the embryo attached to the paper ring, it is then transferred dorsal side up into an imaging plate containing Albumen/Agar. (**d**) Selected embryos (up to 6) are then transferred to the culture chamber for imaging. (**e**) The culture chamber is finally sealed with insulation tape. Scale bar = 0.5 cm

to avian embryos [14, 30, 31]. This technique allows any cells undergoing SN to be manipulated in vivo in order to investigate the MET process, and it permits cell dynamics to be followed in vivo. Briefly, stage HH9 chick embryos are electroporated in ovo to transform the preSN region, injecting the DNA onto the surface of the concave cavity that exists at the posterior end of the embryo, where the primary NT is still open. The electrodes are then positioned carefully above and below the embryo (Fig. 2a, b), and an electrical current is applied so that the posterior cells incorporate the plasmid DNA. Subsequently, the eggs are sealed and incubated for 24 h, until the embryos reach stage HH15 (Fig. 2d–g). The next day the eggs are opened again and the embryos are removed using a filter paper ring (Fig. 3a, b), which not only facilitates their manipulation but also provides the mechanical support essential to generate the correct tensions and deformation that occur during normal embryo elongation [29]. Embryos attached to these rings are then cleaned to avoid any yolk and debris interfering with their visualization, and they are transferred to pre-prepared imaging

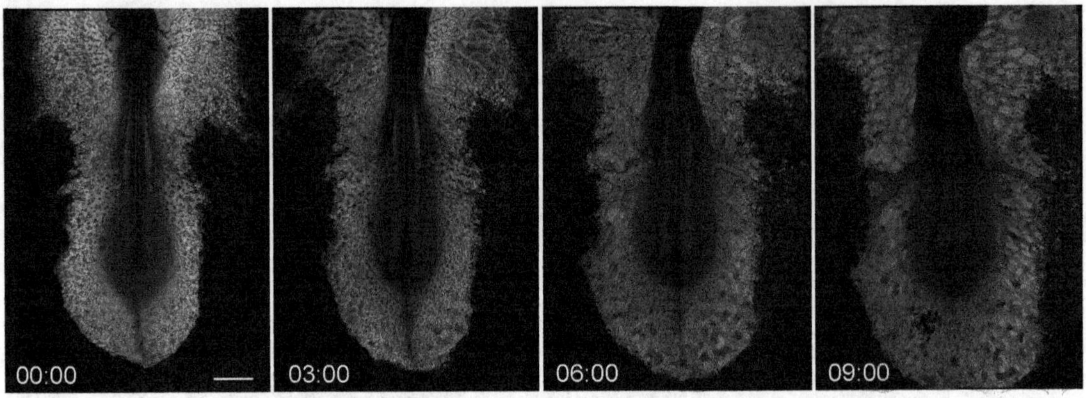

Fig. 4 Normal chick embryo development. Frames from a 5× WT video showing normal chick embryo tail bud elongation and blood vessel formation (hh:mm) (*see* Movie S1). The embryo is at HH13 when the video begins and it develops to HH16. Scale bar = 500 μm

plates containing Agar/Albumen culture media (Fig. 3c). Finally, the embryos are placed in a culture chamber (Fig. 3d, e) and examined under an upright wide-field microscope. The system we describe here allows six specimens to be visualized simultaneously and they can be analyzed over long periods of time, these embryos developing at approximately the same rate as they do *in ovo* (Fig. 4, Movie S1). In the videos generated with this system, the cells undergoing MET can be visualized and tracked during the process of SN (Fig. 5, Movies S2 and S3).

2 Materials

2.1 Electroporation of Chick SN Cells In Ovo

1. Fertilized eggs are obtained from the White-Leghorn chicken strain.
2. Electroporation mixture containing the plasmid DNA. Here we use plasmids containing pSox2:eGFP [32–34] and the Top-Flash:d2eGFP [35, 36] reporters as examples. Before injection, the DNA plasmids are diluted to 2 μg/μL in 60% sucrose prepared in H_2O, adding 1:10 of Fast Green FCF.
3. TSS-20 Ovodyne Electroporator operated by a footswitch or equivalent equipment generating square electrical pulses.
4. Electrodes (Fig. 2c). We separated a pair of commercial platinum electrodes (CUY610P1.5-1, Nepagene or equivalent) and only used one side as the positive electrode. We incorporated a sharpened and bent 90° tungsten needle (Fine Science Tools) into a holder and used it as the negative "microelectrode" [37, 38].

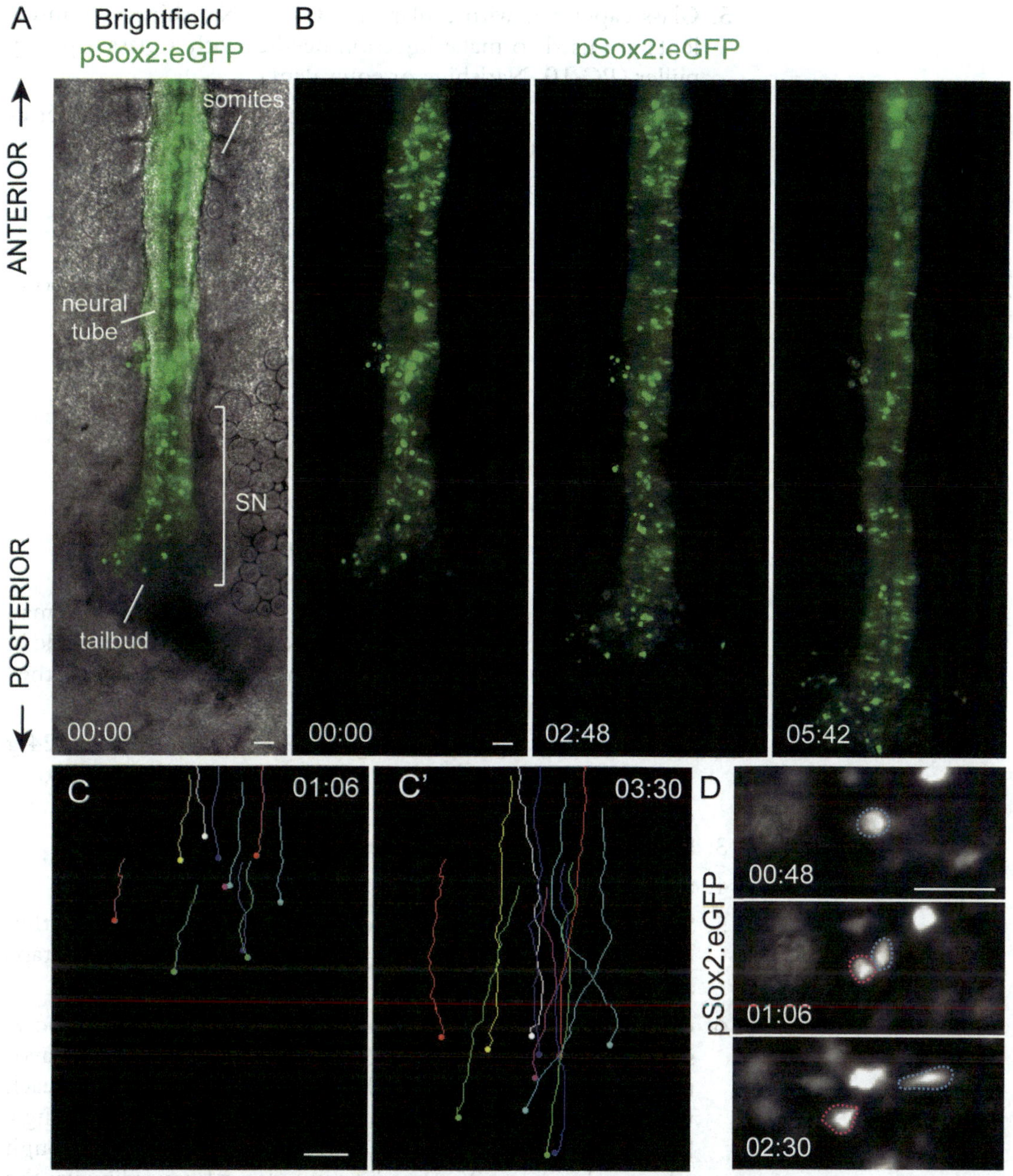

Fig. 5 In vivo analysis of SN cells. (**a**) First frame from a 20× video showing a pSox2:eGFP electroporated embryo (*see* Movie S2). The somites, neural tube, SN region and tail bud are indicated. (**b**) Frames of the same movie as development proceeds, with the tail bud growing caudally and SN advancing. (**c**) Tracks obtained from the embryo in (**a**) and (**b**) using the Manual Tracking plugin of the ImageJ software (*see* Movie S3). (**d**) Frames from the video in (**a**) and (**b**) at a higher magnification. Single cells are easily followed over time. The cells enclosed by dotted lines correspond to the light blue and magenta tracks in (**c**). The blue cell divides and generates two daughter cells. Scale bars = 50 μm

5. Glass capillaries with a filament (GD-1, Narishige or equivalent) are used to make injection needles with a glass capillary puller (PC-10, Narishige or equivalent).
6. Aspirator tube assemblies for calibrated microcapillary pipettes (Sigma-Aldrich, A5177-5EA or equivalent).
7. 1% Penicillin/Streptomycin (P/S) in MilliQ H_2O.
8. Syringe, thin forceps, curved scissors and plastic tape.

2.2 Imaging Plate Preparation

1. Millicell cell culture plate inserts (0.4 mm: Millipore, PICMORG50).
2. 25 mL of thin albumen from eggs incubated overnight.
3. 10% glucose in MilliQ H_2O.
4. 0.6% Granulated Agar (Difco, 0145-17-0) in MilliQ H_2O.
5. 5 M NaCl in MilliQ H_2O.
6. P/S, undiluted.
7. Water bath, 50 mL Falcon tubes and Pasteur pipettes.

2.3 Mounting and In Vivo Time-Lapse Imaging

1. Filter paper rings prepared from 2 × 2 cm squares of Whatman grade 1 filter paper in which a cloverleaf shaped hole is made in the center with a paper punch, cutting the corners so that they fit in the round imaging plates (Fig. 3a).
2. PBS 1× (1 L): 8 g NaCl, 0.2 g KCl, 1.44 g Na_2HPO_4, 0.24 g KH_2PO_4, 800 mL of MilliQ H_2O, adjusted pH to 7.4 and to 1 L, sterilized by autoclaving and stored at room temperature.
3. 5 mL of thin albumen from eggs incubated overnight.
4. 5 mL of 123 mM NaCl in MilliQ H_2O.
5. Soft tissues (Kimtech Science Kimwipes, or equivalent), thin forceps, fine scissors, Pasteur pipettes, electrical insulation tape and 100 mm petri dishes.
6. Culture chamber [14] created from a Corning® Costar® polystyrene 6-well plate (Sigma, CLS3736 or equivalent). To favor the optics, the plastic in the lid is replaced with glass. In each well, a 23 mm hole is made in the centre of the lid by pushing a heated cork borer through the plastic, smoothing the rough edges and sealing a 25-mm-diameter glass #1 coverslip over the hole using Marine Adhesive (Zolux Silicone "SA 500") (Fig. 3d).

2.4 Image Processing and Analysis

1. ZEN software (Zeiss) Version 2.3 blue edition with the experiment designer option or any equivalent option allowing time-lapse imaging with multiple *XYZ* positions.
2. Image J/Fiji software [39, 40].
3. BioFormat plugin [41] (https://imagej.net/Bio-Formats).

4. Image5D (Joachim Walter—https://imagej.nih.gov/ij/plugins/image5d.html) plugins.
5. Stack focuser (Michael Umorin–https://imagej.nih.gov/ij/plugins/stack-focuser.html) plugin.
6. MultiStackReg plugin (Brad Busse and Kota Miura—http://bradbusse.net/downloads.html).
7. StackReg plugin [42] (http://bigwww.epfl.ch/thevenaz/stackreg/).
8. Grid/Collection stitching plugin [43] (https://imagej.net/Image_Stitching#Grid.2FCollection_Stitching).
9. Manual Tracking plugin (Fabrice P. Cordelières—https://imagej.nih.gov/ij/plugins/track/track.html).

3 Methods

3.1 Electroporation of Chick SN Cells *In Ovo*

1. Incubate eggs horizontally at 38.5 °C in an atmosphere of 70% humidity until stage HH9.
2. Remove 5 mL of albumen from the egg with a syringe.
3. Open a window at the top of the shell with curved scissors to visualize the embryo.
4. Make a small hole with thin forceps in the posterior region of the area opaca, just outside of the area pellucida. Avoid touching the embryo (Fig. 2a).
5. Pour 200 mL of 1% P/S onto the embryo to improve electrode conductivity (*see* **Note 1**).
6. Inject the DNA solution onto the epiblast with a glass capillary by blowing air through the aspirator tube. Introduce the DNA into the small concave region at their posterior end of the stage HH9 embryo, where the neural tube is still open (Fig. 2a, b; *see* **Notes 2** and **3**).
7. Carefully insert the platinum electrode connected to the positive lead (+) below the embryo through the hole made previously, parallel to its anteroposterior axis (Fig. 2a, b).
8. Position the tungsten microelectrode connected to the negative lead (−) on top of the embryo, also parallel to the embryo's anteroposterior axis (Fig. 2a, b).
9. Deliver five 50 ms square pulses of 5 V at intervals of 50 ms with the electroporator (*see* **Note 4**).
10. Seal the window in the shell with tape and incubate embryos until they reach stage HH15 (+24 h).

3.2 Imaging Plate Preparation

1. Heat a water bath to 50 °C. Collect 25 mL of thin albumen from the eggs incubated overnight in a sterile 50 mL tube and stir it for 15 min at RT (*see* **Note 5**).
2. Add 0.75 mL of 10% glucose to the albumen and place the mixture in the water bath.
3. Boil 25 mL of 0.6% BactAgar in MilliQ H_2O and add 0.615 mL of 5 M NaCl. Transfer this solution into another sterile 50 mL tube and place it into the water bath to equilibrate to 50 °C.
4. Mix the solutions of Albumen/Glucose and Agar/NaCl in a sterile 50 mL tube and add 100 μL of P/S (*see* **Note 6**).
5. Use a sterile Pasteur pipette to pour 1.5 mL of the Albumen/Agar mix onto the cell culture inserts on a flat surface (*see* **Note 7**), leaving the Albumen/Agar imaging plates at RT to cool until the gel becomes solid and then storing them at 4 °C (*see* **Note 8**).

3.3 Mounting and In Vivo Time-Lapse Imaging

1. Reopen the tape-sealed window in the egg, and carefully remove the thick albumen surrounding and covering the embryo with a soft tissue (*see* **Note 9**).
2. Place a paper ring on top of the vitelline membrane so that the embryo is located in the center of the clover-shaped hole (Fig. 3a, b), and cut through the vitelline membrane and around the whole perimeter of the filter paper ring with a small pair of scissors, carefully pulling the filter with the embryo attached away from the yolk (*see* **Note 10**).
3. Place the embryo ventral side up in a petri dish containing 1× PBS, and clean the remaining yolk and debris by blowing streams of PBS over the embryo with a Pasteur pipette (*see* **Note 11**).
4. Select the embryos with the best overall morphology and the greatest level of transgene expression for imaging (*see* **Note 12**).
5. Transfer the selected embryos to an Agar/Albumen imaging plate dorsal side up (Fig. 3c), and fill each well of the culture chamber with 1.5 mL of a solution of 5 mL thin albumen and 5 mL of 123 mM NaCl.
6. Transfer the embryos in the imaging plates to the wells of the culture chamber (*see* **Note 13**) and add 1× PBS between the wells to maintain a moist environment inside the culture chamber. Seal the culture chamber with electrical insulation tape so that up to six embryos can be imaged at the same time (Fig. 3d, e: *see* **Note 14**).
7. Visualize the embryos under an upright wide-field microscope Axio Imager 2 (Zeiss) equipped with a motorized stage and an

incubation chamber. Set the temperature to 39.5 °C so that the temperature at the level of the embryo is around 37.5 °C (*see* **Note 15**).

8. Use the Experiment designer module of the ZEN software to setup the acquisition parameters, creating one experimental block for each embryo (i.e., one 6-well plate = 6 blocks). Define the acquisition parameters for each block after focusing and positioning the embryo in the centre of the field of view (*see* **Note 16**).
9. Define a last "blank" block situated in the middle of the imaging plate with the objective in a higher position.
10. Create a "delay block" as the final block and define the delay, which will be the time interval in the time-lapse movie. This delay is synchronized with the preceding blocks and it can be used to pause the process in order to correct drift or loss of focus (*see* **Note 17**).
11. Define the number of acquisition loops, which will correspond to the number of time points in the time-lapse video. The motorized stage is now set automatically to move to each previously defined block position, where the given numbers of tiles, channels and slices will be acquired, a process that will be repeated for each loop in every given time interval.
12. With a 5× objective, acquire 10 *z* images every 10 min at a resolution of 1024 × 1024, binning 4 × 4 (Fig. 4, Movie S1). With a 20× objective, acquire 10 *z* images every 6 min at a resolution of 1024 × 1024, binning 4 × 4 (Fig. 5: *see* **Note 18**).

3.4 Image Processing and Analysis

1. Time stitch the images of each embryo acquired with the ZEN software.
2. Open the Tiff file in ImageJ/Fiji using the BioFormat Importer plugin.
3. Convert each tile to the Image5D format (In ImageJ: Plugin tab/Image5D/Stack to Image5D) and apply the Stack focuser plugin, which allows the best focused parts to be selected and projected onto the same plane (*see* **Note 19**). Then convert the best focused projection in the Image5D format to the hyperstack format (In ImageJ: Plugin tab/Image5D/Image5D to stack and then Image tab/Hyperstack/Stack to Hyperstack and redefine the dimensions). Repeat the process for each time point of the time-lapse.
4. Stitch together the best focused tiles using the Grid/Collection stitching plugin in order to reconstruct the whole best focused time-lapse movie.

5. Finally, correct the drift along the time-lapse video using the MultiStackReg plugin that is based on the StackReg plugin (*see* **Note 20**). The best focused time-lapse movie is now aligned (Fig. 5a, b, Movie S2).
6. Individual cells are followed spatiotemporally using the Manual Tracking plugin of the Image J software (Fig. 5c, d, Movie S3).

4 Notes

1. It is best to perform **steps 4** and **5** before placing the electroporation mixture on the epiblast to avoid its diffusion.
2. Repeated injection of DNA through the membranes may block the needle, in which case, clean the tip with a wet tissue or cotton swab.
3. The electroporation mixture can be passed over the top of the epiblast by blowing air through the aspirator tube or with the help of a syringe.
4. Maintain the electrodes moist with 1% of P/S in MilliQ H_2O to diminish the resistance between the electrodes.
5. To prepare the imaging plates, collect the thin albumen at the time of electroporation with a syringe, as both procedures are performed on the same day. Thin albumen can also be collected by cracking an egg incubated overnight into a petri dish and recovering it with a Pasteur pipette.
6. Wait for the two solutions to equilibrate to the same temperature as the hot agar could "cook" the thin albumen if they are mixed too early.
7. Try to fill the culture inserts in a way that the culture medium dries as flat as possible, avoiding the introduction of bubbles.
8. Imaging plates should be prepared fresh, no more than 2 days in advance. The plates can be stored at 4 °C in sterile 35 mm petri dishes.
9. The albumen prevents the membranes from attaching to the filter paper so try to remove as much as possible. However, if the embryo is already dry and it is touched again with the tissue, the blastoderm may attach and break and the embryo will be lost.
10. Pull the filter paper with the embryo away from the yolk obliquely. It is best to pull in the direction of the yolk flow produced by the cutting of membranes or along the anteroposterior embryonic axis.

11. Do not blow streams of PBS directly onto the embryo. Clean it by blowing away from the centre where the embryo is located, toward the sides of the filter paper.
12. Electroporated embryos can be screened and selected in the egg, in PBS or in 35 mm petri dishes embedded in the Agar/Albumen mix. These can be done when preparing the imaging plates.
13. Dip the bottom of the plates in PBS before transferring them to the culture chamber to avoid forming bubbles under the imaging plates.
14. Before sealing the culture chamber and to avoid any condensation during image acquisition, the coverslip should be sprayed with an antifog solution typically used for scuba diving masks.
15. Place the 6-well plate into the incubation chamber of the microscope but do not start the video recording straight away. Leave the embryos to recover for at least 1 h.
16. Define the number of slices, number of tiles, number of channels, laser power and exposure time. Images are acquired in a mosaic format, with different tiles representing each field of view.
17. Frequently check the time-lapse image acquisition during the first hours of imaging as the focus may be lost due to the embryo settling down and due to its elongation. Refocus as often as necessary, during the pausing time.
18. In vivo time-lapse imaging of SN cells can be performed using other systems using some mounting modifications. For example, for confocal imaging we mount the embryos in Lab-Tek® 2-well glass chamber slides.
19. The Stack focuser plugin detects the sharpest details of each image in a *z*-stack.
20. Registration is calculated on the channel displaying the highest contrast and it is then applied to the other channels.

References

1. Gouignard N, Andrieu C, Theveneau E (2018) Neural crest delamination and migration: looking forward to the next 150 years. Genesis 56 (6–7):e23107. https://doi.org/10.1002/dvg.23107
2. Mayor R, Theveneau E (2013) The neural crest. Development 140(11):2247–2251. https://doi.org/10.1242/dev.091751
3. Theveneau E, Mayor R (2012) Neural crest migration: interplay between chemorepellents, chemoattractants, contact inhibition, epithelial-mesenchymal transition, and collective cell migration. Wiley Interdiscip Rev Dev Biol 1(3):435–445. https://doi.org/10.1002/wdev.28
4. Harrington MJ, Hong E, Brewster R (2009) Comparative analysis of neurulation: first impressions do not count. Mol Reprod Dev 76(10):954–965. https://doi.org/10.1002/mrd.21085
5. Lowery LA, Sive H (2004) Strategies of vertebrate neurulation and a re-evaluation of teleost neural tube formation. Mech Dev 121

(10):1189–1197. https://doi.org/10.1016/j.mod.2004.04.022
6. Smith JL, Schoenwolf GC (1987) Cell cycle and neuroepithelial cell shape during bending of the chick neural plate. Anat Rec 218 (2):196–206. https://doi.org/10.1002/ar.1092180215
7. Colas JF, Schoenwolf GC (2001) Towards a cellular and molecular understanding of neurulation. Dev Dyn 221(2):117–145. https://doi.org/10.1002/dvdy.1144
8. Nikolopoulou E, Galea GL, Rolo A, Greene ND, Copp AJ (2017) Neural tube closure: cellular, molecular and biomechanical mechanisms. Development 144(4):552–566. https://doi.org/10.1242/dev.145904
9. Saitsu H, Shiota K (2008) Involvement of the axially condensed tail bud mesenchyme in normal and abnormal human posterior neural tube development. Congenit Anom (Kyoto) 48 (1):1–6. https://doi.org/10.1111/j.1741-4520.2007.00178.x
10. Saitsu H, Yamada S, Uwabe C, Ishibashi M, Shiota K (2004) Development of the posterior neural tube in human embryos. Anat Embryol (Berl) 209(2):107–117. https://doi.org/10.1007/s00429-004-0421-2
11. O'Rahilly R, Muller F (2002) The two sites of fusion of the neural folds and the two neuropores in the human embryo. Teratology 65 (4):162–170. https://doi.org/10.1002/tera.10007
12. O'Rahilly R, Muller F (1994) Neurulation in the normal human embryo. Ciba Found Symp 181:70–82; discussion 82–9
13. Hamburger V, Hamilton HL (1992) A series of normal stages in the development of the chick embryo. 1951. Dev Dyn 195(4):231–272. https://doi.org/10.1002/aja.1001950404
14. Rupp PA, Rongish BJ, Czirok A, Little CD (2003) Culturing of avian embryos for time-lapse imaging. Biotechniques 34(2):274–278. https://doi.org/10.2144/03342st01
15. Dady A, Havis E, Escriou V, Catala M, Duband JL (2014) Junctional neurulation: a unique developmental program shaping a discrete region of the spinal cord highly susceptible to neural tube defects. J Neurosci 34 (39):13208–13221. https://doi.org/10.1523/JNEUROSCI.1850-14.2014
16. Criley BB (1969) Analysis of embryonic sources and mechanims of development of posterior levels of chick neural tubes. J Morphol 128(4):465–501. https://doi.org/10.1002/jmor.1051280406
17. Shum AS, Tang LS, Copp AJ, Roelink H (2010) Lack of motor neuron differentiation is an intrinsic property of the mouse secondary neural tube. Dev Dyn 239(12):3192–3203. https://doi.org/10.1002/dvdy.22457
18. Schoenwolf GC (1984) Histological and ultrastructural studies of secondary neurulation in mouse embryos. Am J Anat 169(4):361–376. https://doi.org/10.1002/aja.1001690402
19. Nievelstein RA, Hartwig NG, Vermeij-Keers C, Valk J (1993) Embryonic development of the mammalian caudal neural tube. Teratology 48 (1):21–31. https://doi.org/10.1002/tera.1420480106
20. Schoenwolf GC, Delongo J (1980) Ultrastructure of secondary neurulation in the chick embryo. Am J Anat 158(1):43–63. https://doi.org/10.1002/aja.1001580106
21. Schoenwolf GC, Kelley RO (1980) Characterization of intercellular junctions in the caudal portion of the developing neural tube of the chick embryo. Am J Anat 158(1):29–41. https://doi.org/10.1002/aja.1001580105
22. Shimokita E, Takahashi Y (2011) Secondary neurulation: fate-mapping and gene manipulation of the neural tube in tail bud. Develop Growth Differ 53(3):401–410. https://doi.org/10.1111/j.1440-169X.2011.01260.x
23. Catala M, Teillet MA, De Robertis EM, Le Douarin ML (1996) A spinal cord fate map in the avian embryo: while regressing, Hensen's node lays down the notochord and floor plate thus joining the spinal cord lateral walls. Development 122(9):2599–2610
24. Catala M, Teillet MA, Le Douarin NM (1995) Organization and development of the tail bud analyzed with the quail-chick chimaera system. Mech Dev 51(1):51–65
25. Le Douarin NM, Teillet MA, Catala M (1998) Neurulation in amniote vertebrates: a novel view deduced from the use of quail-chick chimeras. Int J Dev Biol 42(7):909–916
26. Le Douarin NM (2001) Early neurogenesis in Amniote vertebrates. Int J Dev Biol 45 (1):373–378
27. Voiculescu O, Papanayotou C, Stern CD (2008) Spatially and temporally controlled electroporation of early chick embryos. Nat Protoc 3(3):419–426. https://doi.org/10.1038/nprot.2008.10
28. Hatakeyama J, Shimamura K (2008) Method for electroporation for the early chick embryo. Develop Growth Differ 50(6):449–452. https://doi.org/10.1111/j.1440-169X.2008.01040.x
29. Chapman SC, Collignon J, Schoenwolf GC, Lumsden A (2001) Improved method for chick whole-embryo culture using a filter paper carrier. Dev Dyn 220(3):284–289.

https://doi.org/10.1002/1097-0177(20010301)220:3<284::AID-DVDY1102>3.0.CO;2-5

30. Benazeraf B, Beaupeux M, Tchernookov M, Wallingford A, Salisbury T, Shirtz A, Shirtz A, Huss D, Pourquie O, Francois P, Lansford R (2017) Multi-scale quantification of tissue behavior during amniote embryo axis elongation. Development 144(23):4462–4472. https://doi.org/10.1242/dev.150557
31. Benazeraf B, Francois P, Baker RE, Denans N, Little CD, Pourquie O (2010) A random cell motility gradient downstream of FGF controls elongation of an amniote embryo. Nature 466 (7303):248–252. https://doi.org/10.1038/nature09151
32. Saade M, Gutierrez-Vallejo I, Le Dreau G, Rabadan MA, Miguez DG, Buceta J, Marti E (2013) Sonic hedgehog signaling switches the mode of division in the developing nervous system. Cell Rep 4(3):492–503. https://doi.org/10.1016/j.celrep.2013.06.038
33. Uchikawa M, Ishida Y, Takemoto T, Kamachi Y, Kondoh H (2003) Functional analysis of chicken Sox2 enhancers highlights an array of diverse regulatory elements that are conserved in mammals. Dev Cell 4 (4):509–519
34. Le Dreau G, Saade M, Gutierrez-Vallejo I, Marti E (2014) The strength of SMAD1/5 activity determines the mode of stem cell division in the developing spinal cord. J Cell Biol 204(4):591–605. https://doi.org/10.1083/jcb.201307031
35. Rios AC, Denans N, Marcelle C (2010) Real-time observation of Wnt beta-catenin signaling in the chick embryo. Dev Dyn 239 (1):346–353. https://doi.org/10.1002/dvdy.22174
36. Serralbo O, Marcelle C (2014) Migrating cells mediate long-range WNT signaling. Development 141(10):2057–2063. https://doi.org/10.1242/dev.107656
37. Momose T, Tonegawa A, Takeuchi J, Ogawa H, Umesono K, Yasuda K (1999) Efficient targeting of gene expression in chick embryos by microelectroporation. Develop Growth Differ 41(3):335–344
38. Yasuda K, Momose T, Takahashi Y (2000) Applications of microelectroporation for studies of chick embryogenesis. Develop Growth Differ 42(3):203–206
39. Rueden CT, Schindelin J, Hiner MC, DeZonia BE, Walter AE, Arena ET, Eliceiri KW (2017) ImageJ2: ImageJ for the next generation of scientific image data. BMC Bioinformatics 18 (1):529. https://doi.org/10.1186/s12859-017-1934-z
40. Schindelin J, Arganda-Carreras I, Frise E, Kaynig V, Longair M, Pietzsch T, Preibisch S, Rueden C, Saalfeld S, Schmid B, Tinevez JY, White DJ, Hartenstein V, Eliceiri K, Tomancak P, Cardona A (2012) Fiji: an open-source platform for biological-image analysis. Nat Methods 9(7):676–682. https://doi.org/10.1038/nmeth.2019
41. Linkert M, Rueden CT, Allan C, Burel JM, Moore W, Patterson A, Loranger B, Moore J, Neves C, Macdonald D, Tarkowska A, Sticco C, Hill E, Rossner M, Eliceiri KW, Swedlow JR (2010) Metadata matters: access to image data in the real world. J Cell Biol 189 (5):777–782. https://doi.org/10.1083/jcb.201004104
42. Thevenaz P, Ruttimann UE, Unser M (1998) A pyramid approach to subpixel registration based on intensity. IEEE Trans Image Process 7(1):27–41. https://doi.org/10.1109/83.650848
43. Preibisch S, Saalfeld S, Tomancak P (2009) Globally optimal stitching of tiled 3D microscopic image acquisitions. Bioinformatics 25 (11):1463–1465. https://doi.org/10.1093/bioinformatics/btp184

Chapter 17

Multiscale In Vivo Imaging of Collective Cell Migration in *Drosophila* Embryos

Gordana Scepanovic, Alexandru Florea, and Rodrigo Fernandez-Gonzalez

Abstract

Coordinated cell movements drive embryonic development and tissue repair, and can also spread disease. Time-lapse microscopy is an integral part in the study of the cell biology of collective cell movements. Advances in imaging techniques enable monitoring dynamic cellular and molecular events in real time within living animals. Here, we demonstrate the use of spinning disk confocal microscopy to investigate coordinated cell movements and epithelial-to-mesenchymal-like transitions during embryonic wound closure in *Drosophila*. We describe image-based metrics to quantify the efficiency of collective cell migration. Finally, we show the application of super-resolution radial fluctuation microscopy to obtain multidimensional, super-resolution images of protrusive activity in collectively moving cells in vivo. Together, the methods presented here constitute a toolkit for the modern analysis of collective cell migration in living animals.

Key words *Drosophila melanogaster*, Live imaging, Mounting methods, Quantitative microscopy, Spinning disk confocal microscopy, Super-resolution radial fluctuation microscopy, Wound healing

1 Introduction

Collective cell movements are essential for embryonic development. Morphogenetic processes such as gastrulation [1] and limb elongation [2] are mediated by coordinated cell movements and failure of cells to organize cell migration can lead to congenital birth defects. Coordinated cell movements also contribute to the spread of disease during cancer invasion [3]. Therefore, understanding the mechanisms by which cells coordinate their movements may facilitate the development of therapeutic interventions that promote or prevent collective cell migration.

Electronic supplementary material: The online version of this chapter (https://doi.org/10.1007/978-1-0716-0779-4_17) contains supplementary material, which is available to authorized users.

Kyra Campbell and Eric Theveneau (eds.), *The Epithelial-to Mesenchymal Transition: Methods and Protocols*, Methods in Molecular Biology, vol. 2179, https://doi.org/10.1007/978-1-0716-0779-4_17,

Embryonic wound repair provides an excellent model to investigate collective cell movements. Embryonic wound healing is driven by the coordinated migration of the cells around the wound into the damaged region, in a process conserved across species, from fruit flies to mice [4–8]. Wound healing in embryos does not involve cell division, and occurs in the absence of an inflammatory response. Thus, the study of embryonic wound repair allows researchers to specifically address and understand the mechanisms that drive coordinated cell movements. Embryonic wound closure can be imaged live in several organisms, provides an easy way to register experiments based on the time of wounding, and occurs rapidly (a few minutes to 1–2 h). Together, all these features facilitate the use of quantitative methods to analyze embryonic wound healing and increase the statistical power of the analyses.

The collective cell movements that drive wound re-epithelization are associated with a partial epithelial-to-mesenchymal transition (EMT). EMT is characterized by changes in molecular composition and cell behavior that are also present during embryonic wound repair. For example, in classical models of EMT, such as gastrulation in mouse embryos, or cancer cell invasion, E-cadherin-based adherens junctions are dramatically reorganized to facilitate cell movements [9, 10]. Similar junctional rearrangements also occur during embryonic wound closure [11–13]. EMT events are often partial, with cells retaining some degree of collectiveness. Thus, groups of cells undergoing EMT can establish a front-back axis of polarity within the group. The cells at the front exhibit increased protrusive activity in comparison to "follower cells" [14]. The front-to-back axis of polarity is also found in embryonic wound healing, with cells at the wound edge displaying an increased number of actin-based protrusions [15, 16]. Collective invasion by squamous cell carcinoma cells is associated with the assembly of a cable formed by actin and the molecular motor non-muscle myosin II around the group of invading cells [17]. Similarly, actin and myosin become polarized in the cells at the wound edge and accumulate at the interface with the wounded cells [15, 18], assembling an actomyosin cable that coordinates cell movements to drive rapid wound closure [15, 19, 20]. Thus, elucidating the mechanisms of coordinated cell migration during wound closure will increase our understanding of how EMT and collective cell movements contribute to embryonic development and the spread of disease in vivo.

Advances in microscopy have pushed the limits of our understanding of coordinated cell movements to shorter time and length scales. Spinning disk confocal microscopy, for example, overcomes the speed limitations of laser scanning confocal microscopy by illuminating multiple points in the field of view simultaneously [21]. Photons emitted by the sample can be collected more effectively with charge-coupled device cameras than by photomultiplier tubes. With greater detection sensitivity, exposure time can be reduced without affecting the signal to noise ratio. Therefore, spinning disk confocal microscopy allows for faster imaging for longer periods of time and with lower levels of photobleaching. A disadvantage of spinning disk microscopy is the pinhole crosstalk effect, which can affect the axial resolution of the image. A limited axial resolution is problematic to image developmental processes deep within the specimen. To overcome this and other limitations of spinning disk confocal microscopy, including the potential for photobleaching or phototoxicity, other techniques such as light-sheet microscopy have been developed [22, 23].

Super-resolution microscopy enables the visualization of small molecular complexes and single fluorophores, surpassing the diffraction limit. However, the application of super-resolution microscopy often requires special, expensive equipment; and results in slow image acquisition and/or increased phototoxicity, making super-resolution microscopy often not amenable for application in vivo. A novel approach, super-resolution radial fluctuation microscopy (SRRF) was recently developed [24], providing a balance between resolution, cost and speed of image acquisition. SRRF localizes individual fluorophores based on the radial symmetry and persistence in time of their emission (in contrast with noise, which is neither symmetric nor temporally correlated). Using this approach, SRRF can achieve lateral resolutions of 80–100 nm using conventional fluorophores, with temporal resolutions of only a few seconds. Imaging with SRRF does not require special optics or high-power laser excitation. Therefore, SRRF can be used to conduct super-resolution imaging of dynamic molecular events in living cells [25].

In this chapter, we describe methods to image collective cell movements with high spatial and temporal resolution in the context of developing *Drosophila* embryos. The *Drosophila* embryo is amenable to genetic, pharmacological and biophysical manipulations, thus rendering it an excellent system to probe the mechanisms of collective cell migration. Furthermore, transgenic fluorescent lines useful to study molecular dynamics during coordinated cell movements are freely available (*see* Table 1 for examples). In this chapter, we illustrate how to prepare embryos for time-lapse spinning disk confocal microscopy. We introduce methods to wound the embryonic epidermis, and we demonstrate how to use quantitative metrics to measure the efficiency of collective cell movements. Finally, we demonstrate how to apply SRRF to image protrusive activity in cells that move in a coordinated manner. Overall, the methods presented here enable multi-scale, in vivo imaging of cellular and molecular dynamics during collective cell migration.

Table 1
Some publicly available fluorescent lines used for live imaging of cellular and molecular dynamics of collective cell movements in *Drosophila* embryos

Line	Bloomington Stock Center stock number	Protein	Reference
endo-shg:tdTomato	58789	E-cadherin	[40]
endo-shg:GFP	60584	E-cadherin	[40]
UASp-shg:GFP	58445	E-cadherin	Donated by Pernille Rorth
sqh-sqh:GFP	57145	Regulatory light chain of the non-muscle myosin II	[41]
sqh-sqh:mCherry	59024	Regulatory light chain of the non-muscle myosin II	[42]
sqh-GFP:MoesinABD	59023	Moesin actin binding domain (a filamentous actin reporter)	[43]
rho1-GFP:rho1	9527	Rho1	[44]
arm-arm:GFP	8555	β-catenin	Derivative of the *arm* minigene described in [45]
UAS-nls:GFP	4776	Nuclear localization signal	[46]
UAS-mCherry:mito	66533	Miro transmembrane domain (labels the outer mitochondrial membrane)	[47]

2 Materials

2.1 Embryo Staging and Collection

1. Stereomicroscope.
2. Embryo collection cages (VWR, #25384-152 or equivalent). A fine pin can be used to drill ventilation holes around the cage (Fig. 1a, left). A commercial solution with a fine mesh for ventilation at the top of the cage is also available (Diamed, #GEN59-100 or equivalent; Fig. 1a, right).
3. 60 mm Petri dishes for embryo collection (Fig. 1b).
4. Medium to fill embryo collection plates (makes approximately 100 plates) (*see* **Note 1**): 20.25 g agar powder, 675 mL tap water, 22.5 g sucrose, 1.35 g nipagin, and 225 mL apple juice.
5. Yeast and tap water.
6. A spatula to mix the yeast paste.
7. 50 mL tubes.

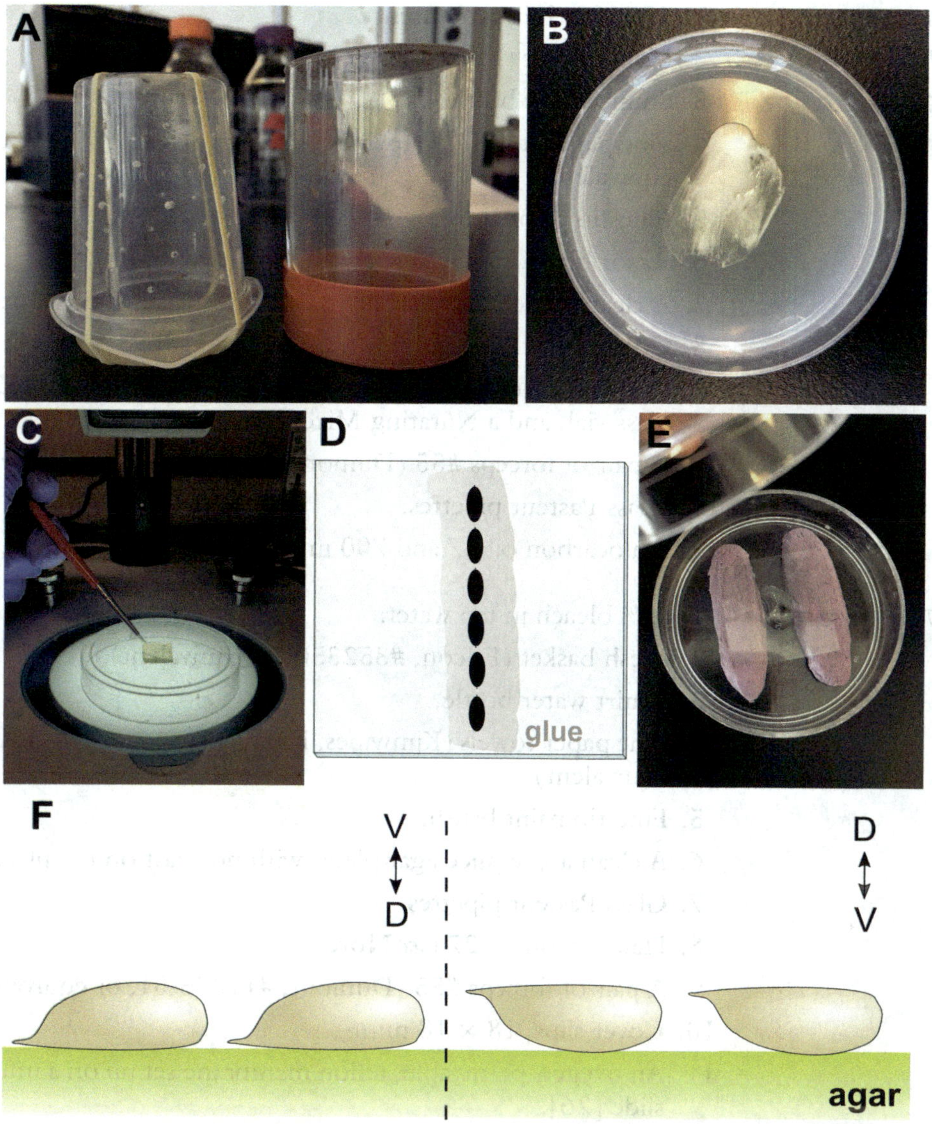

Fig. 1 Collecting and mounting embryos for live imaging. (**a**) Embryo collection cages with ventilation holes manually drilled and rubber bands holding the agar plate in place (left), or with a wire mesh at the top and a plastic insert holding the plate (right). (**b**) Yeasted agar plate for embryo collection. (**c**) Embryos transferred onto a block of agar using a paint brush. (**d**) Schematic showing aligned embryos glued onto a coverslip for imaging. (**e**) Plasticine blocks supporting a coverslip with glued embryos covered in oil. (**f**) Cartoon showing embryos lined up on an agar block with their ventral side up to image the ventral side of the animal (left), or with their ventral side down to image the dorsal side (right)

2.2 Mounting Embryos

2.2.1 Glue Method

1. 50% bleach in tap water.
2. Mesh basket (Falcon, #352350, or equivalent).
3. Squirt water bottle.
4. Fine paper towels (Kimwipes, Kimtech Science, #KC34120, or equivalent).
5. Fine-tip paint brush.
6. Razor blade.
7. A clean apple juice agar plate (with no yeast on the plate).
8. Cover slips, 18 × 18 mm.
9. Heptane glue (*see* **Note 2**): heptane, double-sided tape, 20 mL glass vial, and a Nutating Mixer.
10. A pair of forceps #55 (Dumont, #11295-51 or equivalent).
11. Glass Pasteur pipettes.
12. Halocarbon oil 27 and 700 mixed 1:1 in a 50 mL tube.

2.2.2 Teflon Membrane Method

1. 50% bleach in tap water.
2. Mesh basket (Falcon, #352350, or equivalent).
3. Squirt water bottle.
4. Fine paper towels (Kimwipes, Kimtech Science, #KC34120, or equivalent).
5. Fine-tip paint brush.
6. A clean apple juice agar plate (with no yeast on the plate).
7. Glass Pasteur pipettes.
8. Halocarbon oil 27 (*see* **Note 3**).
9. A pair of forceps #55 (Dumont, #11295-51, or equivalent).
10. Cover slips, 18 × 18 mm.
11. An oxygen-permeable, teflon membrane set up on a mounting slide [26].

2.3 Microinjection

1. 100 mm Petri dishes for embryo storage (Fig. 2a).
2. Small container filled with silica beads, covered with a small cardboard platform (Fig. 2b).
3. Glass Pasteur pipettes, 150 mm.
4. Halocarbon oil 27 and 700 mixed 1:1 in a 50 mL tube.
5. 4-in., thin wall glass needle (WPI, #08E, or equivalent).
6. Needle puller (e.g., Sutter Instrument, Flaming/Brown Micropipette Puller, #P-97 or equivalent).
7. 0.5–10 μL pipette.
8. Microloader tips.

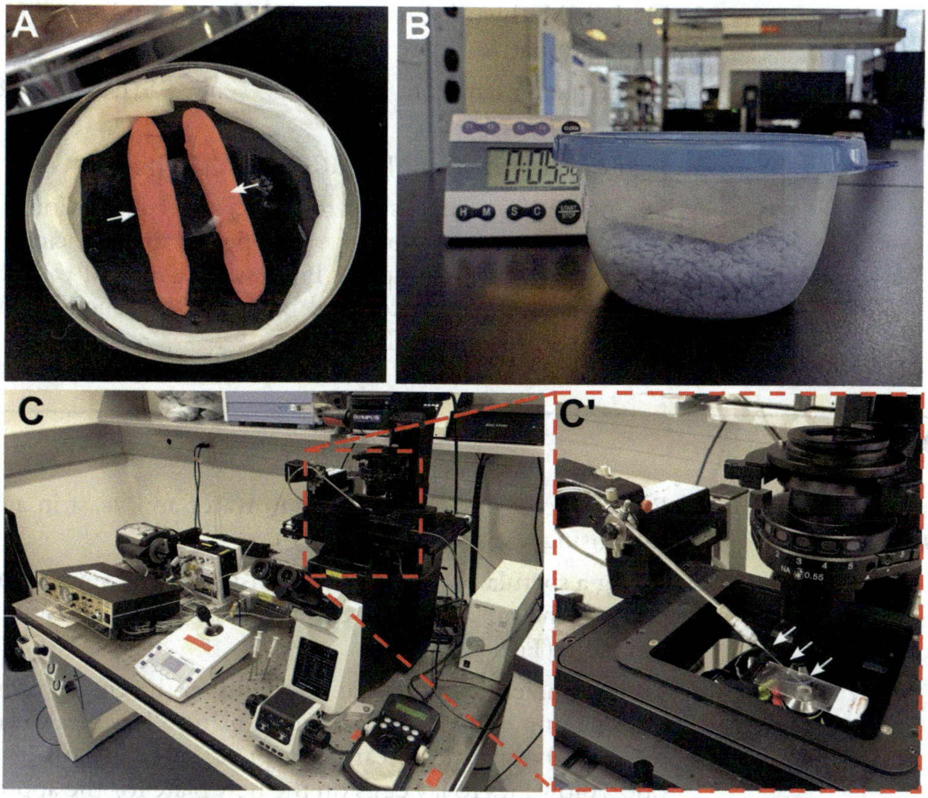

Fig. 2 Setup for embryo microinjection. (**a**) Embryo incubation chamber to age embryos glued onto a coverslip. Arrows indicate the position of the coverslip. Pink plasticine support posts can be seen below. (**b**) Tupperware container filled with silica beads for embryo dehydration. (**c**) Microinjection system coupled to a confocal microscope (**c**) and a zoom-in showing a microneedle inserted into the micromanipulator (arrows) and resting on a needle-breaking slide (**c′**)

9. A slide to break the tip of microneedles (*see* **Note 4**).
10. Micromanipulator.
11. Pneumatic pump.
12. Inverted microscope with transmitted light illumination and a 10× or 20× air objective, coupled to the micromanipulator and the microinjector (Fig. 2c).

2.4 Time-Lapse Imaging of Collective Cell Movements

1. Spinning disk confocal microscope (*see* **Note 5**) equipped with a 10× air lens to find the embryos, a 60× or 100× oil-immersion lens, and a high-power laser (e.g., Andor Technology Micropoint or equivalent) for tissue ablation. Laser ablation is often conducted at 365 nm, but other wavelengths—e.g., 405 and 440 nm—are also effective for ablation.
2. Metamorph software (Molecular Devices) or an alternative image acquisition software, such as Micromanager [27].

3. Matlab (Mathworks) using the DIPImage toolbox (TU Delft) or an alternative image analysis software.
4. Image segmentation software (e.g., SIESTA [28] or equivalent).
5. Fiji [29] and the NanoJ-SRRF plugin [24].
6. For processing and visualization of super-resolution microscopy images, a computer equipped with sufficient RAM (at least 32 GB). A Graphics Processing Unit (GPU) is recommended, as it will accelerate image reconstruction through parallel processing (*see* **Note 6**).

3 Methods

3.1 Embryo Staging and Collection

1. Prepare yeast paste by mixing tap water and yeast in a 50 mL tube using a spatula (*see* **Note** 7).
2. Using a spatula or your finger, transfer a small amount of yeast paste to the surface of the embryo collection plates and spread the paste (Fig. 1b).
3. To maximize the number and viability of the embryos used for live imaging, replace the collection plate, allow flies to lay eggs for 30–60 min, and discard that plate, replacing it with a new one. Allow flies to lay eggs on the new plate for the appropriate time (*see* **Note 8**).

3.2 Mounting Embryos

For live imaging, embryos can be glued to a coverslip, or immobilized between a coverslip and an oxygen-permeable teflon membrane. The glue method facilitates proper embryo orientation and minimizes embryo compression, while maintaining embryos accessible for additional manipulations such as microinjection. However, the glue method is technically hard for beginners, and it limits the number of embryos that can be mounted per coverslip, as embryos are susceptible to dehydration during the alignment process. Finally, the glue method reduces the size of the field of cells that can be imaged, due to the curvature of the embryonic surface. In contrast, the membrane mounting method is easy to use, provides lots of embryos ready for live imaging, and a large field of cells for imaging, as the surface of the embryo is flattened during the mounting procedure. Unfortunately, membrane mounting makes it difficult to orient the embryos, microinjections are not feasible with this method, and embryo compression can cause changes in gene expression [30], and affect the dynamics of morphogenetic processes [16], so caution should be applied when choosing the membrane method.

3.2.1 Glue Method

1. Using a razor blade, cut out a rectangular piece of agar from a clean apple juice agar plate. Leave the agar block on the covered plate so that it does not dry out.
2. Using a glass pipette, place a drop of heptane glue on a coverslip. Tilt the coverslip so that the glue forms a thin layer along a stripe on the surface of the coverslip. Store the coverslip glue-side up on a Petri dish so that dirt does not stick to the glue (*see* **Note 9**).
3. Retrieve the plate containing the embryos from the collection cage and replace it with a new plate.
4. Pour 50% bleach in the plate containing the embryos and incubate for 90–120 s. During the incubation, gently shake the plate with circular movements to dislodge embryos from the yeast and agar.
5. Pour the bleach and embryos into a wire mesh basket.
6. Thoroughly wash embryos in the mesh basket with tap water using a squirt bottle. Continue to wash embryos for 1–2 min. Dab the bottom and sides of the basket with a fine paper towel once during the process to ensure maximum bleach removal.
7. Using a clean, fine paper towel, dab the bottom and sides of the basket to dry out any remaining bleach and water.
8. Place the agar block cut out in **step 1** on a Petri dish cover under the stereoscope (Fig. 1c).
9. Using a paintbrush, gently transfer embryos from the mesh basket onto the agar block (Fig. 1c).
10. Working under the stereoscope, line up and orient the embryos using a pair of forceps (*see* **Notes 10** and **11**). Epi-illumination makes it easier to visualize embryonic morphology.
11. Transfer embryos onto the coverslip by gently pressing the glue-painted coverslip surface against the agar block. For embryo injection proceed to the next section without covering the embryos in oil. If ready to image, use a glass pipette to cover the embryos with three drops of halocarbon oil mixture, tilting the coverslip as necessary to facilitate the spread of oil (Fig. 1d, e).

3.2.2 Teflon Membrane Method

1. Retrieve the plate containing the embryos from the collection cage and replace it with a new plate.
2. Pour 50% bleach in the plate containing the embryos and incubate for 90–120 s. During the incubation, gently shake the plate with circular movements to dislodge embryos from the yeast and agar.
3. Pour the bleach and embryos into a wire mesh basket.

4. Thoroughly wash embryos in the mesh basket with tap water using a squirt bottle. Continue to wash embryos for 1–2 min. Dab the bottom and sides of the basket with a fine paper towel once during the process to ensure maximum bleach removal.
5. Using a clean fine paper towel, dab the bottom and sides of the basket to dry out any remaining bleach and water.
6. Using a paintbrush, transfer embryos from the basket on to the surface of the clean apple juice agar plate.
7. Using a glass pipette, cover embryos in halocarbon oil.
8. Add three drops of halocarbon oil on a coverslip using a glass pipette.
9. Working under the stereoscope, use forceps (or a paintbrush) to transfer embryos from the agar plate and into the oil drop on the surface of the coverslip. Epi-illumination makes it easier to visualize embryonic morphology.
10. Place the oxygen-permeable membrane on the coverslip, taking care to minimize embryo compression and bursting (*see* **Note 12**).

3.3 Microinjection

1. Begin with embryos glued to a coverslip, but not covered in oil.
2. Dehydrate embryos by placing the coverslip with the glued embryos in a sealed container full of silica beads for 5–15 min (*see* **Note 13**).
3. Remove embryos from the container and cover them immediately with three drops of halocarbon oil mixture (*see* **Note 14**), tilting the coverslip as necessary to spread of the oil.
4. Prepare a pulled microneedle (*see* **Note 15**) by loading 1–2 μL of the substance to be injected (typically small molecule inhibitors, dyes, or dsRNA) into the intact needle.
5. Place the slide to break the microneedle on the microscope stage, coverslip-side up. Add a drop of halocarbon oil on the edge of the coverslip.
6. Carefully load the needle onto the pneumatic pump nozzle, mounted on the micromanipulator (Fig. 2c′).
7. Using the micromanipulator, carefully lower the needle toward the slide until the tip of the needle is inside the drop of oil at the edge of the coverslip. Focus the microscope on the needle tip. Adjust the position of the needle using the micromanipulator until both the needle tip and the edge of the coverslip are on the same plane.
8. Try to eject liquid with the pump to make sure that the needle is not broken. No bubbles should come out of the tip of the needle.

9. Using the micromanipulator, gently move the needle tip laterally, and touch the side of the coverslip (Movie 1). Minimal contact between the needle tip and the coverslip, while ejecting liquid from the needle, will suffice to break the needle tip. Move the needle back and ensure that it was broken by trying to eject some solution. A bubble (not a stream of solution) is indicative of a successful needle break. From this point on, minimize the time that the tip of the needle is not immersed in oil to prevent clogging.
10. To prepare for embryo injection, slowly move the needle upwards and replace the needle-breaking slide with the coverslip with glued embryos.
11. Lower the needle slowly until the tip is inside the oil that covers the embryos. Focus the microscope on the needle tip. Make sure that the embryo cross-section and the needle tip are in focus on the same plane by adjusting the position of the needle.
12. Gently push the needle against the vitelline membrane (Movie 2). The vitelline membrane should display some light resistance and will eventually be pierced by the pressure from the needle. For injections in the perivitelline space (e.g., membrane-permeable dyes or small molecule inhibitors), slowly pull back the needle (Movie 2). The vitelline membrane will retract with the needle, forming a pocket where solution can be ejected. Membrane-impermeable molecules can be injected in the yolk before cellularization is complete. Embryos can then be aged until the appropriate stage (*see* **Note 16**). If the embryo bursts upon injection, reduce the ejection pressure of the pneumatic pump, use a needle with a thinner tip, or increase the embryo dehydration time.
13. If embryos need to be moved to an incubation chamber for aging or to a different microscope for imaging, make sure to place the slide used to break the needle back on the stage and lower the needle tip into the oil to prevent needle clogging.
14. If injections were conducted on the same microscope in which imaging will occur, leave the needle tip inside the oil that covers the embryos to prevent needle clogging.

3.4 Time-Lapse Imaging of Collective Cell Movements

3.4.1 Live Imaging Using Spinning Disk Confocal Microscopy

1. Startup Metamorph.
2. If using the Micropoint for laser ablation, make sure that the laser is calibrated with the objective lens required for ablation (*see* **Note 17**).
3. Place the coverslip containing the embryos on the microscope stage, and find an embryo using the 10 or 20× objective and bright field illumination.

4. Switch to a higher magnification objective. For oil-immersion lenses only, remember to add a drop of oil on the lens surface.
5. Focus on the embryo using the eyepieces and bright field illumination (*see* **Note 18**).
6. Use the live imaging mode in the image acquisition software with the appropriate laser illumination to visualize fluorescent proteins and focus on the correct plane (*see* **Notes 19** and **20**).
7. Adjust the laser power and exposure time for optimal imaging (Fig. 3, *see* **Notes 21** and **22**).
8. To acquire a movie, select the range of depths in which images must be acquired, the spacing of slices within that range, and the temporal resolution of image acquisition (Movie 3, *see* **Note 23**).
9. Depending on the microscope set up, multiple stage positions can be acquired for imaging larger samples or multiple samples in parallel.

3.4.2 Wound Healing Assays

For image acquisition:

1. Focus the objective on the plane in which ablation should occur.
2. Select the number of pulses to use for ablation (*see* **Note 24**). Test the laser to make sure that it can sever the tissue of interest (Movie 4, *see* **Note 25**).
3. Find a new embryo and acquire at least two-to-four images of the cells that will be wounded to be able to reliably calculate their area before wounding.
4. Fire the laser on one or more spots (*see* **Note 26**).
5. Continue imaging and monitoring the progress of wound closure until the wound is closed (Movie 3).

For image analysis:

5. Using ImageJ (or other software) obtain maximum intensity projections for each of the stacks by opening you image and selecting "Image > Stacks > Z-project . . .". Concatenate each of the resulting images into a time-sequence of the wound healing response by using "Image > Stacks > Z-project > Tools > Concatenate . . .".
6. Segment the wound margin, for instance, using the semi-automated LiveWire algorithm in SIESTA [28] based on Dijkstra's optimal path search [31] (under "Annotations > LiveWire") (Fig. 4a). SIESTA also enables automated delineation of the wound edge using active contours, a region growing algorithm [32] (under Annotations > MEDUSA). Equivalent segmentation algorithms can be found in other software tools.

7. The segmentation results can be used to quantify the cellular and molecular dynamics of cell movement during wound repair, for instance by measuring the area of the wound over time (Fig. 4b), the rate of wound closure (Fig. 4c, *see* **Note 27**), or fluorescence dynamics at the leading edge of the advancing cells (Fig. 4d, *see* **Notes 28** and **29**).

3.4.3 Super-Resolution Radial Fluctuation Microscopy of Living Embryos

For image acquisition:

1. Prepare embryos as indicated in Subheading 3.4.1.
2. Initialize Micromanager 1.4, nightly build November 20, 2018 or later.
3. Select camera configuration and illumination settings in the "Configuration Settings" panel in the top right corner of the main Micromanager window. We recommend having a SRRF-specific "Group" or "Preset" to ensure that the appropriate camera and illumination settings are used consistently.
4. Check that the camera parameters in the "Device Property Browser" ("Tools > Device Property Browser") are set to optimize the speed of image acquisition (Table 2).
5. Open the SRRF acquisition script (https://github.com/RFGLabAtUofT/SRRF) in the script panel of Micromanager ("Tools > Script Panel").
6. Edit the parameter values for time-lapse, 3D SRRF image acquisition: image name, root directory for image storage, number of time points to be acquired, number of slices per time point, *Z* step size, time interval, number of images per SRRF slice, and exposure time (*see* **Note 30**).

Table 2
Camera settings in Micromanager for an iXon Ultra 897 to enable rapid image acquisition. Settings may differ for different camera types

Property name	Description	Value
Frame transfer	Allows the camera to initiate the acquisition of a new image while the previous one is being read out	On
Vertical clock voltage	The voltage of the clock pulses required to shift rows of pixels toward the readout register. A greater voltage results in faster readout, but introduces noise	+4
Vertical speed	The speed in microseconds at which rows of pixels move toward the readout register	0.30
Trigger	Internal triggering ensures that the camera does not wait for a signal from the image acquisition software before acquiring the next image, but rather begins as soon as it is ready (as determined by the parameters above)	Internal

7. Save the edited script.
8. Manually adjust the focal plane to the bottom of the stack.
9. From the "Script" panel in Micromanager, select "Run" (*see* **Note 31**).

For image reconstruction:

10. SRRF sequences can be processed using the Fiji plugin developed by Ricardo Henriques' lab (https://bitbucket.org/rhenriqueslab/nanoj-srrf/wiki/Home) [24]. The plugin can be used to convert a raw image sequence into a single SRRF image (Fig. 5a, b), or to process a group of SRRF sequences. To do this, open the raw data and initialize the plug-in by selecting "Plugins > NanoJ-SRRF > SRRF Analysis."
11. Select the desired SRRF reconstruction settings (*see* **Notes 32** and **33**) and click "OK" to run the script.
12. We developed a Fiji macro to reconstruct multidimensional SRRF images (time-lapse, *Z*-slices) automatically. To use it, open the macro "SRRF_Processing_Timelapse MM1.4_20190711.ijm" (https://github.com/RFGLabAtUofT/SRRF) in Fiji from the "Plugins > Macros > Edit" option.
13. Edit the values for the following parameters: number of slices per stack; number of images per SRRF reconstruction; root directory containing the raw data (which will contain folders *time0–timeN*, corresponding to time points 0 to *N*); output directory to save the SRRF image reconstructions; output directory to save average intensity projection images (for comparison with SRRF reconstructions); and the values for the three SRRF processing parameters (*see* **Notes 32** and **33**).
14. Save the edited script.
15. Run the script by selecting "Run" in the Fiji macro window. Each raw burst of images will be automatically reconstructed into individual SRRF images. SRRF images corresponding to different *Z* slices of a specific time point will be saved as independent files in a common subfolder. At the end, all the slices corresponding to a time point will be combined and saved as a multipage tiff file, and all the multipage tiff files corresponding to all the movie time points will saved in the same folder. At this point, the individual time point folders can be deleted, and the results are opened as a hyperstack in Fiji (Movie 5).

4 Notes

1. To make medium for embryo collection, mix water and agar in a glass beaker. Combine the rest of the reagents in a different beaker. Heat water and agar in a microwave for 5 min, pausing periodically (approximately every 30 s) to stir gently. The rest of the reagents should be microwaved for 2.5 min. The two mixtures can then be mixed together, and poured into Petri dishes while still hot.
2. To make heptane glue, place double-sided tape into a glass vial (use a glass pipette to push tape into the vial), and fill the vial with heptane. Leave the vial on a nutator for 12–24 h to dissolve the glue from the tape into the heptane. Use a glass pipette to transfer the heptane glue into a new vial. The longer double-sided tape is left in heptane, the thicker the glue and the more auto-fluorescent. Therefore, it is best to transfer out the glue or remove the double-sided tape within 24 h of mixing.
3. As microinjections are not feasible using the membrane method, it is possible to use the less viscous and cheaper halocarbon oil 27 for mounting, without mixing with halocarbon oil 700. However, for wound healing studies, it is recommended to mix halocarbon oil 27 and 700 to create a more viscous medium that minimizes embryo leakage upon wounding.
4. To make a slide to break needles, adhere an 18 × 18 mm cover slip to a 25 × 75 mm microscope slide using a drop of water to create a seal between the two surfaces.
5. For super-resolution microscopy using SRRF, an acquisition of >50 FPS for live samples is recommended to minimize artifacts due to sample movement. Slower acquisition rates can be used for fixed samples. High magnification (60–100×) and a small pixel size (100–150 nm) should also be used.
6. NanoJ-SRRF is written in Java using Aparapi (http://aparapi.com/), a library that enables running Java code on GPUs (with support for AMD, Intel, and NVIDIA GPUs). A workstation equipped with an Intel Core i7-9700K, 64 GB of RAM, and an NVIDIA Titan Xp can reconstruct one SRRF image from 100 512 × 512 images within 4 s. Without the GPU, the total processing time for the same image is 146 s. For comparison, an iMac 2011 with a 2.5 GHz Intel Core i5-2400S processor, 32 GB of RAM and an AMD Radeon HD 6750M graphics card with 512 MB of dedicated memory requires 200 s to reconstruct the same image, and 520 s if the graphics card is disabled.

7. Combine water and yeast in small quantities in a tube, while continuously mixing with a spatula until the consistency resembles that of smooth peanut butter. Store the tube on a tray (to prevent spills) at 4 °C. The lid of the Falcon tube should only be partially closed, for as the yeast ferments, carbon dioxide is produced, and the tube can burst if the gas cannot be released.
8. Embryos can be collected throughout the day or overnight, and selected under the stereomicroscope based on their morphology [33]; or collections can be timed and embryos can be aged until they reach the appropriate developmental stage. Development is 1.8 times slower at 18 °C than at 25 °C [34], a fact that can be used to facilitate experiments. For example, to select embryos at stage 14–15 of embryonic development for wound healing assays, collect embryos for 2 h and age the plate for 10 h at 25 °C or for 18 h at 18 °C. dsRNA injections must be conducted in syncytial embryos, before cells form, and thus, embryos must be collected for 30–60 min and immediately prepared for injection.
9. Two strips of plasticine can be used as supports to facilitate retrieval of the coverslip from the Petri dish (Fig. 1e).
10. To avoid desiccation, do not spend more than 5 min aligning embryos. At the end, use the back of the forceps to remove all the non-aligned embryos from the agar block so that they are not transferred to the coverslip.
11. To image the ventral epidermis in an inverted microscope (e.g., for wound healing assays), orient embryos with their convex (ventral) side up (Fig. 1f, left). To image dorsally (e.g., to image dorsal closure or cardiac morphogenesis), the convex side of the embryo should face the agar block (Fig. 1f, right).
12. If embryos burst when the oxygen-permeable membrane is placed on top of the coverslip, or if the oil does not spread to cover the entire coverslip, repeat using an additional drop of oil. On the contrary, if embryos float and move in between the membrane and the coverslip, reduce the amount of oil.
13. Dehydration times will be longer when humidity is high. Begin with 5 min. If embryos are plump and the vitelline membrane is hard to pierce with a microneedle, increase the dehydration time in 1-min increments. If the vitelline membrane forms folds or does not slightly resist deformation by the needle, decrease the dehydration time.
14. For microinjections, it is often preferable to use only halocarbon oil 700 without mixing with halocarbon oil 27. The greater viscosity of halocarbon oil 700 helps preserve the integrity of the embryo and prevent "leaks" during the injection procedure.

15. Needles can be prepared for microinjection using a needle puller. It is important to use needles with thin, long tips that do not break upon contact with the embryo. For the Sutter Instrument Flaming/Brown Micropipette Puller, we use the following parameters: pressure 250, heat 492, pull 110, velocity 110, and delay 200. A longer tip can be obtained by reducing the temperature of the filament, by pulling more slowly, or by increasing the pressure value.
16. To age embryos, use a 100 mm Petri dish as an incubation chamber. Place two strips of plasticine in the dish to be used as supports for the coverslip (Fig. 2a). Arrange wet, fine paper towels around the circumference of the dish to prevent dehydration. Cover the Petri dish and incubate for the time necessary (*see* **Note 8**).
17. To calibrate the Micropoint, place the mirror slide on the microscope stage, and focus on the slide using the objective lens that needs to be calibrated (do not forget to add oil on the objective if it is an oil immersion lens). Find an area of the mirror slide with no scratches or previous calibration patterns. Use the calibration routine provided by the software to ensure accurate laser targeting with the objective lens to be used for ablation. The Micropoint calibration routine in Metamorph fires the laser in a 3 × 3 grid. The software tries to automatically detect the point where the laser hit the mirror slide based on the change in contrast between images acquired before and after the laser is fired. Both the exposure time used to collect the images in the calibration routine and the laser power used to fire the laser will determine the success of the automated spot detection. Aim to generate small spots, so that their position is well defined, by using short exposure times (e.g., 50 ms with bright field illumination) and low Micropoint powers. The Micropoint laser power can be controlled with a neutral density filter that has 30 positions (we typically set the neutral density filter to positions 20–25, with position 30 being that of minimal attenuation), and with a motorized attenuator controlled from the software (we often use transmittance values of 5–30% for calibration). If the automated routine fails, it is possible to manually calibrate the Micropoint by clicking on the spots generated by the laser on the mirror slide during the calibration routine. When calibration is complete, a red rectangle will be displayed on the screen, indicating the region where the laser is properly calibrated. If a trapezoid rather than a rectangle is displayed, the calibration has not worked and should be repeated.
18. As the lens approaches the coverslip, transition from the coarse to the fine focus adjustment controls on the microscope to avoid hitting the coverslip with the objective.

19. Typically, 458, 477, 488, or 514 nm lasers are used for green fluorescent proteins; 543, 561, and 568 nm for red; and 633 or 647 for far red.
20. If imaging a sample with more than one fluorescent marker, use a dichromatic beamsplitter to acquire multiple wavelengths without having to alter the illumination path, thus optimizing the temporal resolution of the acquisition.
21. Be mindful of laser power and exposure time, as excessive laser exposure can cause photobleaching (loss of fluorescence) and phototoxicity (cell damage) to the sample.
22. Set laser power and exposure time using the image histogram as a reference. Ideally, the histogram of the images should fill the dynamic range of the camera with a minimal number of saturated or underexposed pixels (aim for less than 5% in each case) (Fig. 3).
23. Finer *Z* and temporal resolutions can provide valuable information about cell and protein dynamics, but they also increase photobleaching. Depending on the fluorescent protein, imaging tissue slices 10-μm-thick, at 0.5 μm intervals every 15–30 s, can provide relatively high *Z* and temporal resolutions with minimal photobleaching for imaging sessions with durations of approximately 1 h.
24. Typically, using a Micropoint laser with the neutral density filter in positions 20–25, and the mechanical attenuator between 50–90% transmittance, ten pulses on a single spot can be used to sever a junction in approximately 670 ms. If rapid laser ablation is required, use fewer pulses at a higher laser power.
25. Laser ablation of *Drosophila* embryos using ultraviolet light causes photoactivation of the vitelline membrane.
26. If wounding along a line or within a shape, it will be necessary to create discrete spots representing those geometries. Metamorph has an option ("segment region") to break any shape into discrete spots with specific inter-spot distances.
27. To calculate the wound closure rate constant, *k*, we use nonlinear least-squares fitting by the Gauss-Newton method to fit the area of the wound from the time of its maximum value, using an exponential of the form:

$$a(t) = Ae^{-kt}, \tag{1}$$

where $a(t)$ represents the area of the wound at time t with respect to the time of maximum wound area, and A is the maximum area of the wound. The decay constant of the exponential, k, is a rate constant indicative of the efficiency of collective cell movement.

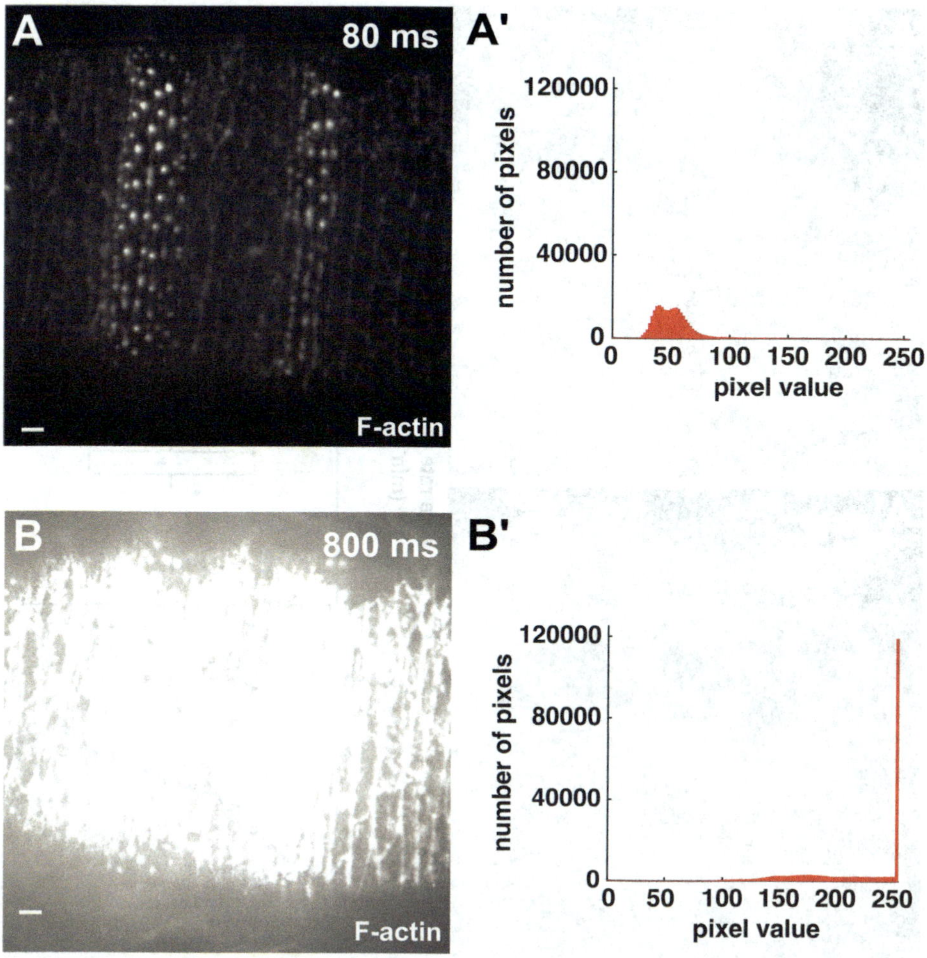

Fig. 3 Using histograms to set up live imaging parameters. (**a**, **b**) Epidermal cells in a *Drosophila* embryo expressing GFP:UtrophinABD, a reporter of filamentous actin [37]. Images were acquired using an 8-bit camera, with moderate exposure time (80 ms, **a**), or high exposure time (800 ms, **b**). Anterior left, dorsal up. Bars, 5 μm. (**a′**, **b′**) Histograms corresponding to the images shown in (**a**) and (**b**). The saturation of the pixel values in (**b**) is obvious from the large fraction of pixels with the maximum possible value

28. To calculate fluorescence at the wound margin, the mean fluorescence of the pixels under the LiveWire annotation (we typically use 0.6-μm-wide mask) is calculated. Images should be corrected for photobleaching by dividing by the mean image intensity at each time point. Background subtraction should also be used when quantifying fluorescence. For cortically enriched signals, the image mode (the most frequent pixel value) can be used as the background value, as there are many more cytoplasmic pixels than cortical ones.
29. Other features of the fluorescence distribution at the wound edge can be useful to determine the efficiency of collective cell movements. For instance, we previously showed that heterogeneity in the distribution at the wound edge of the

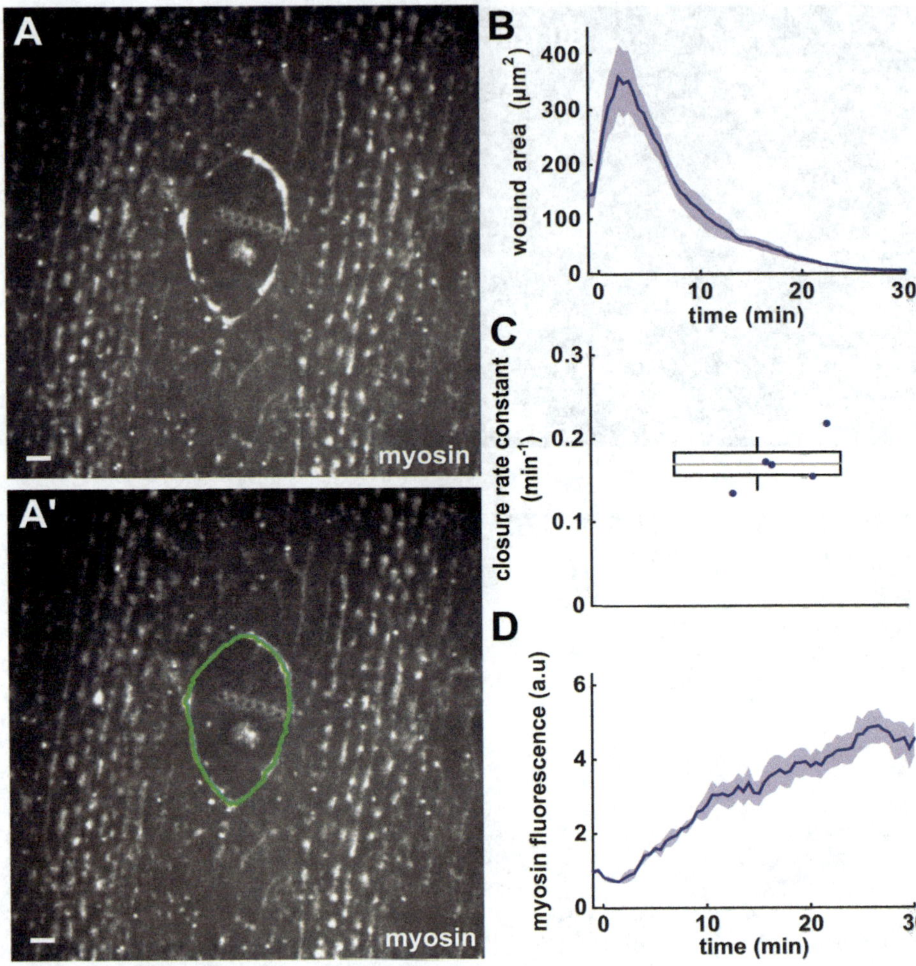

Fig. 4 Analyzing embryonic wound closure from time-lapse data. (**a**) Wound in the epidermis of a *Drosophila* embryo expressing myosin:GFP (**a**) and segmentation of the wound edge using the LiveWire algorithm (**a′**). Anterior left, dorsal up. Bars, 5 μm. (**b–d**) Wound area over time (**b**), wound closure rate constant (**c**), and mean myosin fluorescence at the wound edge (**d**) in control embryos ($n = 5$ embryos). (**b**, **d**) Error bars, standard error of the mean. (**c**) Error bars, standard deviation; box, standard error of the mean; gray line, mean. (**d**) Myosin was quantified in a 0.6-μm-wide mask. Mean values were background-subtracted and corrected for photobleaching (*see* **Note 28**). The resulting curves were normalized to the mean pre-wound values

cytoskeletal protein actin and the molecular motor non-muscle myosin II is critical for rapid cell movements [20, 32]. Cytoskeletal heterogeneity can be quantified as:

$$heterogeneity(t) = sd(t)/mean(t), \qquad (2)$$

where $sd(t)$ represents the standard deviation of the pixel values under the mask at time t, after background subtraction and correction for photobleaching (*see* **Note 28**), and $mean(t)$ indicates the mean pixel value under the mask at time t, also after background subtraction and photobleaching correction.

30. Use short exposure times and low laser power to minimize photobleaching. Also, ensure that the camera is configured to maximize the rate of image acquisition (Table 2). There is a tradeoff between spatial resolution and temporal resolution using SRRF: the more images in a SRRF sequence, the better the reconstruction, but the longer the acquisition time and the likelihood that cellular or molecular dynamics in the sample will interfere with the reconstruction. In addition, longer SRRF sequences will increase sample photobleaching. Typically, 50–100 images can be used to reconstruct a single SRRF image from a live sample [24, 35]. Thus, single-plane time-lapse SRRF can be performed with a temporal resolution of under 4 s, if one accounts not only for the image acquisition time, but also for the time necessary to save the images and clear them from RAM. If acquiring *Z*-stacks, a temporal resolution of 1–2 min is possible.
31. If the name for an image acquisition already exists in the selected directory, the new acquisition will be stored in a folder with the same name and an incremented index at the end to avoid overwriting the original data set. Each time point in the SRRF acquisition will be saved in a different subfolder (*time0* to *timeN*) within the acquisition folder. And each image in a SRRF sequence will be automatically saved as an individual file, using the following naming convention: "img_XXXXXXXXX_Default0_YYY.tif", where XXXXXXXXX is the *Z* slice number, padded with zeros, and YYY is image position within a specific SRRF sequence.
32. Default SRRF parameter values for radiality magnification, ring radius, and axes can be used to start [24] (Fig. 5a, c). Radiality magnification determines the number of subpixels that each pixel is split into. For example, if this value is set to 5, each pixel will be split into a 5 × 5 grid of subpixels, and a 512 × 512 raw image sequence will lead to a 2560 × 2560 super-resolution image. Greater magnification values provide better final resolution, but increase the computation time necessary to reconstruct an image. The ring radius (in pixels) describes the local area in which the convergence of the image gradient is quantified to determine radial symmetry and the presence of a fluorophore. A smaller ring can provide better resolution, but will also increase noise of the reconstruction (Fig. 5c, f). Finally, the axes value indicates the number of axes used to evaluate gradient convergence. A higher value (up to 8) is preferred for a better assessment of radiality, but again, will increase the processing time during reconstruction.
33. The quality of SRRF images is affected by both acquisition settings (number of images per burst, laser power, exposure time) and image reconstruction ones (ring size, radiality

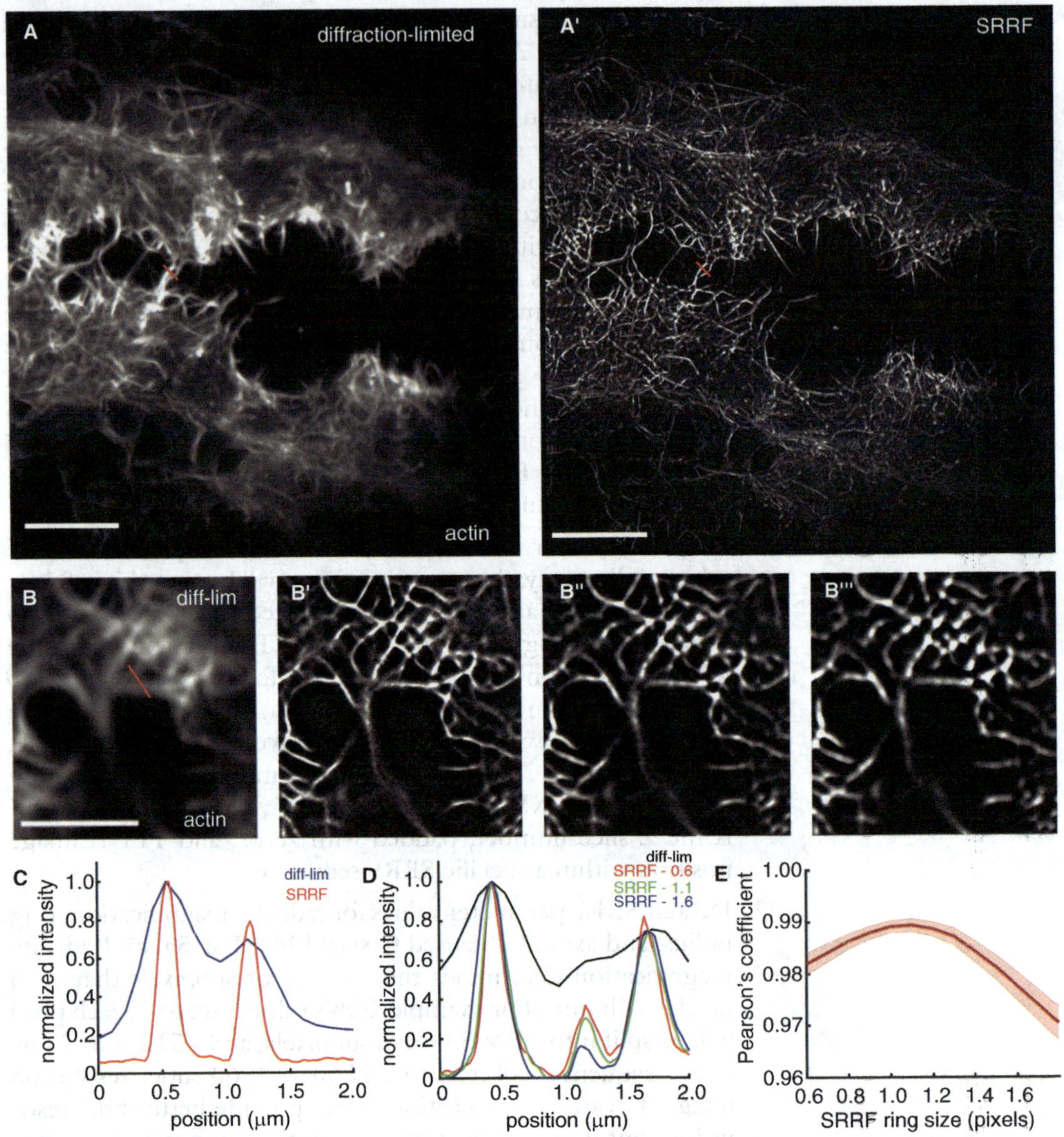

Fig. 5 Super-resolution microscopy to visualize and quantify protrusive activity. (**a**) Maximum intensity projections of diffraction-limited (**a**) and corresponding SRRF (**a′**) *Z*-stacks showing protrusions in cardiac progenitors (cardioblasts) in *Drosophila* embryos expressing GFP:MoesinABD, a marker for filamentous actin [38]. Cardioblasts migrate collectively during the formation of the heart tube, and display both epithelial and mesenchymal characteristics [39], thus constituting an excellent system for the in vivo analysis of the epithelial-mesenchymal continuum and coordinated cell movements. Bars, 10 μm. (**b**) Diffraction-limited (**b**) and SRRF (**b′**–**b‴**) single *Z*-plane images reconstructed with different ring radius values (0.6, 1.1, and 1.6, respectively). Bar, 5 μm. Anterior, left. (**c**) Line intensity profile along the red lines in (**a**, **a′**) for diffraction-limited (blue) and SRRF (red) images, normalized to the maximum intensity for each image. (**d**) Line intensity profile along the red line in (**c**) for diffraction-limited (black) and SRRF reconstructions (red, green, blue) with different ring radii. (**e**) Correlation between experimental diffraction-limited images (**a**) and theoretical ones generated by applying SQUIRREL to SRRF reconstructions using different ring radii (**a′**)

magnification, number of axes, processing method). To optimize the values for these settings, the software SQUIRREL (Super-resolution Quantitative Image Rating and Reporting of Error Locations) can be used [35]. SQUIRREL estimates a resolution scaling function (RSF), which converts the point-spread function (PSF) of the super-resolved image to the PSF of the diffraction-limited image. The RSF can also be manually entered by the user as a Gaussian function, calculated from the PSF of the diffraction-limited image and the PSF of the super-resolution image [36]. SQUIRREL applies the RSF to the SRRF image to generate a theoretical diffraction-limited image. An experimental diffraction-limited image is generated by averaging all the raw images in a SRRF burst. The theoretical diffraction-limited image is quantitatively compared to the experimentally obtained one by calculating the absolute difference, the Pearson Correlation Coefficient, and the root-mean-square error [35]. Since the experimental and the theoretical diffraction-limited images are generated from the same raw data, any differences can be attributed to artifacts caused by the SRRF reconstruction. Thus, it is possible to vary both image acquisition and SRRF reconstruction parameters and use the results of SQUIRREL to determine the parameter values that optimize image reconstruction (Fig. 5d).

Acknowledgements

We thank Katheryn Rothenberg for assistance with Fig. 3, and Christopher McFaul and Eric Théveneau for comments of the manuscript. G.S. is supported in part by an Ontario Graduate Scholarship. Work in our lab is funded by grants to R.F.G. from the Natural Sciences and Engineering Research Council of Canada (418438-13), the Canada Foundation for Innovation (30279), the Ontario Ministry of Economic Development and Innovation (ER14-10-170), the Ted Rogers Centre for Heart Research TBEP Seed Program, and the Canadian Institutes of Health Research (156279). RFG is the Tier II Canada Research Chair in Quantitative Cell Biology and Morphogenesis.

References

1. Yang X, Dormann D, Munsterberg AE, Weijer CJ (2002) Cell movement patterns during gastrulation in the chick are controlled by positive and negative chemotaxis mediated by FGF4 and FGF8. Dev Cell 3:425–437
2. Wyngaarden LA, Vogeli KM, Ciruna BG, Wells M, Hadjantonakis A-K, Hopyan S (2010) Oriented cell motility and division underlie early limb bud morphogenesis. Development 137:2551–2558. https://doi.org/10.1242/dev.046987
3. Gaggioli C, Hooper S, Hidalgo-Carcedo C, Grosse R, Marshall JF, Harrington K, Sahai E (2007) Fibroblast-led collective invasion of carcinoma cells with differing roles for

RhoGTPases in leading and following cells. Nat Cell Biol 9:1392–1400

4. Martin P, Lewis J (1992) Actin cables and epidermal movement in embryonic wound healing. Nature 360:179–183. https://doi.org/10.1038/360179a0
5. McCluskey J, Hopkinson-Woolley J, Luke B, Martin P (1993) A study of wound healing in the E11.5 mouse embryo by light and electron microscopy. Tissue Cell 25:173–181. https://doi.org/10.1016/0040-8166(93)90017-F
6. Davidson LA, Ezin AM, Keller R (2002) Embryonic wound healing by apical contraction and ingression in *Xenopus laevis*. Cell Motil 53:163–176. https://doi.org/10.1002/cm.10070
7. Clark AG, Miller AL, Vaughan E, Yu H-YE, Penkert R, Bement WM (2009) Integration of single and multicellular wound responses. Curr Biol 19:1389–1395. https://doi.org/10.1016/j.cub.2009.06.044
8. Hunter MV, Willoughby PM, Bruce AEE, Fernandez-Gonzalez R (2018) Developmental cell oxidative stress orchestrates cell polarity to promote embryonic wound healing. Dev Cell 47:377–387.e374
9. Ciruna B, Rossant J (2001) FGF signaling regulates mesoderm cell fate specification and morphogenetic movement at the primitive streak. Dev Cell 1(1):37–49
10. Tikhmyanova N, Golemis EA (2011) NEDD9 and BCAR1 negatively regulate E-cadherin membrane localization, and promote E-cadherin degradation. PLoS One 6(7): e22102. https://doi.org/10.1371/journal.pone.0022102
11. Carvalho L, Jacinto A, Matova N (2014) The Toll/NF-kappaB signaling pathway is required for epidermal wound repair in *Drosophila*. Proc Natl Acad Sci U S A 111:E5373–E5382. https://doi.org/10.1073/pnas.1408224111
12. Hunter MV, Lee DM, Harris TJC, Gonzalez RF (2015) Polarized E-cadherin endocytosis directs actomyosin remodeling during embryonic wound repair. J Cell Biol 210:801–816. https://doi.org/10.1083/jcb.201501076
13. Matsubayashi Y, Gilmer CC, Millard TH (2015) Endocytosis-dependent coordination of multiple actin regulators is required for wound healing. J Cell Biol 210:5–7
14. Campbell K, Casanova J (2016) A common framework for EMT and collective cell migration. Development 143:4291–4300. https://doi.org/10.1242/dev.139071
15. Wood W, Jacinto A, Grose R, Woolner S, Gale J, Wilson C, Martin P (2002) Wound healing recapitulates morphogenesis in *Drosophila* embryos. Nat Cell Biol 4:1–7. https://doi.org/10.1038/ncb875
16. Abreu-Blanco MT, Verboon JM, Liu R, Watts JJ, Parkhurst SM (2012) *Drosophila* embryos close epithelial wounds using a combination of cellular protrusions and an actomyosin purse string. J Cell Sci 125:5984–5997. https://doi.org/10.1242/jcs.109066
17. Hidalgo-Carcedo C, Hooper S, Chaudhry SI, Williamson P, Harrington K, Leitinger B, Sahai E (2011) Collective cell migration requires suppression of actomyosin at cell-cell contacts mediated by DDR1 and the cell polarity regulators Par3 and Par6. Nat Cell Biol 13 (1):49–58. https://doi.org/10.1038/ncb2133
18. Brock J, Midwinter K, Lewis J, Martin P (1996) Healing of incisional wounds in the embryonic chick wing bud: characterization of the actin purse-string and demonstration of a requirement for rho activation. J Cell Biol 135:1097–1107
19. Ducuing A, Vincent S (2016) The actin cable is dispensable in directing dorsal closure dynamics but neutralizes mechanical stress to prevent scarring in the *Drosophila* embryo. Nat Cell Biol 18:1149–1160. https://doi.org/10.1038/ncb3421
20. Zulueta-Coarasa T, Fernandez-Gonzalez R (2018) Dynamic force patterns promote collective cell movements during embryonic wound repair. Nat Phys 14:750–758. https://doi.org/10.1038/s41567-018-0111-2
21. Oreopoulos J, Berman R, Browne M (2014) Spinning-disk confocal microscopy: present technology and future trends. In: Quantitative imaging in cell biology, vol 123, 1st edn. Elsevier Inc., Amsterdam, pp 153–175. https://doi.org/10.1016/B978-0-12-420138-5.00009-4
22. Huisken J, Swoger J, Del Bene F, Wittbrodt J, Stelzer EH (2004) Optical sectioning deep inside live embryos by selective plane illumination microscopy. Science 305 (5686):1007–1009. https://doi.org/10.1126/science.1100035
23. Schmied C, Tomancak P (2016) Sample preparation and mounting of *Drosophila* embryos for multiview light sheet microscopy. In: *Drosophila*: methods and protocols. Springer New York, New York. https://doi.org/10.1007/978-1-4939-6371-3_10
24. Gustafsson N, Culley S, Ashdown G, Owen DM, Pereira PM, Henriques R (2016) Fast live-cell conventional fluorophore nanoscopy with ImageJ through super-resolution radial fluctuations. Nat Commun 7:12471

25. Culley S, Tosheva KL, Matos Pereira P, Henriques R (2018) SRRF: universal live-cell super-resolution microscopy. Int J Biochem Cell Biol 101:74–79. https://doi.org/10.1016/J.BIOCEL.2018.05.014
26. Kiehart DP, Montague RA, Rickoll WL, Foard D, Thomas GH (1994) High-resolution microscopic methods for the analysis of cellular movements in *Drosophila* embryos. In: Methods in cell biology, vol 44. Academic Press, New York. https://doi.org/10.1016/S0091-679X(08)60929-2
27. Edelstein A, Amodaj N, Hoover K, Vale R, Stuurman N (2010) Computer control of microscopes using μManager. Curr Protoc Mol Biol 92:14.20.11–14.20.17. https://doi.org/10.1002/0471142727.mb1420s92
28. Fernandez-Gonzalez R, Zallen J (2013) Wounded cells drive rapid epidermal repair in the early *Drosophila* embryo. Mol Biol Cell 24:3227–3237. https://doi.org/10.1091/mbc.E13-05-0228
29. Schindelin J, Arganda-Carreras I, Frise E, Kaynig V, Longair M, Pietzsch T, Preibisch S, Rueden C, Saalfeld S, Schmid B, Tinevez J-Y, White DJ, Hartenstein V, Eliceiri K, Tomancak P, Cardona A (2012) Fiji: an open-source platform for biological-image analysis. Nat Methods 9:676–682. https://doi.org/10.1038/nmeth.2019
30. Farge E (2003) Mechanical induction of Twist in the *Drosophila* foregut/stomodeal primordium. Curr Biol 13:1365–1377. https://doi.org/10.1016/s0960-9822(03)00576-1
31. Dijkstra EW (1959) A note on two problems in connexion with graphs. Numer Math 1:269–271. https://doi.org/10.1007/BF01386390
32. Zulueta-Coarasa T, Tamada M, Lee EJ, Fernandez-Gonzalez R (2014) Automated multidimensional image analysis reveals a role for Abl in embryonic wound repair. Development 141:2901–2911. https://doi.org/10.1242/dev.106898
33. Sullivan W, Ashburner M, Hawley SR (2000) *Drosophila* protocols. Cold Spring Harbor Laboratory Press, Cold Spring Harbor
34. Kuntz SG, Eisen MB (2014) *Drosophila* embryogenesis scales uniformly across temperature in developmentally diverse species. PLoS Genet 10:e1004293. https://doi.org/10.1371/journal.pgen.1004293
35. Culley S, Albrecht D, Jacobs C, Pereira PM, Leterrier C, Mercer J, Henriques R (2018) Quantitative mapping and minimization of super-resolution optical imaging artifacts. Nat Methods 15(4):263–266. https://doi.org/10.1038/nmeth.4605
36. Cole RW, Jinadasa T, Brown CM (2011) Measuring and interpreting point spread functions to determine confocal microscope resolution and ensure quality control. Nat Protoc 6:1929–1941
37. Rauzi M, Lenne PF, Lecuit T (2010) Planar polarized actomyosin contractile flows control epithelial junction remodelling. Nature 468:1110–1115. https://doi.org/10.1038/nature09566
38. Haack T, Schneider M, Schwendele B, Renault AD (2014) *Drosophila* heart cell movement to the midline occurs through both cell autonomous migration and dorsal closure. Dev Biol 396:169–182. https://doi.org/10.1016/j.ydbio.2014.08.033
39. Frémion F, Astier M, Zaffran S, Guillèn A, Homburger V, Sémériva M (1999) The heterotrimeric protein G is required for the formation of heart epithelium in *Drosophila*. J Cell Biol 145:1063–1076. https://doi.org/10.1083/jcb.145.5.1063
40. Huang J, Zhou W, Dong W, Watson AM, Hong Y (2009) Directed, efficient, and versatile modifications of the *Drosophila* genome by genomic engineering. Proc Natl Acad Sci U S A 106:8284–8289. https://doi.org/10.1073/pnas.0900641106
41. Royou A, Field C, Sisson JC, Sullivan W, Karess R (2003) Reassessing the role and dynamics of nonmuscle myosin II during furrow formation in early *Drosophila* embryos. Mol Biol Cell 15:838–850. https://doi.org/10.1091/mbc.e03-06-0440
42. Martin AC, Kaschube M, Wieschaus EF (2009) Pulsed contractions of an actin–myosin network drive apical constriction. Nature 457:495–499
43. Kiehart DP, Galbraith CG, Edwards KA, Rickoll WL, Montague RA (2000) Multiple forces contribute to cell sheet morphogenesis for dorsal closure in *Drosophila*. J Cell Biol 149:471–490
44. Rosales-Nieves AE, Johndrow JE, Keller LC, Magie CR, Pinto-Santini DM, Parkhurst SM (2006) Coordination of microtubule and microfilament dynamics by *Drosophila* Rho1, Spire and Cappuccino. Nat Cell Biol 8:367–376. https://doi.org/10.1038/ncb1385
45. Orsulic S, Peifer M (1996) An in vivo structure-function study of armadillo, the beta-catenin homologue, reveals both separate and overlapping regions of the protein required

for cell adhesion and for wingless signaling. J Cell Biol 134:1283–1300. https://doi.org/10.1083/jcb.134.5.1283

46. Shiga Y, Tanaka-Matakatsu M, Hayashi S (1996) A nuclear GFP/β-galactosidase fusion protein as a marker for morphogenesis in living *Drosophila*. Develop Growth Differ 38:99–106. https://doi.org/10.1046/j.1440-169X.1996.00012.x

47. Vagnoni A, Bullock SL (2016) A simple method for imaging axonal transport in aging neurons using the adult *Drosophila* wing. Nat Protoc 11:1711–1723. https://doi.org/10.1038/nprot.2016.112

Part V

Ex Vivo/In Vitro Methods to Study EMT/MET

Chapter 18

Methods to Generate Tube Micropatterns for Epithelial Morphogenetic Analyses and Tissue Engineering

Minerva Bosch-Fortea and Fernando Martín-Belmonte

Abstract

Cells live in a highly curved and folded 3D microenvironment within the human body. Since epithelial cells in internal organs usually adopt a tubular shape, there is a need to engineer simple in vitro devices to promote this cellular configuration. The aim of these devices would be to investigate epithelial morphogenesis and cell behavior—leading to the development of more sophisticated platforms for tissue engineering and regenerative medicine. In this chapter, we first explain the need for such epithelial tubular micropatterns based on anatomical considerations and then survey methods that can be used to study different aspects of epithelial tubulogenesis. The methods examined can broadly be divided into two classes: conventional 2D microfabrication for the formation of simple epithelial tubes in substrates of different stiffness; and 3D approaches to enable the self-assembly of organoid-derived epithelial tubes in a tubular configuration. These methods demonstrate that modeling tubulogenesis in vitro with high resolution, accuracy, and reproducibility is possible.

Key words 3D microenvironment, Epithelial morphogenesis, 2D microfabrication, Tissue engineering

1 Introduction

Simple epithelial monolayers cover most internal surfaces of the organism with epithelial cells being the fundamental building blocks of many organs. Epithelial organs accomplish a plethora of different functions: from digestion in the intestine to excretion through the kidneys; from lactation in the mammary glands to breathing through the lungs. Indeed, most of our internal organs are made of polarized epithelial cells forming tubes, which organize as an intricate network mostly in charge of transporting and distributing metabolites throughout the body [1]. The fine orchestration of the described morphogenetic processes is paramount for the coordination and synchronization of the cells that would form these tubular structures.

Kyra Campbell and Eric Theveneau (eds.), *The Epithelial-to Mesenchymal Transition: Methods and Protocols*, Methods in Molecular Biology, vol. 2179, https://doi.org/10.1007/978-1-0716-0779-4_18,

Interestingly, the variability of tubular organs derives from the broad diversity of different strategies to form epithelial tubes during development [2]. This variability generates a remarkable structural and cellular diversity: different tube sizes, shapes, and connecting patterns [2–4]. The establishment and control of a correct tubular architecture are fundamental, not only for the understanding of the development of organs but also because many diseases are often related with the loss of any of these epithelial characteristics. For instance, in some cancers, cells undergo an epithelial-to-mesenchymal transition (EMT) that causes loss of cell adhesion and polarity markers, which results in the activation of migratory phenotypes [5]. This loss of epithelial integrity is associated with tumor progression and poor prognosis [6] also leading to invasion and metastasis [7]. Besides cancer, defects in epithelial morphogenesis during development also lead to other diseases like polycystic kidney disease, atherosclerotic heart disease, or faciogenital dysplasia, among others. Therefore, understanding how an epithelium can self-organize and create a tube at a molecular level, is paramount to develop new tools and strategies for diagnosis and treatment.

The control of the cellular microenvironment allows a deeper understanding of the different cellular mechanisms that underlie cell–cell and cell–matrix interactions. In traditional 2D cultures, cells are randomly seeded and lack a specific organization as they are maintained on flat and homogeneous substrates. Major drawbacks of these adherent cultures are the difficulty to control environmental parameters as well as the impossibility to recapitulate multicellular architectures, tissue–tissue interfaces, and the physicochemical microenvironment found in vivo. This absence of a physiologically relevant cell conformation strongly impacts on cell function and behavior. However, micropatterning technology allows the mechanical interaction with the cells by controlling the geometry of cell adhesion and substrate rigidity. Thus, micropattern-based substrates have helped to gain insights into how the extracellular environment influences processes such as the orientation of the cell division, organelle positioning, cytoskeleton rearrangement, cell differentiation, and directionality of cell migration and lumen formation among others [8–13].

In this chapter, we show that micropatterning technology can be used to fabricate a system able to recapitulate epithelial tube morphogenesis in vitro. As there is no universal solution to produce micropatterns, a compromise between simplicity, reproducibility and excellence of patterning, together with an appropriate optical quality of the substrate has to be found for each specific application [14]. Each of the platforms presented in this chapter has different features and serves different purposes as well as provides different rigidity cues. However, all of the developed micropatterned systems rely on the supplementation with Matrigel to provide apicobasal polarization cues to allow cells to attain a 3D architecture.

2 Material

Prepare all solutions using ultrapure water and analytical grade reagents. Prepare and store all reagents as specified by the manufacturer. For cell culture purposes, keep sterile conditions. Carefully follow the waste disposal regulations.

2.1 Fabrication of Flat Micropatterns

1. Photolithographed silicon wafer with the desired microfeatures.
2. Trichloro(1H,1H,2H,2H-perfluorooctyl)silane.
3. Oven (should reach 80 °C).
4. Desiccator.
5. Vacuum pump.
6. Hood.
7. UV/O_3 cleaner.
8. Spin coater.
9. Borosilicate glass Petri dishes.
10. 18 mm coverslips.
11. 6-Well plates.
12. 1.5 ml tubes.
13. PDMS (Sylgard 184).
14. Laminin from Engelbreth-Holm-Swarm murine sarcoma basement membrane.
15. Human plasma fibronectin purified protein or Rhodamine-Fibronectin.
16. Pluronic F127.
17. MilliQ water.
18. Phosphate-buffered saline (PBS).
19. Compressed air duster.
20. Sharp scalpel and single edge razor, curved forceps.

2.2 Fabrication of the 3D Microwells

1. Photolithographed silicon wafer or counter-mold with the desired microfeatures.
2. Oven (should reach 80 °C).
3. Desiccator.
4. Vacuum pump.

5. Hood.
6. UV/O_3 cleaner.
7. Spin coater.
8. Borosilicate glass Petri dishes.
9. 60 mm Bacteriological Petri dishes.
10. 18 mm coverslips.
11. 6-Well plates.
12. PDMS (Sylgard 184).
13. Rhodamine-fibronectin.
14. MilliQ water.
15. Phosphate-buffered saline (PBS).
16. Sharp scalpel and single edge razor, curved forceps.
17. Fluorescence microscope.

2.3 Culture of MDCK Cells as 3D Tubes

1. Tissue-culture hood.
2. Flat patterned surface (either CYTOO chip or μCPrinted chip) or 3D substrates with microchannels.
3. Centrifuge.
4. MDCK II cells.
5. Ice bucket.
6. MEM medium supplemented with 10% fetal bovine serum (FBS), 2 mM L-glutamine, 100 U/ml penicillin, and 100 mg/ml streptomycin for cell maintenance.
7. 0.25% (w/v) trypsin/1 mM EDTA.
8. Tissue-culture dishes.
9. Hemocytometer.
10. 15 ml conical tube.
11. Matrigel.

2.4 Culture of Mice Mammary Gland Organoids on Line Microchannels

1. Tissue-culture hood.
2. ECM-coated PDMS substrates with microwells.
3. Centrifuge suitable for plates.
4. Isolated mice mammary gland organoids.
5. DMEM/F12.
6. Fetal bovine serum.
7. Bovine serum albumin.
8. Ice bucket.
9. DispaseII.
10. DNaseI.

11. 0.25% (w/v) trypsin/1 mM EDTA.
12. Insulin-transferrin-selenium-X.
13. 40-μm cell strainer filter.
14. 6-Well plates.
15. 12-mm coverslips.
16. Tissue-culture dishes.
17. Hemocytometer.
18. 15 and 50 ml conical tubes.
19. Matrigel growth factor reduced basement membrane.

3 Methods

3.1 Fabrication of Flat Surfaces Bearing Line-Shaped Microfeatures

For the design of flat surfaces, different adhesive micropattern prototypes were designed using either photopatterning or μCP techniques. Photopatterning allows printing of the microfeatures on glass substrates, which provides an excellent optical quality suitable to perform imaging experiments. This micropatterning technique has been extensively revised elsewhere [11]. However, we have developed a commercial device in collaboration with CYTOO SA (Fig. 1a, b), with the optimized parameters to build renal epithelial tubules in vitro [14]. When designing the photomask, several factors need to be taken into account. The first of them is the shape of the pattern. The most apparent shape to achieve a tubular structure is a line. However, given the wide diversity of growing patterns occurring in vivo, other shapes can also be taken into consideration like circular lines and S-shaped lines (Fig. 1c). The second issue is the length of the patterns, corresponding to the anteroposterior axis of the tube. For the CYTOO chip, we selected three different measures based on the capability of adapting them to an automated quantification procedure and on the sizes able to fit into a microscopy image taken at high magnification (40–63×). Patterns of 100, 200, and 300 μm were selected (Fig. 1c). Third, the width of the patterns that would be the dorsoventral axis of the tube should be chosen. We designed patterns of 15, 20, 30, and 40 μm in width (Fig. 1c). Taking into account the fact that the basal size of cells decreases through polarization process, we expected these patterns to provide adhesion to 2–6 cells in a *Y*-cross-section of a fully polarized tube.

The second technique used to generate line micropatterns is Microcontact printing (μCP), which allows patterning on other substrates with lower rigidity than glass (like silicone or hydrogels) and can be helpful to assess the role of matrix stiffness in tube formation. The principles of μCP consist of the transfer of the microfeatures to an activated surface by direct contact with an

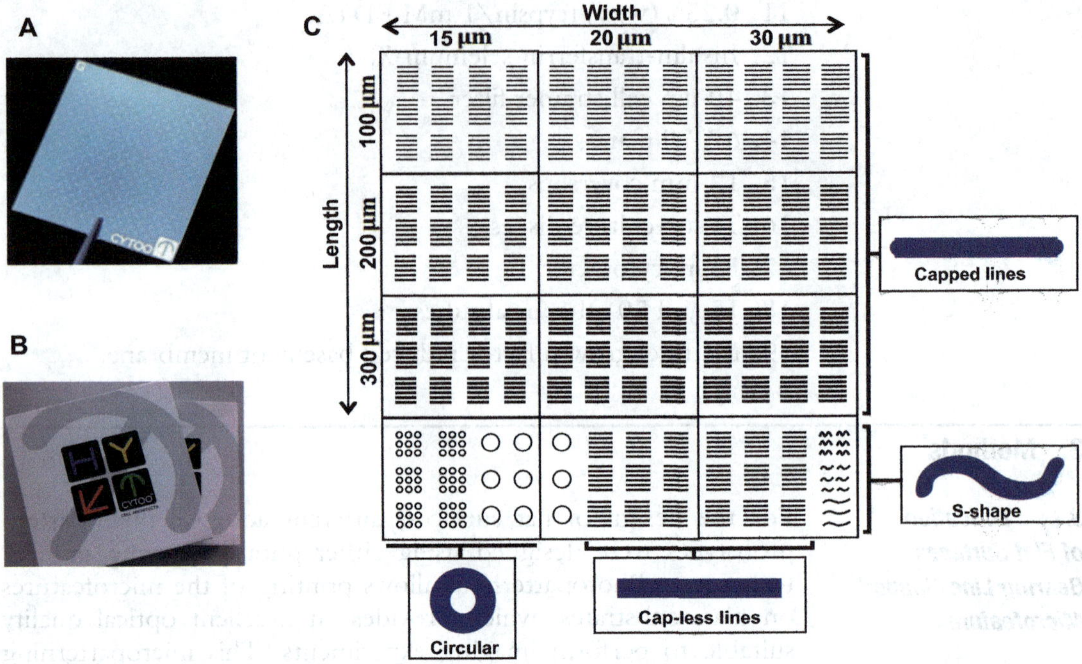

Fig. 1 Micropatterning on borosilicate glass with deep UV light. (**a**) Commercial CYTOO chips are 19.5 × 19.5 mm and made of high quality and low fluorescence borosilicate glass. Grid coordinates are also printed on the underside of the chip to provide orientation and location if required. (**b**) Selling format for CYTOO chips. (**c**) The designed mask has different shapes (green) and different sizes. Line lengths (upper) range from 100 to 300 μm and tube widths comprised between 15 and 30 μm. Lines are spaced with a 40 μm gap in the vertical axis and with an 80 μm gap in the horizontal axis. Canonical lines are capped at the ends providing a rounded end. Cap-less lines (lower center) have sharper squared ends and 200 μm length and 20 μm width. S-shapes (lower right) vary between 200 to 500 μm in length while S-shape patterns are 20 μm in diameter. Circular patterns (lower left) are 20 μm wide and have a radius between 50 and 100 μm. Repetitions of the same patterns are separated from the others through a micropatterned grid composed of 40-μm-width continuous line

ECM-coated stamp. To perform μCP, a quartz photomask similar to that used for photopatterning should be designed. For the fabrication of photomasks and silicon wafers, please refer to [15] (Fig. 2a). When designing the photomask, similar principles to those used for the design of the photopatterning photomask should be considered (Fig. 2b–d).

1. Use PDMS Sylgard184 to produce the stamps. Thoroughly mix the elastomer and cross-linker components in a 1:10 (cross-linker: elastomer) ratio and keep under vacuum using a vacuum bell or desiccator for 20–30 min to degas the mixture. The amount of PDMS should be determined depending on the number of stamps to fabricate.
2. Pour the PDMS over the silicon wafer and bake at 80 °C for 1–2 h in the oven (*see* **Notes 1** and **2**).

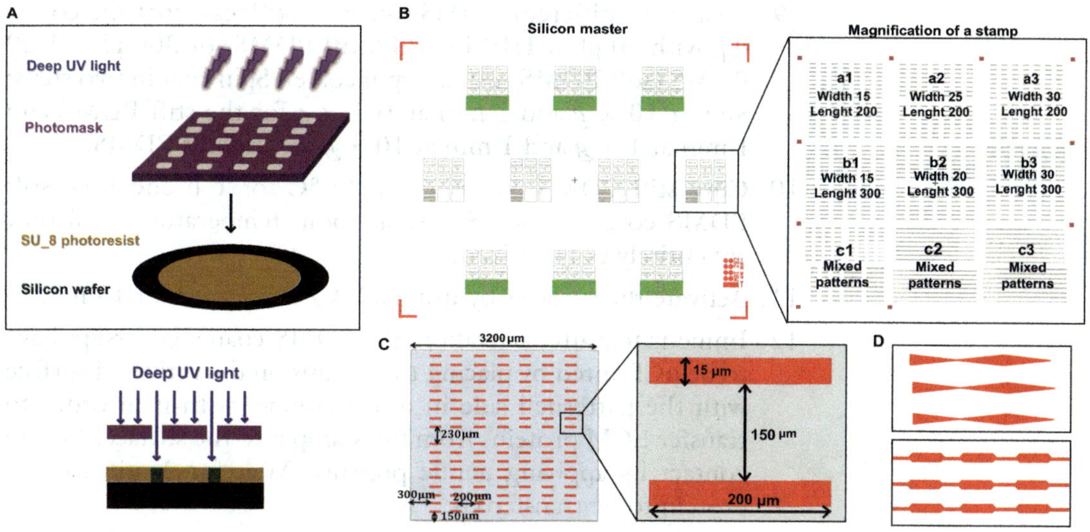

Fig. 2 Microfabrication of the wafer. (**a**) Photolithography procedure was used to fabricate the silicon wafer. It uses UV light that passes through a photomask and etches the microfeatures on the wafer. Lower panel shows how three-dimensional features are generated by photolithography when using a SU_8 negative photoresist (yellow). When exposed to 365 nm light through a photomask, the exposed part crosslinks and becomes hard (green). After exposure, the wafer is developed, and the not-crosslinked SU_8 photoresist is dissolved. (**b**) Schematic representation of the silicon wafer is shown on the left panel. Wafer contains ten stamps each of them measuring 1 cm × 1 cm with a spacing of 1 cm from the adjacent stamps. Green boxes represent variable patterns that differ between stamps. The right panel shows a magnification of a stamp. A letter and a number identify each set of patterns. Patterns in the upper row are 200 μm in length and 15, 20, and 30 μm wide. Patterns in the middle row are 300 μm in length and 15, 20, and 30 μm wide. Patterns in the lower row are composed of distinct features different from the canonical ones. (**c**) A magnification of a set of patterns from the a1 region. The dimensions of the patterns and the spacing regions are represented. (**d**) Examples of mixed patterns contained in the mixed regions c1, c2, and c3 of the stamps

3. Peel-off the cured PDMS and cut it into small squared pieces containing the patterned regions. Keep them on a 6-well plate with the patterned region up-side.
4. Prepare ECM solutions in PBS using sterile conditions. Use either 20 μg/ml Laminin, 20 μg/ml fibronectin, or 20 μg/ml Rhodamine-fibronectin depending on the cell type or experimental approach.
5. Coat the stamps with ECM proteins by covering the patterned with a 300 μl drop of the solution prepared in **step 4**. Incubate for 1 h at RT.
6. Rinse the stamps two times with sterile PBS and make the last wash with milliQ water.
7. Let the stamps dry and keep them at 4 °C for a maximum of 24 h.
8. Clean 18 mm borosilicate coverslips by sonication and isopropanol wash and dry using compressed air.

9. To achieve different PDMS substrate stiffness, coat the coverslip with 50 μl of 1:10 PDMS (Stiff PDMS) or 300 μl of 1:20 PDMS (soft PDMS) using a spin coater. Spin coat in two steps: 30 s at 10 × g and 1 min at 161 × g for the stiff PDMS and 1 min at 1 × g and 1 min at 10 × g for the soft PDMS.
10. Cure stiff PDMS coverslips at 80 °C for 1 h and keep soft PDMS coverslips for 5 days at room temperature until they are entirely polymerized.
11. Activate the surfaces by using a UV/O_3 cleaner for 15 min.
12. Immediately after activation, the PDMS-coated coverslips have to be μCPrinted by placing the stamps on the activated surface with the patterned side in direct contact with it in order to transfer ECM proteins from the stamp into the surface. Ensure contact by applying gentle pressure. Wait for 2 min for the protein to be transferred.
13. Carefully remove the stamps and discard them.
14. Treat the surfaces with 2% Pluronic F127 diluted in water at room temperature for 40 min–1 h or overnight at 4 °C.
15. Rinse the patterned surface thoroughly with PBS to prevent cell toxicity. Patterned surfaces can be stored in PBS for 2–3 days at 4 °C.

3.2 Culture of MDCK Tubes on Flat Line Micropatterns

The vast majority of studies using micropatterns have been done either with single cells or with a group of cells growing in 2D conditions. However, building a tube in vitro requires a third dimension that, in this case, was provided by the supplementation with MG [13]. Taking this into consideration, we established a culturing protocol for MDCK cells (Fig. 3). When using a different cell line, in addition to adjusting the concentration of MG, other parameters were optimized such as the medium to be used, the FBS concentration, the number of cells to seed, and the ECM coating. The timing of the protocol should also be determined as cell lines displayed different growing dynamics.

1. To prepare MDCK tubes (Fig. 3), place patterned surface in a 6-well plate and rinse with PBS.
2. Incubate with culture medium for 1 h at 37 °C.
3. Trypsinise an 80% confluent dish of MDCK cells to a single cell suspension by incubating with sterile PBS for 20 min and Trypsin solution for 5 min at 37 °C.
4. Seed 6 × 10^4 cells per coverslip MEM culture medium and incubate for 1 h at room temperature and 2 h at 37 °C in a humidified atmosphere containing 5% CO_2 until cells entirely adhere to the patterns.

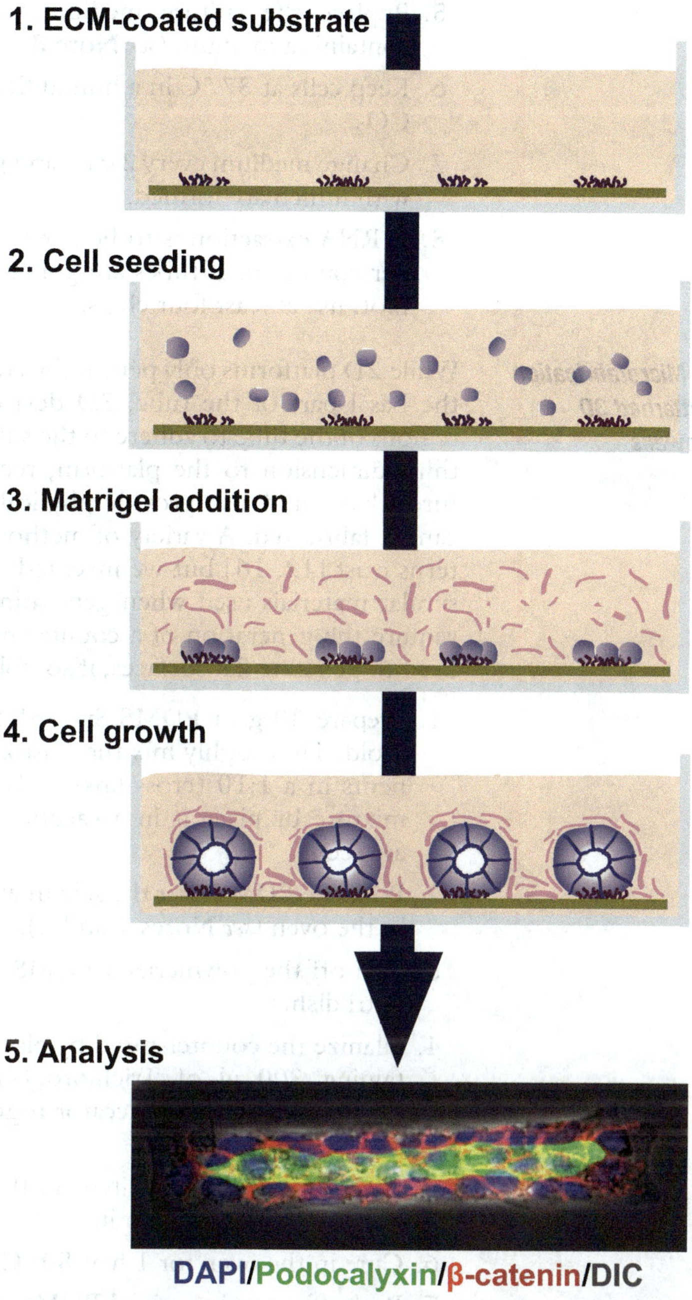

Fig. 3 Protocol to grow MDCK cells on ECM-coated flat micropatterns. Chips are coated with a solution of purified ECM components (1), and cells are seeded on top (2). Once cells adhere to the micropatterns, MG solution is added to medium (3). Micropatterns are then placed in the incubator until cells reach a tubular structure (4). At that point, they are suitable for cell imaging, or for biochemical studies (5)

5. Replace the culture medium with 3% FBS, 3% Matrigel-containing medium (*see* **Note 3**).
6. Keep cells at 37 °C in a humidified atmosphere containing 5% CO_2.
7. Change medium every 2 days and grow for 1–4 days until tubes with lumen are formed.
8. If RNA extraction is to be performed, use at least eight chips per condition. If tubes are going to be analyzed by immunoblot, use at least four chips.

3.3 Microfabrication of Patterned 3D Microwells

While 2D platforms only permit the adhesion of the cells located in the basal part of the tube, 3D devices allow cells at the laterals regions of the tube to adhere to the substrate (Fig. 4a). By adding a third dimension to the platform, rectangular well-like microfeatures that could also provide physical confinement to the system can be fabricated. A variety of methods to fabricate 3D micropatterns exist [12, 16] but we invented a simple technique requiring similar materials used when generating flat micropatterns. It may require the generation of a counter-mold out of the silicon wafer used to generate flat surfaces, if so, follow **steps 1–8**.

1. Prepare 30 g of PDMS Sylgard184 to fabricate the counter-mold. Thoroughly mix the elastomer and cross-linker components in a 1:10 (cross-linker: elastomer) ratio and degas the mixture by place it in a vacuum desiccator until no bubbles are seen.
2. Pour the PDMS over the silicon wafer and bake at 80 °C for 2 h in the oven (*see* **Notes 1** and **2**).
3. Peel-off the polymerized PDMS and place it in a borosilicate Petri dish.
4. Silanize the counter-mold by placing an open Eppendorf containing 300 μl of Trichloro(1H,1H,2H,2H-perfluorooctyl) silane in a vacuum desiccator together with the counter-mold (*see* **Note 2**).
5. Clean the silane deposited on the counter-mold by casting a new PDMS mixture on it.
6. Cure in the oven for 1 h at 80 °C.
7. Peel off the polymerized PDMS and discard.
8. The counter-mold will have the complementary patterns of the silicon wafer and can be reused many times. Silanization process should be performed every 20–30 uses again (*see* **Note 2**).
9. Use PDMS Sylgard184 to produce the 3D microwells (Fig. 4b). Thoroughly mix the elastomer and cross-linker components in a 1:10 (cross-linker: elastomer) ratio and keep under

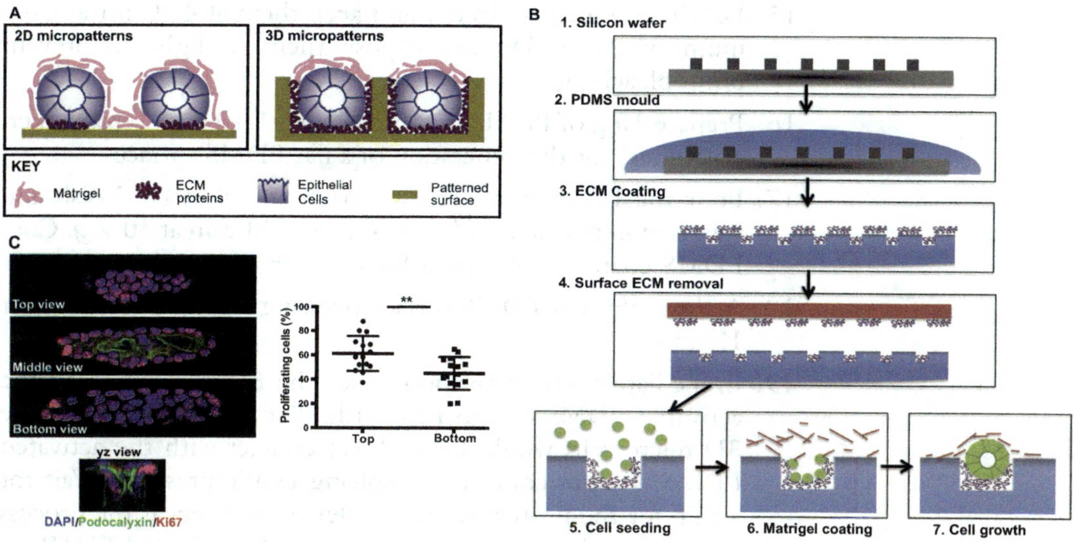

Fig. 4 Fabrication of microwells and culture of MDCK cells as tubes. (**a**) Comparative analysis of 2D and 3D devices. While 2D devices (upper) only provide adhesion to the cells on the basal part of the tube, 3D platforms (lower), which have a well-like shape, provide adhesion also to the lateral parts of the tube. (**b**) Schematic showing fabrication of 3D microwells. PDMS casting on the silicon wafer (1) allowed the generation of the 3D PDMS substrate (2). This last substrate was coated with ECM proteins that adhered to the whole PDMS surface (3). To remove the ECM proteins adhered outside of the micropatterned wells, we performed μCP repeatedly on an activated PDMS flat surface (4). Cells were then seeded (5), and Matrigel was added after adhesion (6). Cells were then grown for 3–6 days (7). (**c**) Various confocal sections of an MDCK tube growing on a 3D micropattern. Sections are from the upper part of the tube exposed to the culture medium, the middle, and the bottom part where cells attach to the ECM. Tubes are stained with a proliferation marker (Ki67) and *yz* view (right): Note, Ki67 staining changes depending on the region of the tube. Quantification shows the percentage of proliferating cells in the top and the bottom slices of tubes growing on 3D micropatterns. Values are mean ± SD from 15 different tubes ($n > 50$ cells/tube; $**P < 0.01$)

vacuum using a vacuum desiccator for 20–30 min to degas the mixture. The amount of PDMS should be determined depending on the number of substrates to fabricate.

10. Pour the PDMS over the silicon wafer or counter-mold and bake at 80 °C for 2 h in the oven.
11. Peel off the cured PDMS and cut it into small squared pieces containing the patterned regions. Keep them on a 6-well plate with the patterned region up-side.
12. Prepare 20 μg/ml Rhodamine-fibronectin solutions in PBS using sterile conditions.
13. Coat the 3D microwells with the ECM solution by covering the patterned region with a 300 μl drop. Incubate for 1 h at RT in the dark.
14. Rinse the 3D microwells two times with sterile PBS and make the last wash with milliQ water.

15. Let the 3D microwells dry and keep them at 4 °C for a maximum of 24 h. Do not expose them to light to prevent photobleaching.
16. Prepare 10 g of PDMS mixture using a 1:10 ratio (crosslinker: elastomer) for the generation of a flat PDMS surface.
17. Pour the PDMS mixture on a bacteriological Petri dish and spin coat in two steps: 30 s at 10 × *g* and 1 min at 40 × *g*. Cure PDMS-coated coverslips at RT for 24 h.
18. Activate this flat PDMS surface by using a UV/O_3 cleaner for 15 min.
19. Immediately after activation, place the 3D microwells on the activated PDMS-coated Petri dish. The patterned side of the 3D microwells would be in direct contact with the activated PDMS. Ensure contact by applying gentle pressure. Wait for 2 min for the protein to be transferred and repeat the process three more times in a clean region of the activated PDMS to ensure ECM removal from the surface.
20. Carefully remove the 3D microwells and discard the PDMS-coated Petri dish.
21. Visualize the 3D microwells under a fluorescence microscope, the absence of ECM proteins outside of the wells should be confirmed.
22. 3D microwells can be stored in PBS for 2–3 days at 4 °C in the dark.

3.4 Culture of MDCK Tubes on Line Microwells

3D-microwells prevent MDCK cells from growing out of the patterns and better confine them into a tube-like shape with cells enclosing a central lumen (Fig. 4c). By using these microwells to culture MDCK cells, we assessed the proliferation rates at the different sections of the tubes. Results indicate that the basal part of the tube provides higher confinement as fewer cells were proliferating when compared to the top part, where the absence of physical constraint allowed cells to grow in a less compressive environment (Fig. 4c). MDCK cells on microchannels are grown as follows. Make sure to use ice and cold instruments when working with Matrigel.

1. To prepare MDCK tubes on microchannels (Fig. 4b), place patterned surface in a 6-well plate with the patterns facing up and rinse with PBS.
2. Incubate with 2 ml of culture medium for 1 h at 37 °C.
3. Trypsinize an 80% confluent dish of MDCK cells to a single cell suspension by incubating with sterile PBS for 20 min and trypsin solution for 5 min at 37 °C.

4. Resuspend 10^5 cells in 250 μl of culture medium and placed it on the 3D surface. Incubate for 3–6 h at 37 °C in a humidified atmosphere containing 5% CO_2 until cells adhere inside the microwells. Bubbles can form inside the microwells, pipetting medium every 1–2 h might be necessary to break the bubbles.
5. Once cells adhere, remove the drop and gently wash with 2 ml of medium. Replace the culture medium with 3% FBS, 3% Matrigel-containing medium (*see* **Note 3**).
6. Keep cells at 37 °C in a humidified atmosphere containing 5% CO_2.
7. Change medium every 2 days and grow for 1–4 days until tubes with lumen are formed (*see* **Note 4**).

3.5 Culture of Mice Mammary Gland Organoids on Line Microchannels

Nowadays, many laboratories are beginning to develop a promising cell culture system, denominated organoids. These organoids are formed from isolated stem cells or a combination of stem cells with animal explants or cell lines. Together, these cells create an organ-like structure in vitro under specific culture conditions, sharing many similarities with the organ they derive from [17–19]. Although studies using mammary gland organoids have helped to gain insights into the complexity of morphogenesis, mammary gland organoids have a rounded shape and do not acquire the tubular architecture present in mammary gland epithelial ducts. Thus, they cannot fully recapitulate the complexity of tube formation in physiological conditions. To solve this problem, the microwell technology can also be used to fabricate microchannels to culture mouse mammary gland organoid-derived cells in a tubular shape. Mammary gland organoids were isolated following the protocol previously described for FVB mice (Fig. 5a) [20].

1. Purify organoids from three different individuals by following the method described elsewhere [20] (*see* **Note 5**). After purification, a white pellet containing the organoids is obtained. To prepare smaller organoids, gently pipette the pellet for 2 min in 2 ml of trypsin-EDTA 0.25%.
2. Centrifuge for 5 min at 10 × *g*, and then pipette for 2 min in 2 ml of DMEM/F12 containing DispaseII (5 mg/ml) and DNaseI (1 mg/ml).
3. Centrifuge for 5 min at 10 × *g* and then resuspend in 5 ml DMEM/F12 containing 2% FBS.
4. Filter clumps by using a 40-μm cell strainer filter.
5. If isolating organoids from three mice, about eight different microwells can be used for cell culture.
6. Count cells and centrifuge for 5 min at 10 × *g*.
7. Resuspend the total amount of cells in the appropriate amount of DMEM/F1 medium containing 1% penicillin/streptomycin and 1% insulin-transferrin-selenium-X (Organoid medium). To

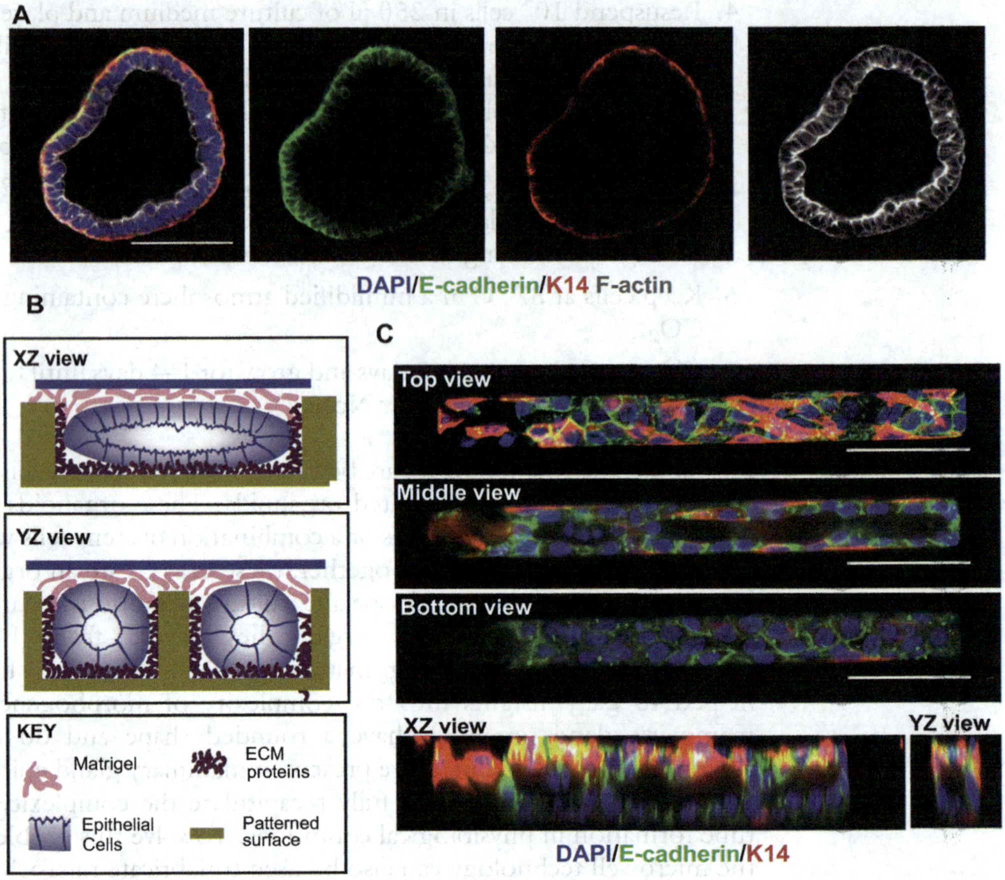

Fig. 5 Culturing mammary gland organoids on microchannels. (**a**) Three days mouse mammary gland organoid. Note that luminal epithelial cells (green) are surrounded by myoepithelial cells (red). Scale bar, 100 μm. (**b**) Schematic images showing cells cultured on microwells growing on Matrigel supplemented medium (upper) or in microchannels done by covering the cells with a Matrigel-coated coverslip. (**c**) Different confocal sections of mammary gland organoid-derived cells growing on a 3D channel. Sections are from the upper part of the tube exposed to the MG-coated coverslip, the middle, and the bottom part where cells attach to the ECM. Note, *xz* and *yz* views show cells enclosing the central lumens

calculate the appropriate volume, take into account that 250 μl of cell-containing solutions are used for each microwell and that cell numbers should be not less than 2×10^5 cells and not more than of 10^6 cells per microwell.

8. Stick the appropriate number of PDMS microwells (depending on cell yields) to the bottom of a 6-well plate by using a PBS drop.
9. Place 250 μl of cell solution on the microwells covering the whole surface and centrifuge the plate for 2 min at $10 \times g$.
10. Add 3 ml of organoid medium to the wells without disturbing the PDMS surface.

11. Incubate for 3–6 h at 37 °C in a humidified atmosphere containing 5% CO_2 until cells adhere inside the microwells.
12. Prepare sterile 18 mm coverslips coated with 50 μl of growth factors-reduced Matrigel (*see* **Note 3**). Keep them on a cold Petri dish (on ice) to prevent Matrigel from polymerizing.
13. Once the cells are attached to the wells, remove culture medium, and stick the Matrigel-coated coverslip on top of the microwells (Fig. 5b).
14. Incubate for 1 h at 37 °C in a humidified atmosphere containing 5% CO_2.
15. Add 2 ml of Organoid medium containing 5% Matrigel (*see* **Note 3**).
16. Keep cells at 37 °C in a humidified atmosphere containing 5% CO_2. Change medium every 2 days and grow for 5–9 days until tubes with lumen are formed (*see* **Note 4**) (Fig. 5c).

4 Notes

1. Photomask should be silanized with Trichloro (1H,1H,2H,2H-perfluorooctyl) silane for 1 h in a vacuum desiccator when used for the first time [15]. Keep it in a glass Petri dish.
2. The silanization process is very toxic so a fume hood and appropriated protection must be used.
3. Use ice and cold instruments when working with Matrigel.
4. If fluorescent microscopy is to be performed, use phenol red--free medium.
5. When working with organoids, coat every plastic consumable (tips, conical tubes, etc.) with sterile 5% BSA solution in PBS to prevent organoids from sticking to the walls.

References

1. Hogan BL, Kolodziej PA (2002) Organogenesis: molecular mechanisms of tubulogenesis. Nat Rev Genet 3(7):513–523. https://doi.org/10.1038/nrg840
2. Lubarsky B, Krasnow MA (2003) Tube morphogenesis: making and shaping biological tubes. Cell 112(1):19–28
3. Datta A, Bryant DM, Mostov KE (2011) Molecular regulation of lumen morphogenesis. Curr Biol 21(3):R126–R136. https://doi.org/10.1016/j.cub.2010.12.003
4. Bryant DM, Mostov KE (2008) From cells to organs: building polarized tissue. Nat Rev Mol Cell Biol 9(11):887–901. https://doi.org/10.1038/nrm2523
5. Barriere G, Fici P, Gallerani G, Fabbri F, Rigaud M (2015) Epithelial mesenchymal transition: a double-edged sword. Clin Transl Med 4:14. https://doi.org/10.1186/s40169-015-0055-4
6. Martin-Belmonte F, Perez-Moreno M (2011) Epithelial cell polarity, stem cells and cancer. Nat Rev Cancer 12(1):23–38. https://doi.org/10.1038/nrc3169
7. Wei SC, Fattet L, Tsai JH, Guo Y, Pai VH, Majeski HE, Chen AC, Sah RL, Taylor SS,

Engler AJ, Yang J (2015) Matrix stiffness drives epithelial-mesenchymal transition and tumour metastasis through a TWIST1-G3BP2 mechanotransduction pathway. Nat Cell Biol 17(5):678–688. https://doi.org/10.1038/ncb3157

8. Azioune A, Carpi N, Tseng Q, Thery M, Piel M (2010) Protein micropatterns: a direct printing protocol using deep UVs. Methods Cell Biol 97:133–146. https://doi.org/10.1016/S0091-679X(10)97008-8
9. Leight JL, Liu WF, Chaturvedi RR, Chen S, Yang MT, Raghavan S, Chen CS (2012) Manipulation of 3D cluster size and geometry by release from 2D micropatterns. Cell Mol Bioeng 5(3):299–306. https://doi.org/10.1007/s12195-012-0236-9
10. Purrucker O, Fortig A, Ludtke K, Jordan R, Tanaka M (2005) Confinement of transmembrane cell receptors in tunable stripe micropatterns. J Am Chem Soc 127(4):1258–1264. https://doi.org/10.1021/ja045713m
11. Thery M, Piel M (2009) Adhesive micropatterns for cells: a microcontact printing protocol. Cold Spring Harb Protoc 2009(7):pdb.prot5255. https://doi.org/10.1101/pdb.prot5255
12. Yilmaz CO, Xu ZS, Gracias DH (2014) Curved and folded micropatterns in 3D cell culture and tissue engineering. Methods Cell Biol 121:121–139. https://doi.org/10.1016/B978-0-12-800281-0.00009-9
13. Rodriguez-Fraticelli AE, Auzan M, Alonso MA, Bornens M, Martin-Belmonte F (2012) Cell confinement controls centrosome positioning and lumen initiation during epithelial morphogenesis. J Cell Biol 198(6):1011–1023. https://doi.org/10.1083/jcb.201203075
14. Bosch-Fortea M et al (2019) Micropattern-based platform as a physiologically relevant model to study epithelial morphogenesis and nephrotoxicity. Biomaterials 218:119339. https://doi.org/10.1016/j.biomaterials.2019.119339
15. Ghibaudo M, Di Meglio JM, Hersen P, Ladoux B (2011) Mechanics of cell spreading within 3D-micropatterned environments. Lab Chip 11(5):805–812. https://doi.org/10.1039/c0lc00221f
16. Li Q et al (2016) Extracellular matrix scaffolding guides lumen elongation by inducing anisotropic intercellular mechanical tension. Nat Cell Biol 18:311–318. https://doi.org/10.1038/ncb3310
17. Eiraku M, Takata N, Ishibashi H, Kawada M, Sakakura E, Okuda S, Sekiguchi K, Adachi T, Sasai Y (2011) Self-organizing optic-cup morphogenesis in three-dimensional culture. Nature 472(7341):51–56. https://doi.org/10.1038/nature09941
18. Sato T, Vries RG, Snippert HJ, van de Wetering M, Barker N, Stange DE, van Es JH, Abo A, Kujala P, Peters PJ, Clevers H (2009) Single Lgr5 stem cells build crypt-villus structures in vitro without a mesenchymal niche. Nature 459(7244):262–265. https://doi.org/10.1038/nature07935
19. Xia Y, Nivet E, Sancho-Martinez I, Gallegos T, Suzuki K, Okamura D, Wu MZ, Dubova I, Esteban CR, Montserrat N, Campistol JM, Izpisua Belmonte JC (2013) Directed differentiation of human pluripotent cells to ureteric bud kidney progenitor-like cells. Nat Cell Biol 15(12):1507–1515. https://doi.org/10.1038/ncb2872
20. Nguyen-Ngoc KV, Shamir ER, Huebner RJ, Beck JN, Cheung KJ, Ewald AJ (2015) 3D culture assays of murine mammary branching morphogenesis and epithelial invasion. Methods Mol Biol 1189:135–162. https://doi.org/10.1007/978-1-4939-1164-6_10

Chapter 19

CAFs and Cancer Cells Co-Migration in 3D Spheroid Invasion Assay

Sefora Conti, Takuya Kato, Danielle Park, Erik Sahai, Xavier Trepat, and Anna Labernadie

Abstract

In many solid tumors, collective cell invasion prevails over single-cell dissemination strategies. Collective modes of invasion often display specific front/rear cellular organization, where invasive leader cells arise from cancer cell populations or the tumor stroma. Collective invasion involves coordinated cellular movements which require tight mechanical crosstalk through specific combinations of cell–cell interactions and cell–matrix adhesions. Cancer Associated Fibroblasts (CAFs) have been recently reported to drive the dissemination of epithelial cancer cells through ECM remodeling and direct intercellular contact. However, the cooperation between tumor and stromal cells remains poorly understood. Here we present a simple spheroid invasion assay to assess the role of CAFs in the collective migration of epithelial tumor cells. This method enables the characterization of 3D spheroid invasion patterns through live cell fluorescent labeling combined with spinning disc microscopy. When embedded in extracellular matrix, the invasive strands of spheroids can be tracked and leader/follower organization of CAFs and cancer cells can be quantified.

Key words 3D spheroid invasion, Cancer Associated Fibroblasts, Epithelial cancer cells, Collective migration, Leader/follower cells

1 Introduction

Metastatic dissemination is responsible for most cancer-related deaths. In the early stages of cancer spreading, tumor cells escape from the primary site, breach through the basement membrane (BM), infiltrate into the stroma and eventually enter the blood circulation [1]. Increasing evidence suggests that a large number of solid tumors invade as multicellular units within the peritumoral stroma and in the bloodstream [2–4]. These modes of collective migration generally involve an invasive front presenting specific combinations of cell–cell interactions and cell–matrix adhesion

Sefora Conti and Takuya Kato are co-authors.

Kyra Campbell and Eric Theveneau (eds.), *The Epithelial-to Mesenchymal Transition: Methods and Protocols*, Methods in Molecular Biology, vol. 2179, https://doi.org/10.1007/978-1-0716-0779-4_19,

[5, 6]. Several reports support a cooperative model for cancer invasion where distinct cell populations interact, resulting in enhanced invasiveness and survival of the moving cell cluster [5–7]. Cancer Associated Fibroblasts (CAFs) are one of the most abundant cell types in solid tumors, and they play an active role in promoting tumor invasion and metastasis. Recent studies reported that CAFs drive the initiation of collective dissemination of epithelial tumors through ECM remodeling and intercellular interactions mediated by adherent junction proteins [8, 9]. These findings highlight that the stromal contribution to metastasis is not limited to chemical crosstalk but can also be mediated by direct mechanical interactions [9, 10].

Here we provide a three-dimensional spheroid invasion assay to assess the invasive pattern of cocultured CAFs and cancer cells in organotypic ECM. This method enables multicellular coculture and is compatible with time-lapse fluorescent imaging using spinning disc microscopy. Collective invasion patterns can be monitored over a period of 24–72 h and leader/follower behavior of fibroblasts and distinct tumor cell subpopulations can be tracked within the invasive strands. We describe how to assess the role of CAFs in the collective invasion of epithelial cancer cells. Analysis of spheroid invasion patterns in 3D revealed that CAFs positioned at the leading front within the invading strands and triggered cancer cell dissemination of otherwise poorly invasive epithelial cancer cells through specific cell–cell contacts [11]. Together with other studies, these findings highlight the importance of developing models of tumor invasion that integrate cancer cell heterogeneity and tumor–stroma interactions.

2 Materials

2.1 Cell Culture

All cell culture procedures should be performed under sterile conditions in a biological safety cabinet class II. In the following protocol, human CAFs and HCC1806 cells were used as a model for collective cell invasion but all the steps can be adapted to different types of fibroblast and cancer cell if spheroid formation and coculture are feasible (*see* **Note 1**).

1. Cells: Human Invasive Ductal Breast Carcinoma Cancer Associated Fibroblasts were collected from the BCI Breast Tissue bank and immortalized by pBABE-Hygro-HTERT retroviral transfection (Courtesy of Eric Sahai, Francis Crick Institute, UK). HCC1806 cells (Human breast ductal cancer cell line).
2. Culture medium: Dulbecco's modified Eagle's medium (DMEM) high glucose, supplemented with 10% fetal bovine serum (FBS), 1% insulin-transferrin-selenium, 100 U/mL penicillin, 100 μg/mL streptomycin, and 29.2 mg/mL glutamine.

HCC1806 cells were grown in RPMI 1640 Medium supplemented with 10% FBS, 100 U/mL penicillin, 100 μg/mL streptomycin and 1% of MEM Nonessential Amino Acid Solution (100×).

3. Trypsin EDTA solution 1×.
4. Phosphate buffered saline sterile.

2.2 Hanging-Drop Method for Spheroid Self-Assembly

1. Sterile syringe filter with a 0.22 μm pore size hydrophilic Polyethersulfone (PES) membrane, 33 mm diameter.
2. 1.2% Methylcellulose solution: Weight 0.6 g Methylcellulose and transfer into a 50 mL tube with a magnetic stirring bar. Make up to 50 mL with serum-free DMEM high glucose. Stir until fully dissolved (it might take several minutes). Filter with a 0.22 μm filter, aliquot and store at 4 °C (*see* **Note 2**).
3. CAFs culture medium (*see* **Note 3**), warmed at 37 °C before use.
4. 0.5% Trypsin EDTA, warmed at 37 °C before use.
5. PBS buffer, sterile.
6. Cell labeling: In order to visualize the different population of cell in the invasion assay, several strategies of cell labeling by fluorescent dyes can be used. In this protocol we use CellTracker™ Green CMFDA Dye and Deep Red Dye staining solution (Invitrogen) to stain the CAFs and cancer cells respectively (*see* **Note 4**).
7. Sterile Petri dishes (100 mm).
8. 1.5 mL sterile centrifuge tubes.
9. Hemocytometer chamber for cell counting.
10. Cell mixture: *see* Table 1.

2.3 Embedding Spheroids into Matrix

1. 5×Col solution (*see* **Note 5**): Weigh 2.5 g of DMEM powder, high glucose, pyruvate, and 1 g Sodium Bicarbonate ($NaHCO_3$) transfer into a 50 mL tube. Add 5 mL of 1 M Hepes pH 7.5 and make up to 50 mL with milli-Q water. Stir until the powder is completely dissolved. Filter with a 0.22 μm filter and store at 4 °C.
2. Rat tail collagen type I. Store at 4 °C. The stock concentration ranges between 8 and 11 mg/mL (*see* **Note 6**).
3. Matrigel matrix (*see* **Note 7**). Make aliquots and store at −20 °C. Avoid multiple freeze-thaws. Thaw overnight at 4 °C in ice one day before use.
4. Fetal bovine serum (FBS), filtered with a 0.22 μm filter. Keep at 4 °C until use.

Table 1
Cell suspension solution for hanging-drop culture

	Mix 1:1 ratio CAFs:CC	Mix 2:1:1 ratio CAFs:CC_1:CC_2
1.2% Methylcellulose solution	200 μL	200 μL
CAF suspension (750,000 cells/mL)	83.5 μL	83.5 μL
Cancer cell suspension (750,000 cells/mL)	83.5 μL	CC_1 41.75 μL CC_2 41.75 μL
CAF culture medium	633 μL	633 μL

CC cancer cell, *CC_1* cancer cell type 1, *CC_2* cancer cell type 2

Table 2
Collagen I/Matrigel gel mixture

	Volume (μL)
Matrigel	225
Rat tail collagen type I	421
5×Col	84
FBS	100
Culture medium	120
Pelleted spheroid (remaining volume)	50

5. CAFs culture medium, warmed at 37 °C before use. Keep a 500 μL aliquot at 4 °C.
6. 12-Well glass-bottomed cell culture plates, glass thickness #0 (0.085–0.115 mm) (MatTek) (*see* **Note 8**).
7. ECM mixture (*see* Table 2):
   5×Col solution.

   Matrigel. The aliquot should be thawed the day before the experiment in a box of ice at 4 °C.

   FBS, filtered.

   CAFs medium (*see* **Note 3**).

   Rat tail collagen type I.

2.4 Image Acquisition and Analysis

1. Confocal microscope with and environmental chamber for temperature and CO_2 control and a motorized stage for multiple *XYZ* positions.

2. Software to program multiple positions, *z*-stack and multiple wave lengths acquisition.
3. 10× objective (e.g., Plan Fluor, NA 0.30, Ph1) (*see* **Note 9**).
4. CO_2 chamber and temperature control (set at 37 °C).
5. Image analysis software (e.g., ImageJ/FIJI, Imaris or equivalent).

3 Methods

3.1 Hanging-Drop Method for Spheroid Self-Assembly

1. Trypsinize HCC1806 cells and CAFs from a 60% confluent flask to single-cell suspension. Remove the cell culture medium, wash with PBS and add 1 mL of Trypsin EDTA 1× for a 75 cm^2 flask or 0.5 mL for a 25 cm^2 flask. Incubate at 37 °C until the cells detach (approximately 2 min incubation). Neutralize the trypsin by adding 10 mL culture medium and transfer to a 15 mL centrifuge tube. Mix well and keep 10 μL for counting cells (can be added directly to hemocytometer chamber).
2. Pellet cells by centrifugation. Aspirate the supernatant and resuspend the cell pellet with 3 mL of culture medium containing 2 μL of CellTracker staining solution (10 mM). Incubate each cell type with the chosen CellTracker Dye for 30 min at 37 °C (*see* **Note 4**).
3. Count HCC1806 cells and CAFs. Determine the resuspension volume needed to reach a concentration of 750,000 cells/mL.
4. After incubating for 30 min, dilute the CellTracker staining solution with the appropriate culture medium for each cell type up to 15 mL.
5. Centrifuge both tubes containing HCC1806 cells and CAFs and aspirate the supernatant.
6. Resuspend the cells with previously calculated volumes.
7. In a 1.5 mL centrifuge tube, prepare the cell suspension solution for spheroid preparation (*see* Table 1). The detailed volumes refer to 1:1 and 2:1:1 ratio between CAFs and cancer cells respectively. These proportions and the number of different cell types added may be adjusted according to the requirements of each experiment.
8. Fill the bottom of a sterile Petri dish with 10 mL sterile PBS to create a humid chamber (PBS will prevent the suspended drops from drying during spheroid formation) (Fig. 1).
9. Mix thoroughly the cell suspension solution and plate 20 μL drops on the lid of the Petri dish. Avoid bubble formation

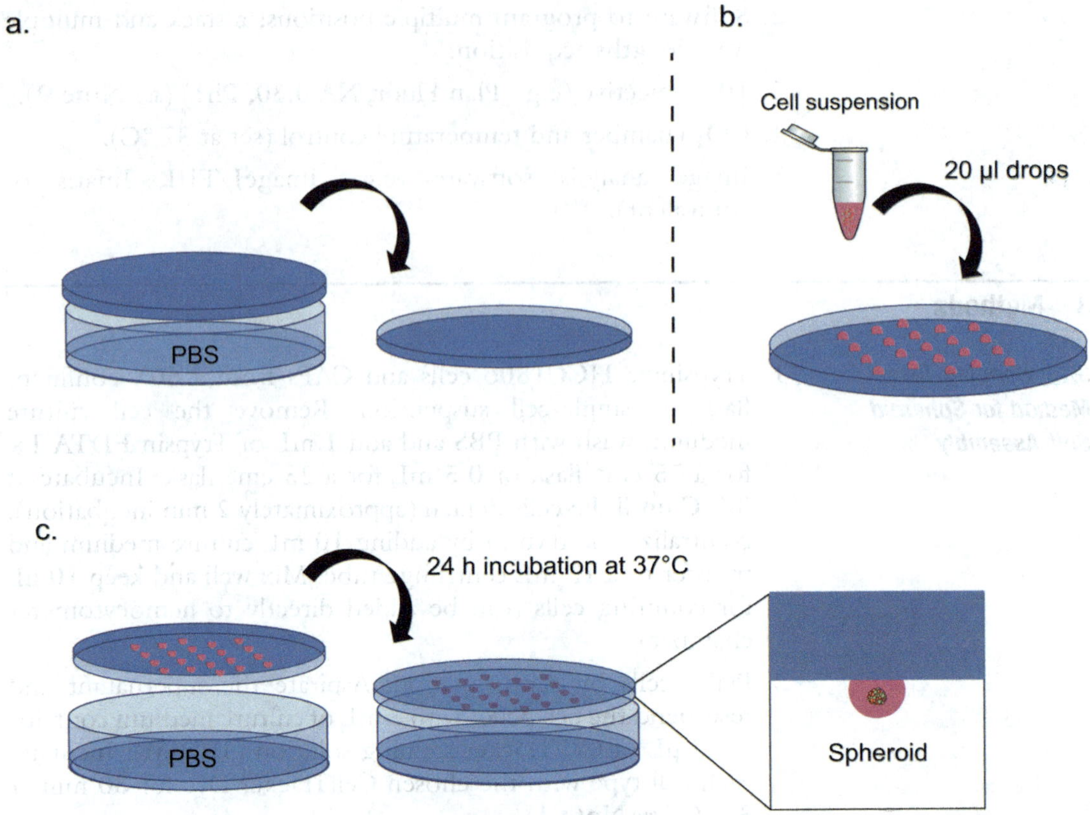

Fig. 1 Hanging-drop spheroid preparation method. (**a**) The bottom of a sterile Petri dish is filled with 10 mL sterile PBS. (**b**) 20 µL drops of the cell mixture are plated on the internal surface of the Petri dish. (**c**) The lid is then carefully placed on the Petri dish bottom (drops are positioned upside-down) and incubated at 37 °C, 5% CO_2 for at least 24 h to obtain one spheroid in each drop

within the drops. 1 mL of cell suspension should yield 50 spheroids of 2500 cells each (Fig. 1).

10. Gently invert the lid and place it on top of the Petri dish containing PBS.
11. Carefully transfer the plate into the incubator (37 °C, 5% CO_2).
12. After 24 h each drop should contain a single spheroid (it can take longer depending on the cell types) (Fig. 1).

3.2 Embedding Spheroids into Matrix

All steps should be carried out on ice unless specified otherwise. This protocol describes the 3D embedding of multiple spheroids into a Matrigel/collagen I gel (2 and 4 mg/mL final concentration, respectively).

1. In a 15 mL tube, kept on ice, add first the FBS, the culture medium, and the 5×Col solution (*see* **Note 10**), add then Matrigel and finally the collagen. While pipetting highly

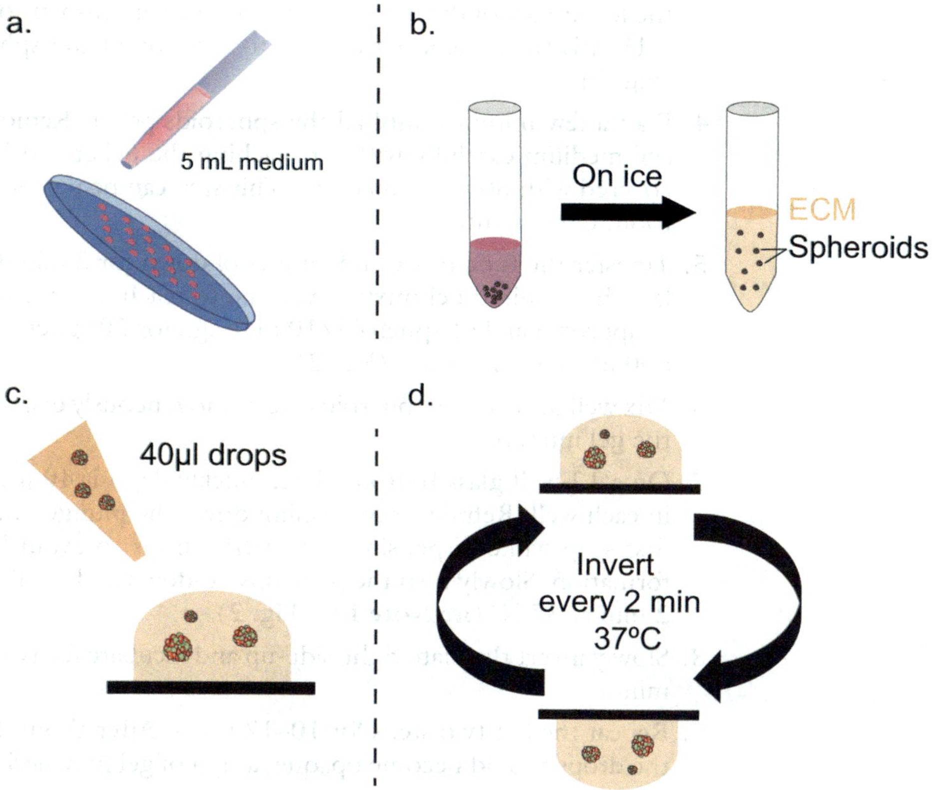

Fig. 2 Embedding the spheroids in the ECM gel mixture. (**a**) Spheroids are collected into a 15 mL tube by washing out the drops with 5 mL of cell culture medium. (**b**) After the spheroids pellet, the culture medium is removed. The tube is placed on ice and the gel mixture is added. (**c**) A 40 μL drop of ECM containing the spheroids is deposited in each well of a 12-well glass-bottom plate. (**d**) The plate is inverted upside-down every 2 min maintaining the plate at 37 °C, 5% CO_2. This step is repeated at least ten times

viscous solutions such as Matrigel and collagen, leave the tip immersed in the liquid for few seconds after releasing the piston of the pipette allowing the proper aspiration of the desired volume. Moreover, refrain from pushing down the plunger after the first stop when dispensing the viscous fluids to avoid the formation of air bubbles. Table 2 describes volumes adjusted for a 9.5 mg/mL collagen I and 8.9 mg/mL Matrigel stock solutions, respectively for a total of 1 mL gel volume (*see* **Notes** 7 and **8**).

2. With a 1 mL pipette, mix thoroughly avoiding bubble formation. Keep mixing slowly until the color of the solution becomes salmon pink. Always work on ice to avoid premature gel polymerization. Keep the solution on ice during the spheroid collection step.
3. Spheroid collection: tilt the lid and slowly flush 5 ml of culture medium with a 5 mL pipette to concentrate the spheroids in

the lower part of the lid. Transfer the spheroid suspension into a 15 mL tube. Repeat this step to retrieve all the spheroids (Fig. 2).

4. Wait a few minutes until all the spheroids pellet. Remove the cell medium carefully without reaching the pellet and leave it covered with 50 μL of medium. This step can be performed at room temperature.
5. Transfer the tube to ice and let it cool down for 2 min. Calculate the required gel mixture volume to reach a concentration of approximately 1 spheroid/10 μL (e.g., for 50 spheroids add 500 μL of gel mixture) (Fig. 2).
6. Mix well until all the spheroids are homogeneously dispersed in the gel mixture.
7. On a 12-well glass-bottom plate, quickly deposit 40 μL drops in each well. Refrain from pushing down the plunger after the first stop while dispensing the matrix drops to avoid bubble formation. Slowly turn the plate upside-down and incubate for 2 min at 37 °C (*see* **Note 11**) (Fig. 2).
8. Slowly invert the plate right-side-up and incubate for two more minutes.
9. Repeat the last two steps for 10–12 times. After about 20 min the drops should become opaque, a sign of gel polymerization.
10. Incubate the plate 20 min at 37 °C and 5% CO_2 while keeping it upside-down (*see* **Note 12**).
11. Add 1 mL of culture medium in each well dispensing it slowly on the border of the well to avoid detachment of the gel drops.
12. At this step, proper 3D embedding can be checked using a standard inverted phase contrast microscope. The optimal position of the spheroid is in the center of the drop, as spheroids located too close to the upper surface or to the glass bottom may migrate in 2D. Incubate at 37 °C and 5% CO_2 for 48–72 h until invasive strands are formed (incubation time can vary depending on the cell type).

3.3 Image Acquisition

Microscopy acquisition is performed at 37 °C in a CO_2 humidity-controlled chamber. Acquisition can be performed either semiautomatically or sequentially, recording a *z*-stack of each spheroid separately. Here we describe a semi-automated procedure where all positions are set prior to acquisition. This option can also be used to run time-lapse imaging.

1. Insert 10× objectives (e.g., Plan Fluor, NA 0.30, Ph1) (*see* **Note 9**).
2. Carefully fix the plate inside the plate holder and cover with the CO_2 chamber.

3. Turn on the microscopy software.
4. Using bright-field live imaging, find a spheroid and assess optimal laser power and exposure time for each fluorescent laser channel.
5. In the multi-acquisition tab, select *z*-stack option, relative *z*, start with 0 and end with 200–250 μm with a step of 1–10 μm depending on the desired resolution and the spheroid size. In the described experimental setting, spheroid diameter varies from 200 to 250 μm approximately, but this depends on the cell type used. Set acquisition channels with the optimal exposure times established in **step 4**.
6. Set the position focus on the lowest *z* plane of the spheroid. Stack acquisition will start from that plane, moving upward to the uppermost plane of the spheroid. When running time-lapse acquisition we used an automated focus system to avoid focus drift. If your microscope is not equipped with such system, you may need to manually adjust the focus at regular intervals during acquisition and/or set-up wider *z*-stacks to avoid the sample to go out of the field of view during the acquisition (*see* **Note 9**).
7. Select destination folder for saving images.
8. Once all positions are set, run the experiment.

3.4 Image Analysis

Image analysis is performed using the open source ImageJ/FIJI [12]. Open a stack in ImageJ and using the ImageJ *z*-projection function, under the "Stacks" tab, process all stacks to obtain maximum-intensity projections of the saved *z*-series (acquired with the 10× objective). From each *z*-stack, the number and the length of invading strands as well as the identification of leader/follower cells within each strand can be assessed as follows (Fig. 3) (*see* **Note 13**):

1. Calibrate spatially your image setting appropriate image scale based on the camera pixel-size value and magnification (in this experiment, camera pixel-size value: 0.645 μm/pixel using a 10× magnification objective).
2. Select the straight-line tool. Draw a line starting from the strand base, at the spheroid circumference, to its tip. Measure its length. Repeat this step for all strands, evaluating both length and number of invading strands (Fig. 3a).
3. From time-lapse images, identify leader/follower dynamics (Fig. 3b).

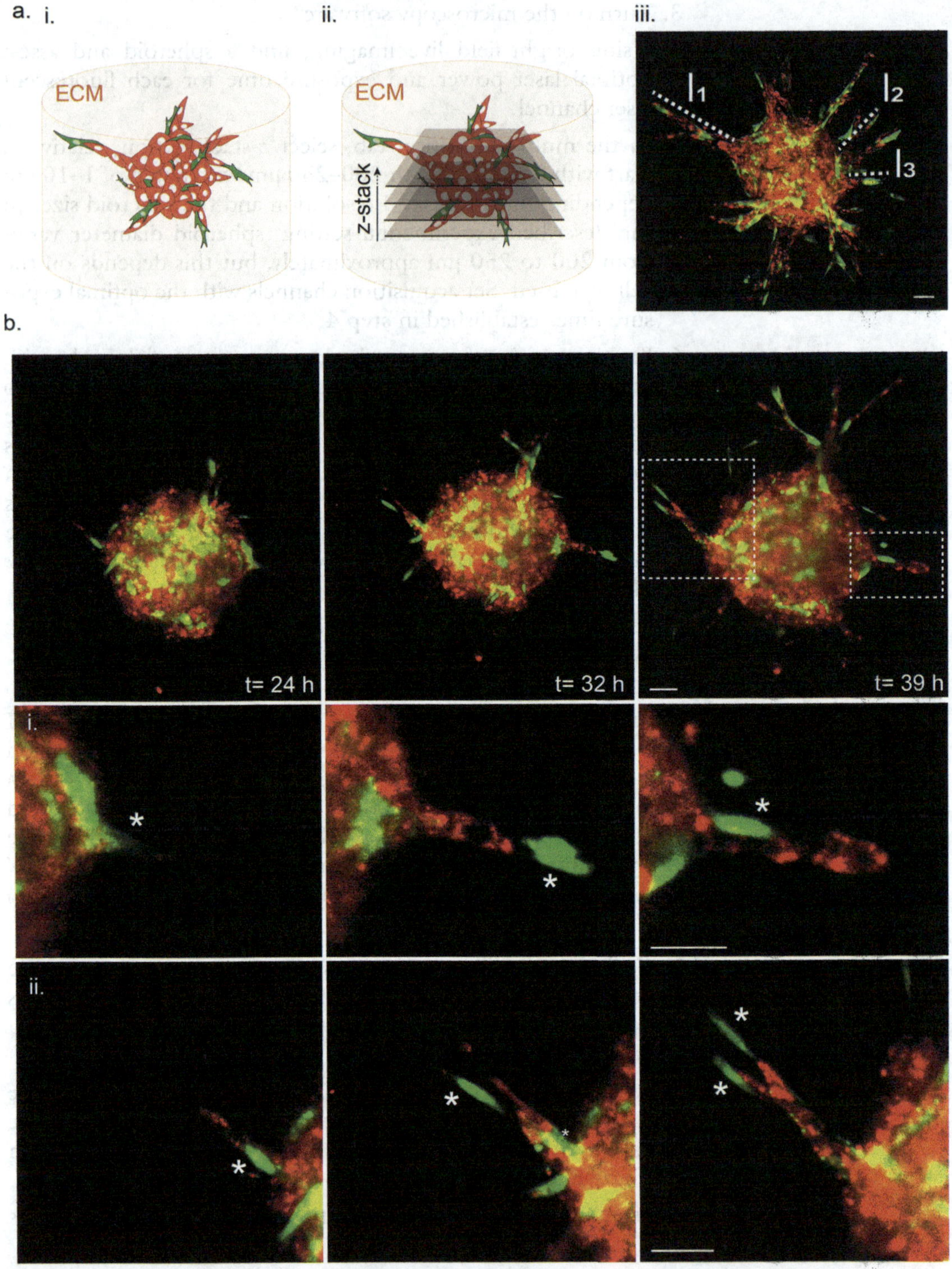

Fig. 3 Image acquisition of the 3D spheroid invasion. (**a**(i) and **a**(ii)) Illustrations of the acquisition process. (**a**(iii)) Representative maximum fluorescence intensity *z*-projected images of spheroid invasion 48 h after embedding in ECM (taken with 10× objective). White dashed lines depict some measurement of the lengths of invasive strands (l_1, l_2 and l_3 respectively). Cancer cells (red), CAFs (green) were mixed in 1:1 ratio. Scale bar:

4 Notes

1. This technique can be applied to different cell types and multicellular models. Depending on the chosen cell type, optimization of the spheroid assembly step may be required. In fact, differences in cell–cell adhesions can result in loose aggregates or nonhomogeneous spheroids. Several strategies may help to promote cell clustering including use of ultralow attachment culture plates, higher densities of methylcellulose or using other additives to modulate viscosity and compactness [13].
2. Even when completely dissolved, methylcellulose solution is extremely viscous and difficult to filter. We suggest filtering small volumes at a time and changing filters often to prevent clogging. Nevertheless, it is likely that filtering might result into a slightly lower percentage of methylcellulose in the final stock. Alternatively, the solution can be autoclaved.
3. This protocol involves at least two different cell types. This implies having to choose the optimal culture medium for both cell types. If both cell types require specific media, the two can be mixed in equal proportion. It is then recommendable to test the medium compatibility for all cell types prior coculturing the cells.
4. Cell labeling with CellTracker: 2–3 days before the experiment, HCC1806 cells and CAFs can be passed to a 25 cm^2 flask so they will reach 70% confluency the day of spheroid preparation step. Cell labeling can be performed directly in the flask, removing culture medium and adding the CellTracker staining solution.
5. Collagen concentration, polymerization temperature, and pH strongly affect the mechanical and structural properties of the obtained hydrogel. Therefore, it is important to maintain stable conditions of temperature across experiments to ensure reproducibility. 5×Col functions as a buffer increasing the pH of the ECM mixture until it becomes salmon-colored, indicating that a neutral pH has been reached. Consequently, it is of critical importance to adjust the volume of the 5×Col solution accordingly to stock concentrations of collagen I and Matrigel, to maintain comparable gel porosity after polymerization while using different stock solutions [14].

Fig. 3 (continued) 50 μm. (**b**) Representative maximum fluorescence intensity *z*-projected images of three time points (time = 24, 32 and 39 h) extracted from time-lapse imaging of spheroid invasion acquired 24 h after embedding in ECM (taken with 10× objective). Images were taken every 30 min during 15 h. Images showed in **b**(i) and **b**(ii) correspond to the dashed box depicted in (**b**). Asterisks show the position of CAFs over time within the invasive strands. Scale bars: 50 μm

6. Keep Collagen Rat tail type I on ice anytime it is not stored at 4 °C to avoid polymerization. Different batches may have different pH and concentration. Anytime a new batch is used, calculate volume needed to reach 4 mg/mL concentration and verify that the gel mix solution becomes salmon colored once properly mixed.
7. Matrigel concentration varies from one batch to another, consider adjusting all volumes of the ECM mixture when changing batch. The final concentration should be 2 mg/mL.
8. To image thick samples such as 3D embedded spheroids with high magnification and high numerical aperture objectives (e.g., 40× NA 0.75) it is important to use bottom glass dishes with glass thickness #0 (0.085–0.115 mm) for optimal microscopy imaging.
9. The choice of the objective will depend on the quality of the fluorescent signal and on the purpose of the assay. If the signal to noise ratio of the fluorescent signal is high enough and the migration features are easy to describe (e.g., cell strands, chains, or clusters) then the positioning of cells in strand can be assessed with a 10× objective. To assess more precisely cellular or subcellular events (e.g., cell–cell interactions, actin dynamic) or if the fluorescent signal to noise ratio is deem, the use of 20–40× with high numerical aperture (e.g., 20×, NA 0.75; 40×, NA 0.75) might be required.
10. During gel mixture preparation, slowly pipette the highly viscous collagen and Matrigel solution, allowing it to flow into the 1 mL pipette tip after fully releasing the plunger button. While dispensing, do not immerge the tip into the solution (causing bubbles). Instead, pipette viscous reagents to the sidewalls of the tube allowing the solution to flow to the bottom of the tube. Avoiding bubbles is critical in this step. We emphasize the importance of preventing drops containing air bubbles. Indeed, they easily detach from the glass bottom and reduce the quality of the acquisitions.
11. This step should be performed rapidly to avoid the polymerization of the first drops at room temperature while dispensing new drops. Rapid deposition will also prevent the embedded spheroids from precipitating into the drop, reaching the glass bottom of the plate. This could result in a 2D migration of the spheroids within the drop. To overcome this problem, we suggest keeping the plate on ice while seeding the drops. Alternatively, less drops can be plated for each experiment using 6-well glass-bottom plates or single glass-bottom dishes.
12. Do not leave the ECM drops without media more than 40 min in total (including the step of plate inversion and the following 20-min incubation) to prevent drying.

13. Quantification of the length of the strands based on a z-projected image is affected by the orientation of the strand toward the z plane. This method of quantification gives an estimation of the overall invasive capacity of the spheroid and z-projection could result in an underestimation of the real length values. For more exact quantification, measurement of invasive strands can be performed through 3D reconstruction of the spheroid with smaller z steps.

Acknowledgements

This study was supported by MINECO/FEDER (PGC2018-099645-B-I00 to X.T.), the Generalitat de Catalunya (SGR-2017-01602 to X.T. and CERCA Program), the European Research Council (CoG-616480 to X.T.), the European Commission (Grant Agreement SEP-210342844 to X.T.). A.L. has received financial support through the Junior Leader Postdoctoral Fellowship Programme from "la Caixa" Banking Foundation (LCF/BQ/PR18/11640001). T.K. is funded by Marie-Curie action (HeteroCancerInvasion # 708651) and the Japanese Strategic Young Researcher Overseas Visits Program for Accelerating Brain Circulation. D.P is supported by Breast Cancer Now (BCN) (2013NovPR182/Breast Cancer Now (BCN)). S.C is supported by the 2017 FPI MINECO doctoral fellowship programme (BES-2017-079847). E.S. is supported by the Francis Crick Institute, which receives its core funding from Cancer Research UK (FC001144), the UK Medical Research Council (FC001144) and the Wellcome Trust (FC001144). The authors wish to acknowledge the role of the Breast Cancer Now Tissue Bank in collecting and making available the samples used in the generation of this publication. IBEC is recipient of a Severo Ochoa Award of Excellence from the MINECO.

References

1. Chambers AF, Groom AC, MacDonald IC (2002) Dissemination and growth of cancer cells in metastatic sites. Nat Rev Cancer 2:563–572. https://doi.org/10.1038/nrc865
2. Cheung KJ, Ewald AJ (2016) A collective route to metastasis: seeding by tumor cell clusters. Science 352:167–169. https://doi.org/10.1126/science.aaf6546
3. Konen J, Summerbell E, Dwivedi B et al (2017) Image-guided genomics of phenotypically heterogeneous populations reveals vascular signalling during symbiotic collective cancer invasion. Nat Commun 8:15078
4. Cheung KJ, Padmanaban V, Silvestri V et al (2016) Polyclonal breast cancer metastases arise from collective dissemination of keratin 14-expressing tumor cell clusters. Proc Natl Acad Sci U S A 113:E854–E863. https://doi.org/10.1073/pnas.1508541113
5. Mayor R, Etienne-Manneville S (2016) The front and rear of collective cell migration. Nat Rev Mol Cell Biol 17:97–109
6. Cheung KJ, Gabrielson E, Werb Z, Ewald AJ (2013) Collective invasion in breast cancer requires a conserved basal epithelial program. Cell 155:1639–1651

7. Zhang J, Goliwas KF, Wang W et al (2019) Energetic regulation of coordinated leader–follower dynamics during collective invasion of breast cancer cells. Proc Natl Acad Sci U S A 116:7867–7872. https://doi.org/10.1073/PNAS.1809964116
8. Gaggioli C, Hooper S, Hidalgo-Carcedo C et al (2007) Fibroblast-led collective invasion of carcinoma cells with differing roles for RhoGTPases in leading and following cells. Nat Cell Biol 9:1392–1400
9. Labernadie A, Kato T, Brugués A et al (2017) A mechanically active heterotypic E-cadherin/N-cadherin adhesion enables fibroblasts to drive cancer cell invasion. Nat Cell Biol 19:224–237. https://doi.org/10.1038/ncb3478
10. Glentis A, Oertle P, Mariani P et al (2017) Cancer-associated fibroblasts induce metalloprotease-independent cancer cell invasion of the basement membrane. Nat Commun 8(1):924. https://doi.org/10.1038/s41467-017-00985-8
11. Theveneau E, Linker C (2017) Leaders in collective migration: are front cells really endowed with a particular set of skills? F1000Res 6:1899. https://doi.org/10.12688/f1000research.11889.1
12. Schindelin J, Arganda-Carreras I, Frise E et al (2012) Fiji: an open-source platform for biological-image analysis. Nat Methods 9:676–682. https://doi.org/10.1038/nmeth.2019
13. Leung BM, Lesher-Perez SC, Matsuoka T, Moraes C, Takayama S (2015) Media additives to promote spheroid circularity and compactness in hanging drop platform. Bio Sci 3 (2):336–344
14. Antoine EE, Vlachos PP, Rylander MN (2014) Review of Collagen I Hydrogels for Bioengineered Tissue Microenvironments: Characterization of Mechanics, Structure, and Transport. Tissue Engineering Part B: Reviews 20 (6):683–696

Chapter 20

Using Xenopus Neural Crest Explants to Study Epithelial-Mesenchymal Transition

Nadège Gouignard, Christian Rouvière, and Eric Theveneau

Abstract

The epithelial-mesenchymal transition (EMT) converts coherent epithelial structures into single cells. EMT is a dynamic cellular process that is not systematically completed (not all EMTs lead to single cells) and reversible (cells can re-epithelialize). EMT is orchestrated at multiple levels from transcription, to post-translational modifications, to protein turnover. It involves remodeling of polarity and adhesion and enhances migratory capabilities. During physiological events such as embryogenesis or wound healing EMT is used to initiate cell migration, but EMT can also occur in pathological settings. In particular, EMT has been linked to fibrosis and cancer. Neural crest (NC) cells, an embryonic stem cell population whose behavior recapitulates the main steps of carcinoma progression, are a great model to study EMT. In this chapter, we provide a fully detailed protocol to extract NC cells from Xenopus embryos and culture them to study the dynamics of cell–cell adhesion, cell motility, and dispersion.

Key words Neural crest, Epithelial-mesenchymal transition, Cell migration, Adhesion, Polarity, Dispersion

1 Introduction

In the 1960s, during chick embryo gastrulation, epiblast cells were shown to dissociate from the epithelium, lose their epithelial morphology, and become mesoderm and endoderm cells [1, 2]. The concept of Epithelial-Mesenchymal Transformation was then introduced and defined as a complete loss of epithelial traits, including apicobasal polarity and cell–cell adhesion, accompanied by total acquisition of mesenchymal characteristics (e.g., front-back polarity, cell motility) [1, 2]. E-Cadherin, a cell–cell junction protein, and vimentin, an intermediate filament protein, were soon proposed as makers for epithelial and mesenchymal cells, respectively.

Electronic supplementary material: The online version of this chapter (https://doi.org/10.1007/978-1-0716-0779-4_20) contains supplementary material, which is available to authorized users.

Kyra Campbell and Eric Theveneau (eds.), *The Epithelial-to Mesenchymal Transition: Methods and Protocols*, Methods in Molecular Biology, vol. 2179, https://doi.org/10.1007/978-1-0716-0779-4_20,

However, a decade later, the term transition was preferred over transformation to reflect the dynamics, the complexity, and the plasticity of this mechanism. Indeed, cells were described in various stages of EMT co-expressing markers of epithelial and mesenchymal states or neither. For a historical perspective on the field of EMT we recommend the review by Angela Nieto in Annual Reviews of Cell and Developmental Biology published in 2011 [3].

EMT can be initiated by various signaling pathways, including the Bone Morphogenetic Protein (BMP), Wnt, Fibroblast Growth Factor (FGF), Notch and hypoxia pathways, which induce the expression of transcription factors from the Twist, Snail, SoxE, FoxD, Ets, and Zeb families. These transcription factors modulate the expression of the proteins involved in cell–cell junctions (type I and type II cadherins) to reduce cell–cell junction strength and promote the loss of apicobasal polarity. They also induce the expression of cell-matrix adhesion proteins (e.g., Integrins), extracellular matrix components (e.g., Fibronectin, Collagen, Vitronectin, Laminin) and remodeling factors (e.g., ADAMs, MMPs) as well as cytoskeleton proteins and regulators (e.g., Keratins, Vimentin, small GTPases) necessary for front-back polarity and motility. In the 1990s, a parallel was drawn between tumor progression/metastasis and EMT, identifying common actors and signaling pathways (reviewed in [4, 5]). It was suggested that cancer cells were hijacking the embryonic EMT programs. Therefore, a better understanding of the molecular mechanisms regulating EMT was believed to be of primary importance to understand tumor progression.

During the development of vertebrate embryos, EMT is essential for the delamination and long-distance migration of neural crest (NC) cells, a transient multipotent population of cells that arises at the neural plate border and gives rise to a wide variety of derivatives [6]. Over the years, NC cells became a well establish model for physiological EMT, delamination and migration mechanisms, notably for its accessibility and ease of manipulation both in vivo and ex vivo.

In particular, Xenopus NC development recapitulates the main step of EMT and long-distance migration observed in most epithelial cancers. Xenopus NC cells can be defined as the cell population initiating the expressing of a repertoire of EMT transcription factors (Twist1, Snail1/2, Ets1, Zeb2) at the border of the neural plate, at the end of neurulation. Expression of these genes is controlled by the BMP, Wnt, FGF, Notch and hypoxia pathways [7–9]. Xenopus NC cells then undergo a typical case of cadherin switching, going from E-cadherin-dependent cell–cell junctions to N-cadherin ones [10]. This switch is associated with a change of cell polarity from apicobasal to front-rear, and is directly linked to the acquisition of motility and dispersion, by endowing NC cells with the ability to perform contact-inhibition of locomotion [10]. Importantly, EMT in Xenopus NC cells is also linked with the acquisition of chemotactic capabilities, since the Hif pathway co-regulates Twist, an E-cadherin repressor, and the expression of CXCR4, the main receptor for CXCL12/Stromal cell-derived factor 1 [9, 11]. Finally, Xenopus NC cells express a wide range of matrix remodeling factors including several MMPs and ADAMs [12]. Altogether, this makes the Xenopus NC cells an extremely relevant model system to study EMT, from its upstream regulators to its downstream effectors and their impact on cell motility.

From a practical point of view, Xenopus produce a high number of large embryos, developing externally, allowing complete access at all times to developmental stages which are hardly accessible in other models. Furthermore, Xenopus embryos allow for gain and loss of function studies, by simple microinjection of Morpholino oligonucleotides, CRISPR/gRNA or mRNA in blastomeres during segmentation. The use of nontoxic fluorescent dyes permits lineage tracing during developmental processes. Using this technique, a cell fate map was generated, allowing reliable targeting of subpopulations of cells, including NC, by injecting specific blastomeres. Finally, NC are easily extracted from developing embryos to be grafted into host embryos for in vivo studies, or cultured ex-vivo in coated Petri dishes in a simple culture medium up to differentiation stages. If dissected at late neurula stage, NC are fully induced and will spontaneously undergo EMT-driven dispersion and migration during the first few hours following grafts or culture. EMT and migration of NC cells can be monitored in vivo or ex vivo using time-lapse cinematography.

In this book chapter, we describe the protocol to extract NC cells for ex vivo culture and discuss how to monitor and analyze NC dispersion and migration from time-lapse images.

2 Materials

2.1 Ex-Vivo Culture of Xenopus Neural Crest cells

2.1.1 Extracellular Matrix Preparation

1. Petri dishes (any dish from 35 to 90 mm made of untreated plastic) or multi-well dishes (e.g., LabTek, Ibidi µSlides 80.821 or equivalent).
2. Purified Fibronectin in solution (Sigma F1141 or equivalent reference from other providers).
3. Phosphate Buffer Saline (PBS) 1×: Start with 800 mL of distilled water add 8 g of NaCl, 0.2 g of KCl, 1.44 g of Na_2HPO_4, 0.24 g of KH_2PO_4, adjust the pH to 7.4 with HCl, add distilled water to a final volume of 1 L.
4. Bovine Serum Albumin (BSA).
5. Heating block.

2.1.2 Microdissection (Fig. 1)

1. Hair knife. Dip the tip of a short glass Pasteur pipette into melted wax and insert the root of a hair (eyebrow, eyelash or else) in the tip of the pipette before the wax has set.
2. A dish filled with modelling clay.
3. Glass ball. Prepared by melting the tip of a short glass Pasteur pipette.
4. Fine forceps.
5. P20 micropipette.
6. A small uncoated petri dish filled with culture medium to collect NC explants.

2.1.3 Amphibian Media and Culture Media

1. Normal Amphibian Medium (NAM) stock solutions, must be autoclaved and can be stored at room temperature. In distilled water, NAM A (NaCl, 1.1 M; KCl, 20 mM; $NaHCO_3$, 10 mM), NAM B ($MgSO_4$, 10 mM; $CaCl_2$, 10 mM), NAM C (EDTA, 1 mM), NAM D (Na_2HPO_4, 16 mM; NaH_2PO_4, 4 mM).
2. Working solutions: NAM 0.1× (for 1 L add 10 mL of each NAM stock solutions and add 960 mL of distilled water), NAM 0.25× (for 1 L add 25 mL of each NAM stock solutions and add 900 mL of distilled water).
3. Culture medium—Danilchik's for Amy (DFA): 53 mM NaCl, 5 mM Na_2CO_3, 4.5 mM KGluconate, 32 mM NaGluconate, 1 mM $MgSO_4$ ($7H_2O$), 1 mM $CaCl_2$, 0.1% BSA; adjust pH to 8.3 with 1 M Bicine. Store at −20 °C. After thawing, add 1000 U penicillin and 100 μg/mL streptomycin.

2.2 Time-Lapse Imaging

1. Microscopes: inverted, upright microscope, dissecting microscope (*see* **Note 1**).
2. Camera and software.
3. Temperature controlled room (*see* **Note 2**).
4. Formaldehyde.

2.3 Analysis of Cell Migration and Dispersion Dynamics

1. A computer equipped with a modern CPU and with sufficient amount of RAM and storage space to process the data. The total RAM available should be twice the size of your largest dataset to allow opening and processing of the data.
2. A free (FIJI/Image J) or commercial software (e.g., Imaris/Bitplane) for automated detection of nuclei, tracking, and dispersion analysis.

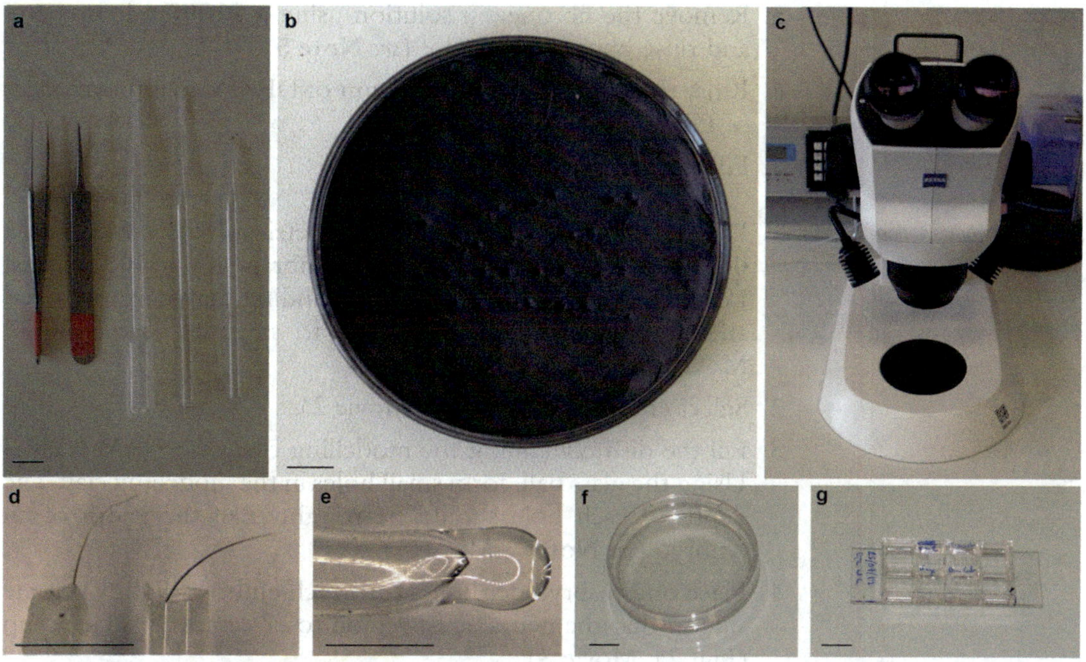

Fig. 1 Tools for microdissection of Neural Crest explants. (**a**) Tools for microdissection: from left to right, fine forceps to remove the vitelline membrane, plastic Pasteur pipette for transferring embryos, hair knives and glass ball for digging holes in the modelling clay. (**b**) Dish filled with modelling clay to hold embryos during dissection. (**c**) Stereomicroscope. (**d**) Zoom-in on the tips of the hair knives. (**e**) Zoom-in on the tip of the glass ball. (**f**) A small dish containing DFA1× to collect explants. (**g**) A fibronectin-coated dish to culture the explants. All scale bars are 1 cm, except in (**d**) and (**e**), 500 μm

3 Methods

3.1 Ex-Vivo Culture of Xenopus Neural Crest Cells

3.1.1 Extracellular Matrix Preparation

1. Aliquot the Fibronectin solution as a 10× stock by 20 or 50 μL in 500 μL centrifuge tubes and store at −20 °C. For coating on glass coverslips, the 10× stock is at 1 mg/mL. For coating on plastic petri dishes, the 10× stock is at 100 μg/mL. When making the 10× stock for plastic, use PBS1× containing 0.1% BSA for long-term stability of the solution.
2. On the day of the experiment, select a dish appropriate to your microscope (*see* **Note 3**).
3. For glass dishes/coverslips dilute your Fibronectin in PBS1× to a final concentration of 100 μg/mL, for plastic, to 10 μg/mL. Directly thaw the frozen aliquot of Fibronectin by adding the required amount of PBS1× at room temperature in the Fibronectin aliquot. (e.g., add 450 μL of PBS1× on top of a 50 μL aliquot of 10× Fibronectin to obtain 500 μL of Fibronectin 1× for coating) (*see* **Note 4**).
4. Place the dish on a heating block for 1 h at 37 °C.

5. Remove the fibronectin solution using a P1000 micropipette and rinse once with PBS1× (*see* **Note 5**).
6. Rinse the dish with a small amount of DFA1× to remove traces of PBS1×.
7. Fill the dish with DFA1×.

3.1.2 Microdissection

1. Place embryos at stage 15/16 in a petri dish filled with NAM 0.25× and remove the vitelline membrane using fine forceps. Let them recover as superficial wounding may have occurred and let embryos further develop to stage 18 (*see* **Note 6**, Movie 1).
2. Select embryos at stage 18 (Movie 2).
3. Fill the dish containing the modelling clay with NAM 0.25×. Using the glass ball, form small holes in the modelling clay. The depth of each hole should be roughly half the width of an embryo (*see* **Note** 7, Movie 3).
4. Place one embryo per hole. Turn each embryo slightly on one side so that the neural crest region to be removed is facing up (Fig. 2a, Movie 3).
5. Using the glass ball, move gently the clay toward the embryo until it touches the epidermis making sure the orientation is preserved (Fig. 2b) then apply pressure all around the embryo by pushing the clay onto its sides to firmly secure the embryo (Fig. 2c, Movie 3, *see* **Note 8**).
6. Using the hair knife make a small incision posteriorly and dorsally to the neural crest area (Fig. 2c, step 1, Movie 3).
7. Insert the tip of the hair knife in that initial hole, keeping the hair parallel to the tissue. The hair knife should be in between the pigmented layer and the neural tissue (Fig. 2c, step 2, Movie 3).
8. Gently remove the pigmented layer from medial to lateral to uncover the neural crest region (Fig. 2c, step 3, Movie 3).
9. Place the hair knife parallel to the ventral side of the neural crest region (dotted line on Fig. 2d). Apply a gentle pressure (do not attempt to cut the tissue) and move from lateral to medial repeatedly. The neural crest region will detach from its surroundings (Fig. 2e, e′). Continue until the whole region detaches by itself (Movie 2). After dissection there should be a hole in which the underlying mesoderm can be seen (Fig. 2f). When a dissected embryo is released from the clay the actual size of the wound should be relatively small and the neural tube on the dissected side should be intact (Fig. 2g, g′). Look out for contamination from other cell types (Fig. 3, *see* **Note 9**). An example of a rough dissection including mesoderm is shown at the end of Movie 3.

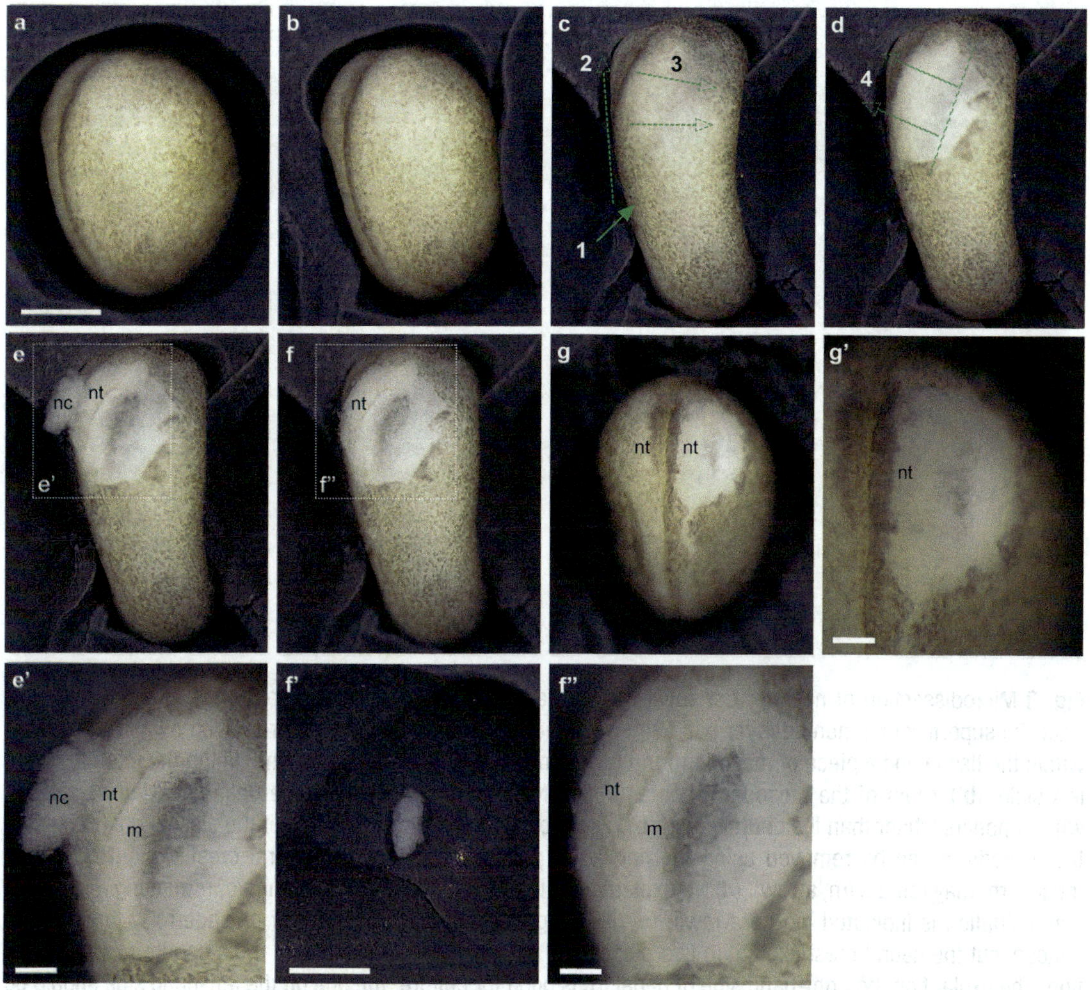

Fig. 2 Microdissection of neural crest cells: an ideal case. (**a**) Stage 18 embryo slightly turned on its left side to place the right NC region up for dissection. (**b**) The clay has been gently pushed to make contact with the embryo and to lock the orientation. (**c**) The clay was further pushed toward the embryo to firmly maintain the embryo prior to dissection. Three steps are necessary to remove the overlying pigmented layer. Step 1, make a superficial incision using the tip of the hair knife. Step 2, insert the hair knife underneath the skin, parallel to the embryo and move anteriorly. Do not stab the embryo toward the inside. Step 3, gently move from medial to lateral to remove the pigmented layer. Avoid using the tip of the hair knife, rather work with the side of the hair. (**d**) The embryo after the superficial pigmented layer has been removed. The green dotted line indicates where the hair knife should be positioned to remove the neural crest. Slightly press down using the side of the hair knife and move from lateral to medial (step 4, green arrows). (**e**) The NC explant (nc) has been lifted, the neural tube (nt) and mesoderm (m) are visible, *see* the zoom-in in (**e′**). (**f**) The NC explant has been removed, seen in (**f′**), the dissected region is shown in (**f″**). (**g**, **g′**) The embryo after dissection and freed from the clay. Scale bars: (**a–g**) 1 mm; (**e′**, **f″**, **g′**) 150 μm; (**f′**) 500 μm

10. Using a P20 micropipette set to 10 μL rapidly transfer the explant into a dish containing DFA1× (Movie 3).
11. Repeat **steps 3–10** to collect multiple explants.

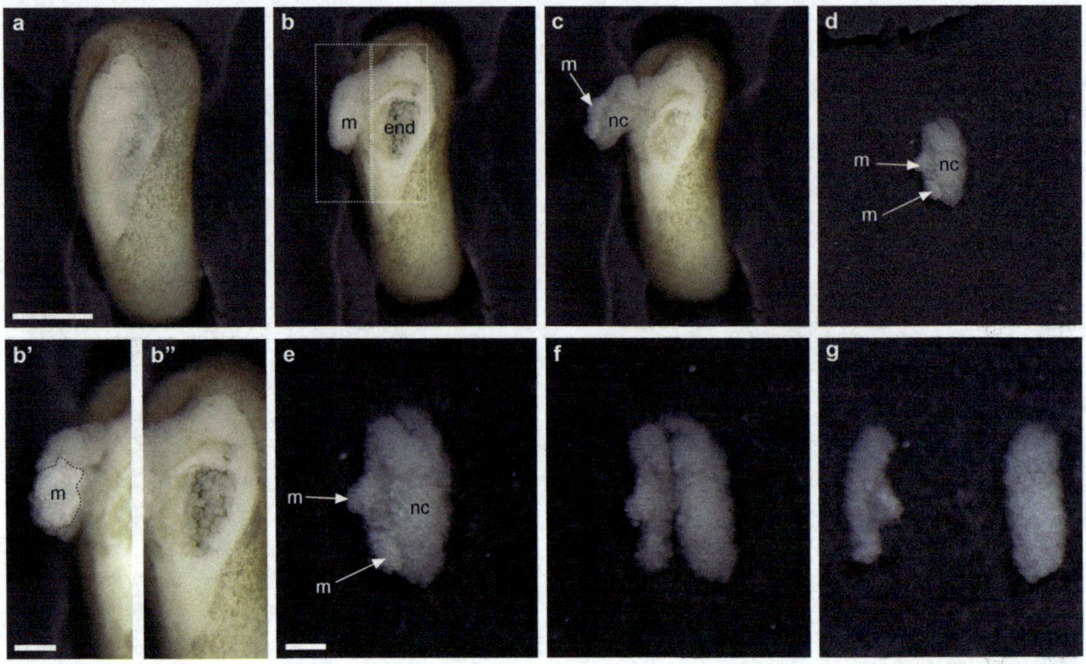

Fig. 3 Microdissection of neural crest cells: how to handle mesoderm contamination. (**a**) Stage 18 embryo after the superficial pigmented layer has been removed. (**b**) The NC explant was removed by going too deep within the tissue and a piece of mesoderm (m) is coming with it. The endoderm (end) lining the internal cavity is visible. (**b′**) Zoom of the mesoderm-contaminated NC explant. The dotted line delineates the mesoderm which appears whiter than the neural crest. (**b″**) Zoom of the endoderm at the site of the dissection. (**c**) Most of the mesoderm can be removed using the hair knife prior to detaching the neural crest explant but some mesoderm may remain (m, arrow). (**d**) NC explant after being detached from the embryo, remaining mesoderm contamination is indicated by the arrows. (**e–g**) Trying to clean the small bits of mesoderm is too difficult. Instead, cut the neural crest explant in two pieces and discard the half containing the mesoderm contamination. The explant on the right-hand side of panel (**g**) is good for culture, the one on the left-hand side should be discarded

12. Cut the explants in small pieces using the hair knife and fine forceps (Movie 4). A complete unilateral neural crest region can be cut into 4–5 pieces each containing ~100–200 cells.
13. Transfer the explants onto the Fibronectin-coated dish.
14. Array the explants to facilitate monitoring of the explants under the microscope (*see* **Note 10**).
15. Let the explants adhere to the substrate for at least 15 min before moving the dish.

3.2 Time-Lapse Imaging

1. Gently place your dish onto the microscope.
2. Check explants viability and fluorescence if embryos were microinjected with a tracer prior to dissection.
3. Adjust lighting conditions (light intensity, exposure time).
4. Record the *XY* positions of each explants.

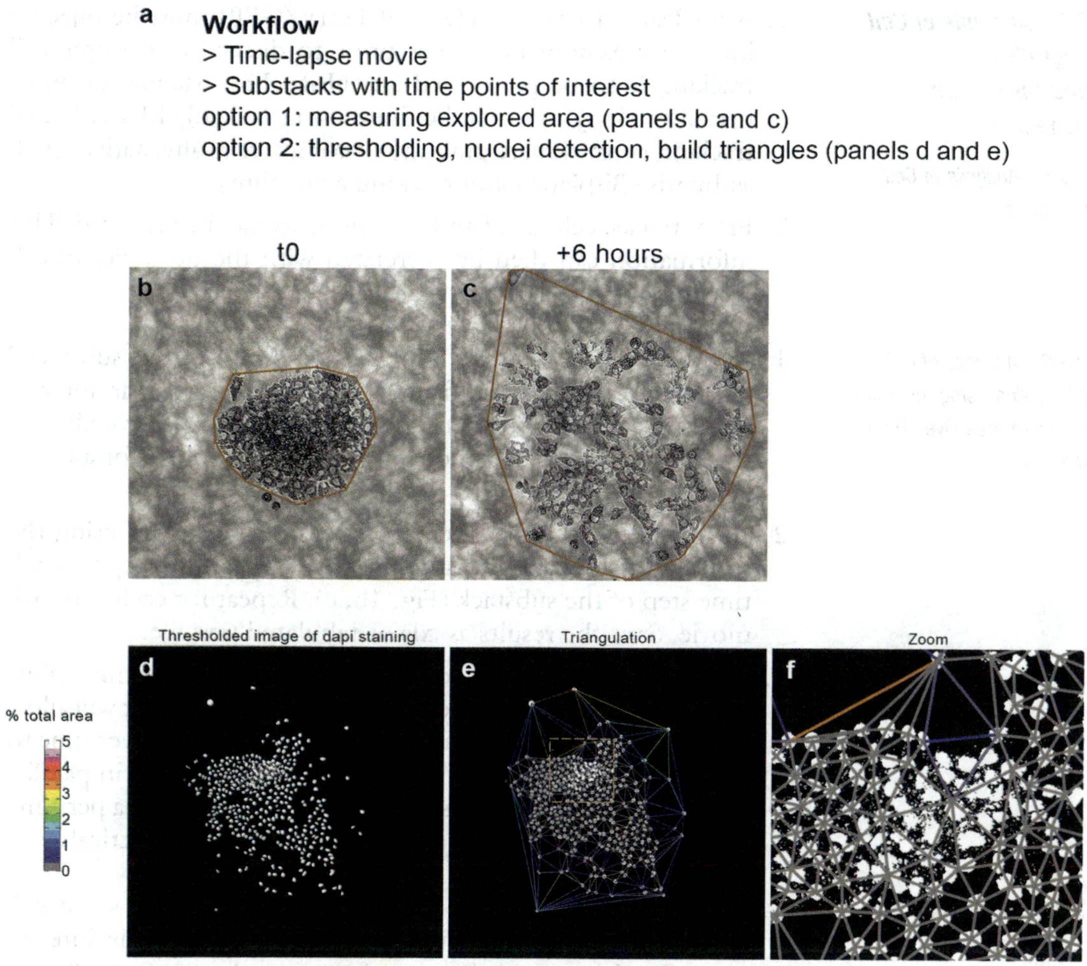

Fig. 4 Analyzing the dynamics of cell dispersion. (**a**) Workflow from cell culture to substacks for dispersion analyses. (**b**, **c**) Total explored area is retrieved by linking all external cells per explant, per time point. (**d–f**) Triangulation analysis. (**d**) Thresholded image of DAPI staining. (**e**) Triangles obtained after performing particles detection and Delaunay/Voronoi triangulation. Each triangle is color-coded according to the following ratio (area of the triangle/total area). (**f**) Example of artifacts in areas of high cell density due to poor particle detection. Steps on panels (**d**) and (**e**) are repeated for each time point of interest

5. Set the time interval and total duration of the movie (*see* **Note 11**).
6. Start the acquisition.
7. At the end of the movie, cells can be fixed in formaldehyde 4% for further analysis and staining (*see* **Note 12**). Typical dispersion of control neural crest explants is shown in Movie 4 and Fig. 4.

3.3 Analysis of Cell Migration and Dispersion Dynamics

3.3.1 Analysis of Cell Migration

1. A nuclear tracer (e.g., H2B-mCherry/GFP) must be injected into early Xenopus embryos prior to dissection if single cell tracking is to be performed (*see* **Note 13**). Manual or automated tracking can then be done using ImageJ/FIJI (Manual tracking or trackmate plug-in) or commercial alternatives such as Imaris/Bitplane (spot tracking algorithm).
2. From tracks, cell speed and directionality can be retrieved. This information can then be correlated with the dynamics of cell dispersion (*see* **Note 14**).

3.3.2 Analysis of Cell Dispersion Using the Total Explored Area Over Time (Fig. 4)

1. In ImageJ/FIJI use the "image\stacks\tools\make substack" tool to extract images from each movie at regular interval (e.g., first image and then one image per hour). Save all substacks (one per *XY* position) into a separate folder for analysis (Fig. 4a).
2. In ImageJ/FIJI, link all external cells of an explant using the broken line tool and measure the area. Repeat this step for each time step of the substack (Fig. 4b, c). Repeat for each substack movie. Save the results as .xls or tabulated text file.
3. If only one experimental condition is analyzed over time, plotting the data as box whiskers plots for each time point will allow variability at each time point and overall dispersion over time to be assessed. If multiple conditions are to be plotted in parallel on the same graph, curves representing the mean area per time point alongside the standard deviation are more practical.

3.3.3 Analysis of Cell Dispersion Using Triangulation Between Nearest Neighbors

1. In ImageJ/FIJI use the "image\stacks\tools\make substack" tool to extract images from each movie at regular interval (e.g., first image and then one image per hour). Save all substacks (one per *XY* position) into a separate folder for analysis.
2. In imageJ/FIJI, go to "analyze/set measurements" and make sure "centroid" and "area" are selected.
3. Open one of the substacks and estimate the mean area occupied by one nucleus in your samples by measuring about 20 different nuclei using the elliptical or polygon selection tools in ImageJ/FIJI.
4. Go to "analyze/analyze particles." Set the minimum and maximum area occupied by a nucleus as measured in **step 3** and press OK. This will produce a list of *XY* coordinates of all nuclei detected.
5. Use the Delaunay/Voronoi plug-in in ImageJ/FIJI to build triangles from coordinates produced by the "analyze particles" tool. The options "infer selection from particles" and "make Delaunay ROI" must be selected.

6. Go to "analyze/tools/ROI manager" and do the following steps: "Add", "Split", "multi-measures."
7. Save the results table containing the areas of all triangles. It is important to note that the first value of that list is the cumulated area occupied by all triangles (*see* **Note 15**).
8. If this has to be done on multiple images, a macro can be used. A macro for batch processing and a custom-made Look-up table (LUT) for color-coding of triangles were developed in the Theveneau lab and are available upon request.

The respective merits of the two methods to analyze dispersion are discussed in **Note 16**.

4 Notes

1. In Xenopus NC cells the dynamics of dispersion depends on the timing of cell–cell dissociation, which is linked to the E-to-N-cadherin switch. Thus, the timing (how long it takes) and intensity (are all cells becoming individual?) of dispersion are a proxy for EMT dynamics. This can be assessed on low magnification images even without time-lapse if several conditions are plated at the same t0 and compared at a similar time point after the beginning of the cell culture. Therefore, monitoring simultaneously several explants at low magnification on a dissecting microscope could be sufficient for a first screening of experimental conditions. However, if dispersion analysis is to be performed using single cell detection by nuclear staining or if cell tracking is to be done in parallel, then higher magnifications (in *XY* and time) are required and a motorized stage is necessary. A wide field microscope with a 10× objective will allow visualization of NC explants the size of a fifth of a complete unilateral NC region, while providing enough *XY* resolution to track single cells accurately using a time interval of 3 min. If cells are labelled with a membrane tracer and the dynamics of cell–cell contacts is to be analyzed, 20× or 40× objectives are needed and shorter time intervals are necessary (1 min or below). Inverted microscopes are the usual choice for cell culture assays. However, we obtained very good results using upright microscopes with dry lenses. In this case, cells need to be cultured in a dish that can be filled entirely with culture medium, without any air bubbles, and flipped upside down. Several providers have culture dishes with tight lids in their catalogue. If your cell culture device of choice does not exist in a version with a tight-fit lid, it is possible to seal the lid using high vacuum silicone grease. This type of upside-down cell culture removes all dead and/or non-adherent cells,

keeping the substrate clean for the healthy cells. However, this set-up is not appropriate if experimental conditions impairs cell–matrix adhesion as cells would detach and thus be lost.

2. Xenopus embryos can develop between 12 and 28 °C, the optimal range being 14–25 °C. NC explants are a bit less flexible and usually survive and migrate better between 18 and 21 °C. Below 18 °C cell–cell dispersion is affected and explants usually remain as pseudoepithelial sheets for much longer than they would in vivo. Above 21 °C, cell survival is lower. Between 18–21 °C, explants roughly the size of a fifth of one unilateral NC region dissected at stage 18, spread extensively during the first 2 h and then disperse as small clusters and eventually as single cells.
3. On upright microscopes, use tight-fit lid petri dishes that can be completely filled with culture medium and closed hermetically or open dishes with a water-immersion lens. For inverted microscopes, any type of multi-well dishes can be used depending on the type of motorized stage. In our hands, the 8-well Ibidi μSlides (80821) gave the best results. They allow culture in small volumes (300 μL) suitable for drug screening. They are made of a polymer that allows low concentration of Fibronectin coating for plastic to be used while being compatible with confocal imaging.
4. When coating large dishes do not cover the entire dish with Fibronectin. Use a marker pen to delimitate an area where Fibronectin will be coated. A circle with a diameter of 1 cm is enough to plate five rows of five explants and can be coated with a 500 μL drop of Fibronectin 1× solution. Always cover the dish during Fibronectin incubation at 37 °C to avoid evaporation. If very large dishes are used (e.g., 90 mm bacteria plates), place a piece of wet paper inside the dish to keep it moist.
5. A successful Fibronectin coating can be assessed by eye. When the Fibronectin solution is removed at the end of the 1-h incubation, the part of the dish or the well that is coated remains slightly wet and appears shiny when exposed to light whereas a similar incubation with a drop of water or PBS1× will leave the dish dry as soon as the liquid is removed. The area that received water or PBS cannot be distinguished from the rest of the dish.
6. Dissection of Xenopus NC explants is very easy when the embryos are at the right developmental stage. Just prior to migration, NC cells are not significantly attached to their surrounding tissues (ectoderm, mesoderm, ectodermal placodes). Yet, they strongly attach to each other. This allows the whole NC region to be removed, as it is an epithelial structure only

loosely connected to the other tissues. The best stage is 18. At this stage the neural folds nearly finished closing (Movie 2). Only the most anterior part of the neural plate is still open and has the shape of a tear drop. The NC region appears as a bump lateral to the anterior neural plate and extends roughly to the middle of the anteroposterior axis. Before that stage, the NC region is still attached to the neural plate and one would need to arbitrarily cut the explant out of the embryo without any clear landmark to use as reference. After stage 18, cells initiate migration and start adhering strongly to the mesoderm and the ectoderm. First, many cells are lost while peeling the superficial ectoderm. Second, it increases the probability of mesoderm contamination (Figs. 3 and 5, last part of Movie 3). Dissections at early (prior to stage 18) or late stages (after stage 18) are still technically possible, but they require an extensive amount of training. If such stages are absolutely needed for a given project, one should train dissecting by placing embryos at similar stages that have been stained with a neural crest marker (e.g., slug, twist) next to the embryos to be dissected, for reference.

7. Embryos should not be put completely inside the clay. It is important that roughly a third of the embryo still pops out of the modelling clay after the embryo has been immobilized. Otherwise, the region of interest will be difficult to access for dissection with the hair knife.

8. Once the embryo has been immobilized by applying pressure with the modelling clay, all previous morphological landmarks are gone. Thus, it is critical that the orientation of the embryo to be dissected is correct before applying pressure.

9. At the end of neurulation (stage 18), NC cells adhere to each other more than they adhere to their surrounding tissues (*see* **Note 6**) making dissection relatively easy. Yet, contamination from adjacent cell types can sometimes occur. Mesoderm and endoderm are richer in vitellus and appear whiter than NC cells, which are comparatively translucent and gray. The last part of Movie 3 shows an example of a rough dissection during which a bit of mesoderm is taken out together with the neural crest. *See* also Fig. 3 and its legend to see how to proceed when mesoderm is mistakenly dissected alongside NC cells. Superficial ectoderm may be detected by the presence of dark spots corresponding to the pigments. Explants should always be screened for contamination at the end of the culture and explants with contamination from other cell types should be excluded. In 2D-culture, ectoderm, neural plate, neural crest, mesoderm and endoderm can be easily recognized morphologically using several criteria such as the shape of the cells and nuclei or the size of vitellus platelets (Fig. 5). Nonneural ectodermal cells flatten extensively onto the substrate and form a

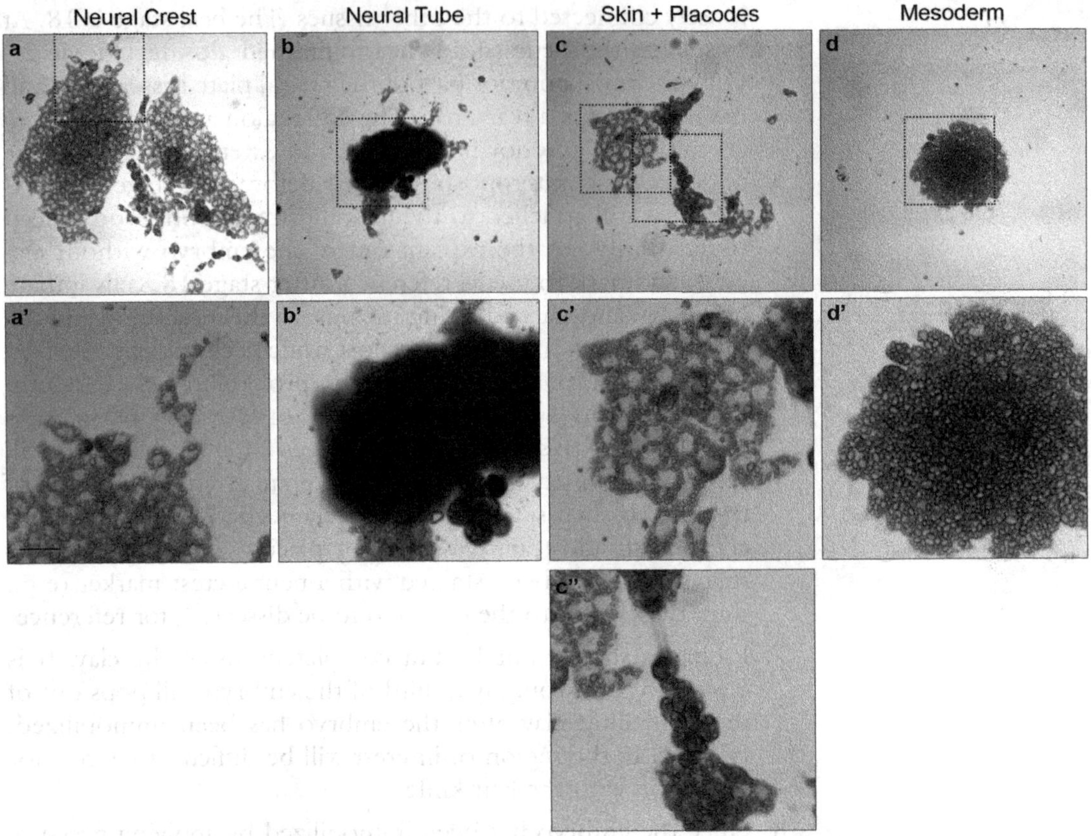

Fig. 5 How to distinguish neural tube, skin, placodes and mesoderm from neural crest cells. (**a**) Typical NC culture observed using bright field. After a few hours, the explant should have flattened on the substrate and cells should start dispersing. NC cells have small vitellus platelets and their nuclei should be visible as a translucent circle. (**b**) Neural plate/tube explants do not flatten on the substrate. They form a solid block from which few cells may migrate out. Often axons are projected directly from the explant onto the matrix. (**c**) Nonneural ectoderm and placodes. Even if the superficial pigmented layer has been properly removed the deep layers of the ectoderm are still present and may be taken out together with neural crest. The deep layers of the ectoderm surrounding the NC contain prospective skin cells and placodes. Skin cells flatten even more than NC and as a consequence their nuclei appear very large. They form a coherent sheet without any visible gaps (**c′**). Placodes produces neurons of the cranial ganglia; therefore, placode contamination often leads to few grapes of round cells projecting axons after an overnight culture (**c″**). (**d**) Mesoderm. Mesodermal cells are rich in vitellus and the vitelline platelets are larger than those observed in crest cells. Also, the mesoderm underlying the neural crest at stage 18 flattens on the substrate but does not disperse extensively and rarely produces single cells. Dotted lines on panels **a** to **d** indicate the position of the zooms for the different cell types that are presented in panels (**a′**–**d′**)

continuous monolayers with large nuclei. Neural plate explants remain as a tight cluster and project axons within few hours of cell culture. Mesodermal and endodermal cells are significantly larger and whiter than neural crest cells. They contain large vitelline platelets.

10. When plating explants next to each other onto Fibronectin, do not put explants to close to one another as NC cells can attract each other via a process called co-attraction or mutual attraction. This would lead neighboring explants to merge. A rule of thumb is to make sure that each explant is at least at five times its own diameter away from the nearest explants in the dish.
11. If only overall explant dispersion is to be analyzed, time-lapse imaging is not necessary. Explants can be fixed after several hours and the dispersion can be analyzed using triangulation, total explant area or nearest neighbors analysis. However, having the dynamics of the population and the movement of single cells can be very informative. For movies done with a 10× objective at a temperature between 18–21 °C, cells migrate at around 1–3 μm/min and a time interval of 3 min is optimal for cell tracking (either manual or automated). Eight-hour movies at 21 °C are enough to see the main steps of dispersion: adhesion, spreading, break-up of the explants as clusters and single cell dispersion. If under control conditions single cell dispersion does not occur within this time frame it can be due to several possibilities: (i) the temperature is too low, (ii) cells were extracted at earlier stages than 18 and have not yet completed their cadherin switch (there is a sharp decrease of the amount of E-cadherin after stage 15), (iii) the substrate is not optimal for migration and cells have problems moving away from each other.
12. To fix cells at the end of a culture experiments do not remove the DFA1×, instead add directly formaldehyde into the dish. In 8-well slides, if each well contains 300 μL of DFA1X, simply add 100 μL of formaldehyde 16% in each well. In larger dishes, add locally formaldehyde 16% atop the explants to pre-fix the cells, wait 5 min, remove the medium, fill with formaldehyde 4%, and further fix for 30 min.
13. To label NC cells prior to cell culture, one can inject at either 2, 4, 8, 16, or 32-cell stage. If simple tracers to label nuclei and/or membranes are used, the easiest way is to inject two blastomeres at 2-cell stage and select the brightest embryos at stage 16 prior to removing the vitelline membrane. If nuclear/membrane tracers are to be used in combination with other treatments that may affect early development, it is recommended to inject at 8-cell stage or later. From 8-cell stage the animal blastomeres only contribute to the ectoderm (neural plate, NC, placodes, epidermis), therefore injections at 8-cell stage or later will leave the mesoderm and endoderm intact, reducing possible side effects on gastrulation.
14. In control conditions, cell–cell dissociation leads to a progressive shift from collective/pseudoepithelial-like behavior, to a collective/mesenchymal behavior, to single cell behavior.

Individual cell directionality decreases in single cells compared to cells within a group, whereas individual cell speed increases. This, in control conditions, typically correlates with the rate of expansion of the explant. The initial slow-paced directional radial expansion turns into a fast disorganized dispersion of single cells. Overall dispersion comes to a halt when local cell density is such that cells have an equal probability of moving in any direction. However, under experimental conditions, one can interfere with cell–cell or cell–matrix adhesion and affect the relationship between explant dispersion and single cell behavior. For instance, reducing cell–cell adhesion may promote single cell behavior while overall radial expansion of the explant is impaired. Thus analyzing single cell behavior and explant dispersion in parallel can be very informative.

15. Triangles' areas follow an exponential distribution, thus the usual statistical tests for normal/Gaussian datasets do not apply if one wants to compare the distribution of triangles' areas from one explant to another. This problem is easily solved when comparing populations of explants. One should plot the mean triangle area of each explant per experimental condition. That way each explant has the same weight in the mean of a given condition (otherwise explants with more cells contribute more triangles and can skew the mean of the dataset). In addition, the distribution of mean areas under control condition follows a Gaussian distribution and allows simple statistical tests to be performed.

16. Analyzing explant dispersion using the total explored area is the simplest and most effective way. This technique does not require pre- or post-labeling of cell's nuclei. It enables the inclusion of the impact of all cells, and can be done on simple low resolution bright-field or phase-contrast images. Triangulation on the other hand has several caveats. First, one needs to detect single cell nuclei. This can be done by labeling NC cells with a nuclear tracer or post-staining them with DAPI. Automatic detection of nuclei is possible (e.g., analyze particles plug-in in Image J/FIJI) but weaker nuclei are often not detected, artificially increasing the area of some triangles. Similarly, areas with high cell density cannot be properly segmented and this results in artifacts (Fig. 4). This later problem can be partially solved by acquiring images on a confocal microscope, although high resolution images of nuclei often confuses the automated detection of particles, since pixel intensity within each nucleus can vary substantially, and nuclei no longer appear as solid structures. Alternatively, one can mark each cell's nucleus manually. The first option, using confocal imaging, will never be a realistic routine procedure, even if one owns a dedicated confocal microscope for time-lapse imaging. The

second option, manual marking of individual nuclei, is extremely time consuming, especially if one wants to perform triangulation at multiple time points for multiple explants to assess the dynamics of dispersion. A typical explant has between 100 and 200 cells. If one wants to perform triangulation on 20 explants at 5 time points each, it means marking manually between 10,000–20,000 nuclei. This increases exponentially if one needs to do experiments in triplicate and has to compare multiple experimental conditions. Thus, if one absolutely needs to use triangulation (e.g., local differences in terms of distance between neighbors within each explant are important), we recommend using the automated detection of nuclei. However, it is important to bear in mind that some nuclei will not be detected regardless of the technique used to mark them, or to detect them. It is important to first establish the conditions for automated detection on a small subset of data and to compare the overall output with the one obtained when marking nuclei manually. A difference of more than 5% should not be considered acceptable and image acquisition and nuclei detection would need to be improved before embarking on large-scale automated detection of nuclei.

Acknowledgements

Eric Theveneau and Christian Rouviere are permanent CNRS staff. Work in the Theveneau lab is supported by the Region Midi-Pyrenees (Installation Grants for Excellent Researchers, 13053025), the Fondation pour la Recherche Medicale (AJE201224), the CNRS and the Université Paul Sabatier. Nadège Gouignard is the recipient of an individual fellowship from FRM (ARF20150934153) and the Marie Curie Prestiges Program (PRESTIGES 2015-4-007).

References

1. Hay ED (1995) An overview of epithelio-mesenchymal transformation. Acta Anat 154 (1):8–20. https://doi.org/10.1159/000147748
2. Hay ED, Zuk A (1995) Transformations between epithelium and mesenchyme: normal, pathological, and experimentally induced. Am J Kidney Dis 26(4):678–690. https://doi.org/10.1016/0272-6386(95)90610-x
3. Nieto MA (2011) The ins and outs of the epithelial to mesenchymal transition in health and disease. Annu Rev Cell Dev Biol 27:347–376. https://doi.org/10.1146/annurev-cellbio-092910-154036
4. Thiery JP, Acloque H, Huang RY, Nieto MA (2009) Epithelial-mesenchymal transitions in development and disease. Cell 139 (5):871–890. https://doi.org/10.1016/j.cell.2009.11.007
5. Thiery JP (2002) Epithelial-mesenchymal transitions in tumour progression. Nat Rev Cancer 2(6):442–454. https://doi.org/10.1038/nrc822
6. Gouignard N, Andrieu C, Theveneau E (2018) Neural crest delamination and migration: Looking forward to the next 150 years. Genesis 56:e23107. https://doi.org/10.1002/dvg.23107

7. Steventon B, Carmona-Fontaine C, Mayor R (2005) Genetic network during neural crest induction: from cell specification to cell survival. Semin Cell Dev Biol 16(6):647–654. https://doi.org/10.1016/j.semcdb.2005.06.001
8. Steventon B, Mayor R (2012) Early neural crest induction requires an initial inhibition of Wnt signals. Dev Biol 365(1):196–207. https://doi.org/10.1016/j.ydbio.2012.02.029
9. Barriga EH, Maxwell PH, Reyes AE, Mayor R (2013) The hypoxia factor Hif-1alpha controls neural crest chemotaxis and epithelial to mesenchymal transition. J Cell Biol 201(5):759–776. https://doi.org/10.1083/jcb.201212100
10. Scarpa E, Szabo A, Bibonne A, Theveneau E, Parsons M, Mayor R (2015) Cadherin switch during EMT in neural crest cells leads to contact inhibition of locomotion via repolarization of forces. Dev Cell 34(4):421–434. https://doi.org/10.1016/j.devcel.2015.06.012
11. Theveneau E, Marchant L, Kuriyama S, Gull M, Moepps B, Parsons M, Mayor R (2010) Collective chemotaxis requires contact-dependent cell polarity. Dev Cell 19(1):39–53. https://doi.org/10.1016/j.devcel.2010.06.012
12. Christian L, Bahudhanapati H, Wei S (2013) Extracellular metalloproteinases in neural crest development and craniofacial morphogenesis. Crit Rev Biochem Mol Biol 48(6):544–560. https://doi.org/10.3109/10409238.2013.838203

Chapter 21

Xenopus Deep Cell Aggregates: A 3D Tissue Model for Mesenchymal-to-Epithelial Transition

Hye Young Kim and Lance A. Davidson

Abstract

Mesenchymal-to-epithelial transition (MET) describes the ability of loosely associated migratory cells to form a more adherent sheet-like assembly of cells. MET is a conserved motif occurring throughout organogenesis and plays a key role in regeneration and cancer metastasis, and is the first step in producing induced pluripotent stem cells (iPSCs). To resolve fundamental biological questions about MET, its relation to epithelial-to-mesenchymal transition, and to explore MET's role in tissue assembly and remodeling requires live models for MET that are amenable to experimentation. Many cases of clinically important MET are inferred since they occur deep with the body of the embryo or adult. We have developed a tractable model for MET, where cellular transitions can be directly observed under conditions where molecular, mechanical, and cellular contexts can be controlled experimentally. In this chapter, we introduce a 3-dimensional (3D) tissue model to study MET using *Xenopus laevis* embryonic mesenchymal cell aggregates.

Key words Mesenchymal-to-epithelial transition, 3D aggregate, Mesenchyme, Immunofluorescence, Dissociation, Microsurgery

1 Introduction

Cells in our body are highly plastic with regard to both their phenotype and their mechanical properties. For instance, a tightly adherent epithelial cell in a sheet may undergo a transition into a loosely bound mesenchymal cell allowing it to migrate into surrounding tissues (e.g., an epithelial-to-mesenchymal transition or EMT). Conversely, motile mesenchymal cells can undergo a mesenchymal-to-epithelial transition (MET) during which they aggregate and form an organized epithelium. METs involve large-scale changes at the level of transcription, the cytoarchitecture, and cell behavior [1]. MET appears to have evolved as a revolutionary strategy in the metazoan lineage [2] where it plays a fundamental role in many tissue-shaping processes including cancer [3], development [4, 5] and stem cell reprogramming [6, 7]. METs are key

Kyra Campbell and Eric Theveneau (eds.), *The Epithelial-to Mesenchymal Transition: Methods and Protocols*, Methods in Molecular Biology, vol. 2179, https://doi.org/10.1007/978-1-0716-0779-4_21,

to organogenesis of the kidney [8] and the heart [5, 9, 10] as clusters of mesenchymal cells migrate great distances in the embryo and transition to epithelial phenotypes to form the simple tubule of the developing kidney or a disk of the heart primordia. The timing and progression of MET during development parallels the events leading to the formation of secondary metastases [11] and suggest these processes share common regulatory mechanisms [12].

Despite the significance of MET in topics ranging from cell biology to human health, little is known about the spatiotemporal dynamics of MET progression or how the physical regulation of MET integrates with conventional intracellular molecular signaling pathways [13]. The limited understanding of the cellular and molecular mechanisms of MET directly results from both the inherent unpredictability of MET emergence deep within multiple layers of cells in vivo and a lack of a tractable 3D model system that can recapitulate the full process of MET. By contrast, the past decade of research has provided great insight into the regulatory mechanisms of EMT due to established in vitro and animal models that permit profiling differential gene expression, tracing cellular transitions, and manipulating signaling factors [14, 15]. MET is often viewed as the reverse of EMT; however, the complex sequential progression of MET may not be reversed by reintroduction of mesenchymal inducing factors alone [16]. These gaps in our understanding of MET indicate a clear demand for a model system that can be used to explore the full progression by which mesenchymal cells initiate and progressively adopt apicobasal polarity, propagate "epithelialness" and stabilize the new architecture of the epithelium [13].

The physical mechanics of the microenvironment changes prior to and concomitant with MET, as dispersed migratory mesenchymal cells first cluster then spontaneously develop cell–cell adhesions. As MET progresses, transitioning tissues often form compact spherical aggregates, disks, or cords. Mechanical tension, driven by cortical actin contraction and cell–cell adhesion, is known to shape cells and tissues during aggregation in vivo [17] and in vitro [18, 19] and has been implicated in other aspects of tissue self-assembly [20], cell rearrangement [21, 22], and cell fate determination [23]. To understand the role of tension in these processes, numerous methods for measuring tension sensed by cells have emerged [24, 25], including FRET-based tension sensors [26] on load-bearing proteins such as cadherin [27], vinculin [28, 29], and talin [30, 31]. Quantitative measures of tension may be complemented by qualitative indicators such as nuclear translocation of YAP [32], an endogenous mechanosensor, that can be used as a proxy to assess tension experienced by a cell. Combining quantitative biomechanical analysis with qualitative tension sensors and molecular genetic approaches can expose the relationship between tension and MET.

Here we describe a reliable, controllable and scalable 3D cell aggregate consisting of mesenchymal cells isolated from the deep ectoderm of *Xenopus laevis* embryos that provides ready access to the processes driving MET [33]. The ectoderm of early *Xenopus* embryos is composed of one or more rows of deep mesenchymal cells that actively express and assemble the extracellular matrix protein fibronectin [34] and are covered by a superficial epithelial sheet with stereotypical tight junctions [35] and distinctive keratin intermediate filaments [36–38]. Upon isolation, deep mesenchymal cells adhere to one another to form a dense aggregate. After only 5 h surface cells on the mesenchymal aggregate progressively transition to adopt an epithelial phenotype which is marked by elevated epithelial gene expression along with newly established apical basal polarity and the formation of tight junctions. The intrinsic nature of 3D culture, self-assembly, rapid development, scalability in size and number, and accessibility for both live imaging and mechanical testing all contribute to the utility of this 3D tissue model of MET.

2 Materials

2.1 Embryos

Xenopus laevis embryos are obtained by general procedure [39]. Embryos may be reliably cultured at a variety of temperatures from 14 to 26 °C and staged by external criteria [40].

2.2 Reagents

1. 1x MBS (Modified Barth's Saline): 88 mM NaCl, 1 mM KCl, 2.4 mM $NaHCO_3$, 0.82 mM $MgSO_4$, 0.33 mM $(CaNO_3)^2$, 0.41 mM $CaCl_2$, 10 mM HEPES in distilled water. Adjust pH to 7.4 with NaOH. Use 1/3× MBS for embryo culture.
2. Danilchik's for Amy (DFA): 53 mM NaCl, 5 mM Na_2CO_3, 4.5 mM potassium gluconate, 32 mM sodium gluconate, 1 mM $CaCl_2$, 1 mM $MgSO_4$ in distilled water. Adjust pH to 8.3 with granular Bicine. Filter with 0.22 μm filter. Just prior to use, add 1% of antibiotic/antimycotic (A5955, Sigma-Aldrich) to prevent bacterial growth. DFA is formulated to match the ionic composition of interstitial media of gastrula stage embryos [41]. Use 1× DFA for aggregate culture.
3. Calcium-magnesium free DFA: DFA made without $CaCl_2$ and $MgSO_4$.
4. 3% Ficoll (Sigma-Aldrich) in 1× MBS. Culture media used during microinjection.
5. Phosphate-buffered saline (PBS).
6. Triton X -100.

7. Murray's clear (also known as BB:BA): 2 volumes of benzyl benzoate and 1 volume of benzyl alcohol: optional for clearing a tissue.

2.3 Tools

1. Hair loop and hair knife (looped hair or eye brow hair fixed to the end of pulled Pasteur glass pipette with melted wax).
2. Disposable plastic pipette—Cut the opening of a disposable pipette to enlarge the opening hole to freely transfer embryos, aggregates, or embryonic cells with minimal shear stresses.
3. Forceps (Dumont #5 stainless steel; Fine Science Tools or equivalent).
4. Cover glass (various sizes; #1 1/2 thickness).
5. Silicon grease (high vacuum, Dow Chemical or equivalent).
6. PCR tubes. To prevent cell adhesions to the wall of PCR tube, add 200 μl of 1% BSA to PCR tube and coat for overnight at 4 °C.

2.4 Microinjection

1. Borosilicate glass capillary tubes (1.0 mm OD × 100 mm).
2. Needle puller (p-97, Flaming/Brown Micropipette Puller, Sutter Instrument Co or equivalent).
3. Microinjector (PLI-100, Harvard Apparatus or equivalent).
4. Micromanipulator (M33, Marzhauser or equivalent).
5. Microloader Pipette Tip.
6. mRNA transcription kit (Epicenter or equivalent).
7. Mineral oil.
8. Horizontal graduated scale eyepiece reticle.

2.5 Imaging

1. Imaging chambers (*see* **Note 1** and Fig. 1).
2. Electron microscope grid (optional): Useful for positioning arrays of aggregates within imaging chambers.
3. Stereoscope/Fluorescent stereoscope: Essential for microinjection and screening injected embryos and aggregates.
4. Confocal microscope. Laser scanning or spinning disk mounted on inverted compound microscope. High numerical aperture objectives such as 63× oil, or 25× water immersion are useful to track protein dynamics. Longer working distance objectives are helpful for live imaging and collecting image stacks for 3D reconstruction. A motorized z-stage is essential while a motorized xy-stage is optional.
5. Image analysis software. Essential for post processing image stacks and quantifying epithelial areas and cell dynamics. FIJI/ImageJ or equivalent.

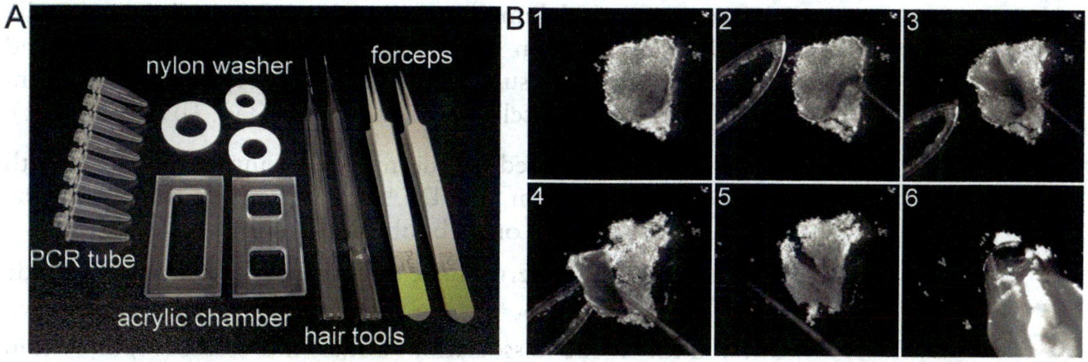

Fig. 1 Assembly of deep ectoderm aggregates. (**a**) Selected tools used to assemble and image the embryonic aggregate from *Xenopus* embryo. (**b**) Steps to isolate deep layer from isolated ectoderm. (1) Isolated ectoderm placed in calcium-magnesium free DFA. (2) Lift the corner of superficial layer (dark pigmented) with a hair knife. (3–5) Carefully peel off the superficial layer. (6) Aspirate the deep cells using pipette and transfer to non-adherent PCR tube

3 Methods

3.1 Isolate the Ectoderm Tissue (Animal Cap) at Stage 10 Xenopus laevis *Embryo*

Dissection of ectoderm is a relatively simple and standard technique [42].

1. Stock microsurgery station with all needed tools (Fig. 1). Prepare dishes with media, tools, imaging chambers, and PCR tubes at the stereomicroscope station within easy reach.
2. Select embryos at stage 10 under a stereoscope based on the external criteria of a visible line or crescent of pigmented cells indicating the early blastopore lip. Early stage 10 ectoderm will have more than 2 layers of deep cells but decreases to a single layer by stage 10½.
3. Transfer selected embryos using a transfer pipette to a dish containing DFA with antibiotic. Do not introduce any air bubbles during transfer.
4. Remove the vitelline membrane using a pair of sharp forceps.
5. Position the embryo to face animal side up.
6. Insert a hair knife into the blastocoel from the side to start an incision. Pull the hair knife out to make the cut.
7. Repeat **step 6** along the margin of ectoderm to excise a circular animal cap tissue from the embryo.
8. Trim the unevenly thick margin of the isolated tissue to eliminate contamination of more vegetal cells such as prospective neural ectoderm, mesendoderm, or mesoderm.

3.2 Separate the Deep Ectoderm and Assemble 3D Aggregates

Brief exposure to calcium-magnesium free DFA media loosens cell–cell adhesions between the deep and superficial cell layers and allows removal of the superficial sheet of the epithelial layer from the multilayered deep cells of the animal cap (Fig. 1b) (*see* **Note 2**).

1. Transfer the isolated ectoderm explant into a dish filled with calcium-magnesium free DFA. Keep explants away from the air–water interface or air bubbles to prevent tissue damage.
2. Position the ectoderm tissue to face the pigmented animal side up using hair tools.
3. Monitor tissue dissociation under the stereoscope. Within 5–10 min you should observe signs that the superficial layer is delaminating from the deep cells as light colored, less-pigmented deep cells move out from the edge of the darker pigmented cells of the superficial layer (*see* **Note 3**).
4. Carefully lift the pigmented superficial layer to detach from the edge of deep cell layers using a hair knife. With practice, the entire sheet of superficial layer can be peeled off in rapid fashion. Discard the superficial layers to prevent further dissociation and reduce chances of contamination with target deep cells.
5. Collect the separated deep cells using 200P pipette and minimize the amount of calcium-magnesium free DFA media transferred with cells to a non-adherent PCR tube filled with 200 μl of DFA with antibiotic (*see* **Note 4**).
6. Stand the PCR tube upright in a rack to induce deep ectoderm cell aggregation (0 h post aggregation, hpa).
7. Deep ectoderm cells form a compact aggregate within 2 hpa and become a spherical aggregate by 5 hpa.
8. To continue the culture of aggregates *en masse*, transfer the aggregate after 5 hpa to a dish filled with DFA with antibiotic. Position aggregate far enough from each other and avoid agitation of the culture dish to prevent conglomeration of individual aggregates.
9. Deep ectoderm aggregates that undergo successful epithelialization and sequential maturation to epidermis can be easily recognized and scored using a simple stereoscope. By 24 hpa the aggregate surface is populated by highly motile multiciliated cells which drive aggregate rotation or movement.

3.3 Imaging the 3D Embryonic Aggregates

Progression of the MET can be visualized using end-point analysis of fixed and immuno-stained aggregates (*see* **steps 1–10**) or can be visualized dynamically from live aggregates expressing epithelial-specific proteins tagged with fluorescent proteins (*see* **steps 11–16**).

1. Fill the glass vial with a fixative (*see* **Note 5**).

2. Transfer aggregates from PCR tube or culture dish to fixative using an end-cut 200P- tip fitted onto a 200P manual pipettor.
3. Fix the aggregate for 15 min at room temperature.
4. Wash 3 times for 15 min each with PBST (PBS + 0.1% Triton X). If using Dent's fixative the embryos must be rehydrated through a series MeOH, dH_2O, and PBST.
5. Block nonspecific antibody binding with 10% goat serum in PBST for 1 h at room temperature.
6. Incubate with first antibody at 4 °C for overnight.
7. Wash three times for 15 min each with PBST.
8. Incubate with 2° antibody at 4 °C for overnight.
9. Wash three times for 15 min each with PBST.
10. (optional) Stained aggregates can be cleared to image inside of the aggregates by following series of dehydration, soaking in Murray's clear until visibly clear under a stereoscope (~5 min).
11. Transfer to a glass-bottomed imaging chamber for high resolution confocal imaging.

    To capture the progression of mesenchymal-to-epithelial transition, live aggregates can be imaged using a confocal microscope by following **steps 12–16** (Fig. 2).
12. Microinject mRNA for protein of interest at one or two cell stage of embryo. For example, injecting mRNA encoding GFP-ZO-1 proteins can reveal progressive changes as cells transition from mesenchymal-to-epithelial phenotypes. To microinject, load the pulled glass capillary with desired mRNA (~3 μl) using a 10P pipettor fitted with a microloader pipette tip. The opening size of injecting glass needle should be minimized to prevent inducing an open hole in embryo during injection and can be adjusted by clipping the pulled end of glass capillary with sharp forceps. To quantify the amount of mRNA, inject small volumes into a drop or dish of mineral oil. The volume of the injected bolus can be measured with a stereomicroscope equipped with a graduated reticle scale from the diameter of the drop of mRNA in mineral oil. To minimize the damage and aid healing after injection, place the embryo in 3% Ficoll/1× MBS during microinjection. Even expression within deep cells can be obtained by injecting 3 to 4 sites across the animal pole of the 1-cell stage embryo (*see* **Note 6**).
13. Just prior to ectoderm dissection, screen and select embryos with optimal expression of fluorescently labeled proteins under the fluorescent stereomicroscope (*see* **Note 7**).
14. Assemble aggregates following the methods above (*see* Subheadings 3.1 and 3.2).

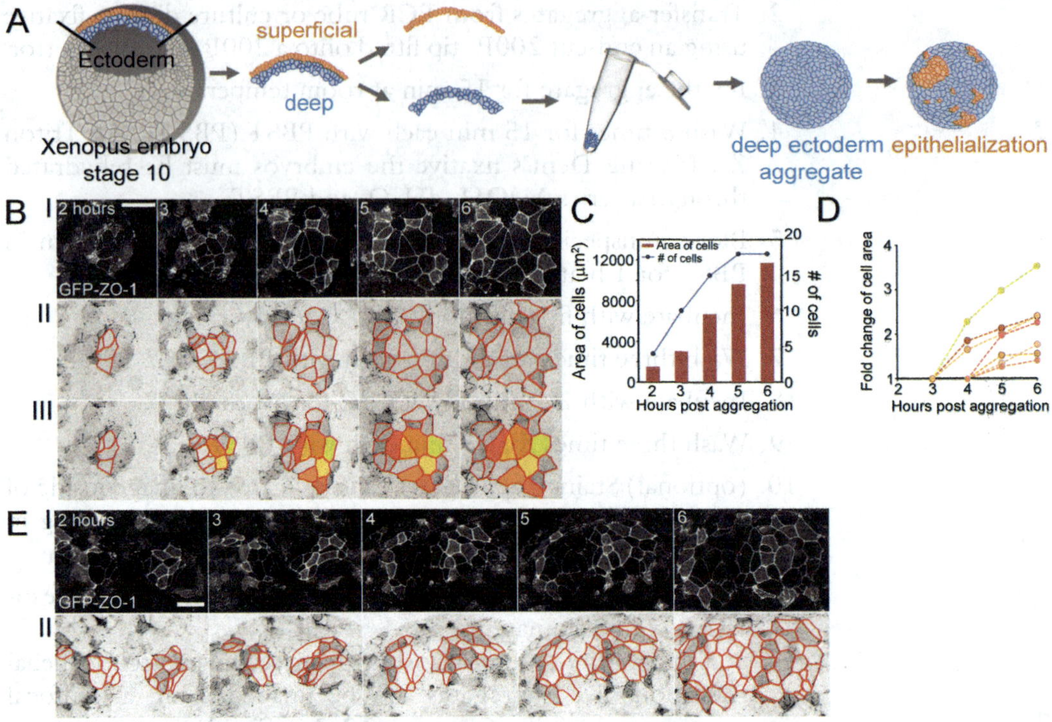

Fig. 2 Imaging mesenchymal-to-epithelial transition in deep ectoderm aggregates (**a**) Schematic representation of steps to make aggregates. Ectoderm explants with superficial epithelial (orange) and deep mesenchymal cell layers (blue) are isolated and separated. After 2 h of aggregation at the bottom of the non-adherent PCR tube, the spheroidal aggregates can be subject to live imaging of the epithelialization. (**b**) (I–III) Representative frames from time-lapse sequences showing expanding epithelial patches from 2 to 6 hpa. Patches expand by recruiting GFP-ZO-1 positive cells to the edge of existing patch. Row I depicts images, row II depicts lookup-table-inverted images where MET cells have been outlined in red, and in row III different colors indicate cell tracking within a growing patch. Scale bar, 50 μm. (**c**) Surface area of the epithelial cell patch (from **b**) continues to increase, although the number of cells within that patch is unchanged after 5 hpa. Note: Patches can continue to add cells beyond this time. (**d**) Surface area of individual cells (from **b** I–III) increase more than twofold over 1–2 h. (**e**) Top row shows representative frames from time-sequences showing MET spreads by joining patches of GFP-ZO-1 positive cells. Lower row indicates MET cells outlined in red on LUT-inverted images. Scale bar, 50 μm

15. Live imaging of aggregates can begin at 2 hpa after all cells in the PCR tube are adherent and form an aggregate.
16. The aggregate can then be mounted in a glass-bottomed culture dish [43]. To reduce drift during imaging, aggregates can positioned either in an EM grid or skewered with a filament cut from a length of heat-pulled plastic pipette tip. The diameter of the filament skewer can range between 50 and 100 μm (*see* **Note 8**).
17. Set the confocal acquisition to collect a *Z*-stack to acquire the images of spherical shaped aggregates over-time. *Z*-intervals, e.g., slice spacing, should be set to allow appropriate 3D reconstruction and image analysis. Live cell imaging allows

tracking of MET in single cells and the emergences of large patches of epithelial clusters (Fig. 2b). Following individual cells over time allow quantitation of cell number and areas as the patches emerge and grow in size (Fig. 2c–d). Live cell ismallermaging also reveals growth of epithelial sheets by fusion of multiple small clusters (Fig. 2e).

4 Notes

1. For live imaging of aggregates, a custom fabricated acrylic well can be glued on to a large cover glass to build a glass-bottomed imaging chamber using silicon grease. Custom sized chamber can be milled by a standard machine shop or fabricated by a 3D FDM printer. The volume of custom imaging chambers can be reduced to limit amounts of small molecule inhibitors or limited availability reagents such as function blocking antibodies. Chambers fashioned from milled acrylic can be reused after thorough washing with 70% ethanol and water. Commercial multi-well chambers are suited for multiple non-imaging experimental conditions. For Imaging of fixed aggregates, inexpensive disposable imaging chambers can be constructed with nylon washer (inner diameter 8.5–13 mm) glued to cover glass with silicon grease (for aqueous mount only) or fingernail polish (required to be resistant to Murray's clear). For the most stable imaging, the top of the chamber should be sealed with an appropriately sized circular cover glass. Variable size and thickness of washers are available through common tool and hardware suppliers (e.g., Grainger Industrial Supply).
2. Cells in the epithelial superficial layer are more tightly bound and are more resistant to dissociation than deep cell layers. This intrinsic property and color of darkly pigmented cells in the superficial layer can be used to distinguish superficial cells from less-pigmented deep cells during preparation of routine aggregates, however, before relying on this routine practice we recommend confirming superficial cells are not contaminating aggregates. To confirm the effectiveness of microsurgery and dissociation we recommend carrying out aggregate assembly using embryos whose surface layers have been labeled with a fluorescent tag. To label the surface layer we take stage 9 embryos and label free prolines on apical proteins with NHS-rhodamine. Incubate embryos in NHS-rhodamine (~1 μg/ml) in elevated pH (>9) 1/3× MBS for 30 min on a slow operating rotator. Wash embryos three times with 1/3× MBS. This produces embryos with a covalent fluorescent label on the apical cell surface can be easily detected within deep ectoderm aggregates if contaminated with cells from superficial layer.

3. During isolation, place the isolated ectoderm apical side up in calcium-magnesium free DFA and monitor under a stereoscope to identify the right time to separate the superficial layer. Tissues cultured too long in CMF-DFA dissociate completely making it impossible to separate superficial from deep ectoderm cells.
4. To transfer aggregates we recommend a manual pipettor with a wide tip opening. We typically cut the end of a 200P tip with sterilized scissors to enlarge the opening. This allows gentle transfer aggregates with minimal transient shear stress. Furthermore, the tip reduces the amount of media transferred to the next chamber or dish.
5. Certain epitopes are sensitive to the type of fixative and duration of the fix. For most studies of the surface layer cells we fix aggregates using 4% fresh paraformaldehyde in PBS for 15 min at room temperature. Longer fixation may be required to preserve deep cells. Some epitopes, notably those found in membrane bound adhesion proteins like cdh1 or cdh3 (E- or C-cadherin in the older cadherin nomenclature) or tight-junction protein TJP1, also known as ZO-1. For these cases we fix aggregates in ice-cold Dent's fixative (80:20 anhydrous methanol (aMeOH) to dimethyl sulfoxide (DMSO) [44] overnight at −20 °C. Often we fix and stain for F-actin to monitor cell phenotypes using 4% fresh paraformaldehyde supplemented with 0.2% glutaraldehyde for 15 min at room temperature and stained with phallacidin or phalloidin. Phalloidin/phallacidin staining for F-actin is not compatible with Methanol. In order to combine F-actin staining with epitopes requiring Dent's fixative we substitute Isopropanol for Methanol. Anhydrous isopropanol must also be used with the dehydration steps in order to preserve phalloidin/phallacidin binding. If co-staining for F-actin, we use complementary fluorescent derivatives of phallacidin or phalloidin (1:800) together with the secondary antibody.
6. Timing is critical to ensure even distribution of injected mRNA. Best results are obtained by restricting mRNA injections to the first 60 min after fertilization. Injections within 15 min of the first cell division, or later stages often result in uneven fluorescence signal.
7. We warn against selecting the "brightest embryos" during screening injected embryos with a fluorescence stereoscope. High levels of overexpression are not needed for imaging in the more sensitive confocal microscope and may interfere with normal protein dynamics or MET. You must use careful titration and expression studies in whole embryos to find expression levels that do not interfere with normal development. These preliminary studies can validate levels of expression for live-aggregate studies.

8. These methods are strongly preferred over conventional explant mounting techniques that involve compression under a glass coverslip bridge since conventional approaches can disrupt MET.

Acknowledgements

We thank members of both the Davidson and Kim groups for their comments and support. This work was supported by grants to LAD from the National Science Foundation (CBET-1547790) and the National Institutes of Health (R01HD044750; R56HL13495). Any opinions, findings, and conclusions or recommendations expressed in this material are those of the authors and do not necessarily reflect the views of the National Science Foundation or the National Institutes of Health. Additionally HYK was supported by a Young Scientist Fellowship from Institute for Basic Science (IBS-R025-Y1).

References

1. Pei D, Shu X, Gassama-Diagne A, Thiery JP (2019) Mesenchymal–epithelial transition in development and reprogramming. Nat Cell Biol 21(1):44
2. Pérez-Pomares JM, Muñoz-Chápuli R (2002) Epithelial–mesenchymal transitions: a mesodermal cell strategy for evolutive innovation in metazoans. Anat Rec 268(3):343–351
3. Sosa MS, Bragado P, Aguirre-Ghiso JA (2014) Mechanisms of disseminated cancer cell dormancy: an awakening field. Nat Rev Cancer 14(9):611–622. https://doi.org/10.1038/nrc3793
4. Stark K, Vainio S, Vassileva G, McMahon AP (1994) Epithelial transformation of metanephric mesenchyme in the developing kidney regulated by Wnt-4. Nature 372(6507):679–683
5. Trinh LA, Stainier DY (2004) Fibronectin regulates epithelial organization during myocardial migration in zebrafish. Dev Cell 6 (3):371–382
6. Li R, Liang J, Ni S, Zhou T, Qing X, Li H, He W, Chen J, Li F, Zhuang Q, Qin B, Xu J, Li W, Yang J, Gan Y, Qin D, Feng S, Song H, Yang D, Zhang B, Zeng L, Lai L, Esteban MA, Pei D (2010) A mesenchymal-to-epithelial transition initiates and is required for the nuclear reprogramming of mouse fibroblasts. Cell Stem Cell 7(1):51–63. https://doi.org/10.1016/j.stem.2010.04.014
7. Samavarchi-Tehrani P, Golipour A, David L, Sung HK, Beyer TA, Datti A, Woltjen K, Nagy A, Wrana JL (2010) Functional genomics reveals a BMP-driven mesenchymal-to-epithelial transition in the initiation of somatic cell reprogramming. Cell Stem Cell 7(1):64–77. https://doi.org/10.1016/j.stem.2010.04.015
8. Dressler GR (2002) Tubulogenesis in the developing mammalian kidney. Trends Cell Biol 12(8):390–395
9. Trinh LA, Yelon D, Stainier DY (2005) Hand2 regulates epithelial formation during myocardial differentiation. Curr Biol 15(5):441–446
10. Jackson TR, Kim HY, Balakrishnan UL, Stuckenholz C, Davidson LA (2017) Spatiotemporally controlled mechanical cues drive progenitor mesenchymal-to-epithelial transition enabling proper heart formation and function. Curr Biol 27(9):1326–1335. https://doi.org/10.1016/j.cub.2017.03.065
11. Chaffer CL, Thompson EW, Williams ED (2007) Mesenchymal to epithelial transition in development and disease. Cells Tissues Organs 185(1–3):7–19. https://doi.org/10.1159/000101298
12. Nieto MA (2013) Epithelial plasticity: a common theme in embryonic and cancer cells. Science 342(6159):1234850. https://doi.org/10.1126/science.1234850

13. Kim HY, Jackson TR, Davidson LA (2017) On the role of mechanics in driving mesenchymal-to-epithelial transitions. Semin Cell Dev Biol 67:113–122. https://doi.org/10.1016/j.semcdb.2016.05.011
14. Yang J, Weinberg RA (2008) Epithelial-mesenchymal transition: at the crossroads of development and tumor metastasis. Dev Cell 14(6):818–829
15. Thiery JP, Acloque H, Huang RY, Nieto MA (2009) Epithelial-mesenchymal transitions in development and disease. Cell 139 (5):871–890
16. Bidarra S, Oliveira P, Rocha S, Saraiva D, Oliveira C, Barrias C (2016) A 3D in vitro model to explore the inter-conversion between epithelial and mesenchymal states during EMT and its reversion. Sci Rep 6:27072
17. Oster GF, Murray JD, Harris AK (1983) Mechanical aspects of mesenchymal morphogenesis. J Embryol Exp Morphol 78:83–125
18. Bhumiratana S, Eton RE, Oungoulian SR, Wan LQ, Ateshian GA, Vunjak-Novakovic G (2014) Large, stratified, and mechanically functional human cartilage grown in vitro by mesenchymal condensation. Proc Natl Acad Sci 111 (19):6940–6945
19. Mammoto T, Mammoto A, Torisawa Y-S, Tat T, Gibbs A, Derda R, Mannix R, de Bruijn M, Yung CW, Huh D (2011) Mechanochemical control of mesenchymal condensation and embryonic tooth organ formation. Dev Cell 21(4):758–769
20. Steinberg MS, Gilbert SF (2004) Townes and Holtfreter (1955): directed movements and selective adhesion of embryonic amphibian cells. J Exp Zool A Comp Exp Biol 301 (9):701–706
21. Maitre JL, Berthoumieux H, Krens SF, Salbreux G, Julicher F, Paluch E, Heisenberg CP (2012) Adhesion functions in cell sorting by mechanically coupling the cortices of adhering cells. Science 338(6104):253–256. https://doi.org/10.1126/science.1225399
22. Harris AK (1976) Is cell sorting caused by differences in the work of intercellular adhesion? A critique of the Steinberg hypothesis. J Theor Biol 61(2):267–285
23. Engler AJ, Sen S, Sweeney HL, Discher DE (2006) Matrix elasticity directs stem cell lineage specification. Cell 126(4):677–689
24. Campàs O (2016) A toolbox to explore the mechanics of living embryonic tissues. Semin Cell Dev Biol 55:119–130
25. Shawky JH, Davidson LA (2015) Tissue mechanics and adhesion during embryo development. Dev Biol 401(1):152–164. https://doi.org/10.1016/j.ydbio.2014.12.005
26. Cost A-L, Ringer P, Chrostek-Grashoff A, Grashoff C (2015) How to measure molecular forces in cells: a guide to evaluating genetically-encoded FRET-based tension sensors. Cell Mol Bioeng 8(1):96–105. https://doi.org/10.1007/s12195-014-0368-1
27. Conway DE, Breckenridge MT, Hinde E, Gratton E, Chen CS, Schwartz MA (2013) Fluid shear stress on endothelial cells modulates mechanical tension across VE-cadherin and PECAM-1. Curr Biol 23 (11):1024–1030. https://doi.org/10.1016/j.cub.2013.04.049
28. Grashoff C, Hoffman BD, Brenner MD, Zhou R, Parsons M, Yang MT, McLean MA, Sligar SG, Chen CS, Ha T, Schwartz MA (2010) Measuring mechanical tension across vinculin reveals regulation of focal adhesion dynamics. Nature 466(7303):263–266. https://doi.org/10.1038/nature09198
29. Leerberg JM, Gomez GA, Verma S, Moussa EJ, Wu SK, Priya R, Hoffman BD, Grashoff C, Schwartz MA, Yap AS (2014) Tension-sensitive actin assembly supports contractility at the epithelial zonula adherens. Curr Biol 24(15):1689–1699
30. Austen K, Ringer P, Mehlich A, Chrostek-Grashoff A, Kluger C, Klingner C, Sabass B, Zent R, Rief M, Grashoff C (2015) Extracellular rigidity sensing by talin isoform-specific mechanical linkages. Nat Cell Biol 17 (12):1597–1606
31. Kumar A, Ouyang M, Van den Dries K, McGhee EJ, Tanaka K, Anderson MD, Groisman A, Goult BT, Anderson KI, Schwartz MA (2016) Talin tension sensor reveals novel features of focal adhesion force transmission and mechanosensitivity. J Cell Biol 213 (3):371–383
32. Aragona M, Panciera T, Manfrin A, Giulitti S, Michielin F, Elvassore N, Dupont S, Piccolo S (2013) A mechanical checkpoint controls multicellular growth through YAP/TAZ regulation by actin-processing factors. Cell 154 (5):1047–1059
33. Kim HY, Jackson TR, Stuckenholz C, Davidson LA (2020) Tissue mechanics drives regeneration of a mucociliated epidermis on the surface of Xenopus embryonic aggregates. Nat Commun 11(1):1–10. https://doi.org/10.1038/s41467-020-14385-y
34. Davidson LA, Dzamba BD, Keller R, Desimone DW (2008) Live imaging of cell protrusive activity, and extracellular matrix assembly and remodeling during morphogenesis in the frog, Xenopus laevis. Dev Dyn 237 (10):2684–2692. https://doi.org/10.1002/dvdy.21600

35. Merzdorf CS, Chen YH, Goodenough DA (1998) Formation of functional tight junctions in Xenopus embryos. Dev Biol 195 (2):187–203
36. Fouquet B, Herrmann H, Franz JK, Franke WW (1988) Expression of intermediate filament proteins during development of Xenopus laevis. III. Identification of mRNAs encoding cytokeratins typical of complex epithelia. Development 104(4):533–548
37. Jonas E, Sargent TD, Dawid IB (1985) Epidermal keratin gene expressed in embryos of *Xenopus laevis*. Proc Natl Acad Sci U S A 82 (16):5413–5417
38. Jamrich M, Sargent TD, Dawid IB (1987) Cell-type-specific expression of epidermal cytokeratin genes during gastrulation of *Xenopus laevis*. Genes Dev 1(2):124–132
39. Sive HL, Grainger RM, Harland RM (2000) Early development of Xenopus laevis: a laboratory manual. CSHL Press, Cold Spring Harbor, NY
40. Nieuwkoop PD, Faber J (1967) Normal tables of *Xenopus laevis* (Daudin). Elsevier North-Holland Biomedical Press, Amsterdam
41. Sater AK, Steinhardt RA, Keller R (1993) Induction of neuronal differentiation by planar signals in *Xenopus* embryos. Dev Dyn 197 (4):268–280
42. Joshi SD, Davidson LA (2010) Live-cell imaging and quantitative analysis of embryonic epithelial cells in Xenopus laevis. J Vis Exp (39). https://doi.org/10.3791/1949; pii: 1949
43. Davidson LA, Wallingford JB (2005) Visualizing morphogenesis in the frog embryo. In: Yuste R, Konnerth A (eds) Imaging in neuroscience and development: a laboratory manual. Cold Spring Harbor Laboratory Press, Cold Spring Harbor, NY
44. Dent JA, Polson AG, Klymkowsky MW (1989) A whole-mount immunocytochemical analysis of the expression of the intermediate filament protein vimentin in Xenopus. Development 105(1):61–74

Chapter 22

Molecular Tension Microscopy of E-Cadherin During Epithelial-Mesenchymal Transition

Helena Canever, Pietro Salvatore Carollo, Romain Fleurisson, Philippe P. Girard, and Nicolas Borghi

Abstract

Molecular Tension Microscopy has been increasingly used in the last years to investigate mechanical forces acting in cells at the molecular scale. Here, we describe a protocol to image the tension of the junctional protein E-cadherin in cultured epithelial cells undergoing Epithelial-Mesenchymal Transition (EMT). We report how to prepare cells and induce EMT, and how to acquire, analyze, and quantitatively interpret FRET data.

Key words FRET Biosensor, E-cadherin, EMT, Microscopy, Mechanotransduction

1 Introduction

Molecular Tension Microscopy (MTM) is the microscopy of molecular tension sensors [1]. Molecular tension sensors are a class of Förster Resonance Energy Transfer (FRET) sensors that are sensitive to molecular tension. They consist in a FRET pair of fluorophores separated by an elastic linker. When a force is exerted on the sensor, the rate of FRET, which is sensitive to the distance and orientation between the donor and acceptor fluorophores, decreases. Knowledge of the relationship between a measure of FRET and the force, from in vitro calibration, allows for the determination of molecular tension [2]. When inserted in a protein of interest, such a sensor can report its tension in live cells.

Molecular tension sensors have been in use for nearly 10 years [3]. A variety of sensors with specific operating ranges have been designed [4, 5], and have been used to measure tension in matrix, cytoskeleton, adhesion, glycocalyx, kinetochore, membrane-cytoskeleton linker or motor proteins [2, 3, 6–9] in cultured cells, *C. elegans*, Xenopus, Zebrafish, or Drosophila [3, 10–12].

Kyra Campbell and Eric Theveneau (eds.), *The Epithelial-to Mesenchymal Transition: Methods and Protocols*, Methods in Molecular Biology, vol. 2179, https://doi.org/10.1007/978-1-0716-0779-4_22,

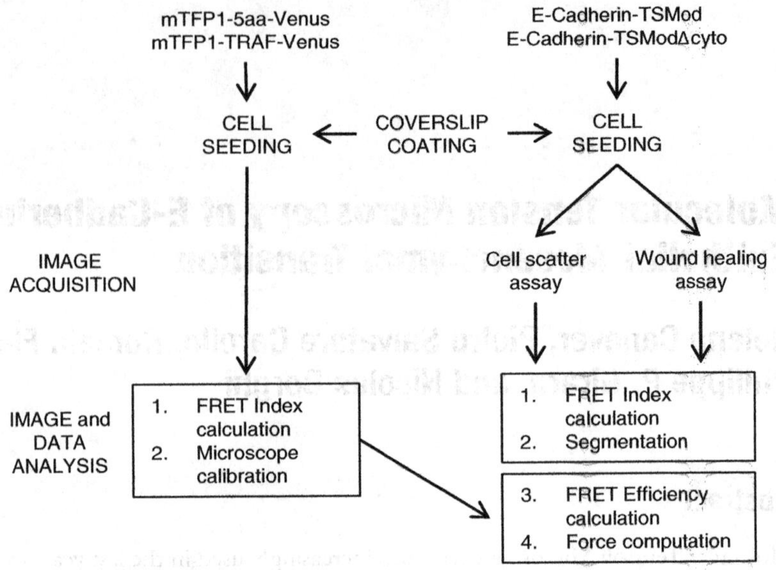

Fig. 1 Experimental workflow for MDCK type II cells sample preparation, EMT induction, image acquisition and image and data analysis

Here, we present a protocol to perform MTM of intercellular adhesion proteins E-cadherins [13] that experience cytoskeleton-generated tension sensitive to external cues [14]. During Epithelial-Mesenchymal Transition (EMT), E-cadherin tension relaxes, which associates with the release of its interactant β-catenin and activation of β-catenin dependent transcription [15].

The protocol describes how to culture epithelial cells (MDCK) stably expressing an E-cadherin tension sensor (E-cadherin-TSMod), induce EMT, and acquire and analyze FRET data using standard confocal microscopy and free image analysis software [16] (*see* Fig. 1). The protocol may be adapted to other proteins or model systems, on different microscopes as well. General considerations about MTM, its strengths and limitations can be found elsewhere [1].

2 Materials

Store DMEM (Dulbecco's Modified Eagle Medium), trypsin, Dulbecco's Phosphate Buffered Saline (DPBS, that is without Calcium and Magnesium), collagen at 4 °C. Store Foetal Bovine Serum (FBS), Penicillin, Streptomycin, and Hepatocyte Growth Factor (HGF) solution at −80 °C.

2.1 Cell Culture

1. Culture medium: low glucose DMEM with phenol red supplemented with 10% v/v Foetal Bovine Serum (FBS) and Penicillin 10 U/mL + Streptomycin 10 μg/mL.

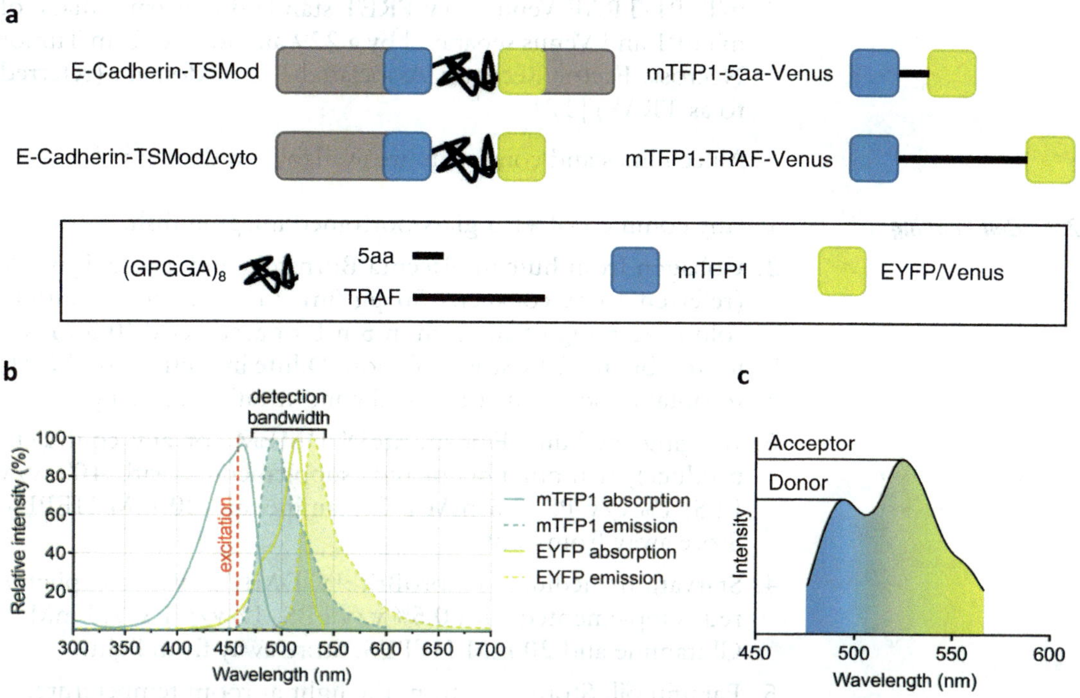

Fig. 2 (**a**) E-Cadherin-TSMod, tension-less control E-Cadherin-TSModΔcyto, and high and low FRET calibration constructs. Blue squares represent the donor fluorophore mTFP1, yellow squares represent the acceptor fluorophore EYFP/Venus. Linkers $(GPGGA)_8$, 5aa, and TRAF are represented in black. (**b**) Absorption and emission spectra of mTFP1 and EYFP/Venus. Laser excitation wavelength is indicated by the red dashed line. (**c**) Spectral emission in the detection bandwidth for mTFP1 and EYFP. Donor and acceptor emission peaks are indicated

2. Washing medium: DPBS.
3. 0.05% trypsin-EDTA solution in DPBS.
4. Madin-Darby canine kidney type II (MDCK II) stable cell lines cultured into 25 cm^2 flasks (T25, TPP) at 37 °C with 5% CO_2 in humidified atmosphere expressing (*see* Fig. 2a):
5. E-cadherin-TSMod: E-Cadherin tension sensor containing the donor protein mTFP1 and acceptor protein EYFP separated by a $(GPGGA)_8$ elastic linker [14].
6. E-Cadherin-TSModΔcyto: tension-less control construct of the E-Cadherin tension sensor, lacking its cytoplasmic tail [14] (*see* **Note 1**).
7. mTFP1-GPGGA-Venus: high FRET standard construct made of mTFP1 and Venus separated by a GPGGA amino acid stretch (referred to as 5aa) [17].

8. mTFP1-TRAF-Venus: low FRET standard construct made of mTFP1 and Venus separated by a 229 amino acid from Tumor Necrosis Factor Receptor Associated Factor domain (referred to as TRAF) [17].

All cell lines and constructs are available on demand.

2.2 Cell Imaging

1. Any commercial #1.5 glass-bottomed imaging dish.
2. Collagen from human placenta Bornstein and Traub Type IV (referred to as collagen) 50 μg/mL in acetic acid solution: solubilize 5 mg of collagen in 5 mL of acetic acid (0.5 M) on ice to obtain a 20× stock solution. Dilute in acetic acid (0.5 M) to obtain a solution at the final concentration of 50 μg/mL.
3. Imaging medium: FluoroBrite™ DMEM (or any equivalent product), without phenol red, supplemented with 10% v/v FBS, 1% v/v PS, 2.5 mM L-Glutamine and 20 mM HEPES. Store away from light.
4. Starvation medium: FluoroBrite™ DMEM, without phenol red, supplemented with 0.5% v/v FBS, 1% v/v PS, 2.5 mM L-Glutamine and 20 mM HEPES. Store away from light.
5. Paraffin oil. Store away from the light at room temperature.

2.3 Cell Scatter Assay

1. Hepatocyte Growth Factor (HGF) 10 ng/μL in BSA/DPBS solution: solubilize 1 mg of BSA (Bovine Serum Albumin) in 1 mL of DPBS. Solubilize 10 μg of HGF in the BSA/DPBS solution. Stock as 10 μL aliquots.
2. Stimulation medium: The day of the experiment, dilute an aliquot in 2 mL of starvation medium for a final HGF concentration of 50 ng/mL.

2.4 Wound Healing Assay

1. 200 μL pipette tips.
2. Washing medium.

2.5 Microscope

1. Commercial confocal inverted microscope equipped with an incubator (37 °C and 5% CO_2 in a humidified atmosphere).
2. High numerical aperture immersion objectives (*see* **Note 2**).
3. Argon laser—the 458 nm line is optimal for mTFP1 excitation—or equivalent, and a beam splitter that transmits above the excitation wavelength.
4. Spectral acquisition: microscopes that use a grating in the emission path (such as Zeiss or Olympus) allow for simultaneous acquisition of the donor and acceptor emission spectra with a spectral resolution below 10 nm and nm precision (*see* Fig. 2b, c).

3 Methods

Carry out all procedures in a cell culture hood unless otherwise specified.

3.1 Cell Culture Maintenance

Pass MDCK II cells three times a week as follows:

1. Prepare a new cell culture flask containing 3 mL of fresh culture medium and put it into the incubator.
2. Remove old culture medium and dead cells from the cell culture flask.
3. Wash once with 1 mL of washing medium and by gently shaking the cell culture flask.
4. Add 1 mL of Trypsin-EDTA solution in DPBS to the cell culture flask.
5. Put the cell culture flask into incubator and wait about 10 min for cells to detach from the surface and from each other.
6. Use a microscope to evaluate cell detachment.
7. Once cells are detached, transfer 1:10 of the collected volume into the new flask.

3.2 Collagen Preparation and Coverslip Coating

1. Take glass-bottomed imaging dishes.
2. Place 1 mL of collagen solution at 50 μg/mL onto the glass-bottomed dish for 10 min at room temperature.
3. Remove collagen solution and let dry for 1 h (*see* **Note 3**).
4. Sterilize the dishes under UV for 20 min (*see* **Note 4**).
5. Before use, wash the sterilized dishes with at least 1 mL of DPBS, twice.

3.3 Cell Seeding for Imaging

Seed cells 24 h before imaging. Prepare cells that express the tension sensor as well as cells that express the tension-less control separately. Follow **steps 1–6** of the Subheading 3.1 and continue as follows:

1. Collect the cell suspension into a sterile centrifugation tube.
2. Pellet the cells by centrifugation at 60–125 × *g* for 5–8 min.
3. Meanwhile, add culture medium onto the collagen coated imaging dishes.
4. Discard the supernatant of the centrifugation tube and re-suspend the pellet with culture medium (*see* **Note 5**).
5. Count cells with a hematocytometer or equivalent device.
6. Plate 1:10 of the collected volume in a new flask with culture medium.

7. Plate cells onto collagen coated imaging dishes: 2×10^5 for a cell scatter assay or 2×10^6 for a Wound Healing assay (*see* **Note 6**).

3.4 Cell Scatter Assay

1. The day after seeding, starve cells for 12 h in 2 mL imaging medium at 0.5% FBS.
2. Wipe the bottom of the glass-bottomed dish once with optical paper and distilled water, and again with ethanol 70%.
3. Install the imaging dish on the stage of the microscope.
4. Acquire pre-treatment images (*see* Subheading 3.6).
5. Replace the imaging medium with stimulation medium (*see* **Note 7**).
6. Add 1 mL paraffin oil at the surface of the imaging medium to prevent evaporation (*see* **Note 8**).
7. Acquire treatment images (*see* Subheading 3.6).

It is recommended to perform a mock experiment in parallel by following **steps 1**–**7** except **step 5** in which the medium does not contain HGF.

3.5 Wound Healing Assay

1. The day after seeding, wash and leave cells in the washing medium for 5–10 min to loosen intercellular contacts (*see* **Note 9**).
2. With a 200 μL pipette tip, perform a straight scratch in the cell monolayer (*see* **Note 10**).
3. Gently wash the dish with imaging medium and leave cells in 2 mL of imaging medium.
4. Any time after the wound, wipe the bottom of the coverslip as in Subheading 3.4, **step 2** and install the imaging dish on the stage of the microscope, add paraffin oil and acquire images (*see* Subheading 3.6).

3.6 Image Acquisition

1. Select an objective and exposure time to ensure sufficient signal to noise ratio, negligible fluorophore photobleaching and no saturation, as for standard fluorescence imaging.
2. Specify multi-position settings: if desired, acquiring images from multiple positions can increase your sample size per experiment. Choose between 5–10 positions considering that the total acquisition time of each position must not exceed the duration of the time interval between frames (*see* **Note 11**).
3. Specify *Z*-stack settings: if desired, this can alleviate focus drift if auto-focus is out-of-order. Make sure to avoid photobleaching.
4. Specify time-lapse settings: cell scattering by HGF and wound healing typically occur through several hours. Choose a time interval between frames of 10 min to 1 h (*see* **Note 12**).

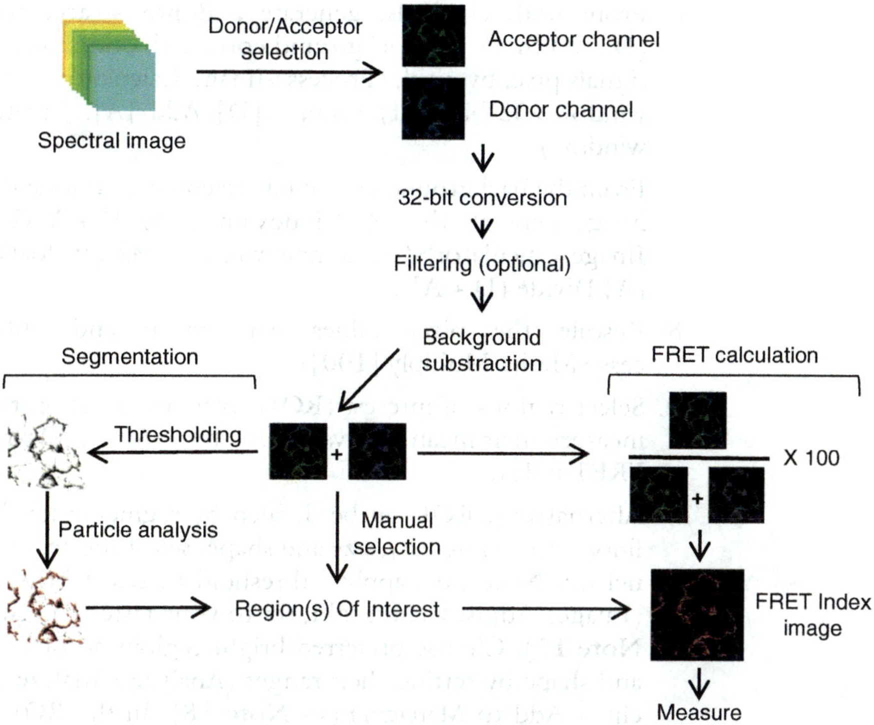

Fig. 3 Image analysis workflow for FRET calculation and image segmentation from a multi-channel spectral image

5. Specify spectral acquisition settings: 10 channels in the 473–561 nm bandwidth will capture both fluorophores. This will generate a multichannel stack.

3.7 Image Analysis

Image analysis can be performed using the free and cross-platform software ImageJ, for instance its FIJI distribution (https://fiji.sc/) (*see* Fig. 3).

1. Make duplicates of the raw data to avoid losses after permanent modifications (*see* **Note 13**). Generate a FRET index image as follows.
2. Open an image file.
3. From the multichannel stack, duplicate the donor and acceptor emission channels (D and A) that correspond to the intensity peaks of the two fluorophores (*see* **Note 14**).
4. On both channels, convert image depth to 32-bit (Image>Type>32-bit).
5. On both channels, subtract background: select a region devoid of cells and measure background level (Analyze>Measure), then subtract the value to the whole image (Process>Math>Subtract [Background value]) (*see* **Note 15**).

6. From both channels, generate a donor + acceptor image (D + A) that sums background corrected, donor and acceptor signals pixel by pixel. (Process>Image Calculator>Create new window>32-bit(float) result >[D] Add [A], Create a new window).
7. From the background corrected, acceptor channel and the sum image, generate the FRET index image A/(D + A) (Process>-Image Calculator>Create new window>32-bit(float) result>[A] Divide [D + A]).
8. Rescale the pixel values between 0 and 100. (Process>Math>Multiply [100]).
9. Select regions of interest (ROI) with any selection tool, then measure their mean gray value (Analyze>Measure) to get their FRET index.
10. Alternatively, ROI can be defined by segmentation based on fluorescence intensity, size and shape: select the acceptor channel (*see* **Note 16**), apply a threshold to select bright regions (Image>Adjust>Threshold, set to your taste then Apply) (*see* **Note 17**). Choose preferred bright regions according to size and shape by setting their ranges (Analyze>Analyze Particles, check Add to Manager) (*see* **Note 18**). In the ROI manager menu, save the ROI set to be able to recall it for future analyses (More>Save). Select/Open the corresponding FRET Index image (*see* Subheading 3.7, **step** 7) and click Measure in the ROI manager menu. You will obtain a measure for each ROI of the set.

3.8 Microscope Calibration

The FRET efficiency E can be related to the FRET index E_R computed as in 3.7 with $E = (1 - a(1 - E_R))/(1 - b(1 - E_R))$, where a and b are instrument-dependent coefficients that account for the donor's spectral bleed-through, the acceptor's direct excitation and differences in donor vs acceptor absorption cross-sections and detection efficiencies [18]. They can be recovered as follows (*see* **Note 19**):

1. Acquire images of the 5aa and TRAF cell lines as in Subheading 3.6. There is no need for time-lapse or Z-stack.
2. Analyze images as in Subheading 3.7.
3. Compute $a = (E_H(1 - E_{R,H}) - E_L(1 - E_{R,L}) + E_H E_L(E_{R,H} - E_{R,L}))/c$ and $b = (E_H(1 - E_{R,L}) - E_L(1 - E_{R,H}) + E_{R,L} - E_{R,H})/c$, with $c = (E_H - E_L)(1 - E_{R,H})(1 - E_{R,L})$, where $E_{R,H}$ and $E_{R,L}$ are the 5aa and TRAF cell lines FRET indices obtained in Subheading 3.8, **step 2**, and E_H and E_L their FRET efficiencies published elsewhere [17].

3.9 Data Interpretation

1. For each experimental condition, compute the difference between the tension sensor and tension-less control FRET efficiencies (*see* **Note 20**).
2. Use previously published FRET efficiency to force calibration [2] to retrieve the force.
3. Mind MTM limitations in your interpretation [1].

4 Notes

1. There exist other strategies for tension-less constructs, which may or may not give equivalent results depending on the protein of interest, as discussed in [1].
2. High magnification is irrelevant but comes with high numerical aperture, which increases resolution and signal intensity, and should also come with apochromatic correction. Choose the highest you can, typically 1.4 for standard objectives.
3. Save used collagen solution back in the final concentration solution for later use. Store at 4 °C indefinitely.
4. Use the UV light of the cell culture hood. Do not cover the dishes with a plastic lid. Multiple dishes can be prepared in advance and stored on the shelf indefinitely.
5. When re-suspending cells, pipette up and down to shear and dissociate cell aggregates. This will facilitate homogenous seeding.
6. When seeding, make sure the cell suspension covers the whole coverslip by gently shaking the dish linearly in X and Y, not circularly as cells would aggregate at the center of the coverslip.
7. Execute with care and promptly directly on the microscope stage to keep focus, temperature and position the same.
8. Make sure there is enough oil to cover the whole surface of the imaging medium to prevent evaporation.
9. Check that the monolayer is confluent before this step.
10. Two perpendicular scratches maximize wound edge length per coverslip. Use/draw fiducial markers on the coverslip/imaging chamber to easily find the wound once on the microscope stage.
11. Selecting positions in close proximity minimizes the spreading and loss of the objective immersion medium.
12. Consider the total acquisition time per time-point (multiple positions and *Z* sections), the total duration of the time-lapse, photobleaching and phototoxicity. 20- to 30-min intervals are adequate for most experiments.

13. Name your processed files consistently with the original files but with a different name to prevent overwriting.
14. Always keep the same donor and acceptor channels throughout all your analyses. You can identify them by plotting the intensity of a region containing fluorophores against the wavelength (or channel number) (select the multichannel stack then Image>Hyperstacks>Hyperstack to Stack, then select a region and Image>Stacks>Plot *Z*-axis profile). The plot should display two maxima (around 490 and 530 nm), if not, do not analyze further: the cell has likely not processed the full construct properly.
15. Before this step, you may filter some noise by local averaging of pixel levels, at the expense of spatial resolution (for instance, Process>Filters>Gaussian Blur).
16. For E-cadherin-TSMod, the donor channel often exhibits intracellular aggregates, probably synthesis or degradation intermediates in which the acceptor fluorophore does not emit. Hence, the acceptor channel is preferred.
17. You can explore the different tools available on Image J to process binary images (Process>Binary>...) to improve segmentation from the thresholded image.
18. More information on how to use the ROI Manager tool can be found here (https://imagej.nih.gov/ij/docs/guide/). Size and shape selection can be very useful to exclude regions too small or round to be cell–cell contacts. In the ROI Manager, you can delete unwanted regions (>Delete) or add missing regions (draw with any selection tool then >Add).
19. This does not need to be done for every experiment but can be performed from time to time to assess the microscope stability.
20. The Ecad-TSModΔcyto is sensitive to intermolecular FRET [14]. This normalization thus corrects for possible FRET changes due to intermolecular FRET among other causes.

Acknowledgements

This work was supported in part by the Centre national de la recherche scientifique (CNRS), the French National Research Agency (ANR) grants (ANR-17-CE13-0013, ANR-17-CE09-0019, ANR-18-CE13-0008), France BioImaging infrastructure (ANR-10-INBS-04), La Ligue contre le Cancer (Allocation de recherche doctorale to HC), the European Union's Horizon 2020 research and innovation programme (Marie Skłodowska-Curie grant agreement No 665850-INSPIRE to PSC). PSC acknowledges the Ecole Doctorale FIRE-Programme Bettencourt. We acknowledge the ImagoSeine facility, member of the France BioImaging infrastructure (ANR-10-INSB-04).

References

1. Gayrard C, Borghi N (2016) FRET-based molecular tension microscopy. Methods 94:33–42. https://doi.org/10.1016/j.ymeth.2015.07.010
2. Grashoff C, Hoffman BD, Brenner MD et al (2010) Measuring mechanical tension across vinculin reveals regulation of focal adhesion dynamics. Nature 466:263–266. https://doi.org/10.1038/nature09198
3. Meng F, Suchyna TM, Sachs F (2008) A fluorescence energy transfer-based mechanical stress sensor for specific proteins in situ. FEBS J 275:3072–3087. https://doi.org/10.1111/j.1742-4658.2008.06461.x
4. Ringer P, Weißl A, Cost A-L et al (2017) Multiplexing molecular tension sensors reveals piconewton force gradient across talin-1. Nat Methods 14:1090–1096. https://doi.org/10.1038/nmeth.4431
5. Brenner MD, Zhou R, Conway DE et al (2016) Spider silk peptide is a compact, linear nanospring ideal for intracellular tension sensing. Nano Lett 16:2096–2102
6. Paszek MJ, DuFort CC, Rossier O et al (2014) The cancer glycocalyx mechanically primes integrin-mediated growth and survival. Nature 511:319–325. https://doi.org/10.1038/nature13535
7. Zhang X, Li G, Guo Y et al (2019) Regulation of ezrin tension by S-nitrosylation mediates non-small cell lung cancer invasion and metastasis. Theranostics 9:2555–2571. https://doi.org/10.7150/thno.32479
8. Hart RG, Kota D, Li F et al (2019) Myosin II tension sensors visualize force generation within the actin cytoskeleton in living cells. bioRxiv 623249. https://doi.org/10.1101/623249
9. Suzuki A, Badger BL, Haase J et al (2016) How the kinetochore couples microtubule force and centromere stretch to move chromosomes. Nat Cell Biol 18:382–392. https://doi.org/10.1038/ncb3323
10. Yamashita S, Tsuboi T, Ishinabe N et al (2016) Wide and high resolution tension measurement using FRET in embryo. Sci Rep 6:28535. https://doi.org/10.1038/srep28535
11. Lagendijk AK, Gomez GA, Baek S et al (2017) Live imaging molecular changes in junctional tension upon VE-cadherin in zebrafish. Nat Commun 8:1402. https://doi.org/10.1038/s41467-017-01325-6
12. Lemke SB, Weidemann T, Cost A-L et al (2019) A small proportion of Talin molecules transmit forces at developing muscle attachments in vivo. PLoS Biol 17:e3000057. https://doi.org/10.1371/journal.pbio.3000057
13. Yoshida C, Takeichi M (1982) Teratocarcinoma cell adhesion: identification of a cell-surface protein involved in calcium-dependent cell aggregation. Cell 28:217–224
14. Borghi N, Sorokina M, Shcherbakova OG et al (2012) E-cadherin is under constitutive actomyosin-generated tension that is increased at cell-cell contacts upon externally applied stretch. Proc Natl Acad Sci U S A 109 (31):12568–12573. https://doi.org/10.1073/pnas.1204390109
15. Gayrard C, Bernaudin C, Déjardin T et al (2018) Src- and confinement-dependent FAK activation causes E-cadherin relaxation and β-catenin activity. J Cell Biol 217:1063–1077. https://doi.org/10.1083/jcb.201706013
16. Schneider CA, Rasband WS, Eliceiri KW (2012) NIH image to ImageJ: 25 years of image analysis. Nat Methods 9:671–675. https://doi.org/10.1038/nmeth.2089
17. Day RN, Booker CF, Periasamy A (2008) Characterization of an improved donor fluorescent protein for Förster resonance energy transfer microscopy. J Biomed Opt 13:031203
18. Lee NK, Kapanidis AN, Wang Y et al (2005) Accurate FRET measurements within single diffusing biomolecules using alternating-laser excitation. Biophys J 88:2939–2953. https://doi.org/10.1529/biophysj.104.054114

Part VI

Methods to Study Epithelial-Mesenchymal Plasticity at the Single Cell, Subcellular, and Molecular Level

Chapter 23

Methodologies for Following EMT In Vivo at Single Cell Resolution

Abdull J. Massri, Geoffrey R. Schiebinger, Alejandro Berrio, Lingyu Wang, Gregory A. Wray, and David R. McClay

Abstract

An epithelial-mesenchymal transition (EMT) occurs in almost every metazoan embryo at the time mesoderm begins to differentiate. Several embryos have a long record as models for studying an EMT given that a known population of cells enters the EMT at a known time thereby enabling a detailed study of the process. Often, however, it is difficult to learn the molecular details of these model EMT systems because the transitioning cells are a minority of the population of cells in the embryo and in most cases there is an inability to isolate that population. Here we provide a method that enables an examination of genes expressed before, during, and after the EMT with a focus on just the cells that undergo the transition. Single cell RNA-seq (scRNA-seq) has advanced as a technology making it feasible to study the trajectory of gene expression specifically in the cells of interest, in vivo, and without the background noise of other cell populations. The sea urchin skeletogenic cells constitute only 5% of the total number of cells in the embryo yet with scRNA-seq it is possible to study the genes expressed by these cells without background noise. This approach, though not perfect, adds a new tool for uncovering the mechanism of EMT in this cell type.

Key words Epithelial-mesenchymal transition, Single cell RNA-sequencing, Sea urchin, Tissue morphogenesis

1 Introduction

Most embryonic mesoderm cells are initially specified when they reside in an epithelium. An epithelial-to-mesenchymal transition (EMT) then removes them from the epithelial layer and they adapt a mesenchymal phenotype. In some cases, these cells again become epithelial and go through additional EMTs. This process of leaving the epithelium also occurs with carcinoma cells. Whether the two EMTs share mechanistic components of the process is a question that has often been asked. Literature reports indicate that they do indeed share multiple properties: they tend to use the same controlling transcription factors (*twist*, *snail*, and *zeb1*), though not always. They appear similar in behavior (the cells become motile,

Kyra Campbell and Eric Theveneau (eds.), *The Epithelial-to Mesenchymal Transition: Methods and Protocols*, Methods in Molecular Biology, vol. 2179, https://doi.org/10.1007/978-1-0716-0779-4_23,

change polarity, invade through the basement membrane, de-adhere from the adherens junction, and the plasma membrane is remodeled), though differences are observed in different model systems. What is clear is that in both the carcinoma and embryo systems, the molecular basis of this complex cellular event called EMT is incompletely understood. Indeed, of the several thousand papers a year on EMT, most focus on the epiphenomenon, that is, does the phenotypic change occur to an epithelial culture, or layer, under applied experimental conditions? Far fewer papers focus on the functional mechanics of that EMT in molecular detail.

A major reason for not understanding the EMT process in greater detail is that most systems are asynchronous, that is, the cells undergoing an EMT are at different states at any given time making it difficult to deduce the precise sequence of molecular events. A few cases of EMT in embryonic systems do offer synchrony, but each of these also has shortcomings. For example, ventral furrow formation in *Drosophila melanogaster (Drosophila)* provides a near synchronous EMT of mesoderm cells, and some genes necessary for the process have been identified. However, the difficulty in that system is that the number of mesoderm cells is small relative to the remaining cells of the embryo, and the EMT occurs relatively early in development, at a time when many maternally expressed genes are still expressed. This makes it difficult to exploit the power of *Drosophila* genetics to discover the genes mechanistically involved specifically in the EMT process [1]. Anchor cell invasion in *Caenorhabditis elegans* is another embryonic EMT in which one cell invades through the basement membrane as part of vulval assembly [2]. In this case the system is genetically tractable and a number of genes involved in the process have been identified. There is no question of synchrony, since only the one cell participates. However, a shortcoming of this system for EMT analysis is that the anchor cell does not complete an EMT. It breaches the basement membrane in a manner similar to that utilized by cells undergoing EMT in other systems, but it does not de-adhere from the epithelium. The sea urchin embryo also has a population of cells that undergo EMT at a precise time in early development and a gene regulatory network of specification is well established for those cells, making this a useful model system for understanding control of the process [3]. Nevertheless, this system also has shortcomings in that the skeletogenic cells that go through the EMT are only 5% of the population of cells in the embryo, making it a challenge to determine the sequence of molecular events in that small population.

Here we describe a method that can be used on any system to at least partially overcome some of the shortcomings possessed by many systems. Single cell RNA-sequencing (scRNA-seq) has advanced to the point where one can obtain a profile of expressed RNA in each cell. Computational approaches along with a temporal trajectory of single cells offers an approach to profile the molecular changes that occur in each cell undergoing the EMT over time. This approach therefore, has the potential of eliminating much of the noise introduced either by asynchrony of the EMT and/or inclusion of noninvolved cells, and the reward is provision of a temporal profile of molecular change.

It should be noted, however, that scRNA-seq is not the perfect solution. Because of the small amount of RNA obtained from each cell, amplification is necessary before sequencing. This and other limitations means that some rare RNA species are less likely to be included in the database than in bulk RNA-seq approaches. Nevertheless, the advances in scRNA-seq approaches provide the investigator with a valuable tool to penetrate EMT mechanisms to a level that heretofore has been unreachable.

2 The Single Cell RNA-Sequencing Approach, a Justification

Next generation sequencing (NGS) platforms increasingly allow in-depth analyses of gene expression and genetic interactions in many biological systems. The approaches allow the investigator unprecedented access to biological questions. The methodology begins with sample preparation, includes library production, sequencing, and data analysis. The latter is most important as software continues to be developed to enable the investigator to gain ever more detail about the biological process in question. As part of the description, the caveats and limitations of these technologies will be discussed. The focus will be on approaches that advance RNA-sequencing technologies and their application to understanding EMTs.

Two methods of RNA-sequencing are currently utilized, single cell RNA-sequencing (scRNA-seq) and bulk RNA-sequencing (RNA-seq). They each have their own individual advantages and disadvantages and are useful for addressing different biological questions. Bulk (whole-tissue) RNA-sequencing has many applications for research including comparative gene expression analyses between samples of various conditions, differential gene expression, identification of mRNA splice variants and small or long noncoding RNAs. RNA material collected from whole-tissue samples requires less or no amplification relative to scRNA-seq and the sample can be more deeply sequenced than that obtained from a single cell. Bulk RNA-seq is also easier: obtaining single cell suspensions from fixed or frozen tissue is nontrivial and may be very difficult for some samples. Thus, bulk RNA-sequencing is a good option in many applications. However, bulk RNA-seq is not as informative for identifying transcriptional differences within heterogeneous cell populations such as in developing and complex tissues because bulk RNA-seq measures the expression level of transcripts across a population of various types of cells, therefore creating an average transcriptomic profile of the tissue. This can become an issue when rare cell types are of interest, because their signal is essentially lost in the noise and more abundant transcripts. One way to get around this issue is by enriching for the population of interest, using a cell surface marker, fluorescence or antibody, however, this will still provide an averaged transcriptome across cells and does not capture heterogeneity at the single cell level. Another way to improve the analysis is to perform a temporal trajectory of the material in question. For embryonic material this can be highly informative because it adds the element of time, although still, within each sample the heterogeneity produces noise.

Single cell RNA-sequencing has the potential to eliminate much of the noise within a mixed population of cells. With a temporal profile it enables the investigator to probe the transcriptional dynamics of heterogeneous cell populations because it measures the distribution of mRNA expression from individual cells. Single cell transcriptomes can be profiled for a number of purposes such as creating cell atlases, mapping cell lineage trajectories [4–10], modeling virtual in situ hybridization [11] and more [12]. Using scRNA, one can capture cell trajectories and developmental processes such as an EMT by applying a scRNA-seq time-course to construct a cell trajectory map [13]. Generating an EMT time-course to capture transient cell states at single cell resolution informs the investigator with information on how this dynamic process occurs over time, providing a resource that is not available in any other known way.

Single cell RNA-sequencing is rapidly becoming a viable alternative to bulk RNA-sequencing, however, there are still some inherent issues with the platform. One challenge is due to the fact that RNA is harvested from only a single cell, and generally needs to be amplified with reverse transcription or PCR. This process of amplification can introduce bias that can lead to an incorrect interpretation. However, this can be overcome during the normalization and computational analysis by using Unique Molecular Identifiers (UMI), to uniquely label individual RNA molecules, greatly reducing amplification bias. Additionally, due to the sparsity of some RNA transcripts present in the cell and the inefficient cell capture process, sometimes a gene may have moderate expression in some cells, but cannot be detected in another cell. These occurrences, known as gene dropouts can be misleading because it is difficult to differentiate between inefficiency of transcript capture and a cell lacking that particular gene expression, or a gene that is expressed intermittently, therefore dimensionality reduction and normalization should to be performed computationally [14, 15].

3 Preparation of Single Cell Suspensions for scRNA-Seq

The key to any scRNA-seq experiment is generating a healthy representative single cell suspension from dissociated tissues or embryos. Therefore, it is imperative to develop a tissue dissociation protocol that properly captures single cells with minimal loss of integrity of the cells and minimal degradation of RNA. To achieve these goals, it is of utmost importance to minimize the time away from a cell's native environment while generating and handling single cell suspensions to accurately capture a cell's RNA identity, before alterations can occur. The transient nature of RNA expression can potentially be fixed in time following dissociation with a proper fixative, such as methanol, and the cells washed and rehydrated in 3× SSC rather than PBS, because rehydration in PBS can cause RNA degradation [16, 17]. Tissue types from various organisms and embryos are highly variable in their composition, therefore to generate a single cell suspension, different tissues require different enzymes, temperatures, salinity, and pH. Many groups have utilized enzymes that degrade extracellular matrix components to facilitate their dissociations. To establish a protocol, single cell preparations should be kept consistent, because altering the method of preparation can introduce a sampling bias. To establish the optimal conditions our single cell dissociation protocol was developed using a pilot study to establish the most reliable approach and as part of that, establish that a fixative such as methanol can be used to stabilize the RNA. The pilot study helped establish optimal scRNA-seq conditions for our system. The details of dissociation and stabilization of RNA are too varied to be covered in this chapter, but in each case the goals outlined above should be sought.

4 Considerations of Approach and Instrumentation Available for Library Preparation from Single Cells

To a research group beginning a scRNA-seq project, the next big question to ask is what platform should be used? Single cell RNA-sequencing has rapidly evolved since it was first used in 2009 [18]. When scRNA-seq was first introduced, it involved manually pipetting single cells into microwells and was relatively low throughput with a considerable amount of work required per cell. Since then, many groups have contributed to making scRNA-seq cost efficient and high throughput, and today many variations of these technologies exist. The introduction of multiplexing in 2011 [19, 20] was a major milestone where they showed many single cells could be sequenced together when UMIs were used. Additionally, in 2013 [21] integrated fluidic circuits, to allow for higher throughput, and more reproducibility. In 2015, [22, 23] introduced droplet-based methods where single cells are placed in droplets using microfluidics and beads with UMIs to uniquely label RNA molecules in each cell.

Currently a number of platforms are available to choose between, each with its advantages and disadvantages. Platforms differ from each other by either method of RNA quantification, or by method of cell capture. RNA expression is quantified by measure of either full length cDNAs or by tag-based UMIs. There are three methods of cell capture, microwell-based, microfluidic-based, and droplet-based. With the various options, it may seem difficult to determine which method is best, and the answer is it depends on the question being asked. Ziegenhain et al. [24] and Svensson et al. [25] realized this and so to assist you in making an informed decision they compare and contrast the common scRNA-seq techniques' accuracy, sensitivity, precision, power, and cost efficiency. Based on their findings, Smart-seq2, had the best sensitivity, accuracy, precision, and the lowest gene dropout rate, however this approach provides relatively low throughput compared to droplet-based methods that are not as sensitive but significantly less costly. Smart-seq2 currently is the best option for increased sequencing depth but for a smaller number of cells, as cost can be quite considerable. If willing to sacrifice some sequencing depth, drop-seq is the most cost efficient of the methods, but requires a tedious multi day protocol to be performed. Labs and sequencing centers also are adapting commercial platforms that include Fluidigm's C1 microfluidic chip, Wafergen ICELL8, BioRad's ddSEQ, and perhaps the most popular, 10× Genomics Chromium. Other alternatives utilize combinatorial indexing such as sciRNA-seq, while SPLiT-seq utilizes a split and pool method of barcoding cells within wells [4, 26]. These allow for higher throughput and cost efficiency than 10× and Drop-seq, however, the sample preparation takes longer, and there is a potential for introduction of sampling bias. In addition, the cell quality reportedly is a bit lower than 10× and Drop-seq. With all these options, it can be difficult to identify which method is best, for your research question. For a process such as EMT which has a temporal component, and for a process that occurs within an in vivo model (in our case, the sea urchin), we sought a method that could process many single cells with the best depth possible. To satisfy such a requirement, 10× Genomics was our choice of platform. Following library construction of single cells via 10× Genomics protocol, cells are sequenced at ~50 k reads per cell and using a 150 bp paired end Illumina run. Similarly, other single cell library preparation protocols utilize Illumina's paired end sequencing but may have different run length of 75, 125, 250 bp and more. Depending on the number of cells and the run length, a variety of options will be available using Illumina. For example, using a total of 1 billion PE reads on the NovaSeq

6000 and 150 bp PE run, roughly, 20 k cells can be sequenced at 50 k reads/cell to generate a single cell atlas capturing EMT. Indeed, the multitude of scRNA-seq techniques and methods are rapidly evolving, and as cost of scRNA-seq decreases, previous technologies will surely become obsolete. Research groups continue to push the limits and cost efficiency of scRNA-seq with methods like cell hashing that allow for "super loading" of cells, and it will only drive the cost down.

5 Bioinformatic Analysis: Overview of Procedure for Analysis of Results

Once single cell libraries are prepared and samples have been sequenced, the first step in analyzing the data is to create an expression matrix from the raw sequencing output. First, your bcl file should be demutiplexed using bcl2fastq to produce fastq files that can be checked for read quality control. A pipeline should be established early on, to identify what type of analyses will be performed (*see* Fig. 1 for a general ScRNA-seq pipeline that can be adapted). Following sequencing, Unique Molecular Identifiers (UMIs) should be extracted and reads demultiplexed using UMI-tools or zUMIs [28, 29]. To perform a quality check on reads, a common tool used is FastQC [30]. Once reads have been checked for quality control, they should be trimmed if a sample has poor base per sequence quality scores below 20, or if any exogenous nucleotides such as adapters were introduced. A few commonly used trimming tools are Trimmomatic, TrimGalore, and Cutadapt [31–33]. Trimmed reads can then be mapped back to a reference genome or transcriptome using a bulk RNA-seq aligner/pseudoaligner such as STAR/Kallisto or an aligner appropriate for your research question [34, 35]. Once reads have been mapped to genes, they are counted on a per gene and per cell basis to generate a single cell gene expression matrix [28, 36]. This matrix has a row for each cell and a column for each gene. The i, j entry encodes the number of molecules of mRNA for gene j in cell i. Therefore, each row encodes the expression profile of a cell as a point in a high-dimensional gene expression space, where there is a dimension for each gene.

With the expression matrix in hand, we are now ready to begin visualizing, exploring and analyzing the data. We begin by visualizing the high-dimensional single cell gene expression profiles in two or three dimensions. Some popular tools for visualizing single cell datasets include force layout embedding (FLE), UMAP, and tSNE [14, 15]. Instead of applying these tools directly to the single cell expression data, it can be helpful to first reduce the dimensionality from 20,000 down to ~1000 by selecting variable genes, and then down to ~100 using principal components analysis (PCA) or diffusion maps. This gradual decrease in dimensionality can help extract

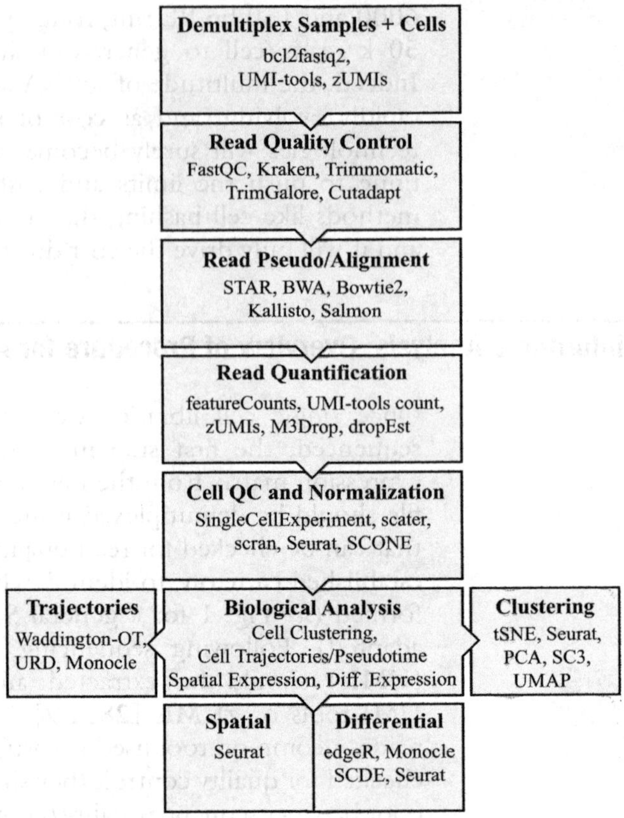

Fig. 1 General scRNA-seq pipeline. Figure adapted from and inspired by the single cell RNA-sequencing course [27]. Bioconductor is a repository that houses toolkits for sequencing and cell quality control, analysis, visualization, exploration, and more. Common packages used for each step in the pipeline are included. Using these methods, each gene's expression during EMT can be quantitatively measured in single cells, allowing for a deeper understanding of the underlying mechanistic structure of EMT

meaningful signals in the visualization. This visualization results in a set of x, y (and maybe z) coordinates that can be used to plot cells as points in 2 or 3 dimensions. Cells can be colored according to time of collection, batch, or expression of individual genes or gene signatures. The second component of exploratory data analysis involves searching for sets of cells with coherent gene expression programs. There are two main ways to do this. The first is to cluster cells (e.g., using Louvain clustering in diffusion component space). The second is to define cell sets according to expression of gene signatures. A gene signature is a list of genes (10 to 100 genes) related to a specific biological process or cell state (e.g., Epithelial Identity). To define an Epithelial cell state, we could select the top 10% of cells with highest expression of the Epithelial Identity gene signature.

In a time-course experiment, an expression matrix is obtained for each time point. The exploratory analysis described above can be applied to all time-points together in order to learn about general trends in expression over time. But, in order to learn about the different developmental trajectories and gene regulatory networks controlling differentiation, we must perform trajectory analysis.

The first goal of trajectory analysis is to infer ancestor–descendant relationships between pairs of time-points. This is crucial because scRNA-seq kills cells; therefore, we cannot use it to directly measure the change in gene expression of any individual cell over time. Live-cell imaging with fluorescent reporters can address this, but only for a handful of genes at a time. Many algorithms have been proposed to recover trajectories from scRNA-seq data. Waddington-OT is the only algorithm developed to date that is capable of modeling cell growth and development in a scRNA-seq time-course. All other algorithms either cannot incorporate known information about time of collection, or assume that all cells grow at the same rate (and therefore give rise to the same number of descendants). Waddington-OT infers ancestor-descendant relationships between pairs of time-points by leveraging a classical mathematical tool called optimal transport (OT). Intuitively, OT is based on the principle that cells can't change expression of all genes by large amounts in a short period of time. Therefore, cells are connected to "putative descendants" in a way that minimizes the total net change in expression over time. Each cell is allocated a certain amount of "descendant mass" according to an estimate of its proliferative ability and apoptosis rate (i.e., more proliferative cells are connected to more descendants). These growth rates are initially based on gene signatures of cell cycle and apoptosis, but are ultimately learned from data. The output of this first step of trajectory analysis is a "transport matrix" connecting each pair of time-points. The transport matrix has a row for each cell at time 1 and a column for each cell at time 2. The entries of the matrix indicate the amount of descendant mass each cell from time 1 gives rise to at time 2 (if we hadn't killed the cells).

After inferring ancestor–descendant relationships, the second goal of trajectory analysis is to infer gene regulatory networks controlling development and differentiation. To do this, Waddington-OT looks for transcription factors that are most predictive of transitions to various cell sets. For example, in iPSC reprogramming which transcription factors are responsible for pushing cells toward the stem cell state? Waddington-OT also allows us to analyze the shared ancestry connecting pairs of cell sets. This allows us to answer—does this pair of cell sets share a common ancestor near the beginning of the time-course and when does the pair diverge? We can then look for transcription factors that explain the bifurcation.

One common drawback of all these techniques is that spatial information is lost when cells are dissociated into suspension, however, the robust characterization of spatial markers within a tissue and developing embryo make it possible to reconstruct spatial patterning in silico. To reconstruct spatial information from dissociated tissues or embryos, Seurat can be employed to estimate a cell's likely position within spatial domains of the original tissue or embryo. As software matures and techniques improve in resolution, spatial transcriptomic technologies like Spatial Transcriptomics, Slide-Seq, and Seurat can provide more accurate spatial transcriptomic distributions [37, 38].

An outcome sought from this long list of computational options is a list of genes to be used in follow-up mechanistic studies. The question of how to reduce the size of that list varies with the goals in the system. In the case of the EMT, one approach might be to eliminate RNAs that are constitutively expressed since the EMT is fundamentally a change. Then, the direction of change and its timing can be considered using trajectories of RNAs and clustering programs. To that, data on perturbations, either based on known transcription factor control or perhaps on known drug effects can provide differential expression data that helps narrow the candidate list. Ultimately the goal is to identify candidates that are essential to the EMT and will help the investigator understand how the process works. To that end scRNA-seq provides an excellent tool.

Acknowledgements

The authors thank members of the McClay and the Wray labs for their critical input. We also appreciate the help provided by the Duke Core facility and the Benfey lab in the Biology Department. Support for this project was provided by NIH to DRM (RO1 HD 14483 and PO1 HD037105), and by NSF to GAW (IOS-1457305) and AJM for his NSF predoctoral fellowship (DGE-1644868). GS is supported by a Career Award at the Scientific Interface from the Burroughs Welcome Fund, and by funds from the Klarman Cell Observatory.

References

1. Schafer G, Narasimha M, Vogelsang E, Leptin M (2014) Cadherin switching during the formation and differentiation of the Drosophila mesoderm - implications for epithelial-to-mesenchymal transitions. J Cell Sci 127 (Pt 7):1511–1522. https://doi.org/10.1242/jcs.139485
2. Schindler AJ, Sherwood DR (2013) Morphogenesis of the Caenorhabditis elegans vulva. Wiley Interdiscip Rev Dev Biol 2(1):75–95. https://doi.org/10.1002/wdev.87
3. Saunders LR, McClay DR (2014) Sub-circuits of a gene regulatory network control a developmental epithelial-mesenchymal transition. Development 141(7):1503–1513. https://doi.org/10.1242/dev.101436
4. Cao J, Packer JS, Ramani V, Cusanovich DA, Huynh C, Daza R, Qiu X, Lee C, Furlan SN,

Steemers FJ, Adey A, Waterston RH, Trapnell C, Shendure J (2017) Comprehensive single-cell transcriptional profiling of a multicellular organism. Science 357 (6352):661–667. https://doi.org/10.1126/science.aam8940

5. Chen J, Renia L, Ginhoux F (2018) Constructing cell lineages from single-cell transcriptomes. Mol Asp Med 59:95–113. https://doi.org/10.1016/j.mam.2017.10.004
6. Fincher CT, Wurtzel O, de Hoog T, Kravarik KM, Reddien PW (2018) Cell type transcriptome atlas for the planarian Schmidtea mediterranea. Science 360(6391). https://doi.org/10.1126/science.aaq1736
7. Plass M, Solana J, Wolf FA, Ayoub S, Misios A, Glazar P, Obermayer B, Theis FJ, Kocks C, Rajewsky N (2018) Cell type atlas and lineage tree of a whole complex animal by single-cell transcriptomics. Science 360(6391). https://doi.org/10.1126/science.aaq1723
8. Tintori SC, Osborne Nishimura E, Golden P, Lieb JD, Goldstein B (2016) A transcriptional lineage of the early C. elegans embryo. Dev Cell 38(4):430–444. https://doi.org/10.1016/j.devcel.2016.07.025
9. Han X, Wang R, Zhou Y, Fei L, Sun H, Lai S, Saadatpour A, Zhou Z, Chen H, Ye F, Huang D, Xu Y, Huang W, Jiang M, Jiang X, Mao J, Chen Y, Lu C, Xie J, Fang Q, Wang Y, Yue R, Li T, Huang H, Orkin SH, Yuan GC, Chen M, Guo G (2018) Mapping the mouse cell atlas by microwell-seq. Cell 172 (5):1091–1107. e1017. https://doi.org/10.1016/j.cell.2018.02.001
10. Wagner DE, Weinreb C, Collins ZM, Briggs JA, Megason SG, Klein AM (2018) Single-cell mapping of gene expression landscapes and lineage in the zebrafish embryo. Science 360 (6392):981–987. https://doi.org/10.1126/science.aar4362
11. Karaiskos N, Wahle P, Alles J, Boltengagen A, Ayoub S, Kipar C, Kocks C, Rajewsky N, Zinzen RP (2017) The Drosophila embryo at single-cell transcriptome resolution. Science 358(6360):194–199. https://doi.org/10.1126/science.aan3235
12. Haque A, Engel J, Teichmann SA, Lonnberg T (2017) A practical guide to single-cell RNA--sequencing for biomedical research and clinical applications. Genome Med 9(1):75. https://doi.org/10.1186/s13073-017-0467-4
13. Griffiths JA, Scialdone A, Marioni JC (2018) Using single-cell genomics to understand developmental processes and cell fate decisions. Mol Syst Biol 14(4):e8046. https://doi.org/10.15252/msb.20178046
14. Becht E, McInnes L, Healy J, Dutertre CA, Kwok IWH, Ng LG, Ginhoux F, Newell EW (2018) Dimensionality reduction for visualizing single-cell data using UMAP. Nat Biotechnol. https://www.ncbi.nlm.nih.gov/pubmed/30531897
15. Van der Maaten L, Hinton G (2008) Visualizing data using T-SNE. J Mach Learn Res 9:2579–2605
16. Juliano C, Swartz SZ, Wessel G (2014) Isolating specific embryonic cells of the sea urchin by FACS. Methods Mol Biol 1128:187–196. https://doi.org/10.1007/978-1-62703-974-1_12
17. Chen J, Cheung F, Shi R, Zhou H, Lu W, Consortium CHI (2018) PBMC fixation and processing for chromium single-cell RNA sequencing. J Transl Med 16(1):198. https://doi.org/10.1186/s12967-018-1578-4
18. Tang F, Barbacioru C, Wang Y, Nordman E, Lee C, Xu N, Wang X, Bodeau J, Tuch BB, Siddiqui A, Lao K, Surani MA (2009) mRNA-Seq whole-transcriptome analysis of a single cell. Nat Methods 6(5):377–382. https://doi.org/10.1038/nmeth.1315
19. Islam S, Kjallquist U, Moliner A, Zajac P, Fan JB, Lonnerberg P, Linnarsson S (2011) Characterization of the single-cell transcriptional landscape by highly multiplex RNA-seq. Genome Res 21(7):1160–1167. https://doi.org/10.1101/gr.110882.110
20. Islam S, Zeisel A, Joost S, La Manno G, Zajac P, Kasper M, Lonnerberg P, Linnarsson S (2014) Quantitative single-cell RNA-seq with unique molecular identifiers. Nat Methods 11(2):163–166. https://doi.org/10.1038/nmeth.2772
21. Brennecke P, Anders S, Kim JK, Kolodziejczyk AA, Zhang X, Proserpio V, Baying B, Benes V, Teichmann SA, Marioni JC, Heisler MG (2013) Accounting for technical noise in single-cell RNA-seq experiments. Nat Methods 10(11):1093–1095. https://doi.org/10.1038/nmeth.2645
22. Macosko EZ, Basu A, Satija R, Nemesh J, Shekhar K, Goldman M, Tirosh I, Bialas AR, Kamitaki N, Martersteck EM, Trombetta JJ, Weitz DA, Sanes JR, Shalek AK, Regev A, McCarroll SA (2015) Highly parallel genome-wide expression profiling of individual cells using nanoliter droplets. Cell 161 (5):1202–1214. https://doi.org/10.1016/j.cell.2015.05.002
23. Klein AM, Mazutis L, Akartuna I, Tallapragada N, Veres A, Li V, Peshkin L, Weitz DA, Kirschner MW (2015) Droplet barcoding for single-cell transcriptomics applied

to embryonic stem cells. Cell 161 (5):1187–1201. https://doi.org/10.1016/j.cell.2015.04.044

24. Ziegenhain C, Vieth B, Parekh S, Reinius B, Guillaumet-Adkins A, Smets M, Leonhardt H, Heyn H, Hellmann I, Enard W (2017) Comparative analysis of single-cell RNA sequencing methods. Mol Cell 65(4):631–643. e634. https://doi.org/10.1016/j.molcel.2017.01.023
25. Svensson V, Natarajan KN, Ly LH, Miragaia RJ, Labalette C, Macaulay IC, Cvejic A, Teichmann SA (2017) Power analysis of single-cell RNA-sequencing experiments. Nat Methods 14(4):381–387. https://doi.org/10.1038/nmeth.4220
26. Rosenberg AB, Roco CM, Muscat RA, Kuchina A, Sample P, Yao Z, Graybuck LT, Peeler DJ, Mukherjee S, Chen W, Pun SH, Sellers DL, Tasic B, Seelig G (2018) Single-cell profiling of the developing mouse brain and spinal cord with split-pool barcoding. Science 360(6385):176–182. https://doi.org/10.1126/science.aam8999
27. Kiselev VY, Kirschner K, Schaub MT, Andrews T, Yiu A, Chandra T, Natarajan KN, Reik W, Barahona M, Green AR, Hemberg M (2017) SC3: consensus clustering of single-cell RNA-seq data. Nat Methods 14(5):483–486. https://doi.org/10.1038/nmeth.4236
28. Smith TS, Heger A, Sudbery I (2017) UMI-tools: modelling sequencing errors in unique molecular identifiers to improve quantification accuracy. Genome Res 27 (3):491–499. https://doi.org/10.1101/gr.209601.116
29. Parekh S, Ziegenhain C, Vieth B, Enard W, Hellmann I (2018) zUMIs - a fast and flexible pipeline to process RNA sequencing data with UMIs. Gigascience 7(6). https://doi.org/10.1093/gigascience/giy059
30. Andrews S (2010) FastQC: a quality control tool for high throughput sequence data
31. Bolger AM, Lohse M, Usadel B (2014) Trimmomatic: a flexible trimmer for Illumina sequence data. Bioinformatics 30 (15):2114–2120. https://doi.org/10.1093/bioinformatics/btu170
32. Krueger F (2012) http://www.bioinformatics.babraham.ac.uk/projects/trim_galore
33. Martin M (2011) Cutadapt removes adapter sequences from high-throughput sequencing reads. EMBnet J 17(1):10–12
34. Dobin A, Davis CA, Schlesinger F, Drenkow J, Zaleski C, Jha S, Batut P, Chaisson M, Gingeras TR (2013) STAR: ultrafast universal RNA-seq aligner. Bioinformatics 29(1):15–21. https://doi.org/10.1093/bioinformatics/bts635
35. Bray NL, Pimentel H, Melsted P, Pachter L (2016) Near-optimal probabilistic RNA-seq quantification. Nat Biotechnol 34 (5):525–527. https://doi.org/10.1038/nbt.3519
36. Andrews TS, Hemberg M (2019) M3Drop: dropout-based feature selection for scRNASeq. Bioinformatics 35(16):2865–2867. https://doi.org/10.1093/bioinformatics/bty1044
37. Eng CL, Lawson M, Zhu Q, Dries R, Koulena N, Takei Y, Yun J, Cronin C, Karp C, Yuan GC, Cai L (2019) Transcriptome-scale super-resolved imaging in tissues by RNA seqFISH. Nature 568 (7751):235–239. https://doi.org/10.1038/s41586-019-1049-y
38. Rodriques SG, Stickels RR, Goeva A, Martin CA, Murray E, Vanderburg CR, Welch J, Chen LM, Chen F, Macosko EZ (2019) Slide-seq: a scalable technology for measuring genome-wide expression at high spatial resolution. Science 363(6434):1463–1467. https://doi.org/10.1126/science.aaw1219

Chapter 24

Isolation and Identification of EMT Subtypes

Robert J. Norgard and Ben Z. Stanger

Abstract

Metastasis and chemoresistance, the most lethal features of cancer progression, are strongly associated with a form of cellular plasticity known as the epithelial-to-mesenchymal transition (EMT). Carcinoma cells undergoing EMT lose their epithelial morphology and become more mobile, allowing them to invade and migrate more efficiently. This shift is also associated with a change in vulnerability to chemotherapeutic agents. Importantly, EMT does not involve a single mechanism, but rather encompasses a spectrum of phenotypes with differing degrees of epithelial and mesenchymal characteristics. These hybrid/partial epithelial-mesenchymal states are associated with other important aspects of tumor biology, such as distinct modes of cellular invasion and drug resistance, illustrating the need to further characterize this phenomenon in tumor cells. Although simple in theory, the identification of tumor cells that have undergone EMT in vivo has proven difficult due to their high similarity to other mesenchymal cells that populate tumor stroma, such as cancer-associated fibroblasts. This protocol describes two methods for isolating epithelial and EMT cancer cell populations from primary murine tumors and cultured cancer cells to identify different EMT subtypes. These populations can then be used for several applications, including, but not limited to, functional studies of motility or invasion, gene expression analysis (RNA sequencing and RT-qPCR), DNA sequencing, epigenetic analysis, tumor subtyping, western blotting, immunohistochemistry, etc. Finally, we describe a flow cytometry-based approach to identify and study tumors cells that are undergoing partial EMT.

Key words Epithelial-to-mesenchymal transition, EMT, Plasticity, Partial EMT, Subtypes, E-cadherin, Flow cytometry-activated cell sorting, Magnetic-activated cell sorting

1 Introduction

In response to harmful stimuli, normal tissues take protective measures to repair resulting damage. Cellular plasticity—which refers to the ability of cells to undergo significant phenotypic and functional changes—is one such protective measure. When aberrantly activated in cancer cells, however, plasticity can contribute to tumor initiation, progression, and resistance to therapy [1]. In

The authors declare no potential conflicts of interest.

Kyra Campbell and Eric Theveneau (eds.), *The Epithelial-to Mesenchymal Transition: Methods and Protocols*, Methods in Molecular Biology, vol. 2179, https://doi.org/10.1007/978-1-0716-0779-4_24,

carcinomas, the most recognized form of cellular plasticity is epithelial-to-mesenchymal transition (EMT), whereby cancer cells lose their epithelial features (such as the adherens junction protein E-cadherin [ECAD]) and acquire the features of mesenchymal lineages (e.g., fibroblasts or leukocytes), thereby facilitating invasion and metastasis. Along with this gain of motile features, EMT has been reported to be associated with the acquisition of stem-cell-like characteristics, including shifts in the vulnerability of these cells to antineoplastic drugs [2]. Work over the past several years has determined that EMT does not involve a single mechanism in either normal physiology or tumors. Rather, it encompasses a phenotypic spectrum with varying degrees of epithelial and mesenchymal characteristics [3–5]. This has led to the recognition of intermediate epithelial-mesenchymal phenotypes, also referred to as partial EMT (P-EMT) states, whereby tumor cells simultaneously express epithelial and mesenchymal features. Importantly, these partial states may be associated with unique functional abilities, absent in either a pure epithelial state or a pure mesenchymal state, which confer a metastatic advantage by promoting collective migration or through other means [6, 7]. The P-EMT state may also allow for the evasion of certain chemotherapies as another mechanisms of resistance [8, 9].

Standard approaches to understand EMT in tumors have relied on in vitro studies and exogenous factors such as Transforming growth factor beta (TGF-β) [3]. A major challenge facing in vivo studies has been the difficulty distinguishing carcinoma cells which have undergone an EMT (therefore exhibit a fibroblast/mesenchymal morphology) from fibroblasts or other mesenchymal cells that infiltrate and populate the tumor microenvironment [10]. We previously developed a lineage-traced mouse model of pancreatic cancer driven by an oncogenic *Kras* and loss of *Trp53* (KPCY) that provide a means of distinguishing these various populations based on a yellow fluorescent protein (YFP) marker carried exclusively by tumor cells [6, 11]. Here we describe a method to isolate these cell populations using fluorescence-activated cell sorting (FACS) by which we isolate YFP+/ECAD+ and YFP+/ECAD- cells to study transcriptional changes that accompany EMT in vivo. These cells can further be used for a plethora of applications, including functional studies of cells in culture, gene expression analysis (RNA sequencing and RT-qPCR), DNA sequencing, epigenetic analysis, tumor subtyping, western blotting, and immunohistochemistry (Fig. 1a). We also highlight another method that can be used to isolate EMT populations by magnetic-activated cell sorting (MACS) and a dual antibody technique to identify intracellular proteins that can identify tumor cells that have undergone partial EMT both in vivo and in vitro. The following techniques are broadly applicable to other tumor models and tumor cells outside of pancreatic cancer that contain a lineage label.

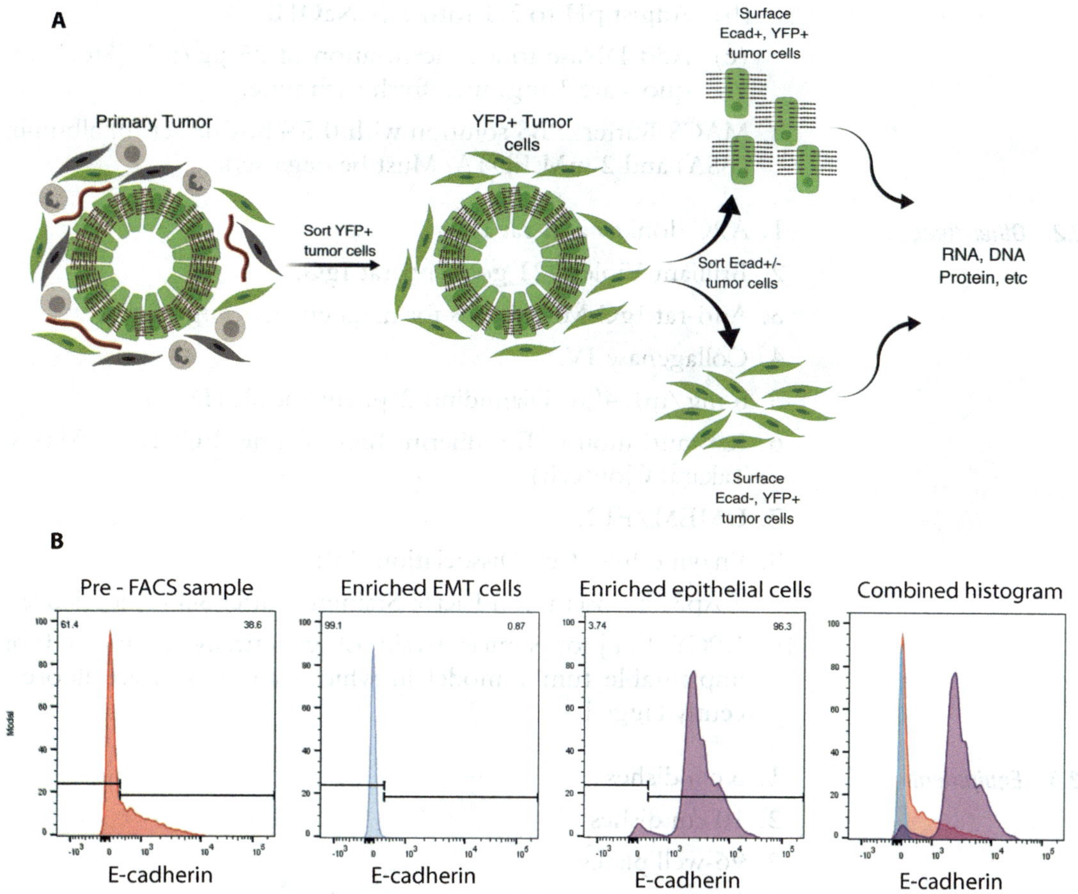

Fig. 1 (**a**) Overall strategy for isolating epithelial (ECAD+/YFP+) and mesenchymal (ECAD-/YFP+) tumor cells by FACS. (**b**) Purity of epithelial and mesenchymal populations after FACS. Left (Red): tumor cells before applying to a MACS column. Left-center (purple) enriched ECAD+ epithelial cells after LS column. Right-center (blue): enriched ECAD- mesenchymal cells after LD column. Right: merge of the three histograms. Upper corners demonstrate percent of purity of E-cadherin

2 Materials

Prepare sorting/staining buffers prior to beginning and store in sterile conditions at 4 °C. This can be stored for several months. Add fresh DNAse to an aliquot just prior to use (*see* **Note 1**).

2.1 Sorting/Staining Buffers

1. FACS Buffer.
   (a) Add the following ingredients to 1 L HBSS without Ca^{2+} and Mg^{2+}: 25 mL HEPES (stock 1 M), 5 mL $MgCl_2$ (Stock 1 M) (*see* **Note 2**), 10 mL Pen/Strep, 5 mL Amphotericin B, 10 ml NEAA (stock 100×), 10 mL L-glutamine (Stock 100×), 3 g Glucose, 10 mL Sodium Pyruvate (Stock 5% FBS).

(b) Adjust pH to 7.4 with 1 N NaOH.

(c) Add DNase to a concentration of 25 μg/mL (Stock aliquots are 1 mg/mL) fresh each time.

2. MACS Buffer: PBS solution with 0.5% bovine serum albumin (BSA) and 2 mM EDTA. Must be degassed.

2.2 Other Reagents

1. APC donkey anti-rat IgG.
2. Brilliant Violet 421 goat anti-rat IgG.
3. Anti-rat IgG Microbeads for magnetic labeling of cells.
4. Collagenase IV.
5. 5 mg/mL 4′,6- Diamidino-2-phenylindole (DAPI).
6. Rat anti-mouse E-cadherin IgG, Clone ECCD-2 (M108, Takara, Clontech).
7. DMEM/F12.
8. Enzyme-free, Cell Dissociation Buffer.
9. Foxp3/Transcription Factor Staining Buffer Set (eBioscience).
10. KPCY [11] or similar strain of genetically engineered or implantable tumor model in which cancer cells are fluorescently tagged.

2.3 Equipment

1. 6 cm dishes.
2. 10 cm dishes.
3. 96-well plates.
4. FACS machine with 100 μm nozzle.
5. Cell strainers (70 μm).
6. Conical tube (15 mL and 50 mL).
7. Transfer pipettes.
8. Fine scissors.
9. Luer Lok syringes (3 mL).
10. Microfuge tubes.
11. Table top microcentrifuge.
12. Needles (18 gauge).
13. Polypropylene round-bottom tubes (5 mL).
14. Polystyrene round-bottom tubes with 35 μm cell strainer cap (5 mL).
15. Benchtop Refrigerator Centrifuge.
16. Vortexer.
17. Water bath.
18. 1.5 mL microcentrifuge tubes.

19. LS Columns (Miltenyi Biotec).
20. LD Columns (Miltenyi Biotec).
21. QuadroMACS separator or suitable magnetic separator.

3 Methods

3.1 Isolation of EMT Populations from YFP-Expressing Tumors by FACS

1. Dissolve collagenase IV in DMEM/F12 at 2 mg/mL and prewarm at 37 °C in a water bath.
2. Remove the tumor-bearing tissue (i.e., pancreas or implantable tumor). Be careful to only remove the tumor portion and not contaminating normal tissue. Transfer to a 6-well dish and wash tissue three times with cold 1× PBS. After the third wash, leave a small amount of liquid to enhance cutting of the tissue.
3. Mince the tissue into small pieces with dissection scissors (alternatively a razor blade and a Petri dish lid can be used) (*see* **Note 3**).
4. Using a transfer pipette, take a small amount of the 2 mg/mL pre-warmed collagenase and transfer the tissue into 10 mL of collagenase in a 15 mL conical tube.
5. Incubate the tissue at 37 °C in a water bath in dark for 20 min. Vortex or shake vigorously every 5 min (*see* **Note 4).**
6. Transfer the now digested tumor mixture into a 50 mL conical tube using a 70 μm cell strainer. Using the plunger of a 3 mL syringe, push the tissue through the filter. Pour several mLs of cold DMEM/F12 and continue to push tissue until large pieces are dissociated.
7. Remove and discard the cell strainer and fill the tube to 50 mL with ice cold DMEM/F12 (*see* **Note 5**).
8. Centrifuge solution at 300 RCF for 5 min at 4 °C.
9. Carefully remove and discard the supernatant, being careful not to disturb or aspirate the mobile cell pellet. Add 500 μL ice cold-staining buffer and transfer cells to a microfuge tube.
10. Centrifuge cells at 4 °C at 300 RCF for 5 min at 4 °C and discard the supernatant.
11. Resuspend cells in 520 μL ice cold-staining buffer and save a small aliquot of cells (~20 μL) as an unstained negative control (this will receive secondary antibody only in the following steps (*see* **Note 6**)). In the remaining 500 μL solution, add 2 μL (1:250) rat anti-ECAD primary antibody and incubate on ice for 15 min.

12. Centrifuge cells at 4 °C at 300 RCF for 5 min and discard supernatant. Wash pellet with 500 μL staining media. Repeat for a total of three times.
13. Resuspend cells in 500 μL ice cold-staining buffer, add 5 μL (1:100) APC donkey anti-rat secondary antibody and incubate on ice for 15 min in the dark. Add 180 μL ice cold-staining buffer to the 20 μL unstained negative control sample, and add 2 μL secondary antibody (1:100).
14. Centrifuge cells at 4 °C at 300 RCF for 5 min and discard supernatant. Wash with 500 μL staining media. Repeat for a total of three times including negative control.
15. Resuspend cells in 1.5 mL sorting/staining buffer with 1.5 μL Dapi (1:1000). Use an 18-gauge needle in a 3 mL syringe to aspirate the cell solution. Remove the needle and then eject the solution into a polystyrene round-bottom tubes with 35 μm cell strainer cap. Depending on the type of FACS machine being used, polypropylene round-bottom tubes may need to be used. The filter cap can be switched onto these tubes or the cells can be transferred into this type of tube. Keep the cells on ice and dark until use.
16. Sort the cells in a FACs machine.
17. When processing using FACS, viable cells are impermeable to DAPI, so DAPI positive cells are no longer viable and excluded from the sort. It is important to use staining buffer without phenol red, as this may interfere with fluorescent signal. Epithelial cells with be positive for YFP and E-cadherin while tumor cells that have undergone EMT will be only YFP positive (Fig. 1b).
18. Centrifuge cells at 4 °C at 300 RCF for 5 min and discard supernatant.
19. Cells can now be processed for analysis to confirm EMT subtype by DNA, RNA, protein, etc. (*see* **Note 7**).

3.2 Isolation of YFP-Positive Mesenchymal Cells from Tissue Culture by FACS (See Note 8)

1. Aspirate media from 10 cm dish and wash the cells twice with ice cold PBS.
2. Add 5 mL of enzyme-free, cell dissociation buffer and incubate for 5–20 min or until cells have dissociated. Vigorously tap every 5 min (*see* **Note 9**).
3. Neutralize media in 5 mL of staining buffer and centrifuge cells at 4 °C at 300 RCF for 5 min.
4. Discard the supernatant and add 500 μL staining buffer and transfer cells to a microfuge tube.
5. Centrifuge cells at 4 °C at 300 RCF for 5 min and discard supernatant.
6. Continue with **step 11** from Subheading 3.1.

3.3 Isolation of Mesenchymal Cells by MACS

For efficient purification of tumor cells that have undergone EMT by MACS, a combination of positive selection (LS columns) and negative selection (LD columns) columns to isolate ECAD+ and ECAD- cells, respectively (Fig. 2a). However, it should be noted that with MACS, it is not possible to sort based on an intracellular lineage label potentially leading to normal cell contamination (*see* Table 1 for advantages and disadvantages).

1. Follow **steps 1–10** from Subheading 3.1 if isolating from primary tumor or **steps 1–5** from Subheading 3.2 if isolating from tissue culture.
2. Count cells and isolate 10^7 cells per column into a 1.5 mL Eppendorf tube. Take a small aliquot of addition cells for a negative control (*see* **Note 10**).

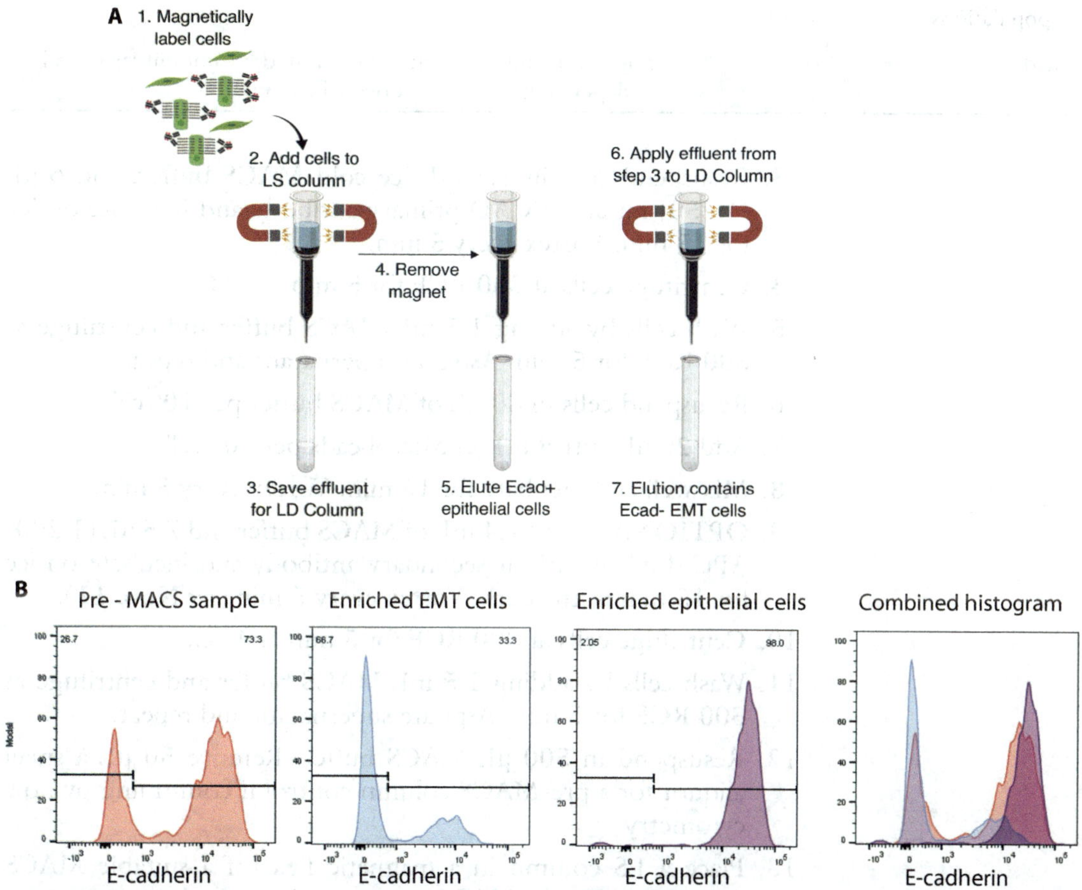

Fig. 2 (**a**) Schematic for isolating epithelial (ECAD+) and mesenchymal (ECAD-) cells by magnetic activated cell sorting (MACS). (**b**) Purity of epithelial and mesenchymal populations after MACS. Left (Red): tumor cells before applying to a MACS column. Left-center (purple) enriched ECAD+ epithelial cells after LS column. Right-center (blue): enriched ECAD- mesenchymal cells after LD column. Right: merge of the three histograms. Upper corners demonstrate percent of purity of E-cadherin. (Figures created with BioRender)

Table 1
Advantages and disadvantages of FACS versus MACS

	FACS	MACS
Technical complexity	High	Low
Time	Short to long	Short
Purity	High (>98%)	Intermed. (80–90%) but dependent on column and effort
Dead cell removal	Simple	Requires additional steps/columns, etc.
Specificity	High	High
Sorting of cells with intracellular fluorescence (i.e., YFP)	Possible	Not possible
Simultaneous sorting of different populations	Possible	Not simultaneously
Yield	High (but time dependent)	High but limited by amount that can be added to column

3. Resuspend cells in 1.5 mL ice cold MACS buffer, add 6 μL (1:250) rat anti-ECAD primary antibody and incubate on ice for 15 min. Vortex every 5 min.
4. Centrifuge cells at 300 RCF for 5 min at 4 °C.
5. Wash cells by adding 1.5 mL MACS buffer and centrifuge at 300 RCF for 5 min. Aspirate supernatant and repeat.
6. Resuspend cells in 80 μL of MACS buffer per 10^7 cells.
7. Add 20 μL anti-Rat IgG Microbeads per 10^7 cells.
8. Mix well and incubate for 15 min. Vortex every 5 min.
9. OPTIONAL: Add 1.4 mL of MACS buffer and 7.5 μL (1:200) APC donkey anti-rat secondary antibody and incubate on ice for 15 min in the dark. Vortex every 5 min (*see* **Note 11**).
10. Centrifuge cells at 300 RCF for 5 min at 4 °C.
11. Wash cells by adding 1.5 mL MACS buffer and centrifuge at 300 RCF for 5 min. Aspirate supernatant and repeat.
12. Resuspend in 500 μL MACS buffer. Remove 50 μL a small aliquot for a pre-MACS column control if confirming by flow cytometry.
13. Place a LS column in a magnetic field of a suitable MACS separator (QuadroMACS separator if multiple columns are being used simultaneously).
14. Prepare LS column by adding 3 mL MACS buffer. All columns are flow stopped and therefore do not run dry. Discard the effluent and change collection tube.

15. Apply cell suspension onto the column and collect flow through into a 15 mL conical tube containing unlabeled or poorly labeled cells (these will be used for the LD column).
16. Wash column 3 × 3 mL with MACS buffer and collect flow through as in step 15.
17. Remove the column from the separator and place it on a suitable collection tube (15 mL conical will suffice).
18. Apply 5 mL MACS column and immediately flush out magnetically labeled cells with the plunger into the column. This is the epithelial ECAD positive or ECAD-high population.
19. Place a LD column in a magnetic field of a suitable MACS separator (QuadroMACS separator if multiple columns are being used simultaneously).
20. Prepare a LD column by rinsing 2 mL of MACS buffer on top of the column. Discard the effluent and change collection tube.
21. Apply the effluent from **steps 15** through **16** and collect the flow through.
22. Wash LD column with 2 × 1 mL MACS buffer and collect flow through (*see* **Note 12**). This is the mesenchymal, EMT ECAD negative or ECAD-low population. These populations can then be confirmed by flow cytometry for purity (Fig. 2b).
23. There are several advantages and disadvantages with FACS versus MACS that are discussed in Table 1.

3.4 Identification of Partial EMT Tumor Cells by Surface/Intracellular Flow for Epithelial Proteins

Partial EMT tumor cells exhibit both epithelial and mesenchymal characteristics. In recent work, we found that some tumor cells re-localize epithelial proteins from the cell surface to the intracellular space. Therefore, it is possible using a dual antibody staining to identify whether the protein has moved from the cell surface to an intracellular compartment.

1. Prepare single cell suspensions from primary tumors (*see* **steps 1–10** from Subheading 3.1) or tissue culture (*see* **steps 1–5** from Subheading 3.2).
2. Resuspend cells in 300 μL ice cold-staining buffer and transfer to the well of a 96-well dish. Save a small aliquot of cells or split sample in half prior to adding the primary as an unstained negative control. This will receive secondary antibody in the following steps (*see* **Note 6**).
3. Centrifuge cells at 4 °C at 300 RCF for 1 min and remove supernatant by inverting the plate with a quick twitch.
4. Resuspend in 100 μL staining buffer with 1:250 rat anti-ECAD primary antibody and incubate on ice for 15 min.
5. Centrifuge cells at 4 °C at 300 RCF for 1 min and remove supernatant by inverting the plate with a quick twitch.

6. Wash with 200 μL staining media. Repeat for a total of three times.
7. Resuspend cells in 100 μL cold-staining buffer, add 1 μL (1:100) Brilliant Violet 421 secondary antibody and incubate on ice for 15 min in the dark. Add secondary antibody to unstained control.
8. Centrifuge cells at 4 °C at 300 RCF for 1 min and remove supernatant. Wash with 200 μL staining media. Repeat for a total of three times.
9. Fix cells by adding 200 μL intracellular fixation buffer (dilute 1-part fixation/permeabilization concentrate with 3 parts fixation/permeabilization diluent) and incubate 30 min at room temperature. Cover plate with aluminum foil or store in drawer.
10. Centrifuge cells at 4 °C at 300 RCF for 1 min and remove supernatant.
11. Add 100 μL 1× Permeabilization Buffer (10× permeabilization buffer diluted tenfold in distilled water).
12. Centrifuge cells at 4 °C at 300 RCF for 1 min and remove supernatant.
13. Resuspend in 100 μL 1× permeabilization buffer with 1:250 rat anti-ECAD primary antibody and incubate on ice for 15 min.
14. Centrifuge cells at 4 °C at 300 RCF for 1 min and remove supernatant. Wash three times with 1× permeabilization buffer.
15. Resuspend cells in 100 μL 1× permeabilization buffer, add 1 μL (1:100) APC anti-rat secondary antibody, and incubate on ice for 15 min in the dark. Add secondary antibody to unstained control. Single color controls should be run as well (*see* **Note 13**).
16. Centrifuge cells at 4 °C at 300 RCF for 1 min and remove supernatant. Wash three times with staining buffer.
17. Resuspend cells with 100 μL staining buffer and analyze by flow cytometry. Tumor cells that are in a partial EMT will be Brilliant Violet E-cadherin negative for surface expression but APC E-cadherin positive for intracellular expression. Strategy for detecting intracellular proteins are shown in Fig. 2d of Aiello et al. [6].

4 Notes

1. DNAse is critical to maintaining single cells suspensions. During the process, tumor and other cell types die and DNA is released into the solution. This results in stickiness or clumping of tumor cells.
2. $MgCl_2$ is necessary for the DNAse to work properly.
3. We are very careful when dissecting tumors. Anatomically (and by use of fluorescent reporters, *see* Maddipati and Stanger [7]), it can be determined whether a few distinct tumors exist within a tumor mass. Therefore, we micro-dissect a tumor and typically cut it into halves or thirds depending on size. One half is used for FACS or other molecular studies and the other half is used for subsequent histological analysis.
4. As much as possible, steps that follow should be done in the dark to preserve fluorescence of tumor cells. Typically, we wrap our tubes or cover plates with aluminum foil.
5. This step is critical to quenching/diluting out collagenase activity. The amount is variable but adding at least up to 30 mL is sufficient.
6. Inclusion of an isotype control primary antibody or secondary antibody alone is best way to identify a true positive population. If performing this protocol with an unconjugated antibody (as in this protocol), use a secondary only control. If performing this protocol with a conjugated antibody, use an isotype control of the same type and fluorescence as the primary antibody.
7. A determination of EMT subtype can be done in several ways (for more information, *see* Aiello et al. [6]). Briefly, RT-qPCR can be used to distinguish EMT subtypes by measuring transcripts for epithelial genes (e.g., ECAD, EpCAM, Cldn7, etc.) and mesenchymal genes (Vimentin, SPARC, FAP, etc.) comparing ECAD+ and ECAD- cells. After sorting tumor cells can also be lysed and western blotting performed in a similar fashion to RT-qPCR.
8. This method is for isolating tumor cells that have undergone EMT within a specific population in vitro or that have undergone EMT spontaneously or that have been induced to undergo EMT by addition of a ligand such as TGF-β. A lineage label is not necessary to separate tumor cells from non-tumor cells in established cancer cell lines.
9. Trypsin is capable of cleaving E-cadherin and other cell surface markers used to ascertain EMT status. Therefore, we use enzyme-free cell dissociation buffer which yields better results and safeguards from false negatives.

10. We have found that 10^7 cells are suitable for efficient isolation. A greater or lesser number of cells can be used, but this may require optimization (e.g., additional washes, more volume, etc.).
11. **Step 9** in Subheading 3.3 is optional for confirmation purposes. Generally, flow cytometry is used to confirm purity, but other methods can be used (i.e., western blotting, microscopy, etc.) (Fig. 2b).
12. The LD column can be removed from the separator to collect the ECAD-intermediate population if desired.
13. Although the unstained control contains both antibody colors, for clarity we typically run single color controls as well as the dual color.

References

1. Yuan S, Norgard RJ, Stanger BZ (2019) Cellular plasticity in cancer. Cancer Discov 9 (7):837–851. https://doi.org/10.1158/2159-8290.CD-19-0015
2. Dongre A, Weinberg RA (2019) New insights into the mechanisms of epithelial–mesenchymal transition and implications for cancer. Nat Rev Mol Cell Biol 20(2):69–84
3. Lamouille S, Xu J, Derynck R (2014) Molecular mechanisms of epithelial-mesenchymal transition. Nat Rev Mol Cell Biol 15:178–196. https://doi.org/10.1038/nrm3758
4. Pastushenko I, Blanpain C (2019) EMT transition states during tumor progression and metastasis. Trends Cell Biol 29(3):212–226
5. Nieto MA, Huang RYYJ, Jackson RAA, Thiery JPP (2016) EMT: 2016. Cell 166:21–45
6. Aiello NM, Maddipati R, Norgard RJ et al (2018) EMT subtype influences epithelial plasticity and mode of cell migration. Dev Cell 45:681–695.e4. https://doi.org/10.1016/j.devcel.2018.05.027
7. Maddipati R, Stanger BZ (2015) Pancreatic cancer metastases harbor evidence of polyclonality. Cancer Discov 5:1086–1097. https://doi.org/10.1158/2159-8290.CD-15-0120
8. Jolly MK, Somarelli JA, Sheth M et al (2019) Hybrid epithelial/mesenchymal phenotypes promote metastasis and therapy resistance across carcinomas. Pharmacol Ther 194:161–184
9. Shibue T, Weinberg RA (2017) EMT, CSCs, and drug resistance: the mechanistic link and clinical implications. Nat Rev Clin Oncol 14 (10):611–629
10. Tarin D (2005) The fallacy of epithelial mesenchymal transition in neoplasia. Cancer Res 65:5996–6000
11. Rhim AD, Mirek ET, Aiello NM et al (2012) EMT and dissemination precede pancreatic tumor formation. Cell 148:349–361. https://doi.org/10.1016/j.cell.2011.11.025

Chapter 25

Studying the Metabolism of Epithelial-Mesenchymal Plasticity Using the Seahorse XFe96 Extracellular Flux Analyzer

Sugandha Bhatia, Erik W. Thompson, and Jennifer H. Gunter

Abstract

The critical role of metabolism in facilitating cancer cell growth and survival has been demonstrated by a combination of methods including, but not limited to, genomic sequencing, transcriptomic and proteomic analyses, measurements of radio-labelled substrate flux and the high throughput measurement of oxidative metabolism in unlabelled live cells using the Seahorse Extracellular Flux (XF) technology. These studies have revealed that tumour cells exhibit a dynamic metabolic plasticity, using numerous pathways including both glycolysis and mitochondrial oxidative phosphorylation (OXPHOS) to support cell proliferation, energy production and the synthesis of biomass. These advanced technologies have also demonstrated metabolic differences between cancer cell types, between molecular subtypes within cancers and between cell states. This has been exemplified by examining the transitions of cancer cells between epithelial and mesenchymal phenotypes, referred to as epithelial-mesenchymal plasticity (EMP). A growing number of studies are demonstrating significant metabolic alterations associated with these transitions, such as increased use of glycolysis by triple negative breast cancers (TNBC) or glutamine addiction in lung cancer. Models of EMP, including invasive cell lines and xenografts, isolated circulating tumour cells and metastatic tissue have been used to examine EMP metabolism. Understanding the metabolism supporting molecular and cellular plasticity and increased metastatic capacity may reveal metabolic vulnerabilities that can be therapeutically exploited. This chapter describes protocols for using the Seahorse Extracellular Flux Analyzer (XFe96), which simultaneously performs real-time monitoring of oxidative phosphorylation and glycolysis in living cells. As an example, we compare the metabolic profiles generated from two breast cancer sublines that reflect epithelial and mesenchymal phenotypes, respectively. We use this example to show how the methodology described can generate bioenergetic results that in turn can be correlated to EMP phenotypes. Normalisation of bioenergetic studies should be considered with respect to cell number, and to potential differences in mitochondrial mass, itself being an important bioenergetics endpoint.

Key words Epithelial-to-mesenchymal plasticity, Glycolysis, Oxidative phosphorylation, Respiration, Metabolism, Metabolic phenotype, Mitochondrial CMxRos, Mitochondrial RedFM, Cellular bioenergetics, Extracellular acidification, Oxygen consumption, Seahorse Extracellular Flux Analyzer

---

Erik W. Thompson and Jennifer H. Gunter shares senior authorship.

Kyra Campbell and Eric Theveneau (eds.), *The Epithelial-to Mesenchymal Transition: Methods and Protocols*, Methods in Molecular Biology, vol. 2179, https://doi.org/10.1007/978-1-0716-0779-4_25,

1 Introduction

Metabolic reprogramming is a hallmark of cancer [1]. Cancer cells utilise a range of nutrient sensing pathways to coordinate the uptake of glucose, lipids and amino acids that are triaged into energy production, biosynthesis pathways and redox control to support cell proliferation and survival. For example, many tumours dramatically increase glucose consumption and this is exploited clinically by fluorodeoxyglucose-positron emission tomography (FDG-PET), to image highly metabolically active cancer lesions.

Glucose is metabolised to pyruvate via the process of glycolysis, where it can either enter the mitochondria for oxidative phosphorylation or generate ATP via its conversion to lactate in a process referred to as aerobic glycolysis. This latter pathway, while producing less ATP, circumvents the product inhibition of biosynthetic pathways and so allows more efficient cell proliferation. Aerobic glycolysis, first observed by Warburg, has been considered the dominant metabolic phenotype of cancer cells [2–4], however evidence now supports an important role of functional mitochondria and oxidative phosphorylation (OXPHOS) in enhancing metastasis [5–9]. Metastatic spread is attributed to dynamic alterations in carcinoma cells associated with the loss of epithelial features and the acquisition of more migratory, mesenchymal properties, referred to as epithelial–mesenchymal transition (EMT). The reverse process, mesenchymal–epithelial transition (MET), is then adopted for seeding into new niches [10–12]. The phenotypic cellular hallmarks that define this epithelial-mesenchymal plasticity (EMP) are well characterised [13], but the underlying metabolism remains to be comprehensively studied.

The interrelationship of EMP and metabolic reprogramming has been identified as a promising therapeutic strategy [14–16]. A genome-wide computational study of glycolytic dependency from NCI-60 cell lines identified metabolic targets to inhibit cellular migration [17]. Using gene signatures in The Cancer Genome Atlas (TCGA) database, the association between downregulation of mitochondrial genes, induction of EMT and metastatic potential was found to be directly associated with patient prognosis [18]. The activation of EMT was also correlated with reduced mtDNA content in prostate [19] and colorectal cancers [20], and a study of 207 primary breast tumour specimens also reflected a direct correlation between low mtDNA content and presence of distant metastases [21]. However, these conclusions were observed using DNA or RNA as reference, so validation studies are needed to correlate EMP with the metabolic phenotypes of cells.

Bioenergetic profiling of cancer cells can be studied on a Seahorse XF Analyzer. This platform contains a single steady-state probe which is lowered over the cell monolayer or spheroids creating a microchamber immediately above the cells. The probe simultaneously monitors extracellular acidification rate (ECAR) by measuring the H^+ in the media, which represents lactate production from glycolysis and the oxygen consumption rate (OCR), by measuring dissolved oxygen which reflects mitochondrial respiration. These key parameters can be used to determine rates of respiration and glycolysis (e.g. Mito Stress Test and Glycolysis Stress Test), the relative contribution of each pathway to total energy production (e.g. ATP Rate Assay and Glycolytic Rate Assay) and the preference of cells for different fuels or substrates such as glucose, glutamine or fatty acids (e.g. Fuel Flex Assay). These assays measure mitochondrial and glycolytic function by the sequential injection of metabolic inhibitors. Here we describe the utilisation of the Mito Stress Test to determine the relative function and dependency of cells with different metastatic potential on OXPHOS and glycolysis. The ATP Rate Assay may also be recommended for this type of analysis.

Until recently, studies have shown that decreased mitochondrial respiration and increased glycolytic activity, associated with EMT and metastasis, may be context-dependent [22]. The concept of a hybrid metabolic phenotype is now evolving, for example metastatic breast cancer cell lines 66 cl4 and 4 T1 have both enhanced ECAR and OCR compared with non-metastatic 67NR cells [23]. Studying metabolism in models of EMP can highlight unique or essential attributes that satisfy energy demands and fuel utilisation to support migration and invasion capabilities, and heralds a major advance in our understanding of the role for metabolic reprogramming in changing cell phenotypes. Additionally, studies of bioenergetic flux of cancer cells can be useful in assessing metabolic drugs that may impact EMP status [24]. Thus, this technology can generate new insights into the dynamics of cancer cell metabolism within the EMP spectrum.

2 Materials

2.1 Equipment

1. Seahorse XF^{e}96 Extracellular Flux Analyzer (Agilent Technologies Inc., Santa Clara, CA, USA).
2. CO_2 incubator for growing cells in an atmosphere of 5–10% CO_2 (*see* **Note 1**).
3. Non-CO_2 incubator to stabilise the cells in atmospheric oxygen immediately prior to analysis.
4. Laminar flow hood.
5. Cell counter.
6. Single channel and multichannel dispensing pipette.
7. pH meter.
8. Water bath.

9. Microplate reader.
10. High-content microscopy imaging system (e.g. InCell Analyzer 2200).
11. Flow cytometer.

2.2 Reagents and Solutions (96-Well Assay Format)

1. Seahorse XFe96 FluxPak containing probe plate, 96-well tissue culture plate and calibrant (Seahorse Bioscience, 102601-100).
2. Seahorse XF Base Medium (DMEM or RPMI without phenol red) or other medium (without glucose, glutamine, sodium bicarbonate, sodium pyruvate, phenol red).
3. Substrate Supplements (final concentration): 2 mM L-glutamine, 1 mM sodium pyruvate, 10 mM glucose.
4. Seahorse XF Cell Mito Stress Kit (Seahorse Bioscience, 103015-100) containing oligomycin, FCCP and Rotenone/Antimycin A. Aliquots are dissolved in media to 100 mM stock concentrations.
5. CellTak™ (Corning, Catalog# 354240), collagen, poly-ornithine or poly-L-Lysine.
6. Hoechst nuclear dye.
7. Syringe Filters, 0.22 μM pore size, 4 mm membrane diameter.
8. Acids and Bases for adjusting pH.
9. Media reservoirs for multichannel dispensing.
10. DNA or protein-based cell counting kits.
11. Trypsin/EDTA Solution.
12. Accutase.
13. Dimethylsulfoxide (DMSO).
14. Fetal Bovine Serum (FBS).
15. Mitotracker dyes (mitochondrial mass and mitochondrial membrane potential).
16. FACS sorting buffer (1× PBS with 0.5% FBS, 1 mM EDTA, 25 mM HEPES pH 7).

3 Methods

3.1 Methodology Overview

The Seahorse assay is performed on live cells. Cells are seeded into specialised 96-well cell culture-treated plates, supplied by the manufacturer. The wells are funnel-shaped which allows the probes to create a snug microchamber of ~2 μL immediately above the cells to allow highly sensitive measurement of dissolved H^+ and O_2. The cells are allowed to adhere, typically overnight. On the day of the assay, the cells are transferred to an unbuffered media with fixed pH equilibrated at temperature and in atmospheric oxygen. During the assay, the metabolic measurements are made with a solid state fluorescent probe attached to a specialised plate, which allows separate and simultaneous measurements of H^+ and O_2 in each well of the plate. Each probe is surrounded by four injection ports, which are loaded with assay-specific inhibitors and programmed to be added to the cells at precise times during the assay in order to assess metabolic function. On the day before the assay, the probe plates are rehydrated in sterile water at 37 °C in atmospheric oxygen. On the day of the assay the inhibitors are added to their respective ports in preparation for the assay and loaded onto the Seahorse machine. The inhibitors used in the Mito Stress Test are *Oligomycin,* which blocks ATP synthase of electron transport chain (ETC) and causes a decrease in oxygen consumed via OXPHOS and usually an increase in glycolysis; *FCCP* (Carbonyl cyanide-4 (trifluoromethoxy) phenylhydrazone), an uncoupling agent that disrupts the mitochondrial membrane potential by collapsing the proton gradient; and a mixture of *Rotenone*, a complex I inhibitor of ETC, and *Antimycin A*, a complex III inhibitor of ETC to inhibit OXPHOS. Following the assay, the plates are removed and data is normalised by the immediate measurement of cell number, DNA or protein content. It is important to consider bioenergetics in the context of cellular mitochondrial mass, which may differ between treatments or phenotypes. A double-staining method for evaluating mitochondrial content and activity is described in this protocol.

3.2 Day 1 – Seeding of the Cells and Probe Rehydration

1. In a laminar flow hood, hydrate the XF assay cartridge plate. Lift the green probe cartridge from the utility plate and place it with probes facing up to avoid damage.
2. Fill each well of the utility plate with 200 μL of ultrapure water and replace the probe cartridge, submerging the sensor sleeves thoroughly in the solution.
3. Place the cartridge at 37 °C in a non-CO_2 incubator overnight along with a falcon tube containing 20 mL of sterile calibration solution in preparation for the following day.
4. Coat the cell culture plate (*See* **Note 2**). Cover the bottom of the plate with matrix and incubate at room temperature for 1–3 h.
5. Rinse the wells at least twice with sterile PBS using a multichannel pipette and reservoir. Allow wells to dry in the laminar flow hood.
6. Select the appropriate cell density. As cells have different proliferation capacities, the number of cells per well should be determined empirically for each cell line. Cells should be seeded so that all the wells are 80–90% confluent on Day 2.

7. Trypsinise the cells, and suspend in appropriate growth media.
8. Mix approximately 10 μL cell suspension in a 1:1 ratio with 0.4% trypan blue and perform cell count.
9. Seed the appropriate number of cells, typically 10,000–20,000 cells in 80 μL of media per well in all but the 4 corner wells. These are used to determine background measurements and must remain cell-free (*see* **Notes 3–5**). Use a multichannel pipette to minimise sampling error.
10. Leave the plate undisturbed for 20–30 min in the hood to assure even cell seeding. If cells are clumped in the middle or edge of the well, a large amount of variation can be observed in the assay.
11. Place the 96-well cell culture plate in a CO_2 incubator at 37 °C, 5% CO_2 and with 95% humidity and monitor after 5–6 h to observe cell adherence. Allow the cells to settle overnight.

3.3 Day 2 – Performing the Mito Stress Test Assay

1. Remove the probe plate from the non-CO_2 incubator.
2. Replace the water in the probe plate with 200 μL of calibration buffer and return to 37 °C non-CO_2 incubator for 1–4 h (*see* **Note 6**).
3. Warm the required volume of unbuffered basal DMEM media at 37 °C and supplement with glucose (10 mM), glutamine (2 mM) and sodium pyruvate (1 mM).
4. Adjust the pH to 7.4 ± 0.05, filter sterilise and warm the assay media to 37 °C.
5. Remove all but 20 μL of the growth media from each well and wash the cells twice gently with 200 μL of assay media (including the background wells A1, A12, H1 and H12) (*see* **Note 7**). After removing the second 200 μL, add 160 μL of media (total volume in well is 180 μL) and place cells in the non-CO_2 incubator to equilibrate to atmospheric oxygen.
6. While the cells equilibrate, prepare the injection compounds in assay media and load them into the ports of the cartridge plate (Fig. 1) as described in Table 1 (*see* **Notes 8** and **9**).
7. Using the XFe96 WaveTM software, create a template file of the protocol by adding details including the name of cell-type(s), experimental conditions or pretreatments, the reagent names and concentrations (*see* **Note 10**).
8. The machine prompts the probe plate to be inserted for calibration. Remove the lid before inserting the plate.
9. After calibration is complete, the sensor plate is retained in the machine and the software prompts the exchange of the utility plate in the loading tray with the cell culture plate.

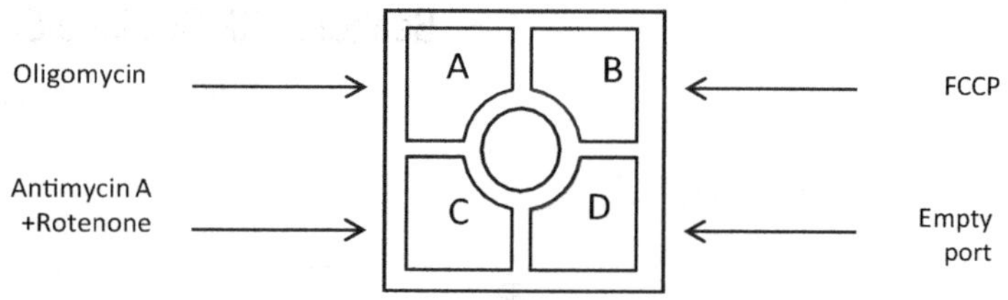

Fig. 1 Schematic drawing of the four injection ports (**a–d**) surrounding the central probe, with compounds loaded for the Mito Stress Test

Table 1
Preparation of injection compounds

Port	Compounds	Stock conc	Port conc stock vol (media vol)	μL loaded per port	Final conc/ well
A	Oligomycin	100 μM	12 μM 360 μL (2640 μL)	20 μL	1.2 μM
B	FCCP	100 μM	15 μM 450 μL (2550 μL)	22 μL	1.5 μM
C	Antimycin A Rotenone	100 μM 100 μM	10 μM 10 μM 300 μL (2700 μL)	25 μL	1 μM 1 μM

10. Run the experiment. The machine allows the cells a chance to equilibrate before recommencing the assay. The default setup is for the probe to mix the well contents for 3 min, measure O_2 and H^+ for 3 min for at least 3 cycles.

3.3.1 A Typical Experiment Will Be

- Mix for 3 min, measure for 3 min. Repeat for 3 measurements.
- Inject port A.
- Mix for 3 min, measure for 3 min. Repeat for 3 measurements.
- Inject port B.
- Mix for 3 min, measure for 3 min. Repeat for 3 measurements.
- Inject port C.
- Mix for 3 min, measure for 3 min. Repeat for 3 measurements.
- End of assay.

The twelve measurements, represented as a line graph in Fig. 2 show the changes in bioenergetic profiles after the injection of mitochondrial inhibitors (sourced with permission from Agilent Technologies Pty Ltd) and explained further in Table 2 at the end of the chapter (*see* **Note 11**).

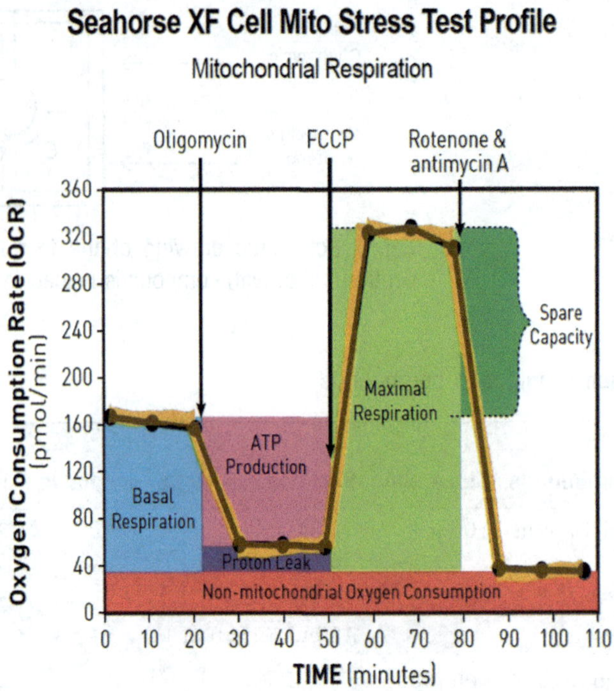

Fig. 2 The numerous parameters of the bioenergetics experiment are calculated from area under the curve (AUC) data obtained from twelve measured values in the Seahorse assay. The non-mitochondrial respiration represents the minimum rate measurements after injection C (rotenone/antimycin A), i.e. mean of the rates 10–12 and represents the non-mitochondrial cellular process that consumes oxygen. The basal respiration (OCR value) is determined from the difference between the mean of the three measurements, prior to injection A (oligomycin), and the OCR values after injection C (rotenone/antimycin A). It represents the actual amount of oxygen that is being consumed to maintain OXPHOS at a level to produce ATP for the growing cells. The proton leak is determined from the difference between the measurement 4, prior to injection B (FCCP), and measurement 10, after injection C. The ATP production is determined from the difference between the measured OCR level and the proton leak. Maximal respiration is the capacity of the cells at which the electron transport chain can supply electrons to complex IV after the addition of FCCP inhibitor, and is determined from the difference between measurement 7 and the measurement 4, and the spare capacity is determined from the differences between the maximal respiration and basal respiration. (Image used with permission from Agilent Technologies)

11. At the end of the assay the machine prompts the removal of the probe and cell plate. Discard the probe plate but retain the cells for normalisation.
12. Calculating the number of cells used in the assay is a crucial step as even cells seeded at the same seeding density might vary in their settling rates as well as rate of cell division (*see* **Note 12**).

Table 2
List of functional parameters calculated using Mito Stress Test

Basal respiration	Oxygen consumption used to meet cellular ATP demand resulting from mitochondrial proton leak. Shows energetic demand of the cell under baseline conditions
ATP production	The decrease in oxygen consumption rate upon injection of the ATP synthase inhibitor oligomycin represents the portion of basal respiration that was being used to drive ATP production. Shows ATP produced by the mitochondria that contributes to meeting the energetic needs of the cell
H+ (proton) leak	Remaining basal respiration not coupled to ATP production. Proton leak can be a sign of mitochondrial damage, or can be used as a mechanism to regulate the mitochondrial ATP production
Maximal respiration	The maximal oxygen consumption rate attained by adding the uncoupler FCCP. FCCP mimics a physiological 'energy demand' by stimulating the respiratory chain to operate at maximum capacity, which causes rapid oxidation of substrates (sugars, fats and amino acids) to meet this metabolic challenge. It shows the maximum rate of respiration that the cell can achieve
Spare respiratory capacity	This measurement indicates the capability of the cell to respond to an energetic demand as well as how closely the cell is to respiring to its theoretical maximum. The cell's ability to respond to demand can be an indicator of cell fitness or flexibility
Non-mitochondrial respiration	Oxygen consumption that persists due to a subset of cellular enzymes that continue to consume oxygen after the addition of rotenone and antimycin A. this is important to get an accurate measure of mitochondrial respiration

13. **Option 1:** assess cell number in each well using a DNA based stain such as CyQuant against a standard curve of known cell numbers.
14. **Option 2:** Alternatively, Hoechst dye can be added to a final concentration of 10 μg/mL for 10–60 min.
15. Image cells using an imaging system such as the InCell Analyzer 2200 high-content microscopy imaging system in the 488 nm channel. Images can be analysed using any software capable of nuclear image segmentation analysis, e.g. Cell Profiler (Broad Institute; https://cellprofiler.org/citations/) can be utilised to determine the number of Hoechst-positive nuclei.
16. Normalise the results by adjusting for the actual cell number.
17. Analyse the data using Wave software (*see* **Note 13**).

3.4 Flow Cytometry Profiling for Mitochondrial Antibodies

1. Use mitochondrial dyes CST Mitotracker Red FM and CMXRos to quantitfy mitochondrial mass and membrane potential, respectively. Concentrations for optimal staining should be optimised using manufacture's recommended dilutions.

2. Lift the cell lines with Accutase and incubate 5×10^5 cells with mitochondrial markers in media for 1 h at 37 °C.
3. Wash the cells once and resuspend in sorting buffer.
4. Strain the samples using cell strainer into 5 mL propylene tubes.
5. Perform quantitative fluorescence analysis using a suitable flow cytometer (*see* **Note 14**).

3.5 Analysis

1. The mitochondrial respiration and extracellular acidification rate, normalised to cell number, are plotted against time and show the metabolic response following the addition of inhibitors (*see* Fig. 3a, b for example).
2. The bioenergetic values of ECAR and OCR can be plotted together for a comparative view. This view can compare intra-experimental phenotypes with low metabolic or dormant cells in the lower left quadrant, more glycolytic cells in bottom right quadrant, more OXPHOS-dependent cells in the top left quadrant and cells utilising both pathways in the most metabolic top right quadrant. *See* Fig. 3c for an example, where the results from an MET model cell line have been summarised.

4 Notes

1. Studies can be performed under hypoxia, but as the Seahorse undertakes measurements under 'atmospheric' oxygen, the Seahorse machine must be placed inside a hypoxia hood.
2. The mixing and movement of the Seahorse probes can disturb cells that are weakly adherent. To avoid this, plates should be precoated to improve cell adherence during the assay. The manufacturer recommends CellTak™ for this. Alternatively, substitutes of ECM matrix, such as collagen, poly-ornithine or poly-L-Lysine can be used. In our experience, coating also improves even cell distribution when plating, and better adherence of cells during media exchanges prior to and during the assay.
3. The 4 corner wells of the 96-well plate (A1, A12, H1 and H12) are temperature correction wells and must not have cells seeded in them. Treat these wells as media blanks.
4. It is recommended to perform a pilot study to assess metabolic potential for range of cell densities, for optimal concentration of inhibitors of oligomycin and FCCP across different cancer cell lines. In our experience we have found a minimum of 4 replicate wells to be required for cell-line observations.

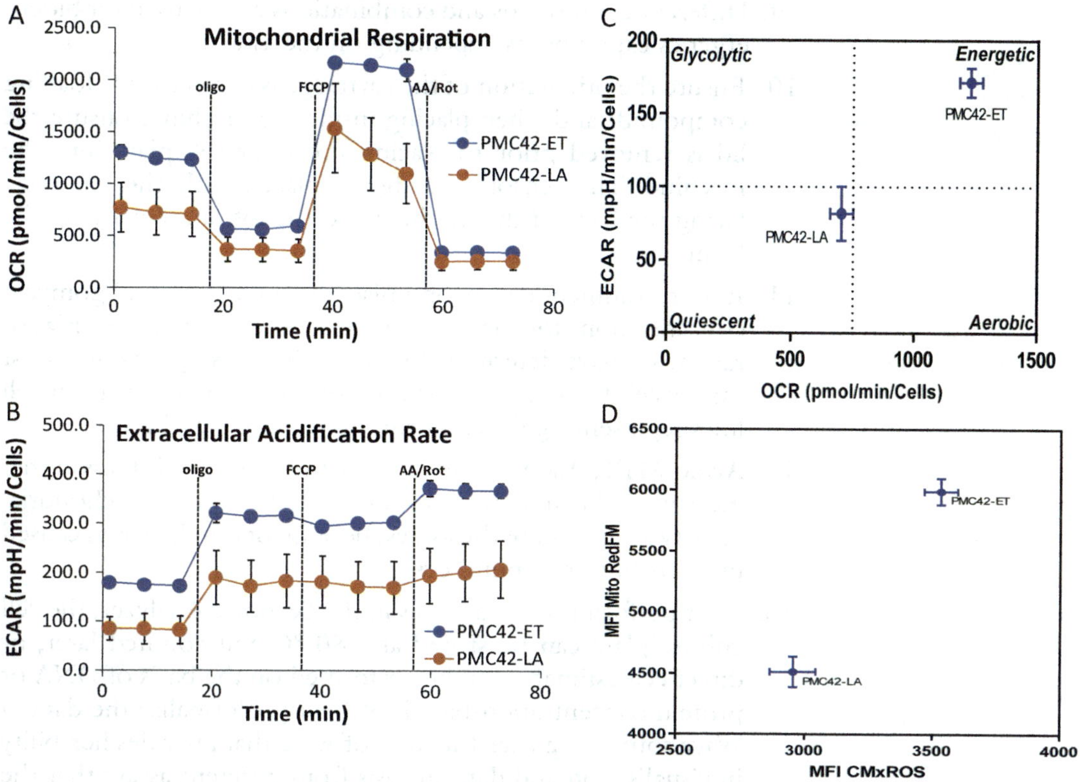

Fig. 3 Examples of real-time measurements of (**a**) OCR (oxygen consumption rate) and (**b**) ECAR (extracellular acidification rate) from the Seahorse Analyzer. (**c**) Profiling of ECAR and OCR measurement values in (parental) PMC42-ET and the epithelially shifted PMC42-LA derivative cells suggests PMC42-ET cells are more glycolytic and energetic as compare to PMC42-LA, and that PMC42-LA cells are more quiescent as compare to PMC42-ET. (**d**) The mitochondrial mass (determined by Mito Red FM) and membrane potential (determined by Mito CMXRos) levels assessed by flow cytometry reflect similar results. *MFI*: Median Fluorescence Intensity

5. These cell numbers are based on epithelial cell lines for a 96-well plate. We have found these to be approximately one third of the number of cells required for XFe24 well plates.
6. Once the probe plate is rehydrated, it is recommended to use plate within 72 h.
7. Be cautious not to remove all media and not to touch the bottom of the wells in order to protect the integrity of the cells. After the last wash, add 160 μL of assay media to the 20 μL of remaining media in wells (final volume 180 μL/well) and keep the plate at 37 °C in a non-CO_2 incubator until ready to load onto the machine.
8. A loading guide, provided with the probe plate, can be used to align to the ports for ease of dispensing. The temperature correction well should be loaded with injection compounds too.

9. Different compounds and combinations can be used for bioenergetics experiments depending on the assay type.
10. Ensure the orientation of the cartridge is correct while loading compounds and when placing inside the machine. Ensure the lid is removed prior to placing probe or cell plate into the machine. The sensor cartridge is placed with the bar code facing the back of the machine and the lot number facing the front.
11. It is recommended to optimise the FCCP and oligomycin concentrations for the bioenergetics experiments. In our laboratory, we have determined that 1–1.5 μM is optimal for most cancer cell lines tested, but this concentration can vary in cell lines representing different tumour subtypes and types.
12. Avoid MTT, Alamar blue or any other proliferation assay that measures cellular viability as survival readout of mitochondrial activity at the end of the assays, because of cellular stress caused by mitochondrial inhibitors.
13. After performing the analysis in the Seahorse Analyzer, the cell culture plate can be stored at −80 °C and counted later, or direct cell estimates can be performed on the basis of DNA or protein concentration-based SRB assay. Normalise the data in Wave software, a user-friendly software that provides flexibility in visualisation and data analysis from different assays that the machine measures, such Mito Stress Test. The generated report produces an excel file with a separate sheet containing an explanation of how parameter calculations were performed/ or how the different values of ECAR, OCR or ATP content measurement were calculated from the assays. ECAR and OCR measurements following inhibitor treatment are used to calculate fundamental parameters from the Seahorse output including non-mitochondrial oxygen consumption (background), basal respiration (prior to first injection), maximal respiration (following uncoupler injection), proton leak, ATP production (following oligomycin injection) and spare respiratory capacity (calculated as the difference between basal and maximal respiration). Figure 2 illustrates how these parameters are calculated from the measurements collected by the Seahorse instrument.
14. Mitochondrial biomass can be affected by your treatments. This can be accounted for by measuring total mitochondrial content.

References

1. Hanahan D, Weinberg RA (2011) Hallmarks of cancer: the next generation. Cell 144 (5):646–674. https://doi.org/10.1016/j.cell.2011.02.013
2. Vander Heiden MG, Cantley LC, Thompson CB (2009) Understanding the Warburg effect: the metabolic requirements of cell proliferation. Science (New York, NY) 324 (5930):1029–1033. https://doi.org/10.1126/science.1160809
3. Menendez JA, Joven J, Cufí S, Corominas-Faja B, Oliveras-Ferraros C, Cuyàs E, Martin-Castillo B, López-Bonet E, Alarcón T, Vazquez-Martin A (2013) The Warburg effect version 2.0: metabolic reprogramming of cancer stem cells. Cell Cycle 12(8):1166–1179. https://doi.org/10.4161/cc.24479
4. Hsu PP, Sabatini DM (2008) Cancer cell metabolism: Warburg and beyond. Cell 134 (5):703–707. https://doi.org/10.1016/j.cell.2008.08.021
5. Maiuri Maria C, Kroemer G (2015) Essential role for oxidative phosphorylation in cancer progression. Cell Metabol 21(1):11–12. https://doi.org/10.1016/j.cmet.2014.12.013
6. Lu CL, Qin L, Liu HC, Candas D, Fan M, Li JJ (2015) Tumor cells switch to mitochondrial oxidative phosphorylation under radiation via mTOR-mediated hexokinase II inhibition—a Warburg-reversing effect. PLoS One 10(3): e0121046. https://doi.org/10.1371/journal.pone.0121046
7. Park JH, Vithayathil S, Kumar S, Sung PL, Dobrolecki LE, Putluri V, Bhat VB, Bhowmik SK, Gupta V, Arora K, Wu D, Tsouko E, Zhang Y, Maity S, Donti TR, Graham BH, Frigo DE, Coarfa C, Yotnda P, Putluri N, Sreekumar A, Lewis MT, Creighton CJ, Wong LC, Kaipparettu BA (2016) Fatty acid oxidation-driven Src links mitochondrial energy reprogramming and oncogenic properties in triple-negative breast cancer. Cell Rep 14 (9):2154–2165. https://doi.org/10.1016/j.celrep.2016.02.004
8. Porporato Paolo E, Payen Valéry L, Pérez-Escuredo J, De Saedeleer Christophe J, Danhier P, Copetti T, Dhup S, Tardy M, Vazeille T, Bouzin C, Feron O, Michiels C, Gallez B, Sonveaux P (2014) A mitochondrial switch promotes tumor metastasis. Cell Rep 8 (3):754–766. https://doi.org/10.1016/j.celrep.2014.06.043
9. LeBleu VS, O'Connell JT, Gonzalez Herrera KN, Wikman H, Pantel K, Haigis MC, de Carvalho FM, Damascena A, Domingos Chinen LT, Rocha RM, Asara JM, Kalluri R (2014) PGC-1alpha mediates mitochondrial biogenesis and oxidative phosphorylation in cancer cells to promote metastasis. Nat Cell Biol 16(10):992–1003, 1001-1015. https://doi.org/10.1038/ncb3039
10. Nieto MA, Huang Ruby Y-J, Jackson Rebecca A, Thiery Jean P (2016) EMT: 2016. Cell 166(1):21–45. https://doi.org/10.1016/j.cell.2016.06.028
11. van Denderen BJ, Thompson EW (2013) Cancer: the to and fro of tumour spread. Nature 493(7433):487–488. https://doi.org/10.1038/493487a
12. Hugo H, Ackland ML, Blick T, Lawrence MG, Clements JA, Williams ED, Thompson EW (2007) Epithelial–mesenchymal and mesenchymal–epithelial transitions in carcinoma progression. J Cell Physiol 213(2):374–383. https://doi.org/10.1002/jcp.21223
13. Bhatia S, Monkman J, Toh AKL, Nagaraj SH, Thompson EW (2017) Targeting epithelial-mesenchymal plasticity in cancer: clinical and preclinical advances in therapy and monitoring. Biochem J 474(19):3269–3306. https://doi.org/10.1042/bcj20160782
14. Lee SY, Ju MK, Jeon HM, Lee YJ, Kim CH, Park HG, Han SI, Kang HS (2018) Oncogenic metabolism acts as a prerequisite step for induction of cancer metastasis and cancer stem cell phenotype. Oxidative Med Cell Longev 2018:1027453–1027453. https://doi.org/10.1155/2018/1027453
15. Li M, Bu X, Cai B, Liang P, Li K, Qu X, Shen L (2019) Biological role of metabolic reprogramming of cancer cells during epithelial mesenchymal transition (review). Oncol Rep 41 (2):727–741. https://doi.org/10.3892/or.2018.6882
16. Sciacovelli M, Frezza C (2017) Metabolic reprogramming and epithelial-to-mesenchymal transition in cancer. FEBS J 284 (19):3132–3144. https://doi.org/10.1111/febs.14090
17. Yizhak K, Le Devedec SE, Rogkoti VM, Baenke F, de Boer VC, Frezza C, Schulze A, van de Water B, Ruppin E (2014) A computational study of the Warburg effect identifies metabolic targets inhibiting cancer migration. Mol Syst Biol 10:744. https://doi.org/10.15252/msb.20134993
18. Gaude E, Frezza C (2016) Tissue-specific and convergent metabolic transformation of cancer correlates with metastatic potential and patient

survival. Nat Commun 7:13041. https://doi.org/10.1038/ncomms13041

19. Koochekpour S, Marlowe T, Singh KK, Attwood K, Chandra D (2013) Reduced mitochondrial DNA content associates with poor prognosis of prostate cancer in African American men. PLoS One 8(9):e74688. https://doi.org/10.1371/journal.pone.0074688
20. Guo J, Zheng L, Liu W, Wang X, Wang Z, Wang Z, French AJ, Kang D, Chen L, Thibodeau SN, Liu W (2011) Frequent truncating mutation of TFAM induces mitochondrial DNA depletion and apoptotic resistance in microsatellite-unstable colorectal cancer. Cancer Res 71(8):2978–2987. https://doi.org/10.1158/0008-5472.CAN-10-3482
21. Weerts MJA, Sieuwerts AM, Smid M, Look MP, Foekens JA, Sleijfer S, Martens JWM (2016) Mitochondrial DNA content in breast cancer: impact on in vitro and in vivo phenotype and patient prognosis. Oncotarget 7 (20):29166–29176. https://doi.org/10.18632/oncotarget.8688
22. Jia D, Park JH, Jung KH, Levine H, Kaipparettu BA (2018) Elucidating the metabolic plasticity of cancer: mitochondrial reprogramming and hybrid metabolic states. Cell 7(3). https://doi.org/10.3390/cells7030021
23. Dupuy F, Tabariès S, Andrzejewski S, Dong Z, Blagih J, Annis Matthew G, Omeroglu A, Gao D, Leung S, Amir E, Clemons M, Aguilar-Mahecha A, Basik M, Vincent Emma E, St.-Pierre J, Jones Russell G, Siegel Peter M (2015) PDK1-dependent metabolic reprogramming dictates metastatic potential in breast cancer. Cell Metabol 22 (4):577–589. https://doi.org/10.1016/j.cmet.2015.08.007
24. Pelicano H, Zhang W, Liu J, Hammoudi N, Dai J, Xu RH, Pusztai L, Huang P (2014) Mitochondrial dysfunction in some triple-negative breast cancer cell lines: role of mTOR pathway and therapeutic potential. Breast Cancer Res 16(5):434. https://doi.org/10.1186/s13058-014-0434-6

Chapter 26

Inducing Sequential Cycles of Epithelial-Mesenchymal and Mesenchymal-Epithelial Transitions in Mammary Epithelial Cells

Cecile Davaine, Eva Hadadi, William Taylor, Annelise Bennaceur-Griscelli, and Hervé Acloque

Abstract

Epithelial-Mesenchymal Transition (EMT) and its reciprocal Mesenchymal-Epithelial Transition (MET) occur naturally as a cycling process during embryonic and foetal development. The capacity of such iterative cycles to drive cell fate and cellular and molecular behaviour in physiology or pathology remains unclear. We describe here a protocol to induce successive cycles of EMT/MET in an untransformed human mammary epithelial cell line (MCF10A) as well as the necessary controls for cycle validation.

Key words EMT, MET, MCF10A, Epithelial-mesenchymal transition, Mesenchymal-epithelial transition, Cycles

1 Introduction

Epithelial-to-mesenchymal transition (EMT) is a process through which cells shift their properties, losing strong cell–cell adherence and gaining migratory capacities [1]. EMT was first described at the onset of gastrulation in the chick embryo, when cells from the epiblast delaminate at the primitive streak to form the mesendoderm [2]. Many other developmental processes in amniotes go through this process, like the prototypic neural crest cell delamination or myotome formation [3–6]. Interestingly, EMT is a reversible process and mesenchymal cells can revert to an epithelial phenotype to produce new epithelial layers. This phenomenon, called mesenchymal-epithelial transition (MET), has been observed during embryogenesis and organogenesis [7]. These round trips between two antagonistic cellular states allow co-opting the properties of mesenchymal cells (motility, directional migration, individual cell migration) or epithelial cells (polarization, mechanical

Kyra Campbell and Eric Theveneau (eds.), *The Epithelial-to Mesenchymal Transition: Methods and Protocols*,
Methods in Molecular Biology, vol. 2179, https://doi.org/10.1007/978-1-0716-0779-4_26,

transduction, ability to form tough and resistant layers) during morphogenesis.

Similarly, succession of EM/ME transitions have also been described during cancer progression [8, 9] and are crucial for cancer cells dissemination and metastasis formation [10]. While the tumoral microenvironment is highly dynamic and can favour either EMT or MET, it is possible that pro-EMT and pro-MET signals are successively received by cancer cells and can eventually modify their biological properties. Indeed, activation of EMT-transcription factors (EMT-TFs) controls a pleiotropic set of cellular functions (cell proliferation, stemness, cell adhesion, cytoskeleton dynamics, cell polarity, etc.) and can lead to a gradation of cellular states [1].

The present protocol allows reproducing in vitro a succession of EMT/MET cycles in human mammary epithelial cells MCF10A. Mesenchymal and epithelial status are confirmed in each cycle by phenotypic and molecular conventional markers (Fig. 1).

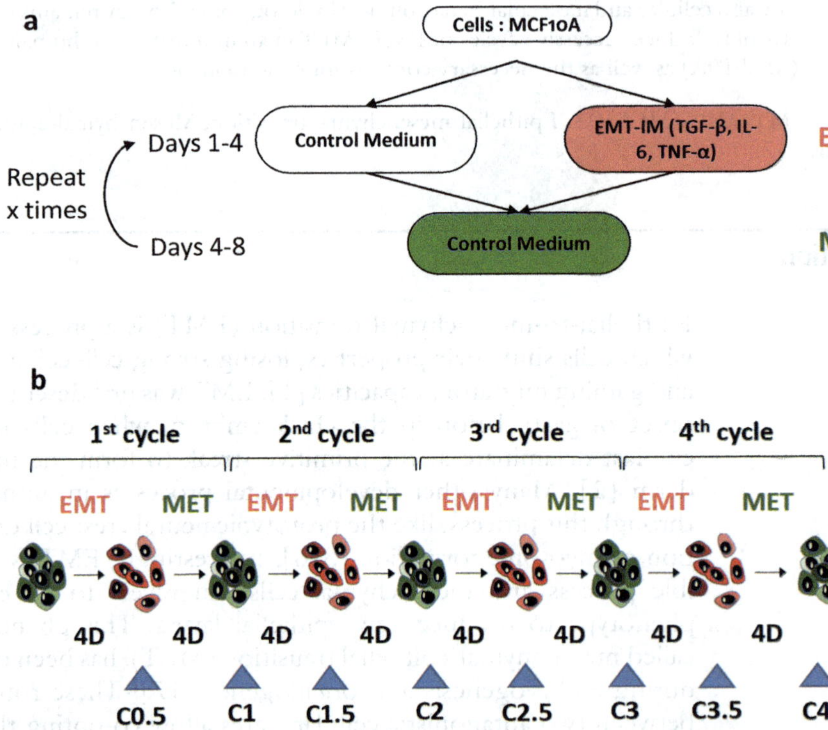

Fig. 1 (**a**) Experimental design for cyclic induction of epithelial-to-mesenchymal and mesenchymal-to-epithelial transitions (EMT-IM: EMT induction medium). (**b**) Timeline of one experiment including 4 EMT/MET cycles. Each cycle lasts for 8 days and cells are dissociated, collected for further analysis and re-plated at the end of each half cycle (blue triangle)

2 Materials

2.1 Cell Culture Equipment

1. Cell incubator allowing CO_2 and temperature control.
2. Laminar flow hood for cell culture.

2.2 Cell Line

Immortalized human breast epithelial cell line MCF10A [11]. Other cell lines may be used (*see* **Note 1**).

2.3 Cell Culture Medium

1. Culture medium: basal medium DMEM/F12 (1:1) (1×) (31330–038 Gibco or any equivalent reference from other suppliers).
2. Horse serum (Merck H1270 or any equivalent reference from other suppliers). Aliquots from the same lot are conserved at −20 °C.
3. Stock solution of cholera toxin (Merck C8052 or any equivalent reference from other suppliers). 0.5 mg are re-suspended in cell culture grade water at a final concentration of 1 mg/mL and filtered using a 0.2 μm membrane. Aliquots are stored at 4 °C (do not freeze).
4. Stock solution of insulin (Merck I9278, 1000× solution or any equivalent reference from other suppliers). The stock solution is stored at 4 °C.
5. Stock solution of hydrocortisone (Merck H0888 or any equivalent reference from other suppliers). 1 g re-suspended in absolute ethanol at a final concentration of 1 mg/mL. Aliquots are stored at −20 °C.
6. Stock solution of Penicillin/Streptomycin (100×), stored at 4 °C.
7. Stock solution of L-Glutamine (100×), stored at −20 °C.
8. Stock solution of 0.1% bovine gelatine in PBS.
9. Accutase solution (1× solution).
10. Stock solution of human recombinant epidermal growth factor (hEGF) (StemCell Technologies or any equivalent reference from other suppliers) at 1 mg/mL prepared following providers' recommendation and stored at −20 °C.
11. Stock solution of human recombinant Interleukin 6 (hIL6) (StemCell Technologies or any equivalent reference from other suppliers) at 10 μg/mL prepared following providers' recommendation and stored at −20 °C.
12. Stock solution of human recombinant Transforming Growth Factor beta (hTGFβ) (StemCell Technologies or any equivalent reference from other suppliers) at 20 μg/mL prepared following providers' recommendation and stored at −20 °C.

13. Stock solution of human recombinant Tumour Necrosis Factor alpha (hTNFα) (StemCell Technologies or any equivalent reference from other suppliers) at 50 μg/mL prepared following providers' recommendation) and stored at −20 °C.
14. Control medium composed of DMEM/F12 supplemented with 5% horse serum, 10 ng/mL cholera toxin, 10 μg/mL insulin, 0.5 μg/mL hydrocortisone, 20 ng/mL human recombinant epidermal growth factor, 1% Penicillin/Streptomycin, 1% L-Glutamine (*see* **Note 2**).
15. EMT inducing medium (EMT-IM): control medium plus TGF-β (5 ng/mL), TNF-α (10 ng/mL), and IL-6 (10 ng/mL) (*see* **Notes 2** and **3**).

2.4 Real-Time PCR

1. Total RNA extraction kit (PureLink™ RNA Mini Kit, ThermoFisher) or any equivalent reference from other suppliers.
2. High-Capacity cDNA Reverse Transcription kit (ThermoFisher) or any equivalent reference from other suppliers.
3. FastStart Universal SYBR Green Master mix (Roche).
4. Real-time PCR primers:

   SNAIL1: FW 5′-GCTGCAGGACTCTAATCCAGA-3′; RV 5′-ATCTCCGGAGGTGGGATG-3′ [11].

   SNAIL2: FW 5′-TTCGGACCCACACATTACCT-3′; RV 5′-TCTCCCCCGTGTGAGTTCTA-3′.

   ZEB1: FW 5′- AGGGCACACCAGAAGCCAG-3′; RV 5′-GAGGTAAAGCGTTTATAGCCTCTATCA-3′ [12].

   TBP: FW 5′-CACGAACCACGGCACTGATT-3′; RV 5′-TTTTCTTGCTGCCAGTCTGGAC-3′ [12].

   36B4: FW 5′-GCTGATGGGCAAGAACACCA-3′; RV 5′-CCGGATATGAGGCAGCAGTT-3′ [11].

   SMA: FW 5′-GAGCGTGAGATTGTCCGGGA-3′; RV 5′-AAGGGAGGATGAGGATGCGG-3′.

5. Real-time PCR thermocycler.

2.5 Flow Cytometry

1. Flow cytometer.
2. LIVE/DEAD Fixable Viability/Cytotoxicity kit for mammalian cells (ThermoFisher).
3. PE-conjugated anti-E-cadherin (Biolegend 324,106).
4. APC-conjugated anti-Fibronectin produced using the Lightning-Link Rapid Conjugation System (Innova Biosciences) and anti-Fibronectin (Agilent A0245).

2.6 Cell Microscopy

1. Microscope equipped with phase-contrast and epifluorescence transillumination.
2. Anti-β-Catenin (Cell Signaling Technology 8480S).

3. Anti-Snail1 (Cell Signaling Technology 3879S).
4. Anti-Fibronectin (Agilent DAKO A0245).
5. Anti-Vimentin (Agilent DAKO M0725).
6. Set of anti-Rabbit and anti-Mouse secondary antibodies coupled with fluorophores.
7. 32% paraformaldehyde aqueous solution (Electron Microscopy Sciences).
8. PBS solution (Gibco or any equivalent reference from other suppliers).
9. Triton X-100 aqueous solution (Merck or any equivalent reference from other suppliers).
10. Foetal Bovine Serum (FBS).
11. Bovine Serum Albumin (BSA).
12. Diamidino-2-phenylindole (DAPI) solution (1 mg/mL) (ThermoFisher) or any equivalent reference from other suppliers.

3 Methods

This protocol consists of the repetition of 4 cycles of EMT/MET (Fig. 1a). At each half cycle, we characterize the cells for epithelial and mesenchymal properties (Fig. 1b).

3.1 Induction of the First EMT

1. For immunofluorescence labelling, place coverslips in 60 mm culture dishes and coat with a 0.1% gelatine/PBS solution. *See* **Note 4** to manage the number of dishes to prepare.
2. Seed MCF10A cells at a density of 400,000 cells per 60 mm culture dish (*see* **Notes 4** and **5**) in control medium and keep in a 5% CO_2 atmosphere at 37 °C.
3. 12 hs after (day 0), replace the control medium by the EMT-IM.
4. After 48 h, replace EMT-IM by the control medium and keep cells in culture for an additional 48 h.

3.2 Induction of the First MET

1. At day 4 (96 h after the beginning of the EMT induction), photograph cells are using a phase-contrast inverted microscope.
2. Recover coverslips and proceed for immunofluorescence labelling.
3. Dissociate the other cells using Accutase, and seed 400,000 cells per 60 mm culture dish in control medium (*see* **Notes 4** and **5**). Keep the remaining cells for gene and protein expression studies, flow cytometry analysis or any other downstream application (*see* **Note 4**).

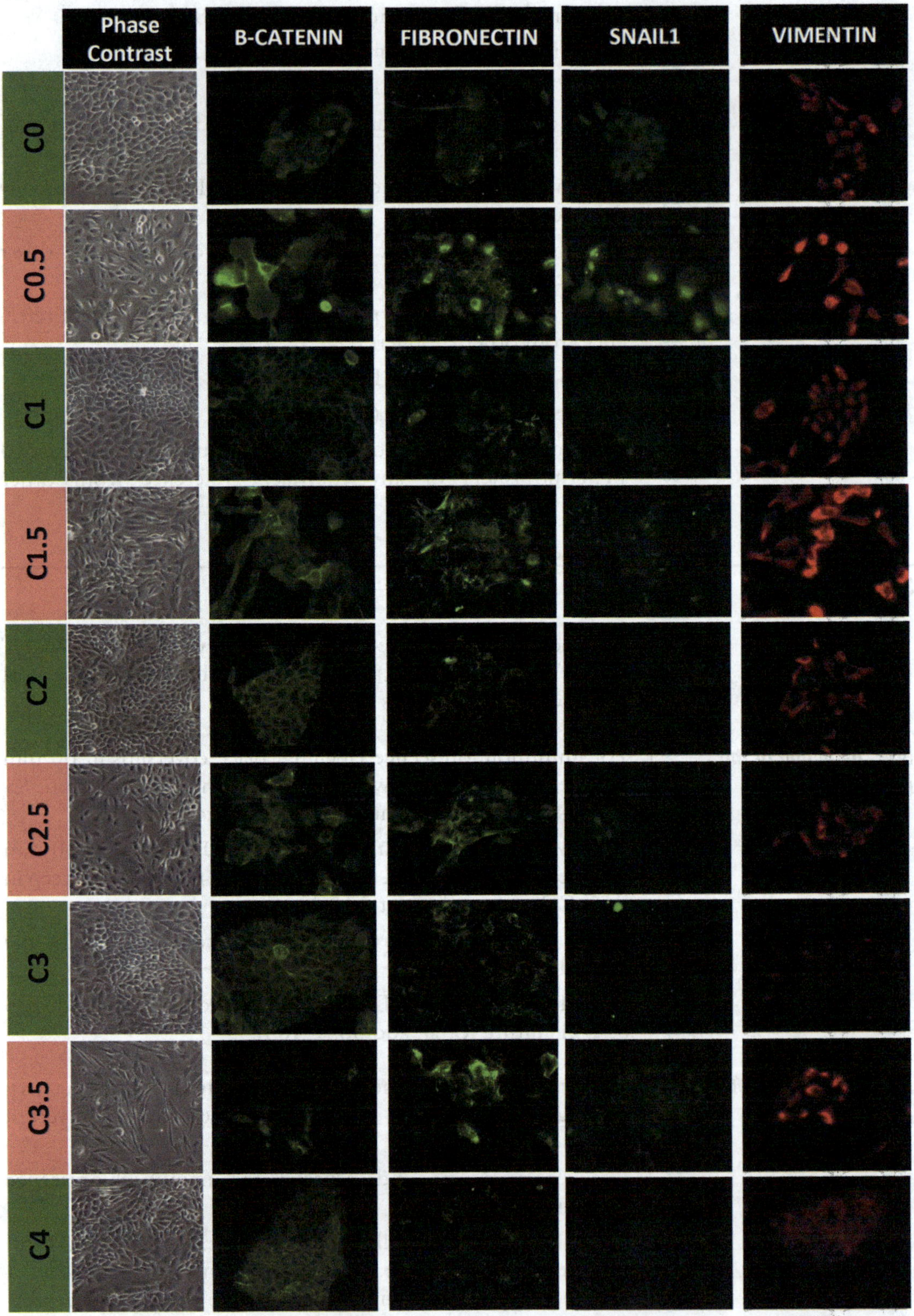

Fig. 2 Efficiencies in EMT and MET are confirmed by following cell morphology using phase-contrast microscopy. While MET is characterized by the formation of typical epithelial layers with cuboid cells, EMT leads to typical mesenchymal-like features with less cell–cell adhesion and cytoplasmic projections.

4. After 48 h, change control medium and maintain cells in culture for an additional 48 h.
5. At day 8, (96 h after the beginning of MET induction), photograph cells using a phase-contrast inverted microscope. Recover coverslips and proceed for immunofluorescence labelling.
6. Dissociate the other cells using Accutase and seed 400,000 cells per 60 mm culture dish in EMT-IM (*see* **Notes 4** and **5**). Keep the remaining cells for gene and protein expression studies, flow cytometry analysis or any other downstream application (*see* **Note 4**).

3.3 Repetitions of EMT/MET Cycles

Repeat steps of Subheadings 3.1 and 3.2 to reach the desired number of EMT/MET cycles (*see* **Note 6**).

3.4 Quality Control of EMT/MET Cycle Using Phase-Contrast and Fluorescent Microscopy

1. At the end of each EMT or MET phases, remove the cover slips and fix cells using a solution of 1% paraformaldehyde/PBS for 10 min at room temperature.
2. Wash cells twice with ice-cold PBS and permeabilize for 20 min at 4 °C using a 0.1% Triton X-100 PBS solution.
3. Block nonspecific antigen binding sites with a 0.1% Triton X-100, 1% FBS PBS blocking solution for 1 h at room temperature.
4. Incubate cells with anti-β-Catenin (using a 1/200 dilution in blocking solution), anti-Fibronectin (using a 1/500 dilution in blocking solution), anti-Snail1 (using a 1/400 dilution in blocking solution) and anti-Vimentin (using a 1/500 dilution in blocking solution) antibodies overnight at 4 °C (*see* **Note** 7).
5. Wash twice with PBS and incubate cells at room temperature with secondary antibodies for 60 min (using a 1/1000 dilution in blocking solution) and stain the nuclei with DAPI (1 μg/mL in PBS) for 10 min then wash with PBS twice.
6. Fix the stained cover slips onto glass slides using Dako fluorescence mounting medium and store away from light at 4 °C.
7. Image cells using an epifluorescence microscope equipped with phase contrast (Fig. 2).

Fig. 2 (continued) Immunofluorescence labelling for prototypic epithelial and mesenchymal markers confirms the transitions between both states. Epithelial cells are characterized by a clear localization of β-CATENIN at adherent junction, low levels of FIBRONECTIN and VIMENTIN and the absence of SNAIL1. Mesenchymal cells are characterized by a cytoplasmic re-localization of β-CATENIN, a strong expression of FIBRONECTIN with an extra-cellular deposition of FIBRONECTIN fibres, a stronger expression of VIMENTIN and nuclear detection of SNAIL1 in some cells

3.5 Flow Cytometry

1. At the end of each EMT or MET phases, recover cells to be analysed by flow cytometry.
2. Stain freshly harvested cells with Live/Dead cell viability solution for 20 min in obscurity. For E-cadherin staining, block cells in 5% FBS/1% BSA/PBS for 30 min and incubate with PE-conjugated anti-E-cadherin (1:10 dilution in 5% FBS/1% BSA/PBS).
3. Fix cells using a solution of 1% PFA/PBS for 30 min, permeabilize with 0.1% Triton-X100/PBS for 20 min and block with 5% FBS/1% BSA/PBS for 30 min.
4. Following PBS washes, incubate cells with APC-conjugated anti-Fibronectin (1:10 dilution in 5% FBS/1% BSA/PBS) for 20 min at 4 °C. All incubations and staining are done at 4 °C.
5. Analyse stained cells by flow cytometry to determine E-cadherin and Fibronectin expression levels (Fig. 3a and **Note 7**). Analyse data with FlowJo software (Tree Star, Ashland, OR).

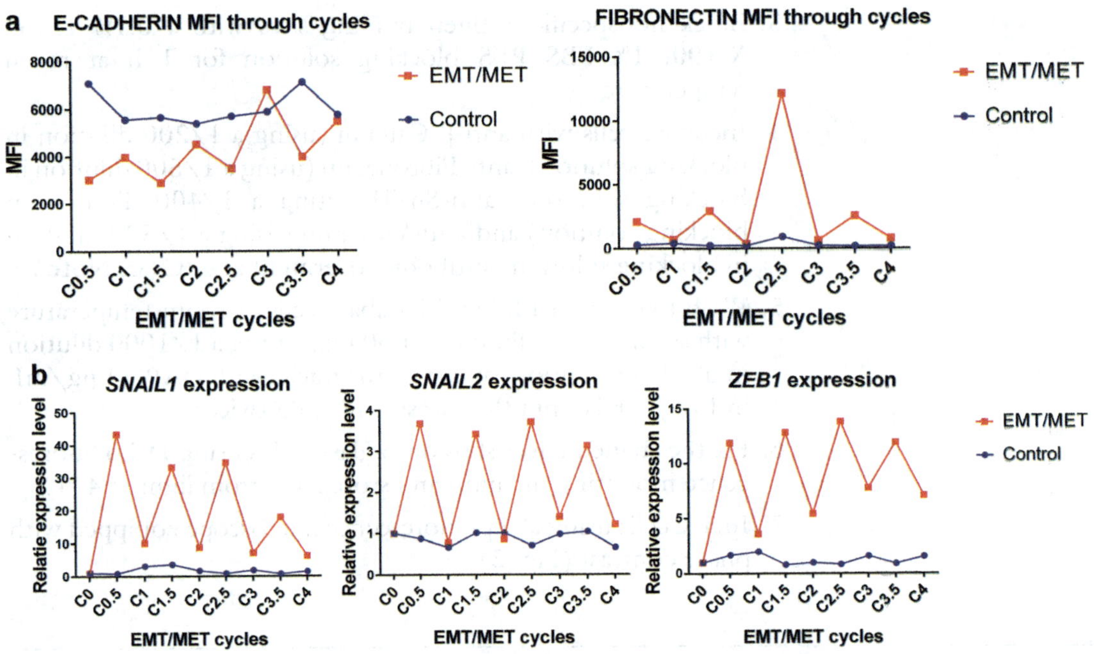

Fig. 3 (**a**) E-Cadherin and Fibronectin levels were quantified by flow cytometry. Mean Fluorescence Intensity (MFI) is shown for each half EMT/MET cycle for control cells (blue circles, not treated) and EMT/MET cycling cells (red square). (**b**) *SNAIL1, SNAIL2* and *ZEB1* expression levels quantified by real-time PCR. Relative expression levels are shown for each half EMT/MET cycle for control cells (blue circles, not treated) and EMT/MET cycling cells (red square). C: cycles

3.6 Real-Time PCR Analysis

1. At the end of each EMT or MET phases, pellet 500,000 cells, snap freeze and store at −80 °C.
2. Extract total RNAs using PureLink RNA kit according to the manufacturer's instructions.
3. Perform cDNA synthesis using High-Capacity cDNA Reverse Transcription kit (Applied Biosystems) with Oligo(dT) primers (ThermoFisher). Use 500 ng of total RNA for transcription. Perform real-time qPCR using FastStart Universal SYBR Green Master mix (Roche) and primers specifically designed to amplify *SNAIL1, SNAIL2* and *ZEB1* (*see* **Note 7**). For each sample, perform qPCRs with experimental replicates, and melting curve analysis is performed for each run. Final results represent relative gene expression values calculated with the ΔΔCt method using the geometric mean of three reference genes for normalization [13] and control MCF10A as the reference sample (Fig. 3b).

4 Notes

1. *Cell types*: these experiments were performed with the MCF10A cell line [11]. The choice of the cell line depends on the biological issue to be addressed, on the ability of the cells to respond to EMT induction and then to return to the epithelial state and on the cell toxicity of EMT inducing molecules.
2. *Culture media*: in order to avoid bias in the reproducibility of the cycles, it is preferable to prepare fresh EMT-IM and control medium for each cycle.
3. *Cytokine choice*: the choice of the cytokines to induce EMT depends on the cells used. The combination of TGF-β, TNF-α and IL-6, as three cytokines known to induce EMT, was a well permissive condition on MCF10A, as well as TGF- β alone.
4. *Estimate of the number of dishes:* usually, for each experiment, each cycle and each experimental condition, we prepare five 60 mm culture dishes: one for performing flow cytometry, one for keeping cell pellets, two (with coverslips) for immunostaining and two to be used to seed cells for the next cycle.
5. *Cell density*: seeding cell density is a crucial point that has to be optimized. A high cell density can inhibit or slow down EMT induction whereas a low cell density can lead to not having enough cells at the end of the cycle to reseed and carry out the various planned analyses. Indeed, EMT induction significantly decreases cell proliferation [14].

6. *End of the experiment:* in this chapter, we describe the realization of four successive cycles of EMT/MET. It is up to the user to define the number of cycles that he/she wishes to achieve according to his hypotheses to be tested. It must be kept in mind that the duration of the experiment is proportional to the number of cycles (4 cycles = 1 month, 8 cycles = 2 months, etc.).
7. *Choice of EMT and MET markers*: we selected well-known EMT and MET markers (EMT-TFs like SNAIL and ZEB, E-cadherin, β-Catenin, Fibronectin and Vimentin) combined with different methods to evaluate their expression level fluctuation. This gave us a fairly robust validation of the effectiveness of the EMT/MET cycles. Interestingly, the oscillations are clearly visible for both proteins and transcripts.

Acknowledgements

We acknowledge all the members of UMRS935 for technical assistance and scientific daily exchanges.

This work was funded by Inserm, University Paris Sud, Association Institut de Cancérologie et d'Immunogénétique (ICIG) and Vaincre le Cancer-NRB. E.H. post-doctoral fellowship was granted by Vaincre le Cancer-NRB and the University Paris-Saclay (Project BioTherAlliance). C.D. Ph.D. fellowship was granted by University Paris-Saclay.

Author contribution: C.D., W.T., E.H. and H.A. performed the experiments.

C.D., E.H., W.T., A.B.G. and H.A. designed the experimental strategy and analysed the results.

C.D., E.H., W.T., A.B.G. and H.A. wrote the manuscript.

References

1. Nieto MA, Huang RY, Jackson RA, Thiery JP (2016) EMT: 2016. Cell 166:21–45
2. Hay ED (1968) Organization and fine structure of epithelium and mesenchyme in the developing chick embryo. In: Fleischmajer R, Billingham RE (eds) Epithelial-mesenchymal interactions. Williams & Wilkins, Baltimore, MD, pp 31–55
3. Acloque H, Adams MS, Fishwick K, Bronner-Fraser M, Nieto MA (2009) Epithelial-mesenchymal transitions: the importance of changing cell state in development and disease. J Clin Invest 119:1438–1449
4. Acloque H, Ocaña OH, Matheu A, Rizzoti K, Wise C, Lovell-Badge R, Nieto MA (2011) Reciprocal repression between Sox3 and snail transcription factors defines embryonic territories at gastrulation. Dev Cell 21:546–558
5. del Barrio MG, Nieto MA (2002) Overexpression of snail family members highlights their ability to promote chick neural crest formation. Development 129:1583–1593
6. Gros J, Scaal M, Marcelle C (2004) A two-step mechanism for myotome formation in chick. Dev Cell 6:875–882
7. Thiery JP, Acloque H, Huang RY, Nieto MA (2009) Epithelial-mesenchymal transitions in development and disease. Cell 139:871–890
8. Ocaña OH, Córcoles R, Fabra A, Moreno-Bueno G, Acloque H, Vega S, Barrallo-

Gimeno A, Cano A, Nieto MA (2012) Metastatic colonization requires the repression of the epithelial-mesenchymal transition inducer prrx1. Cancer Cell 22:709–724

9. Tsai JH, Donaher JL, Murphy DA, Chau S, Yang J (2012) Spatiotemporal regulation of epithelial-mesenchymal transition is essential for squamous cell carcinoma metastasis. Cancer Cell 22:725–736
10. Hanahan D, Weinberg RA (2011) Hallmarks of cancer: the next generation. Cell 144:646–674
11. Soule HD, Maloney TM, Wolman SR, Peterson WD Jr, Brenz R, McGrath CM, Russo J, Pauley RJ, Jones RF, Brooks SC (1990) Isolation and characterization of a spontaneously immortalized human breast epithelial cell line, MCF-10. Cancer Res 50(18):6075–6086
12. Morel AP, Ginestier C, Pommier RM, Cabaud O, Ruiz E, Wicinski J, Devouassoux-Shisheboran M, Combaret V, Finetti P, Chassot C et al (2017) A stemness-related ZEB1-MSRB3 axis governs cellular pliancy and breast cancer genome stability. Nat Med 23:568–578
13. Hadadi E, Souza LEB, Bennaceur-Griscelli A, Acloque H (2018) Identification of valid reference genes for circadian gene-expression studies in human mammary epithelial cells. Chronobiol Int 35:1689–1701
14. Vega S, Morales AV, Ocaña OH, Valdés F, Fabregat I, Nieto MA (2004) Snail blocks the cell cycle and confers resistance to cell death. Genes Dev 18:1131–1143

Chapter 27

Analysis of Cellular EMT States Using Molecular Biology and High Resolution FISH Labeling

Noémie Kempf, Fatima Moutahir, Isabelle Goiffon, Sylvain Cantaloube, Kerstin Bystricky, and Anne-Claire Lavigne

Abstract

Metastasis results from the ability of cancer cells to grow and to spread beyond the primary tumor to distant organs. Epithelial-to-Mesenchymal Transition (EMT), a fundamental developmental process, is reactivated in cancer cells, and causes epithelial properties to evolve into mesenchymal and invasive ones. EMT changes cellular characteristics between two distinct states, yet, the process is not binary but rather reflects a broad spectrum of partial EMT states in which cells exhibit various degrees of intermediate epithelial and mesenchymal phenotypes. EMT is a complex multistep process that involves cellular reprogramming through numerous signaling pathways, alterations in gene expression, and changes in chromatin morphology. Therefore, expression of key proteins, including cadherins, occludin, or vimentin must be precisely regulated. A comprehensive understanding of how changes in nuclear organization, at the level of single genes clusters, correlates with these processes during formation of metastatic cells is still missing and yet may help personalized prognosis and treatment in the clinic. Here, we describe methods to correlate physiological and molecular states of cells undergoing an EMT process with chromatin rearrangements observed via FISH labeling of specific domains.

Key words Epithelial-mesenchymal transition, Hybrid EMT, Plasticity, Cancer, DNA-FISH, High-throughput microscopy, Chromatin reorganization, Genome, Metastasis

1 Introduction

Epithelial-to-Mesenchymal Transition (EMT) is a fundamental development process in which cells lose their epithelial characteristics and acquire mesenchymal features. EMT has long been viewed as a binary process defined by the two extreme phenotypes, epithelial and mesenchymal. It is now clearly recognized that this process is set up gradually with the sequential appearance of different intermediate states (partial or hybrid EMT states) from a morphological, transcriptional, or epigenetic point of view [1]. These partial EMT states have become significant in recent years because of their contribution to carcinogenesis. The EMT process is

Kyra Campbell and Eric Theveneau (eds.), *The Epithelial-to Mesenchymal Transition: Methods and Protocols*, Methods in Molecular Biology, vol. 2179, https://doi.org/10.1007/978-1-0716-0779-4_27,

reactivated in certain cancer cells [2] stimulating tumor cell migration, invasion, and ultimately metastasis to distant organs. During this reactivation, different hybrid EMT states are generated and some of them present the highest metastatic potential [3].

At the molecular level, EMT is a complex multistep process that involves cellular reprogramming through numerous signaling pathways and alterations in gene expression. Epithelial and mesenchymal phenotypes are characterized by highly regulated expression of key proteins such as E-cadherin, N-cadherin, vimentin, or occludin [4].

Furthermore, beyond gene expression and cellular morphology, nuclear structure also reflects cellular activity [5]. Nuclei are highly spatially organized structures where chromatin domains are arranged in discrete functional regions [6]. It is now clear that gene position responds to physiological and pathological changes as it was reported in the case of breast and prostate cancer cells [7–9]. Furthermore, it is well recognized that cells that have undergone an EMT process present an altered nuclear structure [10] and that EMT is characterized by the reorganization of specific chromatin domains across the genome [11]. Undoubtedly, correlating cellular EMT states (complete or partial) with spatial reorganization of genomic regions would be a powerful diagnostic tool but is still in its infancy. Through development of automated methods, artificial intelligence and machine learning, added to improvement of fluorescence microscopy resolution probing the genome may help personalized prognosis and treatment.

Here, we provide detailed protocols including a combination of molecular biology assays to study the genomic localization of EMT-related genes during transforming growth factor beta (TGF-β) dependent EMT induction kinetics.

We use the MCF10A human breast epithelial cell line [12] issued from benign proliferative breast tissue and spontaneously immortalized without defined factors. These non-tumorigenic cells have been extensively used to study EMT because they are capable of undergoing an EMT process upon continuous exposure (one to several weeks) to TGF-β [13]. Interestingly, the MCF10A cell line also responds to short (4 h) transient treatment and rapidly (within 4 days) revert to their initial epithelial phenotype through Mesenchymal-to-Epithelial Transition (MET) [14].

After induction of EMT in the MCF10A cell line with transient or continuous TGF-β treatment, cells were harvested and their physiological state determined using three independent assays: (a) observation of the phenotypic cellular aspect using phase contrast microscopy (b) determination of the EMT-associated gene mRNA expression levels by RT-qPCR and (c) determination of EMT-associated gene protein expression by Western Blotting.

Fluorescence In Situ Hybridization is a commonly used technique to specifically label a genomic region of interest. Unique DNA-FISH-probes targeting EMT-related genes such as CDH1,

CDH2, OCLN or VIM, can be custom-prepared using the genomic DNA fragment of interest cloned into BACs and/or fosmids [15]. DNA-FISH labeled MCF10A cell nuclei are subsequently imaged in three dimensions, and the loci-nuclear localization analyzed. The complete protocol permits defining relative special arrangements of loci and correlating them to the physiological and molecular state of cells undergoing TGF-β induced EMT.

2 Materials

Prepare all solutions using ultrapure water (sensitivity of 18 MΩ cm at 25 °C) and analytical grade reagents. Prepare and store reagents at room temperature (unless indicated otherwise). Follow all safety regulations concerning the handling and waste disposal of chemicals (*see* **Note 1**).

2.1 Material for Cell Culture

1. DMEM/F12, containing phenol red, glutamax and 3.1 g/L D-Glucose.
2. Human epithelial growth factor (hEGF): for stock solution at 7 μg/mL, dissolve 100 μg of commercial hEGF in 14.3 mL sterile water. Store as 500 μL aliquots at −20 °C.
3. 100× Mammary Epithelial Growth Supplement (MEGS).
4. Cholera toxin: for stock solution at 100 ng/μL, dissolve 0.5 mg of cholera toxin in 5 mL sterile water. 0.2 μm filter sterilize and store as 500 μL aliquots at −20 °C.
5. MCF10A cell line growth medium: for 500 mL DMEM/F12, add 500 μL of hEGF at 7 μg/mL, 5 mL MEGS (*see* **Note 2**), 2.5 mL penicillin-streptomycin at 10,000 U/mL, and 500 μL of cholera toxin at 100 ng/μL (100 ng/mL final). Store at 4 °C.
6. 4 mM HCl with 1 mg/mL BSA: dissolve 250 mg of BSA and 83 μL of 37% HCl in 230 mL of H_2O. Adjust the volume up to 250 mL. Vortex to dissolve the BSA and filter the solution using a 0.2 μm filter under a sterile hood. Store at 4 °C.
7. TGF-β stock solution at 20 μg/mL: dissolve 2 μg of human recombinant TGF-β1 in 100 μL of filtrated 4 mM HCl solution with 1 mg/mL BSA. Store at −20 °C.
8. Coverslip #1.5, autoclaved (*see* **Note 3**).
9. 1× Dulbecco's Phosphate Buffered Saline (DPBS), sterile.
10. 1× Trypsin—EDTA solution, sterile.
11. Protease inhibitor (Roche) in 1× PBS: dilute 1 protease inhibitor tablet in 20 mL 1× PBS.
12. Cell spatula.

2.2 Material for Western Blot

2.2.1 Cell Lysis and Loading Buffer

1. 1 M Tris–HCl pH 8.0: in a beaker, dissolve 60.6 g of Tris base in approximately 300 mL H_2O. Add 37% HCl very slowly up to pH = 8.0 (*see* **Note 4**). Complete with H_2O up to 500 mL with a graduated cylinder.
2. Triton-X100.
3. 1 M dithiothreitol (DTT) in H_2O.
4. 0.5 M EDTA pH 8.0: in a beaker, dissolve 146.1 g of EDTA powder in 700 mL H_2O. Add 10 M NaOH very slowly, up to pH = 8.0. Complete with H_2O up to 1 L with a graduate cylinder.
5. 0.2 M Sodium orthovanadate: in a beaker, dissolve 3.68 g of sodium orthovanadate in 80 mL H_2O. To dissolve the powder, heat up the solution up to 50 °C. Complete with H_2O up to 100 mL with a graduate cylinder.
6. Sodium deoxycholate 10% (w/v) in H_2O: For 100 mL, in a beaker, dilute 10 g of sodium deoxycholate in 80 mL of H_2O. Complete with H_2O up to 100 mL with a graduate cylinder.
7. Protease inhibitor.
8. Lysis buffer: for 10 mL solution, combine 200 μL of 1 M Tris–HCl, 10 μL of 1 M DTT, 20 μL of 0.5 M EDTA, 100 μL Triton-X100, 500 μL of sodium deoxycholate 10% (w/v), 50 μL of 0.2 M sodium orthovanadate, ½ protease inhibitor tablet and 9120 μL of H_2O (*see* **Note 5**).
9. 4× Laemmli sample buffer with β-mercaptoethanol: add 100 μL of 99% β-mercaptoethanol (*see* **Note 6**) to 900 μL of 4× Laemmli sample buffer.

2.2.2 Casting of SDS-Polyacrylamide Gels

1. Ammonium persulfate (APS) 10% (w/v) in H_2O: for 100 mL, in a beaker, dilute 10 g of APS in 80 mL of H_2O. Complete with H_2O up to 100 mL with a graduate cylinder (*see* **Note 7**).
2. *N,N,N,N′*-Tetramethyl-ethylenediamine (TEMED) (*see* **Note 8**).
3. 3 M Tris–HCl pH 8.8: For 100 mL, in a beaker, dissolve 181.71 g of Tris base in approximately 450 mL H_2O. Add 37% HCl very slowly until reaching pH = 8.8 (*see* **Note 4**). Complete with H_2O up to 500 mL with a graduated cylinder.
4. 3 M Tris–HCl pH 6.8. For 100 mL: in a beaker, dissolve 36.34 g of Tris base in approximately 60 mL H_2O. Add 37% HCl very slowly up to pH = 6.8 (*see* **Note 4**). Complete with H_2O to 100 mL with a graduated cylinder.
5. Resolving gel: 375 mM Tris–HCl pH 8.8, 10% Acrylamide:Bis-Acrylamide 29:1 (*see* **Note 9**), 0.1% SDS, 0.1% APS, 0.04% TEMED. For one gel of 10 cm × 7.5 cm × 1 mm: mix 4.9 mL H_2O, 1 mL 3 M Tris pH 8.8, 2 mL Acrylamide:Bis-Acrylamide

29:1, and 40 μL 20% SDS. Just before use, add 80 μL 10% APS and 3.2 μL TEMED under a fume hood (*see* **Note 10**).

6. Stacking gel: 375 mM Tris–HCl pH 6.8, 5% Acrylamide:Bis-Acrylamide 29:1 (*see* **Note 9**), 0.1% SDS, 0.1% APS, 0.1% TEMED (*see* **Note 10**). For one gel of 10 cm × 7.5 cm × 1 mm: mix 3.3 mL H_2O, 167 μL 3 M Tris pH 6.8, 0.5 mL 40% Acrylamide:Bis-Acrylamide 29:1 and 20 μL 20% SDS. Just before use, add 40 μL 10%APS and 4 μL TEMED (*see* **Note 10**).

2.2.3 Migration

1. 10× Migration buffer: 0.25 M Tris base, 1.92 M Glycine, 1% SDS. For 1 L: in a beaker, dissolve 144 g Glycine, 30 g of Tris base and 50 mL of 20% SDS in H_2O. Complete with H_2O up to 1 L with a graduated cylinder.
2. 1× Migration buffer: 0.025 M Tris base, 0.192 M Glycine, 0.1% SDS. For 2 L: in a graduated cylinder, measure 200 mL of 10× Migration buffer and complete with H_2O up to 2 L.

2.2.4 Transfer

1. 10× Transfer buffer: 1.92 M Glycine, 0.25 M glycine. For 1 L: in a beaker, dissolve 144 g Glycine and 30 g of Tris base in H_2O. Complete with H_2O up to 1 L with a graduated cylinder.
2. 1× Transfer buffer: 10× transfer buffer, 20% technical EtOH (*see* **Note 11**). For 1 L: mix 100 mL of 10× Transfer buffer with 200 mL technical EtOH and complete with H_2O up to 1 L.
3. Nitrocellulose membrane.
4. Whatman™ Grade 3MM CHR chromatography paper.
5. Red Ponceau S: 0.1% (w/v) Ponceau S in 5% (v/v) acetic acid.

2.2.5 Immunoblotting

1. 10× PBS.
2. PBST: For 1 L, in a beaker, combine 100 mL of 10× PBS into approximately 800 mL of H_2O. Add 2 mL of Tween20 with a serological pipette (*see* **Note 12**) and complete with H_2O up to 1 L in a graduated cylinder.
3. PBSTM (5% (w/v) skim milk in PBST): For 40 mL, dissolve 2 g of skim milk in 40 mL of PBST and mix thoroughly (*see* **Note 13**). Make fresh (can be stored at 4 °C for up to 1 day).

2.3 Material for RT-qPCR

2.3.1 Reverse Transcription

1. Mix 1: per tube, mix 1 μL of 10 mM dNTP (i.e., 0.5 mM final) with 0.25 μL of 100 μM Random hexamer primers (i.e., 1.25 μM final). Prepare freshly for N + 2 tubes and keep on ice.
2. Mix 2: per tube, mix 4 μL of 5× RT buffer with 1 μL Maxima H Minus enzyme (reverse transcriptase). Prepare freshly for N + 2 tubes and keep on ice.

2.3.2 qPCR

1. Primers at 5 μM: For 100 μL, mix 5 μL of forward primer at 100 μM and 5 μL reverse primer at 100 μM to 90 μL ultra-filtrated H_2O.
2. Primers in SYBR green mix: per well, mix 3 μL iTaq Universal SYBR Green supermix with 0.25 μL of the primer mix at 5 μM (*see* **Note 14**).
3. cDNA: dilute cDNA product in RNase-free water to reach 40 ng in a volume of 2.8 μL. Prepare for N + 5 wells and keep on ice.

2.4 Material for DNA-FISH

2.4.1 DNA-FISH Probes Preparation

Our DNA-FISH probes are composed of fluorescently labeled small DNA fragments (from 150 to 1000 bp) custom-prepared using BACs and fosmids identified via the UCSC genome browser.

1. BAC/fosmid DNA at a concentration of 100–200 ng/μL. Stored in H_2O at −20 °C (*see* **Note 15**).
2. 1 M Tris–HCl pH 7.5: for 100 mL, in a beaker, dissolve 12.1 g Tris base in approximately 80 mL H_2O. Add 37% HCl very slowly up to pH = 7.5 (*see* **Note 4**). Complete with H_2O up to 100 mL with a graduated cylinder.
3. 1 M $MgCl_2$: for 100 mL, in a beaker, dissolve 20.3 g $MgCl_2 \cdot 6H_2O$ into H_2O and complete with H_2O up to 100 mL with a graduated cylinder.
4. 10 mg/mL Bovine Serum Albumine (BSA): for 10 mL, dissolve 100 mg BSA in 10 mL H_2O. Store as 1 mL aliquots at −20 °C.
5. 10× Nick Translation Buffer (NTB): 0.5 M Tris–HCl (pH 7.5), 0.05 M $MgCl_2$, 0.5 mg/mL BSA. For 10 mL, combine 5 mL 1 M Tris–HCl pH 7.5, 500 μL 1 M $MgCl_2$ and 500 μL BSA at 10 mg/mL with 4 mL H_2O. Store as 1 mL aliquots at −20 °C.
6. 1 M DTT stock solution: dissolve 1 g DTT into 6.48 mL of H_2O and store at −20 °C. Prepare 0.1 M DTT by diluting tenfold 1 M DTT into H_2O. Store as 50 μL aliquots at −20 °C.
7. 10× dNTP mix: 0.5 mM dATP, 0.5 mM dCTP, 0.5 mM dGTP, 0.125 mM dTTP. For 1 mL, combine 5 μL dATP at 100 mM, 5 μL dCTP at 100 mM, 5 μL dGTP at 100 mM, 1.25 μL dTTP at 100 mM. Store as 50 μL aliquots at −20 °C.
8. DNase I (0.2 U/μL).
9. DNA Polymerase I (10 U/μL).
10. 0.5 mM 5-(3-Aminoallyl)-dUTP: for 500 μL, combine 5 μL of 5-(3-Aminoallyl)-dUTP with 495 μL in H_2O. Store as 20 μL aliquots at −80 °C.
11. 1× TAE: dilute 200 mL of 50× TAE in 10 L H_2O.

12. 2% agarose gel: in a beaker, weigh 2 g of agarose, add 100 mL 1× TAE. Dissolve in microwave and pour in casting system.
13. 3 M sodium acetate pH 5.2. For 100 mL, in a beaker, dissolve 24.6 g NaAc anhydrous in H_2O and adjust pH at 5.2 with glacial acetic acid and complete up to 100 mL with H_2O in a graduated cylinder.
14. 75% ethanol.
15. Cold 70% and 100% ethanol stored at −20 °C.
16. 0.2 M sodium bicarbonate ($NaHCO_3$). For 100 mL: in a beaker, dissolve 1.68 g $NaHCO_3$ in H_2O and complete with H_2O up to 100 mL with a graduated cylinder.
17. 0.5 M Tris–HCl pH 8.5. For 100 mL, in a beaker, dissolve 6.05 g Tris base in approximately 90 mL H_2O. Add 37% HCl very slowly up to pH = 8.5 (*see* **Note 4**). Complete with H_2O up to 100 mL in a graduated cylinder.
18. 10 mM Tris–HCl pH 8.5. For 100 mL, combine 1 mL of 0.5 M Tris–HCl pH 8.5 with 99 mL H_2O.
19. Amine reactive dyes: we used Alexa Fluor Succinimidyl dyes, 488, 555 and 647 from Invitrogen Molecular Probes.
20. Anhydrous DMSO.

2.4.2 DNA-FISH Labeling

1. 4% paraformaldehyde (PFA) in 1× PBS. Always prepare fresh. For 10 mL of fixation solution: combine 2.5 mL 16% paraformaldehyde and 7.5 mL 1× PBS and mix vigorously (*see* **Note 16**).
2. RNase A stock solution at 2 mg/mL in H_2O: for 50 mL, dissolve 100 mg RNase A in 50 mL H_2O. Filter sterilize using a 0.2 μm filter. Store aliquots as 10 mL aliquots at −20 °C.
3. RNase A at 0.2 mg/mL in 2× SSC. Dilute tenfold RNase A stock solution in 2× SSC and store 10 mL aliquots at −20 °C.
4. 20× SSC.
5. 4× SSC in H_2O. For 50 mL: combine 10 mL of 20× SSC stock solution with 40 mL H_2O.
6. 2× SSC in H_2O. Dilute 20× SSC tenfold in H_2O.
7. 0.1× SSC in H_2O. For 50 mL: combine 2.5 mL of 2× SSC with 37.5 mL H_2O.
8. 4× SSC/0.2% Tween20. For 50 mL: dilute 100 μL of Tween20 in 50 mL of 4× SSC (*see* **Note 12**).
9. 0.5% (v/v) Triton X-100 in 1× PBS: for 100 mL, dilute 500 μL Triton X-100 in 100 mL 1× PBS (*see* **Note 17**).
10. 0.01% (v/v) Triton X-100 in 1× PBS: for 100 mL, dilute 2 mL of 0.5% (v/v) Triton X-100 in 198 mL 1× PBS.

11. 0.1 M HCl. For 50 mL: add 425 μL of 37% HCl to 50 mL H_2O.
12. 50% formamide in SSC pH 7–7.2. For 500 mL: combine 250 mL deionized formamide, 50 mL 20× SSC and 120 mL H_2O. Adjust pH with 37% HCl. Store at −20 °C.
13. Human Cot-1 DNA at 1 μg/μL.
14. Single-Stranded-DNA (ssDNA) from salmon testes at 10 mg/mL.
15. Dextran sulfate mix: 20% (w/v) dextran sulfate in 2× SSC. Dissolve 20 g of dextran sulfate powder in 80 mL 2× SSC. Once the powder is dissolved, adjust the volume by adding 2× SSC up to 100 mL using a graduated cylinder. Store at −20 °C (*see* **Note 18**).
16. Glass slides.
17. Kimwipes.
18. Rubber cement.
19. Heating block.
20. Hoechst 33342, trihydrochloride, trihydrate (10 mg/mL solution in water): 1/5000 dilution in 1× PBS. For 10 mL, dilute 2 μL Hoechst in 10 mL 1× PBS.

2.5 Material for Microscopy Acquisitions

To acquire image stacks we recommend using either a spinning disk or a conventional confocal microscope, equipped with a 63X HC PL APO oil immersion objective (NA 1,4–0,7), a Quadrichroic Mirror (405 nm/488 nm/565 nm/640 nm) for simultaneous laser excitation, and a SCMOS camera.

Excitation of the aforementioned dyes (Hoechst, Alexa 488, Alexa 555 and Alexa 647), require lasers with the following wavelengths: 405 nm, 488 nm, 561 nm, and 647 nm.

It is advised to collect fluorescence using single band pass emission filters adapted to each wavelength.

3 Methods

All steps are carried out at room temperature, unless otherwise stated.

3.1 Methods for Cell Culture

Ideally, cells must be cultivated to a maximum of 75–80% confluency.

3.1.1 MCF10A Cell Culture

Cultivate MCF10A with the appropriate growing medium (*see* **Note 19**), in 10 mm cell culture dishes for Western Blot and RT-qPCR analysis and on cleaned coverslips in 6-well plates for

DNA-FISH labeling. For each harvesting day (previously determined), grow three cell samples in parallel: one for the untreated condition (as a negative control), one for the transient TGF-β treatment, and one for the continuous TGF-β treatment (*see* **Note 20**).

3.1.2 TGF-β Treatment

1. For transient TGF-β treatment, add 1.25 μL of TGF-β stock solution (*see* **Note 20**) for 10 mL of growing media (for a final concentration of 2.5 ng/mL) for 4 h, then replace the TGF-β supplemented growing media with non-supplemented MCF10A growing media.
2. For continuous TGF-β treatment, cultivate cells in 2.5 ng/mL TGF-β supplemented growing media (and **Note 20**). Refresh the TGF-β every 48 h.
3. Regularly observe cells by light microscopy to follow phenotypic changes upon transient or continuous TGF-β treatments (*see* Fig. 1a, b and **Note 21**).

3.1.3 Harvesting Cells for Western Blot

Harvest cells for analysis at the desired post- TGF-β treatment time-points (*see* **Note 22**).

1. Put the cell culture dish on ice. We consider that a 10 mm cell culture dish at 80% confluence contains approximately 5 × 10^6 MCF10A cells.
2. Rinse cells with 5 mL of cold 1× PBS containing protease inhibitors.
3. Discard the 1× PBS containing protease inhibitors.
4. Add 2 mL of cold 1× PBS containing protease inhibitors.
5. Scratch cells with a cell spatula making sure that the harvested cells are in the fluids (tilt the cell culture dish).
6. Transfer the harvested cells in a 15 mL tube.
7. Repeat **steps 4–6** once.
8. Centrifuge the 15 mL tube at 1250 × *g* at 4 °C for 15 min.
9. Discard the supernatant.
10. Add 500 μL of lysis buffer.
11. Incubate 30 min at 4 °C.
12. Optional: At this stage, proteins concentrations can be quantified. We use the DC protein assay from Bio-Rad.
13. Add 150 μL of 4× Laemmli sample buffer with β-mercaptoethanol.
14. Store at −20 °C.

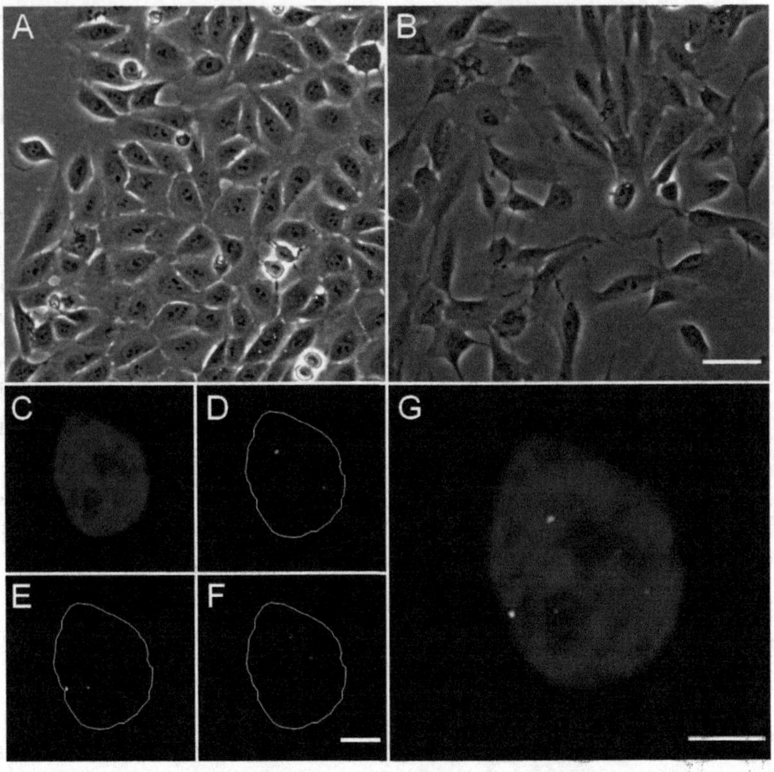

Fig. 1 (**a**, **b**) MCF10A cells observed by light microscopy (**a**) without TGF-β and (**b**) after 9 days of 2.5 nM TGF-β treatment. (**c**–**g**) DNA-FISH labeled MCF10A cell nucleus (untreated; maximum projection of 51 planes). (**c**) The nucleus is stained with Hoechst, (**d**) E-cadherin loci are stained with Alexa 488, (**e**) the control gene loci (here we used the progesterone receptor gene PGR) are stained with Alexa 555 (**f**) and the N-cadherin loci are stained with Alexa 647. (**g**) Overlay of the four channels. Scalebars: 50 μm (**a**, **b**), 5 μm (**c**–**g**)

3.1.4 Harvesting Cells for RT-qPCR

1. Heat 1× DPBS and MCF10A growing media at 37 °C and Trypsin at room temperature.
2. Rinse cells with 10 mL of 1× DPBS.
3. Remove the DPBS and add 1 mL of Trypsin for a 10 mm cell culture dish and incubate 20–25 min at 37 °C.
4. Gently hit the border of the cell culture dish to detach cells.
5. Add 9 mL of growing media and pipette up and down to detach the cells sticking to the container.
6. Count the cells (*see* **Note 23**).
7. Transfer the 10 mL cell solution in a 50 mL tube and centrifuge at 300 × *g* for 5 min.
8. Discard the supernatant, and resuspend cell pellet in a volume of 1× DPBS in order to reach a cell concentration of 1.5 × 10^6 cells/mL.

9. Transfer 1.5×10^6 cells per microtube (prepare as many microtubes as possible) and centrifuge at $9660 \times g$ for 5 min.
10. Discard the supernatant.
11. Flash freeze pellets in liquid nitrogen and store at −80 °C (it is possible to keep pellets prepared as such for months).

3.2 Methods for mRNA and Protein Quantification

By performing Western Blot and RT-qPCR assays on EMT-related genes and proteins, the mRNA and protein expression levels will define the physiological state of the cell at the chosen time-points post TGF-β treatment. Here is a non-exhaustive list of proteins known as EMT markers: E-cadherin and occludin (epithelial phenotype), N-cadherin and vimentin (mesenchymal phenotype), snail1, ZEB1, and twist (Transcription factors). In addition, we also perform every assay on reference genes and proteins (such as RPLP0, H3, or GAPDH), whose expression levels remain invariant upon TGF-β treatment in MCF10A cells.

Western Blot and RT-qPCR procedures are standard excepting cell harvesting.

3.2.1 Western Blot

Cast the gel by following **steps 1–11**. Before starting, make sure that your material is clean and dry, and has been previously rinsed with demineralized water or ethanol.

1. Prepare the casting material as recommended by the supplier.
2. Insert comb into the empty cassette and draw a mark at 0.5–1 cm (*see* **Note 24**) below the bottom of the comb teeth: this will be the level to which the resolving gel will be poured.
3. Remove comb.
4. Use containers that can be closed (Falcon tubes for example), prepare the resolving and stacking mix, without adding the APS and TEMED.
5. Add the appropriate volumes of APS and TEMED to the resolving gel mix, swirl the solution gently but thoroughly and cast gel. Do not exceed the mark indicating the space left for the staking gel, and overlay with isopropanol 100% (*see* **Note 25**).
6. Close the container used for the resolving mix and keep on the side (*see* **Note 26**).
7. Allow the gel to polymerize (*see* **Note 27**), and just before adding the stacking gel, return the gel and absorb isopropanol on a paper towel.
8. Add the appropriate volumes of APS and TEMED in the stacking gel mix, swirl the solution gently but thoroughly and distribute over the resolving gel up to the top of the cassette.

9. Immediately place a comb, taking care not to introduce air bubbles.
10. Allow the gel to polymerize (*see* **Note 28**).
11. Gels can be used immediately or stored at 4 °C (*see* **Note 29**).

    Perform gel loading and migration by following **steps 12–19**.
12. Thaw protein standard at room temperature. It should not be subjected to heat.
13. Thaw samples on ice (*see* **Note 30**).
14. Heat sample 10 min at 95 °C.
15. Vortex in order to homogenize.
16. Spin centrifuge at room temperature (*see* **Note 31**).
17. In the meantime, prepare your gel for electrophoresis: place your cassette into the electrophoresis tank, add migration buffer until it covers the top of your gel and gently take out comb (*see* **Note 32**).
18. Load your samples and the standard into the wells (*see* **Note 33**).
19. Fix voltage at 90 V until the resolving gel has been reached, then set at 150–170 V until the dye front reaches the bottom of the gel.

    Once the migration is terminated, immediately proceed with protein transfer (*see* **steps 20–29**) onto nitrocellulose or PVDF membrane (*see* **Note 34**).
20. Cut one nitrocellulose membrane and 4 × Whatman paper a little bit larger than the size of the gel.
21. In order to orient the membrane (*see* **Note 35**), label the nitrocellulose membrane with an HB pencil or cut-out a corner of the membrane.
22. Prepare the transfer sandwich by soaking the transfer cassette, and the sponges in a bucket containing 1× transfer buffer, and start preparing the sandwich by placing the sponges and the Whatman papers in this order, from the cathode to the anode: black part of the cassette, corresponding to the cathode; two sponges soaked in the transfer buffer; two Whatman papers soaked in transfer buffer.
23. Once the migration is over, remove the gel and separate gel plates with the use of a spatula. The gel will stick to one of the plates.
24. Carefully remove the stacking gel with a spatula and all parts of the gel that are not useful.
25. Pour a small volume of 1× transfer buffer onto the gel, wet your gloves (*see* **Note 36**) and carefully separate gel from the plate and add to the sandwich (*see* **Note 35**).

26. Add the membrane to the sandwich. Verify the orientation to be able to identify the wells later.
27. Add two Whatman papers soaked into 1× transfer buffer onto the sandwich, carefully take out air bubbles by rolling over the sandwich with a roller.
28. Finish the sandwich by adding two sponges soaked into 1× transfer buffer and closing the cassette, carefully taking care not to introduce any air bubbles.
29. Insert the cassette into the system, immerse the system with 1× transfer buffer, transfer to an ice bucket filled up with ice (this will avoid overheating of the cassette) and run in a cold room 60–90 min with the power supply set at 300 mA, 300 W and 300 V (*see* **Note 37**).

    Once the transfer is completed proceed with protein visualization (*see* **steps 30–35**). This step will enable you to confirm the success of the transfer. The gel can be stored at −20 °C (*see* **Note 38**).
30. Remove the sandwich and verify that proteins have nicely transferred onto the membrane (*see* **Note 39**).
31. Transfer the membrane to a small plastic container and wash it in PBST for a few minutes.
32. Trash PBST and replace by Red Ponceau.
33. Slowly agitate on an orbital shaker until colored bands appear.
34. Rinse with demineralized water in order to see well-defined protein bands.
35. Completely destain membrane by incubating it in PBST.

    To prevent unspecific binding of the antibody to lower background, perform membrane blocking (*see* **step 36**).
36. Incubate membrane 1 h in PBSTM at room temperature on an orbital shaker.

    Then, incubate the membrane with the primary and secondary antibodies (*see* **steps 37–44**).
37. Dilute primary antibodies in PBSTM as described in Table 1:
38. For antibodies against N-cadherin, E-cadherin, and occludin, perform incubation overnight at 4 °C on a turning wheel. For antibodies against Vimentin and H3, incubate 1–2 h at room temperature on a turning wheel (*see* **Note 40**).
39. Dilute your secondary antibody and incubate as described by suppliers (*see* **Note 41**).
40. Once the incubation is terminated, recover the antibody for reuse, and place your membrane in a small plastic box having just the size of the membrane.

Table 1
Antibodies used for the detection of EMT-related proteins

Primary antibody	Dilution for immunoblot	Secondary antibody	Supplier	Supplier reference number	Expected molecular weight (kDa)
N-cadherin	1:50	Mouse	BD	610920	130
E-cadherin	1:500	Mouse	BD	610181	120
Occludin	1:125	Rabbit	Thermo	71-1500	65
Vimentin	1:5,000	Mouse	Sigma	V6389	58
H3	1:10,000	Rabbit	Abcam	Ab1791	17

41. Wash 3 × 15 min in PBST on an orbital shaker with slow motion.
42. We perform incubations with 20 mL diluted antibody, within the small plastic box used for the washes.
43. Leave 1 h at room temperature on an orbital shaker with slow motion.
44. Wash 3 × 5 min in PBST before proceeding with revelation.
    Finally, proceed with protein revelation (*see* **steps 45–48**) (*see* **Note 42**).
45. Incubate the membrane with an appropriate volume of the mix: we distribute the mix over a glass plate and dispose the membrane over it, proteins turned toward the mix, taking care not to introduce air bubbles between the membrane and the mix.
46. Incubate for 1 min at room temperature. Absorb the excess of liquid on a paper towel and display on your imaging system.
47. Expose as needed. For our system, usual exposure times required with the antibodies dilutions mentioned above are: 3 min for N-cadherin, 1 min for E-cadherin, 10 s for occludin, 10 s for vimentin, and 5 s for H3.
48. Quantify the protein bands.

3.2.2 RT-qPCR

For all the RT-qPCR steps, we clean surfaces and material with decontaminating solution against RNases such as RNAaseZap (Ambion) or equivalent, we use RNase-free filtered tips to avoid cross-contamination, and ultrapure autoclaved water (resistance at 18.2 MΩ·cm) that is RNase- and DNase-free. Perform RNA extraction following **steps 1–4**.

1. Total RNA from dry cell pellets stored at −80 °C can be extracted by using a ready-to-use kit.

2. Thaw frozen cells at room temperature before proceeding with RNA extraction. We follow the protocol as indicated by the supplier with the following specificities: (a) we do not add β-mercaptoethanol nor DTT into the RLT buffer, (b) we skip the DNase digestion optional step, and (c) we do all centrifugation steps at 9000 × *g*.
3. Elute in 30 μL RNase-free water.
4. You can directly proceed with quantification (*see* **Note 43**) and reverse transcription (recommended) or keep at −20 °C for short storage (up to 1 year). For long storage (more than 1 year), keep at −80 °C. Always thaw on ice and avoid freeze/thaw cycles of the RNA.

   Perform reverse transcription (*see* **steps 5–11**) working on ice, in 250 μL PCR tubes (*see* **Note 44**).
5. Distribute 1500 ng RNA per tube in up to 13.75 μL nuclease-free water from the kit.
6. Add, per tube, 1.25 μL of mix 1, previously prepared and kept on ice. Mix gently and centrifuge briefly.
7. Denature components 5 min at 65 °C in PCR machine. Meanwhile, prepare mix 2 and keep on ice.
8. Chill PCR tube on ice and add 5 μL per tube of mix 2. Mix gently and centrifuge briefly.
9. Incubate 10 min at 25 °C followed by 30 min at 65 °C and 5 min at 85 °C in PCR machine.
10. Chill on ice and directly proceed to qPCR or store at −20 °C for up to 1 week, or at −80 °C for longer storage. Avoid freeze/thaw cycles of the RNA.
11. Proceed with cDNA/RNA hybrid quantification by measuring A_{260} on a spectrophotometer, considering that $A_{260} = 1$ corresponds to a cDNA/RNA hybrid concentration of 50 ng/μL (*see* **Note 45**).

    Perform qPCR following **steps 12–17** (*see* **Note 46**). All volumes indicated are for 384-well qPCR plates. Controls ahead of the qPCR are required (*see* **Note 47**).

    Primers should be designed on exon-exon boundaries in order to avoid genomic DNA amplification (*see* Table 2).
12. Prepare all dilutions and keep on ice (*see* **Note 48**).
13. Use an electronic device and distribute cDNA first, then the primers in SYBR green mix. This will avoid cross-contamination since the SYBR green sticks to the well. From time to time, tap the plate onto the bench to enable droplets to slide to the bottom of the well. The final volume in the wells is 6 μL.

Table 2
qPCR primer sequences

Genes	Primer sense	Primer sequence
RPLP0	Forward Reverse	TGGCAGCATCTACAACCCTGA ACACTGGCAACATTGCGGACA
Vimentin	Forward Reverse	CCAAACTTTTCCTCCCTGAACC GTGATGCTGAGAAGTTTCGTTGA
E-cadherin	Forward Reverse	GAACGCATTGCCACATACAC ATTCGGGCTTGTTGTCATTC
N-cadherin	Forward Reverse	AGCCAACCTTAACTGAGGAGT GGCAAGTTGATTGGAGGGATG
Occludin	Forward Reverse	AGAACTCTCCCGTTTGGATAAAGA TTTGTAATCTGCAGATCCCTTCAC

14. Seal the plate with an adhesive film, using a clean instrument and insisting onto the borders. Only then, take out the protective film. At this stage, it is possible to keep the plate overnight at 4 °C and protected from light.
15. Centrifuge 30 s at 100 × *g*.
16. Enter parameters on the 384-well PCR machine. We use the following cycle:

    95 °C for 1 min.

    95 °C for 15 s.

    60 °C for 30 s—Go to previous step and repeat cycle 40 times.

    Immediately followed by the melting curve parameters:

    95 °C for 15 s.

    60 °C for 15 s.

    95 °C for 15 s.
17. Run the qPCR program.

    Analyze data following **steps 18–23**.
18. Calculate the standard deviation of the triplicates. It should be inferior to 0.1. If not, remove the aberrant value (*see* **Note 49**).
19. Our primers having an efficiency close to 100% (primer efficiency being below 100% is not a problem as far as all primer used have the same efficiency), and considering that the efficiency of both the target and the reference gene are equal to 2 (the amount of the PCR product is doubling at every cycle), we normalize the expression ratio using the "Delta-Delta-C_t" method (or Livak method):

$$\Delta\Delta Ct = 2^{-((Ct\cdot(GOI,T)-Ct(RG,T))-(Ct(GOI,NT)-Ct(RG,NT)))} \quad (1)$$

with:

Ct (GOI,T) being the Ct of the Gene Of Interest (GOI, also called "target" gene) in the test sample (i.e., TGF-β -treated (T) condition).

Ct (GOI,NT) being the Ct of the GOI in the calibrator sample (i.e., non-treated (NT) condition).

Ct (RG,T) being the Ct of the Reference Gene (RG) in the T condition.

Ct (RG,NT) being the Ct of the RG in the NT condition.

In detail:

20. We normalize the Ct of the GOI to that of the RG for both the test sample (T) and the calibrator sample (NT):

$$\Delta\mathrm{Ct(GOI)} = \mathrm{Ct(GOI, T)} - \mathrm{Ct(RG, T)}$$

$$\Delta\mathrm{Ct(NT)} = \mathrm{Ct(GOI, NT)} - \mathrm{Ct(RG, NT)}$$

21. We then normalize the ΔCt of the GOI to the ΔCt of the RG:

$$\Delta\Delta\mathrm{Ct} = \Delta\mathrm{Ct(GOI)} - \Delta\mathrm{Ct(RG)}$$

22. We finally calculate the expression ratio:

$$2^{\Delta\Delta\mathrm{Ct}} = \text{normalized expression ratio}$$

23. Represent data using a spreadsheet program, displaying the normalized expression ratios in a 10-base logarithmic scale.

3.3 Methods for DNA-FISH Labeling

DNA-FISH enables 3D localization of individual genomic loci inside the nucleus at any stage of the cell cycle. To successfully perform DNA-FISH labeling you need (a) to prepare high quality fluorescently tagged DNA-FISH probe and (b) to render DNA accessible for DNA-FISH probes hybridization. Preserving nuclear morphology and chromatin structure despite aggressive cell treatments (fixation, permeabilization, denaturation, hybridization) is essential. It is important to realize that each cell line might need different fixation, permeabilization, denaturation, and hybridization conditions that have to be empirically defined by the user [16]. Optimization of the protocol might be delicate and time consuming, nevertheless, we describe here each parameter that needs to be examined and possibly adjusted.

Here we describe a protocol to perform DNA-FISH in MCF10A cell line to visualize up to three different loci in Hoechst-stained nuclei (*see* Fig. 1c–g).

3.3.1 Fosmid/BAC Preparation

Only a limited number of DNA-FISH probes is commercially available (abnova.com), nevertheless it is possible to custom-prepare specific DNA-FISH probes using genomic DNA fragments of interest cloned in BACs (covers a genomic region of 50–300 kb in length) or fosmids (covers a genomic region of 25–45 kb in length). We select our sequences of interest (*see* Table 3) on the

Human GRCh38/hg38 genome assembly via UCSC website. We prepare high quality BAC or fosmid DNA by using a commercially available kit and we confirm the authenticity of the sequence by PCR followed by sequence-specific restriction digestion [15].

We fluorescently tag the DNA-FISH probes using a two steps protocol. We first perform a nick translation reaction to introduce aminoallyl-dUTP. During the nick translation reaction DNase I will create single strand breaks called "nicks" within the double stranded DNA sequence. In a next step, the Polymerase I will repair these nicks and insert new nucleotides including the modified dUTP. As a result from these reactions we obtain dsDNA fragments (150–1000 bp) containing numerous aminoallyl-dUTP. Finally, we couple fluorophores to the nick-translated modified dsDNA via a simple chemical reaction where Alexa Fluor succinimidyl ester dyes make a covalent bond with the amines of the incorporated aminoallyl-dUTP.

Perform the nick translation reaction following **steps 1–4**

1. Thaw BAC/fosmid DNA, 10× NTB buffer, 0.1 M DTT, 10× dNTP mix, Aminoallyl-dUTP, and DNase A supplier buffer on ice.
2. Dilute 1 μL of DNase I in 49 μL of DNase I supplier buffer to reach the concentration of 0.2 U/μL. Keep on ice (*see* **Note 50**).
3. Set up the following reaction in a PCR tube on ice: mix 5–10 μg of BAC/fosmid DNA (27 μL maximum), 5 μL 10× NTB buffer, 5 μL DTT 0.1 M, 5 μL 10× dNTP mix, 6 μL aminoallyl-dUTP at 0.5 mM, 1 μL DNA polymerase I, 1 μL DNase I at 0.2 U/μL and add water up to 50 μL.
4. Incubate 25 min at 16 °C (*see* **Note 50**). During this time prepare a 2% agarose gel.

Control the nick translation reaction (*see* **steps 5–8**).

5. Following nick translation, immediately transfer the reaction tube on ice.
6. Run 1 μL of this reaction on a 2% agarose gel to verify the size of the obtained fragments. Successful nick translation will result in a smear between 150 and 1000 bp.
7. In the case of unsuccessful reaction, add 1 μL of freshly prepared 0.2 U/μL DNase I dilution and incubate 10 more minutes at 16 °C (*see* **Note50**).
8. Control again the size of the fragments on a 2% agarose gel and repeat this step until you obtain the correct fragment size.

Table 3
References and characteristics of some Fosmids and BACs corresponding to EMT-related genes

References	Covered genes	Protein	Size (bp)	Coordinates (Dec. 2013 (GRCh38/hg38))
WI2-3715L22	CDH1	E-cadherin	45,918	chr16:68,747,043–68,792,960
WI2-1996A10	CDH1	E-cadherin	39,245	chr16:68,792,596–68,831,840
WI2-3409L5	CDH1	E-cadherin	39,413	chr16:68,721,690–68,761,102
WI2-3706F6	CDH1	E-cadherin	38,305	chr16:68,761,988–68,800,292
WI2-2774G15	CDH1	E-cadherin	34,708	chr16:68,796,994–68,831,701
WI2-3194M5	CDH2	N-cadherin	40,532	chr18:27,949,698–27,990,229
WI2-2158H2	CDH2	N-cadherin	36,745	chr18:27,990,131–28,026,875
WI2-3330A4	CDH2	N-cadherin	39,494	chr18:28,028,027–28,067,520
WI2-3413H15	CDH2	N-cadherin	43,383	chr18:28,073,240–28,116,622
WI2-2228E16	CDH2	N-cadherin	44,108	chr18:28,121,814–28,165,921
RP11-643F8	CDH2	N-cadherin	192,389	chr18:27,960,365–28,152,753
WI2-3327C11	OCLN	Occludin	40,741	chr5:69,505,597–69,546,337
WI2-2095C18	OCLN	Occludin	38,977	chr5:69,464,190–69,503,166
RP11-418N14	PGR	PGR	220,426	chr11:101,004,769–101,225,194
WI2-3907J23	VIM	Vimentin	38,741	chr10:17,220,428–17,259,168

PGR is our control gene

Purify the nick-translated DNA by following **steps 9–18**.

9. Inactivate the DNase I by heating at 75 °C for 10 min.
10. Clean-up the amine-modified DNA using a PCR purification kit and elute in 100 μL of water (*see* **Note 51**).
11. Ethanol-precipitate the 100 μL nick-translated DNA by adding 10 μL of 3 M sodium acetate pH 5.2 and 275 μL of cold absolute ethanol.
12. Leave at −20 °C for 1 h minimum (overnight is better).
13. Centrifuge at 20,000 × *g* for 30 min at 4 °C.
14. Discard the supernatant, wash the pellet with 500 μL of 70% cold absolute ethanol and centrifuge 15 min at 20,000 × *g* at 4 °C.
15. Discard the supernatant and let the residual liquid evaporate. The pellet must be dry.
16. As it might stick to the tube wall after desiccation, resuspend the nick-translated DNA pellet in 6 μL of water by vortexing for 30 min at 37 °C.

17. Use 1 μL of the product to determine the concentration using microvolume UV spectroscopy (*see* **Note 52**).
18. Store the clean and nick-translated DNA at −20 °C.

Perform coupling of the fluorescent dye by following **steps 19–28**

19. Adjust 1–3 μg of aminoallyl-modified DNA to a volume of 5 μL (*see* **Note 52**).
20. Seal the tube with parafilm to avoid evaporation and denature the product at 95 °C for 5 min.
21. Immediately transfer the tube on ice for 2 min.
22. Pulse spin and add 3 μL of $NaHCO_3$ at 0.2 M.
23. Dissolve one aliquot of amine reactive dye in 2 μL of anhydrous DMSO.
24. Add the 8 μL of the nick-translated DNA/$NaHCO_3$ mixture, vortex, pulse spin, and incubate at room temperature, in the dark, for 1 h.
25. Add 90 μL of water and purify the labeled DNA-FISH probes using a PCR purification kit (*see* **steps 10–16**). Elute in 60 μL 10 mM Tris-HCl pH 8.5 (*see* **Note 53**).
26. Determine the probe concentration: measure the absorbance of the DNA-dye conjugate at 260 nm (A_{260}) and at the maximum absorbance (λ_{max}) for the used fluorophore (A_{dye}) using microvolume full spectrum spectroscopy. Correct the DNA concentration for the dye absorption according to the correction factor CF_{260} (*see* **Note 54**) in Table 1, using Eq. 2

$$A_{base} = A_{260} - \left(A_{dye} \times CF_{260}\right) \quad (2)$$

27. Determine the labeling efficiency (*see* **Note 55**): Calculate the base to dye molecule ratio using Eq. 3 the extinction coefficient of a DNA base $\varepsilon_{base} = 8919$ and the extinction coefficient of the fluorophore ε_{dye} (*see* Table 4).

$$\text{base} : \text{dye} = \frac{A_{base} \times \varepsilon_{dye}}{A_{dye} \times \varepsilon_{base}} \quad (3)$$

28. The labeled DNA-FISH probe can be stored for months at −20 °C (for DNA-FISH reaction you only need a few microliter of the final product).

3.3.2 Cell Fixation and Permeabilization

1. Cells grown on coverslips are first washed once with 1× PBS.
2. Under a fume hood, fix cells grown on coverslips with 2 mL (*see* **Notes 1**) of 4% paraformaldehyde, without shaking, for 10 min. During the last 3 min, start cell permeabilization by adding 800 μL of 0.5% Triton X-100, gently shake the plate to

Table 4
Absorption characteristics for the used dyes

	λ_{max} (nm)	ε_{dye} (cm^{-1} M^{-1})	Correction factor at 260 nm
Alexa Fluor 488	492	62,000	0.30
Alexa Fluor 555	555	150,000	0.04
Alexa Fluor 647	650	239,000	0.00

homogenize the liquids and leave under the fume hood without shaking until the end of the 10 min (*see* **Note 56**).

3. Rinse cells with 0.01% Triton X-100 in 1× PBS, gently manually shake the plate.
4. Wash with 0.01% Triton X-100 for 3 min. Repeat twice.
5. Incubate with 0.5% Triton X-100 for 10 min (*see* **Note 57**).
6. Incubate with 0.2 mg/mL RNase A solution for 30 min (*see* **Note 58**).
7. Wash with PBS during 10 min. Repeat twice.
8. Incubate in 0.1 M HCl for 5 min (*see* **Note 59**).
9. Wash with 2× SSC for 3 min. Repeat once.
10. Let stand in 50% formamide pH 7–7.2 at room temperature for at least 1 h (*see* **Notes 1** and **60**).

3.3.3 Preparation of the DNA-FISH Probes

1. In a microtube, mix 10–20 ng of labeled DNA-FISH probe (*see* **Note 61**), 1 μL of SS-DNA from salmon testes and 6 μg or 12 μg (for BAC or fosmid probes, respectively) of human Cot-1 DNA (*see* **Note 62**). Adjust the volume to 100 μL with H_2O.
2. Ethanol-precipitate the fluorescent probe mixture by adding 10 μL of 3 M sodium acetate pH 5.2 and 275 μL of cold absolute ethanol.
3. Leave at −20 °C for 1 h minimum (overnight is best). Centrifuge at 20,000 × *g* for 30 min at 4 °C.
4. Discard the supernatant (*see* **Note 63**), wash the pellet with 500 μL of 70% cold ethanol and centrifuge 15 min at 20,000 × *g* at 4 °C.
5. Discard the supernatant (*see* **Note 63**) and evaporate the residual liquid. The pellet must be dry.
6. Resuspend the nick-translated labeled DNA pellet in 5 μL deionized formamide and let for 30 min at 37 °C under an aluminum foil while shaking on a thermomixer at 1000 rpm.

7. Add 5 μL of dextran sulfate mix and shake on a thermomixer at 1000 rpm for further 10 min under an aluminum foil (*see* **Note 64**).

3.3.4 Coverslip Assembly for Hybridization

1. Drop up to 30 μL of the DNA-FISH probe mixture (for localization of up to three different loci) on a glass slide (*see* **Note 65**).
2. Take the glass coverslip with the fixed cells (laying in the 50% formamide solution) using tweezers, quickly but gently wipe the lower side and drain the excess of fluid from the upper side with a tissue (*see* **Note 66**).
3. Return the coverslip on the glass slide to lay the upper side of the coverslip with the DNA-FISH probe mixture droplet and gently dry the excess of fluid around the coverslip.
4. Seal the coverslip with rubber cement (*see* **Note 67**) and let the slide/coverslip set-up in the dark.

3.3.5 Denaturation and Hybridization

1. Humidify the hybridization chamber and place the glass slide/coverslip set-up on the heating block (the coverslip facing up onto the heating block).
2. For MCF10A cells, denaturation conditions are 75 °C for 4 min and hybridization occurs at 37 °C overnight.

3.3.6 Washing FISH Samples

1. Heat up 2× SSC to 45 °C and 0.1× SCC up to 60 °C in water baths.
2. Fill a 6-well plate (or a 35 mm petri dish depending on the number of coverslips to wash), with 2 mL 2× SSC heaten at 45 °C.
3. Gently remove the rubber cement with a tweezer and transfer the coverslip in the well, cells face up (*see* **Note 68**).
4. Remove the supernatant and wash with 2 mL of 45 °C 2× SSC for 3 min. Repeat three times.
5. Wash with 2 mL of 60 °C 0.1× SSC for 3 min. Repeat three times.

3.3.7 Hoechst Staining

1. Add 2 mL of Hoechst 1/5000 dilution in 1× PBS and leave at room temperature and protected from light for 10 min.
2. Wash with 1× PBS for 10 min. Repeat twice.
3. Optional: Sample can be stored in 1× PBS at 4 °C for up to 2 weeks.

3.3.8 Sample Mounting for Microscopy Acquisition

1. Fill the cavity of the concave slide with 80 μL of 1× PBS.
2. Place the coverslip on the cavity.

3. Remove the excess of liquid with a tissue and air bubbles by gently tapping the coverslip with tweezers.
4. Seal the coverslip to the slide with rubber cement or nail polish (*see* **Note 67**).

3.4 Methods for Microscopy Acquisition and Data Analysis

Here we chose to use a spinning disk microscope to perform observation of DNA-FISH labeled nuclei in 3D. Prior to imaging it is assumed that the required laser lines are aligned to the sample using the appropriate dichroic and emission filters. Note that other microscopes (epifluorescence, confocal, high-throughput imaging setup...) may be appropriate as long as the setup is capable to acquire *z*-stacks.

3.4.1 Spinning Disk Image Stacks Acquisitions

Laser intensities and camera integration times were optimized for each channel and each sample. 3D acquisitions were performed in 51 focal planes at 200 nm steps. For each condition, a minimum of 150 fields were acquired to obtain a minimum of 1500 nuclei.

3.4.2 Image Analysis

The DNA-FISH labeled loci of each nucleus appearing on the multichannel image stacks were detected and their position reported using a custom-made script running on ImageJ (https://imagej.nih.gov/ij/; Macro language), enabling automated treatment of the data. Note that commercial scripts are also available, in particular with high-throughput microscopes.

1. Perform nuclei segmentation by doing a *Z-projection* of the image stack (Hoechst staining) followed by (a) a *Smooth* filter of the resulting image (b) a binarization by an automatic intensity threshold (Li methods), (c) a *Fill Holes* function and (d) an *Adjustable Watershed* function (tolerance = 3). We extracted the nuclei morphology parameters (radii length, area, roundness...) with the *Analyze Particles* function (size = 120-Infinity, Circularity = 0.65–1.00).
2. Determination of the nuclei Masks (in 3D) by applying to each stack of the Hoechst channel a smoothing filter, followed by binarization applying an automatic intensity threshold (Li method) and a Fill Holes function. We then extracted the volume of the nuclei with the *3D object counter* plugin [17].
3. Determination of coordinates of the DNA-FISH labeled loci. In each individual nucleus (Region of interest determined **step 1**) we apply on every DNA-FISH channel staining (a) a *Smooth* filter (b) the *Enhance Contrast* function (saturated = 0.0015) (c) the *Subtract Background* function (rolling = 6) and (d) a binarization applying an automatic intensity threshold (RenyiEntropy method). The coordinates of each locus were recorded using the 3D Object Counter plugins.

4. The output of this script is an excel sheet displaying characteristics for each nucleus, such as the dimensions of the nuclei (area, radii length, volume…) loci coordinates, loci-nuclear envelope distances, loci to loci distances… The script can be provided upon request.

Following this protocol (*see* **Note 69**), you will also be able to perform super-resolution 3D-dSTORM acquisitions on your samples to study sub-resolution gene architecture evolution during EMT/MET processes. However, the aim of this chapter is not to develop super-resolution 3D-dSTORM imaging technique, please refer to the relevant references [18] (*see* **Note 70**).

4 Notes

1. Non-exhaustive list: Sodium orthovanadate, TEMED, APS, Acrylamide:Bis-Acrylamide, paraformaldehyde, formamide.
2. Supplementation with MEGS and hEGF brings a final concentration of 0.4% (v/v) of Bovine Pituitary Extract, 0.5 μg/mL hydrocortisone, 10 ng/mL of human EGF, and 1 μg/mL of recombinant human insulin-like growth factor I.
3. We use high precision coverslips (Marienfeld, reference 0107052) in order to lower optical aberrations due to irregularities on the surface of coverslips. Normal coverslips can also be used.
4. This reaction is exothermic. Since pH is temperature-dependent, make sure to measure the pH when the solution is at 25 °C, and with a Tris-compatible electrode (with a double junction). For 100 mL 1 M Tris–HCl pH 8.0, you will need approximately 4.2 mL of 37% HCl to reach the desired pH.

   For 100 mL 3 M Tris–HCl pH 6.8, you will need approximately 24 mL of 37% HCl to reach the desired pH.

   For 500 mL 3 M Tris–HCl pH 8.8, you will need approximately 30 mL of 37% HCl to reach the desired pH.

   For 100 mL 1 M Tris–HCl pH 7.5, you will need approximately 6.5 mL of 37% HCl to reach the desired pH. For more precision, pH can also be adjusted using diluted HCl.

   For 100 mL 0.5 M Tris–HCl pH 8.5, you will need approximately 1.2 mL of 37% HCl to reach the desired pH. For more precision, pH can also be adjusted using diluted HCl.
5. Add protease inhibitors freshly before use.
6. Due to its smell and toxicity, manipulating β-mercaptoethanol under a fume hood is highly recommended.

7. APS is subject to decomposition; therefore, it is best to prepare it freshly. Though, preparing fresh each time can be tedious, and you may therefore prepare 10 mL and store as 1 mL aliquots at −20 °C. In our hands, storing APS as such, and avoiding many thaw/freeze cycles, showed no impact on gel polymerization.
8. Keep TEMED tightly closed as it is subject to oxidation, which leads to gradual loss of reactivity for polymerization. Store at room temperature.
9. Keep the acrylamide solution protected from light.
10. 10% APS and TEMED are responsible for the gel polymerization. Due to its smell and toxicity, manipulating TEMED under a fume hood is highly recommended.
11. In order to avoid evaporation, add EtOH in 1× transfer buffer only.
12. As Tween20 is very viscous, it is highly recommended to follow this procedure in order to have the correct final concentration: prepare a beaker with a magnetic stirrer, add the appropriate volumes of 1× PBS and H_2O and start stirring, slowly withdraw Tween20 with a serological pipette up to 2 mL (do not immerge the pipette too deep into the Tween20 as the Tween will stick to the outside wall of the pipette, increasing the volume of Tween you will be taking). Slowly release the Tween20 into the solution. Leave the serological pipette into the beaker with stirring and release Tween20 from time to time into the solution until the pipette is empty. Then, pipet up and down several times to rinse the pipette.
13. Taking skim milk is mandatory as the fat contained in other milks can induce artifacts. Ensure that the milk is entirely dissolved in the PBST before use, as undissolved milk can provoke background spots on the membrane.
14. Prepare at least 30% more volume than needed in order to account for the viscosity of the SYBR and the dead volume of the electronic pipette used for dispensing.
15. Perform the final elution in water and not in TE buffer because the presence of EDTA will inhibit the DNase I action during nick translation reaction.
16. Paraformaldehyde is toxic and should be disposed using appropriate hazardous waste streams.
17. Triton X-100 is highly viscous and sticks to the pipet tip walls. In order to reach a dilution as precise as possible, cut the bottom end of a 1000 μL-pipette tip before taking 500 μL of Triton X-100 very slowly. Distribute in 20 mL of 1× PBS, wait for the Triton to go down by gravity before releasing again,

then pipet up and down several times to recover all the Triton X-100 sticking to the tip walls.

18. Dextran sulfate can be difficult to get into solution. To dissolve it, vortex the solution and mix on a rocker overnight.
19. MCF10A doubling time is about 24 h.
20. At D0, wait for all cells to be attached before adding TGF-β to the medium. We recommend to wait 24 h when using the MCF10A cell line.
21. Epithelial cells are closely packed and adopt polyhedral shapes (*see* Fig. 1a) while mesenchymal cells are elongated and make few cell-to-cell contacts (*see* Fig. 1b).
22. Wait minimum 48 h after trypsin treatment to harvest the cells.
23. For MCF10A, we use a cell counter with a threshold set-up at 10 μm.
24. To ensure band sharpness, the height of the stacking gel should be at least twice the height of the sample in the well.
25. Isopropanol both prevents contact with oxygen that inhibits acrylamide polymerization and enables to level the resolving gel.
26. The rest of the mix is used to follow the polymerization: if polymerization has occurred in the container, gel will be polymerized in the cassette.
27. Time of polymerization depends on several parameters (thickness of the gel, TEMED and APS concentrations and freshness, acrylamide concentration). If polymerization hasn't occurred after 45 min to 1 h, it has a great chance to never polymerize. In that case, start all over again, with fresh APS and TEMED.
28. Be prompt to use or store the gel once polymerization has occurred as the gel can start drying out with time.
29. If not to be used immediately, wrap gel, with comb still inserted, in a well-humidified paper towel overlaid with a plastic wrap. Store flat at 4 °C for 3 days maximum as pH and ionic strength gradually disappear during storage.
30. If pellet remains undissolved, sonicate your samples. We use a Bioruptor Plus with the following parameters: 3 cycles, 30 s time ON, and 30 s time OFF. We do as many sonicating cycles as needed (number of cycles may vary upon sample). Each sample should be homogeneous.
31. Centrifugating helps recovering any condensates and will pellet insoluble residues.
32. Make sure to pull out slowly and vertically the comb in order to avoid bending the wells, introducing bubbles and gel pieces into the wells. If this occurs, try to take out bubbles and gel pieces by pipetting up and down buffer into the well with a

pipette tip, or by using a Whatman paper. Be careful, as the gel wells are, at this stage, highly sensitive to any movement.

33. We load 7 μL for the standard, and 7–12 μL for samples, depending on the size of the well and the quantity of proteins expected. Do not exceed half of a well volume as sample can pass from one well to the other. In order to obtain nice vertical migrations in all wells, do not leave any empty well, and rather load them with 1× Laemmli sample buffer with β-mercaptoethanol.
34. The choice of the membrane type is personal and both work well. We use nitrocellulose membranes. When using PVDF, the use of methanol instead of ethanol is mandatory, leading to more toxic waste to be treated.
35. If your loading is symmetrical, it is mandatory to give an orientation to the membrane and to be extremely careful on the orientation of the gel and the membrane while assembling the sandwich. If it is asymmetrical, it is still recommended to give an orientation to your membrane.
36. This will keep the gel from sticking to your gloves and will thus minimize the risk of tearing.
37. Run 60 min for low molecular weight proteins such as H3 and up to 90 min for high molecular weight proteins such as E-cadherin and N-cadherin. The time will vary depending on your proteins of interest and your power supply parameters.
38. Storage should be done between two sheets of Whatman paper at −20 °C after the gel has been dried out on a paper tissue at room temperature. Do not over dry.
39. If ladder dyes are visible, it means that the proteins at this size haven't transferred entirely onto the membrane. In a next experiment, leave the transfer longer but do not overrun the transfer as small proteins can also pass through the membrane.
40. We perform incubations with 4–5 mL diluted antibody, within little plastic bags sealed with a plastic sealer. The diluted antibody can be kept at −20 °C for months for reuse. We use them five times maximum or until the signal lowers.
41. We use secondary antibodies coupled to horseradish peroxidase.
42. This step takes advantage of the property of peroxidase to oxidize a peroxide solution.
43. It is important to quantify and ascertain the quality of the RNA extraction before proceeding with the reverse transcription because contaminants (proteins or solvent) could decrease the efficiency of the reverse transcription and/or PCR. This can simply be done with a spectrophotometer by measuring the

optical density at 260 nm (A_{260}: nucleic acids quantification), 280 nm (A_{280}: proteins quantification) and 230 nm (A_{230}: quantification of other contaminants such as guanidine for example), knowing that $A_{260} = 1$ corresponds to an RNA concentration of 40 ng/μL.

The extraction is of high quality if:

- the A_{260}/A_{280} ratio is 2.0–2.2. If below, your sample is contaminated with proteins.
- the A_{260}/A_{230} ratio is 2.2. If below, your sample is contaminated with solvents.

If those ratios are not reached, we advise to discard the sample and repeat the RNA extraction starting from a fresh cell pellet.

44. Prepare sufficient cDNA material depending on the number of genes to be analyzed.
45. A_{260} measures all nucleic acid, which includes, in addition to the cDNA/RNA hybrid, the unused dNTP and hexamer primers from the kit, as well as other RNAs. If the reverse transcription has been a 100% efficient, one can consider that the quantity of cDNA/RNA hybrid equals the quantity of RNA first transferred in the tube (here: 1500 ng). But, if the reverse transcription has not been 100% efficient, one cannot say so. Therefore, in order to be reproducible from one experiment to the other, we prefer taking into account the concentration of the cDNA/RNA hybrid obtained by measurement in the spectrophotometer for the qPCR step.
46. We use the SYBR green technique, more affordable and user-friendly than the TaqMan technology.
47. Primer efficiency and specificity: dilute cDNA to 1/10 steps starting with 40 ng in 2.8 μL of H_2O, and we do 4 dilutions (1; 1/10; 1/100; 1/1000).
    - We use H_2O as a negative control. The resulting regression curve shows the primer's efficiency.
    - Calculate a fusion curve to control the primer specificity.
    - Test at least two reference genes by qPCR (i.e., genes with stable expression upon cell treatments).
    - Include a negative control (i.e., wells without enzyme and/or without cDNA).
48. Prepare triplicates and include reference gene. If several 384-well plates are needed, the reference gene should be included in each of the plates.

49. If two values out of three are close, we consider the latest as aberrant, most probably due to pipetting error.
50. For each fosmid or BAC associated nick translation reaction, the ratio between DNase I nicking and DNA polymerase I translation must be determined. Excess of DNase I will result of short fragment sizes with low reaction yields (due to the excessive DNA digestion) while lack of DNase I will result in a poor aminoallyl-dUTP incorporation rate and large fragment sizes (due to insufficient nicks quantity for the polymerase to initiate translation). Users might want to titrate DNase I and adjust the time to optimize nick translation reaction efficiency.
51. This step removes the trace of amines, which could interfere with the subsequent labeling reaction. To increase the yield of PCR purification, wash with 75% room temperature ethanol (for 100 mL, mix 75 mL of absolute ethanol with 25 mL of water), and elute two times in 50 μL water: put 50 μL of water on the column, let stand for 2 min, centrifuge at the recommended speed for 5 min and repeat once.
52. The final concentration of the product can vary from 600 ng to 3 μg. If the quantity of DNA is below 1 μg, it is possible to pool the product of two reactions and to evaporate the water to reach a final volume of 5 μL.
53. Elution with 60 μL generally gives concentration of 10–15 ng/μL.
54. Correction factor = (A_{260} for the free dye)/(A_{max} for the free dye).
55. The labeling efficiency corresponds to the base to fluorescent dye molecule ratio. Probes containing 16 to 30 bases per fluorescent dye molecule will give a bright FISH signal whereas lower degree of fluorophore incorporation might give undetectable FISH signals.
56. Here volumes are detailed for the use of 6-well plates with 22 × 22 mm coverslips, you may adjust them if you use other materials.
57. Carefully respect the 10 min incubation time as longer incubation may induce nucleus structure alterations.
58. The RNase A treatment will lead to the RNA digestion, which will avoid unspecific hybridization of the probe.
59. HCl treatment will lead to the deproteinization of the DNA, a key step of the DNA-FISH procedure. Times and concentration of the deproteinizing agent are critical and must be observed rigorously. As they can vary depending on cell type users may wish to empirically determine optimal conditions.
60. If cells are not immediately needed, skip the 1 h incubation at room temperature and store the cells in 50% formamide at 4 °C for 1 month maximum. Surround the 6-well plate with parafilm before storing at 4 °C.

61. For DNA-FISH probes from fosmids, the covered genomic region might be too narrow to be clearly visible by fluorescence. To overcome this issue you can combine 2 consecutives fosmids to cover a larger region. In this case combine 10 to 15 ng of each fosmid (obtained during the nick translation step) in the microtube.
62. Cot-1 human DNA is used to block nonspecific hybridization to repetitive sequences in the mammalian genome and is therefore important to reduce background for microscopy.
63. Be careful as the pellet might detach from the tube wall.
64. During this step, avoid the exposure of the labeled DNA-FISH probes to direct light.
65. In the case of multiple FISH labeling, premix all three DNA-FISH probes in a microtube, vortex and pulse spin before dropping the probe mixture on the glass slide.
66. As the cells grew on the upper side of the coverslip, do not directly touch the fixed cells with the tissue.
67. Cover the entire junction between the coverslip and the slide to hermetically seal the coverslip.
68. Should the coverslip stick to the slide, add ~100 μL of 2× SSC onto the coverslip to the slide junction and push very gently the coverslip toward the edges of the slide to avoid sample friction damages.
69. For super-resolution imaging, we advise you to use Alexa 647 dye and to skip the Hoechst staining step to avoid additional background during the acquisition.
70. 3D-dSTORM acquisition was performed on an inverted microscope equipped with a 100X TIRF (NA 1.49) objective, containing a quadrichroic mirror (405 nm/488 nm/565 nm/640 nm) for simultaneous laser excitation. Lasers are delivered via an optical fiber with the following power and wavelengths: 100 mW at 405 nm and 300 mW at 647 nm. Fluorescence is collected using single band pass emission filter. The microscope is equipped with a motorized stage in z and a perfect focus system to correct for drifts. A MicAO 3DSR (Imagine Optic) was used to induce astigmatism to the PSF for acquisitions in 3D.

    We used the SAFe Reagent (Abbelight) as an optimized buffer for dSTORM acquisitions.

References

1. Pastushenko I, Blanpain C (2019) EMT transition states during tumor progression and metastasis. Trends Cell Biol 29:212–226. https://doi.org/10.1016/J.TCB.2018.12.001
2. Hao Y, Baker D, ten Dijke P et al (2019) TGF-β-mediated epithelial-mesenchymal transition and cancer metastasis. Int J Mol Sci 20:2767. https://doi.org/10.3390/ijms20112767

3. Wahl GM, Spike BT (2017) Cell state plasticity, stem cells, EMT, and the generation of intratumoral heterogeneity. NPJ Breast Cancer 3:14. https://doi.org/10.1038/s41523-017-0012-z
4. Lamouille S, Xu J, Derynck R (2014) Molecular mechanisms of epithelial-mesenchymal transition. Nat Rev Mol Cell Biol 15:178–196. https://doi.org/10.1038/nrm3758
5. Shachar S, Misteli T (2017) Causes and consequences of nuclear gene positioning. J Cell Sci 130:1501–1508. https://doi.org/10.1242/jcs.199786
6. Robson MI, Ringel AR, Mundlos S (2019) Regulatory landscaping: how enhancer-promoter communication is sculpted in 3D. Mol Cell 74:1110–1122. https://doi.org/10.1016/J.MOLCEL.2019.05.032
7. Meaburn KJ, Gudla PR, Khan S et al (2009) Disease-specific gene repositioning in breast cancer. J Cell Biol 187:801–812. https://doi.org/10.1083/jcb.200909127
8. Leshner M, Devine M, Roloff GW et al (2016) Locus-specific gene repositioning in prostate cancer. Mol Biol Cell 27:236–246. https://doi.org/10.1091/mbc.e15-05-0280
9. Xu J, Ma H, Ma H et al (2019) Super-resolution imaging reveals the evolution of higher-order chromatin folding in early carcinogenesis. bioRxiv 672105. https://doi.org/10.1101/672105
10. Verdone JE, Parsana P, Veltri RW, Pienta KJ (2015) Epithelial-mesenchymal transition in prostate cancer is associated with quantifiable changes in nuclear structure. Prostate 75:218–224. https://doi.org/10.1002/pros.22908
11. McDonald OG, Wu H, Timp W et al (2011) Genome-scale epigenetic reprogramming during epithelial-to-mesenchymal transition. Nat Struct Mol Biol 18:867–874. https://doi.org/10.1038/nsmb.2084
12. Soule HD, Maloney TM, Wolman SR et al (1990) Isolation and characterization of a spontaneously immortalized human breast epithelial cell line, MCF-10. Cancer Res 50:6075–6086
13. Zavadil J, Böttinger EP (2005) TGF-β and epithelial-to-mesenchymal transitions. Oncogene 24:5764–5774. https://doi.org/10.1038/sj.onc.1208927
14. Tran DD, Corsa CAS, Biswas H et al (2011) Temporal and spatial cooperation of Snail1 and Twist1 during epithelial-mesenchymal transition predicts for human breast cancer recurrence. Mol Cancer Res 9:1644–1657. https://doi.org/10.1158/1541-7786.MCR-11-0371
15. Kocanova S, Goiffon I, Bystricky K (2018) 3D FISH to analyse gene domain-specific chromatin re-modeling in human cancer cell lines. Methods 142:3–15. https://doi.org/10.1016/J.YMETH.2018.02.013
16. Bolland DJ, King MR, Reik W et al (2013) Robust 3D DNA FISH using directly labeled probes. J Vis Exp (78):e50587. https://doi.org/10.3791/50587
17. Bolte S, Cordelières FP (2006) A guided tour into subcellular colocalization analysis in light microscopy. J Microsc 224:213–232. https://doi.org/10.1111/j.1365-2818.2006.01706.x
18. Dempsey GT (2013) A user's guide to localization-based super-resolution fluorescence imaging. Methods Cell Biol 114:561–592. https://doi.org/10.1016/B978-0-12-407761-4.00024-5

Chapter 28

Mathematical Modeling of Plasticity and Heterogeneity in EMT

Shubham Tripathi, Jianhua Xing, Herbert Levine, and Mohit Kumar Jolly

Abstract

The epithelial-mesenchymal transition (EMT) and the corresponding reverse process, mesenchymal-epithelial transition (MET), are dynamic and reversible cellular programs orchestrated by many changes at both biochemical and morphological levels. A recent surge in identifying the molecular mechanisms underlying EMT/MET has led to the development of various mathematical models that have contributed to our improved understanding of dynamics at single-cell and population levels: (a) multi-stability—how many phenotypes can cells attain during an EMT/MET?, (b) reversibility/irreversibility—what time and/or concentration of an EMT inducer marks the "tipping point" when cells induced to undergo EMT cannot revert?, (c) symmetry in EMT/MET—do cells take the same path when reverting as they took during the induction of EMT?, and (d) non-cell autonomous mechanisms—how does a cell undergoing EMT alter the tendency of its neighbors to undergo EMT? These dynamical traits may facilitate a heterogenous response within a cell population undergoing EMT/MET. Here, we present a few examples of designing different mathematical models that can contribute to decoding EMT/MET dynamics.

Key words Mathematical modeling, Epithelial-mesenchymal plasticity, Nongenetic heterogeneity, Multi-stability, Epithelial-mesenchymal heterogeneity

1 Introduction

The epithelial-mesenchymal transition (EMT) is a cellular process involving changes in multiple aspects of cellular behavior, including cell–cell adhesion, cell polarity, cell migration and invasion, and cell shape [1]. EMT and the corresponding reverse process, mesenchymal-epithelial transition (MET), are regulated at multiple levels. These include transcriptional, posttranscriptional, translational, and epigenetic [2] controls, along with non-cell autonomous mechanisms acting through matrix density [3] or cell–cell communication [4–7]. Largely thought of in the past as a binary process, EMT is now known to involve multiple stable intermediates referred to as hybrid epithelial/mesenchymal (hybrid E/M) phenotypes [8]. This updated view of the process has, in part, been

Kyra Campbell and Eric Theveneau (eds.), *The Epithelial-to Mesenchymal Transition: Methods and Protocols*, Methods in Molecular Biology, vol. 2179, https://doi.org/10.1007/978-1-0716-0779-4_28,

driven by predictions made by various mathematical models for the regulatory networks involved in EMT [4, 5, 9–14]. These mathematical models have focused on characterizing the properties of EMT and have predicted that cells can stably maintain one or more hybrid E/M phenotypes [15]. Moreover, these models have also driven insights into how cells may spontaneously switch between various phenotypes due to stochasticity, and thereby determine how cellular plasticity leads to phenotypic heterogeneity associated with EMT as observed experimentally [16, 17]. These models have also offered mechanistic insights into experimental observations showing that EMT and MET are not necessarily symmetric processes [12, 18], i.e., cells may take different paths during EMT and MET in the multi-dimensional landscape of epithelial-mesenchymal plasticity. Finally, these models have helped us gain insights into the interconnection between EMT and other cellular traits such as stemness; for instance, the prediction that a hybrid E/M phenotype is more stem-like and metastatically aggressive than cells exhibiting extremely epithelial or extremely mesenchymal phenotypes [19] was recently confirmed both in vitro and in vivo [20–22]. Here, we introduce a generic framework for developing mathematical models of EMT regulation and share examples of how these models can be used as tools to generate predictions that will guide the next set of experiments.

2 Mathematical Modeling of EMT

The choice of a systems biology approach to study a biological process is highly context-dependent. We here describe a generic procedure for choosing an appropriate approach and detail how this procedure was applied to modeling EMT.

2.1 Identify a Problem that Mathematical Modeling Can Help Address and Form a Team of Experimental and Modeling Researchers

This is a key and probably the most challenging step in modeling studies. There are questions that modeling studies can address and others that they cannot address. It is typically constructive to form a team of experimental and modeling researchers. The team members hold thorough literature review and extensive, in-depth discussions to review existing knowledge and identify open questions regarding the system. One may find it pedagogically illuminating to read accounts of how some successful collaborations were established [23, 24].

2.2 Choose an Appropriate Modeling Framework

Several modeling frameworks have been used to analyze EMT regulatory networks. A Boolean network has dynamics that are discrete in time and involve discrete variable values. The variable values are updated based on a set of Boolean functions that reflect the regulatory relations [25]. Conversely, an ordinary differential equation (ODE)-based model treats time and variables as taking continuous values. Both Boolean network and ODE-based models can be deterministic (meaning that one can precisely predict the temporal evolution of the variables from a set of initial conditions), or stochastic (meaning that the prediction is only probabilistic). There is no best modeling framework for all cases, and one needs to determine what is appropriate and justified for the biological system and process under study. Some general aspects that may be considered include:

1. What is the qualitative and quantitative information available? Compared to a Boolean model, an ODE model typically has more parameters and requires more quantitative data to constrain these model parameters. Therefore, for a large regulatory network without much quantitative data such as the one studied by Steinway et al. [11], the Boolean framework is appropriate. It would be questionable whether an alternative ODE-based model with dozens or even hundreds of free parameters can provide further additional information (but see systematic statistical analyses of model ensembles discussed below).
2. Is the framework sufficient to describe the system dynamics, and does it provide new mechanistic insights that would be unavailable or unclear without the modeling approach? Each framework has its limitations. For example, a Boolean model typically uses some universal parameters and only provides qualitative or at most semiquantitative information. It can be a good starting point to analyze how multiple regulatory factors interact to generate different EMT cell types as demonstrated by Steinway et al. [11]. The model has limited capacity to describe how different time scales of the signal transduction pathways involved in EMT contribute to quantitative detection and encoding of the dose and duration information of the stimulating signals. For the latter purpose, an ODE-based model is a more appropriate choice, as demonstrated by Zhang et al. [26] to show how pathway cross talk leads to a temporal checkpoint mechanism for detecting TGF-β duration information.

As a rule of thumb, one chooses a modeling framework that is simple and sufficient to address the underlying problem. The widely regarded criterion suggested by Einstein for evaluating physics theories also applies here: "Everything should be made as simple as possible, but not simpler." It is possible that for a given problem, initially a coarse-grained framework is appropriate, and as more and more quantitative data becomes available, a different framework becomes necessary to incorporate the new information.

Unfortunately, a commonly held misconception emphasizes that it is always desirable to incorporate additional biological details explicitly into a mathematical model, and this tendency is further reinforced by the expanding computational power. However, abstraction is necessary and is done in all modeling efforts. We want to stress here that the most important reason for using modeling approaches is to provide mechanistic insight buried in the data, and not just to crank machines and obtain some numbers. For this purpose, it is both productive and necessary to perform proper abstraction and idealization as successfully used in theoretical physics [27]. A simple model that only makes qualitative predictions but provides deep mechanistic insight has more value than a complex model that can only "reproduce" experimental data but does not necessarily make a new set of predictions that may be tested experimentally to improve our understanding of the system. To be fair, both detailed and simplified approaches have their merits, and sometimes it is constructive to combine the two strategies https://www.nature.com/articles/s41540-020-0132-1. One may start with detailed models that can reproduce the data, then remove model ingredients step-by-step to identify the minimal components that are essential for recapitulating the key dynamical features of the system.

2.3 Construct a Mathematical Model and Perform Analysis

With the problem identified and an appropriate modeling framework selected, one can follow some generic modeling procedures:

1. Summarize known interacting species into a regulatory network. If there are uncertain interactions, one may construct a set of possible networks for later comparative studies. Figure 1 shows a core EMT regulatory network used in several studies [9, 13, 28].
2. Set up mathematical equations based on the biology. This step is nothing more than translating the relevant biological information into mathematical forms. For example, the equation below governs the temporal evolution of the total level of SNAIL1 mRNA (*[snail1]$_t$*), which is summed over both free (*(snail1)*) and miR-34 bound (*[snail1]$_t$* – *(snail1)*) mRNAs [28].

$$\frac{d[\text{snail1}]_t}{dt} = \underbrace{k_0}_{\text{basal expression}} + k\underbrace{\frac{([\text{TGF}]_t/K_1)^2}{1+([\text{TGF}]_t/K_1)^2}}_{\text{TGF-}\beta\text{ activation}}\underbrace{\frac{1}{1+[\text{SNAIL1}]/K_2}}_{\text{SNAIL1 self-inhibition}} - \underbrace{k_{d0}[\text{snail1}]}_{\text{snail1 basal degradation}} - \underbrace{k_d([\text{snail1}]_t - [\text{snail1}])}_{\text{miR}-34\text{ regulated snail1 degradation}}$$

   Each term on the right-hand side of the above equation corresponds to one of the SNAIL1 related links in Fig. 1.

3. Constrain model parameters using the available quantitative data. Several parameter estimation algorithms are available, from linear regression to the more sophisticated maximum

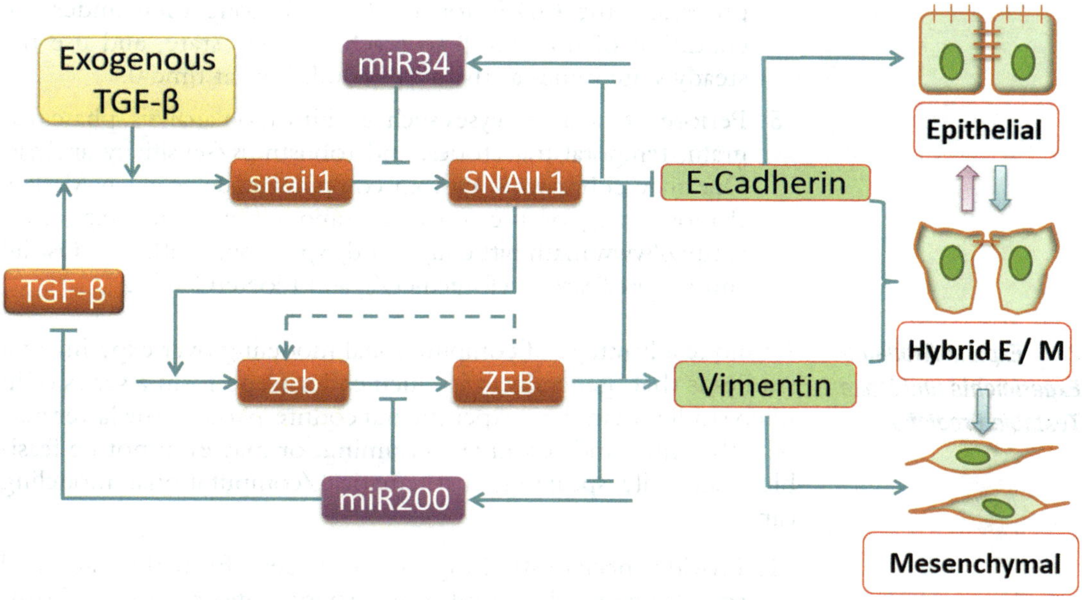

Fig. 1 Core EMT regulatory network that leads to epithelial, hybrid E/M, and mesenchymal phenotypes (adapted from [13]). Pointed arrows represent activation, blunt-end arrows represent inhibition, and the dashed lines represent links first proposed in the modeling study by Lu et al. [9]

likelihood estimation, and Markov chain Monte Carlo methods. Since, in practice, it is rare to have sufficient data for a specific system under study, a commonly adopted practice is to estimate the many parameters based on data from different labs, different cell lines, or cells from different tissues. However, even results from the same cell line can be quantitatively different due to factors such as differences in cell generation, reagent vendors, or even batches. Besides, dynamical parameters such as mRNA turnover rates can differ by orders of magnitude for cells under different conditions. An emerging trend is to collect data from one lab or under the same experimental settings [29], similar to what has been adopted in some large consortiums like ENCODE. Furthermore, instead of using only the best-fit parameter set, one may use an ensemble of model parameters to make model predictions. Zhang et al. [26] adopted such an integrated modeling-quantitative measurement procedure and an ensemble-based approach has been developed previously [30, 31]. Another model ensemble method is discussed in the next section.

4. Specify initial conditions (e.g., initial concentrations of various species) that reflect the experimental setup. For example, if one models cell response after adding TGF-β at time 0, one may first make a rough estimation of the initial concentrations, then

propagate the ODEs for a sufficiently long time under the condition of no TGF-β to reach a steady state, and use the steady state values as the initial conditions at time 0.

5. Perform standard analyses such as bifurcation analysis, phase diagram, temporal trajectories, and robustness/sensitivity analysis. One may either write custom computer code (e.g., in Matlab, Python, etc.), or use available computer packages, e.g., XPP (http://www.math.pitt.edu/~bard/xpp/xpp.html), Oscill8 (http://oscill8.sourceforge.net/), and BioNetGen [32].

2.4 Explain Available Experiments and Make Testable Predictions

A unique advantage of computational modeling over experimental studies is that, generally, it is much easier to perform a series of in silico studies than their experimental counterparts, as the latter may be either time and resource consuming, or may even not be feasible. Generally speaking, mathematical/computational modeling can:

1. Provide mechanistic insights not evident from the data, and sometimes resolve conflicting experimental results or distinguish competing mechanisms. For a given system, data are typically collected from different sources and using different techniques. Each experimental technique or approach can only reveal partial information about the system, and modeling integrates the discrete information. By placing all the experimental results on a common ground, a modeling study allows one to check whether the data are consistent mutually, and with the conceived mechanisms.
2. Make predictions leading to new experimental measurements that might not have been considered otherwise. For example, the modeling study by Tian et al. [13] inspired a subsequent measurement of single cell SNAIL1 expression levels using flow cytometry [28].
3. Identify essential ingredients or missing links necessary to explain the observations. For a given system, there may be too much information, and some of it may not be or may only be marginally relevant to addressing a specific question. By adding or removing individual components and examining the effect on model behavior, one can identify the essential ingredients of a model. In other cases, the available information may be insufficient. In such a scenario, following a similar procedure of systemically adding individual components, one can predict the missing component(s) that are necessary to explain the experimental results. The missing component may then be identified in subsequent experimental studies. For example, the study by Lu et al. [9] suggested the existence of positive feedback in the regulation of ZEB in EMT regulation (dashed line in Fig. 1).

It is important to point out that a model need not necessarily be right in order to be useful. In fact, every model is only an approximation and abstraction of the biological system under study and will be replaced by better approximations when additional information becomes available. "All models are wrong, but some are useful" [33]. Even an eventually falsified model may suggest useful experimental studies that would otherwise not have been performed, and thus help in advancing our knowledge of a biological system. Such models should receive deserved credit.

2.5 Perform Corresponding Experimental Studies

As Katchalsky pointed out [24], "Theory tells us what cannot happen, and it can tell us what could happen. But only experiments tell us what does happen." All model predictions need to be subject to subsequent experimental tests.

2.6 Go Back to Step 2 and Iterate; Expansion of Model (Even After Publishing the Original Work)

It has become more and more common to see studies that have iterations between modeling and experiments. Sometimes the integrated experiment-modeling process may even lead to revisiting step 1 to define new questions and seek expanded collaborations. For example, early modeling studies [9, 13] on EMT focused on the core regulatory network (Fig. 1). Several subsequent studies expanded the network to explore how additional factors contribute to the spectrum of EMT phenotypes [4, 34, 35].

3 Modeling Population Heterogeneity in EMT

Intra-tumoral heterogeneity, wherein cancer cells within the same tumor exhibit different phenotypes, has been reported across multiple cancer types, both in vitro and in vivo [36]. Tumor cell populations in different cancer types including leukemia [37], breast cancer [38], colorectal cancer [39, 40], brain cancer [41], and prostate cancer [42] can consist of subpopulations of cells that exhibit stem-cell-like behavior. Cells in triple-negative breast cancer can exhibit distinct phenotypes including luminal, basal, immunomodulatory, mesenchymal, and stem-like [43]. In small cell lung cancer, tumor cells can exhibit both neuroendocrine and non-neuroendocrine phenotypes [44]. Intra-tumoral heterogeneity has recently been identified as a principal cause for the failure of anticancer therapies [45]. Therefore, characterization of the mechanisms driving this feature of tumor cell populations is key to advancing anticancer therapeutics. In many (perhaps most) cases, genetic heterogeneity does not underlie phenotypic heterogeneity, i.e., tumor cells exhibit different phenotypes in spite of carrying the same genetic alterations. This indicates that nongenetic mechanisms may be the chief driver of intra-tumoral heterogeneity.

Cells within the same tumor can exhibit different EMT-associated phenotypes—an epithelial phenotype, a mesenchymal phenotype, and one or more hybrid E/M phenotypes. This is a canonical example of nongenetic intra-tumoral heterogeneity observed across cancer types including in breast cancer [46], melanoma [47], colorectal cancer [48], and in prostate cancer [49]. Different EMT-associated phenotypes exhibit varying tumor-initiating capabilities [6, 7] and sensitivities to anticancer drugs [50, 51]. How does such epithelial-mesenchymal heterogeneity emerge in a population of cancer cells? How is this heterogeneity maintained and propagated across generations and passages? These are key questions that must be answered if we are to be able to attenuate the role of epithelial-mesenchymal heterogeneity in driving the failure of anticancer therapies.

Multiple nongenetic mechanisms can contribute towards the emergence of phenotypic heterogeneity. The regulatory circuits that govern the phenotypes of different cells often respond differently to the same external cues leading to a phenotypically heterogeneous population. Phenotypes of cells in a population can change stochastically due to the noisy transcription of genes [52] or due to the random partitioning of the parent cell molecules among the daughter cells during cell division [53, 54]. Finally, cell–cell communication can cause cells in a population to acquire distinct phenotypes in a non-cell autonomous manner. Each of these three mechanisms has been implicated in the emergence and maintenance of epithelial-mesenchymal heterogeneity. Mathematical and computational modeling approaches have played a key role in determining how these mechanisms can drive epithelial-mesenchymal heterogeneity in populations of cancer cells. Here, we describe mathematical modeling approaches corresponding to each of the three mechanisms.

4 Heterogeneity from Cell-to-Cell Variation in Regulatory Kinetics

Large and complex gene regulatory networks underlie different cellular functions such as stem cell differentiation [55, 56] and circadian rhythm [57, 58]. The dynamical behavior of such large networks can be understood as being driven by a core regulatory circuit with the remaining genes in the circuit being peripheral to circuit dynamics, acting only to alter the signaling status of the core regulatory circuit [59]. The effects of peripheral genes and exogenous signaling can then be modeled as perturbations to the kinetic parameters governing the dynamics of the core regulatory module. This is the approach underlying the framework known as random circuit perturbation or RACIPE [60]. Here, we describe how to use RACIPE for modeling epithelial-mesenchymal heterogeneity.

While multiple signaling pathways have been implicated in controlling EMT and MET, the activities of many of these pathways converge onto a small set of core regulatory players. This set includes the master regulators such as SNAI1, miR-34, miR-200, and ZEB1 [8, 61]. The effects of different signals modulating EMT and MET can thus be modeled as perturbations to the kinetics of this smaller core regulatory circuit. These perturbations can vary from cell-to-cell, thus representing the differing internal and external signaling states of tumor cells in a population.

We first describe the RACIPE framework using the simple toggle switch as an example. As shown in Fig. 2, the toggle switch consists of two transcription factors, A and B, which form a mutual inhibitory feedback loop. The dynamics of this circuit can be described using a pair of ODEs:

$$\frac{d[A]}{dt} = g_A H^S([B], K_B^A, n_B^A, \lambda_B^A) - k_A[A] \quad (1)$$

$$\frac{d[B]}{dt} = g_B H^S([A], K_A^B, n_A^B, \lambda_A^B) - k_B[B] \quad (2)$$

Here, $[A]$ and $[B]$ are the protein expression levels of genes A and B, respectively. g_A and g_B are the production rates of A and B when no activator or inhibitor is present. k_A and k_B are the inherent degradation rates of the two proteins. The regulatory action of gene B on gene A is modeled via the shifted Hill function:

$$H^S([B], K_B^A, n_B^A, \lambda_B^A) = \lambda_B^A + (1 - \lambda_B^A) H^-([B], K_B^A, n_B^A) \quad (3)$$

$$H^-([B], K_B^A, n_B^A) = \frac{1}{1 + \left(\frac{[B]}{K_B^A}\right)^{n_B^A}} \quad (4)$$

K_B^A is the threshold concentration of B, n_B^A is the Hill coefficient, and λ_B^A is the maximum fold change in the expression level of A that can be caused by the activity of B. If B activates A, $\lambda_B^A > 1$. If B inhibits A, $0 \leq \lambda_B^A < 1$. For the toggle switch, $0 \leq \lambda_B^A, \lambda_A^B < 1$. Thus, there are five types of kinetic parameters in the model. Two of them, g and k, are associated with each gene. The remaining three, K, n, and λ are associated with each regulatory link. Thus, for a circuit with 10 genes and 25 regulatory interactions, the total number of parameters will be $(2 \times 10) + (3 \times 25) = 95$.

RACIPE performs randomization on all five types of circuit parameters to obtain an ensemble of kinetic models for a given circuit topology. The randomization procedure is such that most biologically realizable possibilities are represented by one of the models in the ensemble. RACIPE uses two assumptions to obtain a representative ensemble of models. First, the maximum production rate of each gene is fixed, independent of the number and type of interactions that gene is a target of. For a gene with one activator,

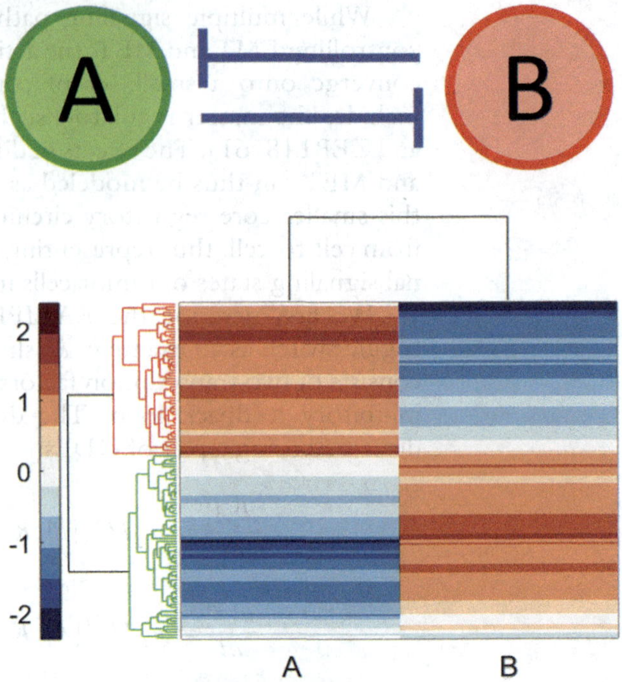

Fig. 2 Transcription factors *A* and *B* with a mutual inhibitory feedback loop (top). RACIPE was used to generate 100 kinetic models corresponding to this topology. A total of 122 distinct steady states were obtained—78 kinetic models exhibited only one steady state while 22 kinetic models exhibited two steady states. Hierarchical clustering of this collection of steady states (bottom) revealed that these steady states can be divided into two phenotypic classes: high A, low B (highlighted in red) and low A, high B (highlighted in green). Thus, in a population wherein each cell carries a copy of this circuit, cells can exhibit two distinct phenotypic states. Hierarchical clustering was carried out using the *Z*-scores of the log2 transformed expression levels

the maximum production rate, *G*, will be obtained when the activator is highly expressed. Thus, the basal production rate of the gene must be $g = \frac{G}{\lambda}, \lambda > 1$. For a gene with only one inhibitor, the maximum production rate will be obtained in the absence of inhibitor expression. Thus, $G = g$ where g is the basal production rate. This approach can easily be generalized to the case when a gene has multiple activators and inhibitors [60]. RACIPE randomizes the maximum production rate (G) and then calculates g using the above-mentioned approach.

The second assumption is that in order for the ensemble of models to be representative of most biological possibilities, each regulatory link in the circuit must have an almost equal chance of being functional and being nonfunctional. To ensure this, RACIPE chooses the threshold parameters in such a manner that the steady state concentration of the corresponding regulator in different models within the ensemble is roughly equally likely to be above the threshold parameter (in which case the interaction is functional) and below the threshold parameter (in which case the interaction is nonfunctional). For a detailed description of how this is achieved, see Huang et al. [60].

In the ensemble generated by RACIPE, all models have the same topology but differ in the values of kinetic parameters governing the model dynamics. The dynamics of each model is then numerically simulated multiple times, each time starting with a different set of initial concentrations of the molecules in the circuit. This allows RACIPE to obtain a set of steady states that a given model can generate. Once this has been done for each model in the ensemble, RACIPE obtains a collection of steady states that the given circuit topology can exhibit. Each model in the ensemble generated by RACIPE may be interpreted as representing a single cell. Thus, the collection of steady states obtained by RACIPE will represent an in silico gene expression profile obtained for a population of cells. One aspect that should be kept in mind is that a model that can exhibit more than one steady states will be counted more often in the collection of steady states generated by RACIPE as compared to a model that can exhibit only one steady state. Nevertheless, this steady state expression data can be analyzed using familiar methodologies including principal component analysis and hierarchical clustering to gain insight into the different classes of steady states that may be exhibited by a given network topology.

The C language computer code implementing the RACIPE framework is available online on GitHub (https://github.com/simonhb1990/RACIPE-1.0). Once the code has been downloaded, change to the folder or directory where the code files are present and use the *make* command to compile the code files for your system. This will generate a single executable named "RACIPE." This executable takes as input a topology file, extension .topo, which describes the topology of the circuit being analyzed. This must be a plain text file with three tab separated columns. The first column ("Source") contains the name of the regulator gene. The second column ("Target") contains the name of the gene being regulated. The third and final column ("Type") describes the interaction type, 1 if the interaction is activating and 2 if the interaction in inhibiting. A sample topology file (TS.topo) is available online with the code. Once the topology file for the circuit of interest has been generated, the RACIPE code can be run as follows:

```
$ ./RACIPE network.topo
```

Additional input options that may be provided to the code are described in the "README.md" file available with the code. Upon execution, the code generates multiple files. Most important among these are:

1. **Parameter ranges file (.prs extension)** This file contains the ranges of different kinetic parameters.
2. **Parameters file (_parameter.dat extension)** This file contains the kinetic parameters for each model in the ensemble along with the number of steady states obtained for that model.
3. **Solutions files** These files contains the gene expression levels in each of the steady states obtained for different models. Steady state expression levels for models exhibiting different numbers of steady states are stored in different files. For models with only steady state, the file extension is "_solution_1.dat." For models with three steady states, the file extension is "_solution_3.dat." All gene expression values reported in these files are log2 normalized.

Descriptions of other output files can be obtained from the "README.md" file available online with the code.

To determine if epithelial-mesenchymal heterogeneity can emerge from cell-to-cell variation in kinetic parameters as simulated using the RACIPE framework, we used a 26-node circuit (Fig. 3; top panel) which was constructed using Ingenuity Pathway Analysis (IPA; QIAGEN Inc.) and literature search [62]. The circuit consists of 17 protein-coding genes and 9 micro-RNAs. The set of protein-coding genes includes transcription factors such as SNAI1, ZEB1, and TWIST1 whose role as master regulators of EMT is well characterized [8]. The set also includes EMT-associated biomarkers such as CDH1 and VIM along with "phenotypic stability factors" [34] such as GRHL2, OVOL2, and $\Delta NP63\alpha$. The collection of steady states that can be exhibited by models with the topology of this EMT circuit was obtained using RACIPE and analyzed using hierarchical clustering (Fig. 3; bottom panel). As mentioned previously, this collection of steady states is representative of the gene expression profile of cells in a tumor. The steady states can be broadly classified into four groups on the basis of expression levels of the 26 proteins and micro-RNAs in the EMT circuit. Group 1 exhibits high levels of expression of epithelial phenotype-associated genes including CDH1 along with high levels expression of EMT inhibitors such as GRHL2 and miR-200. This group thus represents cells that exhibit an epithelial phenotype. In group 4, EMT drivers such as SNAI1 and ZEB1 are highly expressed along with high expression of the mesenchymal marker VIM. This group represents cells that exhibit a mesenchymal phenotype. Groups 2 and 3 consist of steady states with co-expression of both epithelial and mesenchymal-associated factors. The expression of epithelial factors in these groups is lower than the expression of these factors in the epithelial group (group 1) and the expression of mesenchymal factors is lower than that in the mesenchymal group

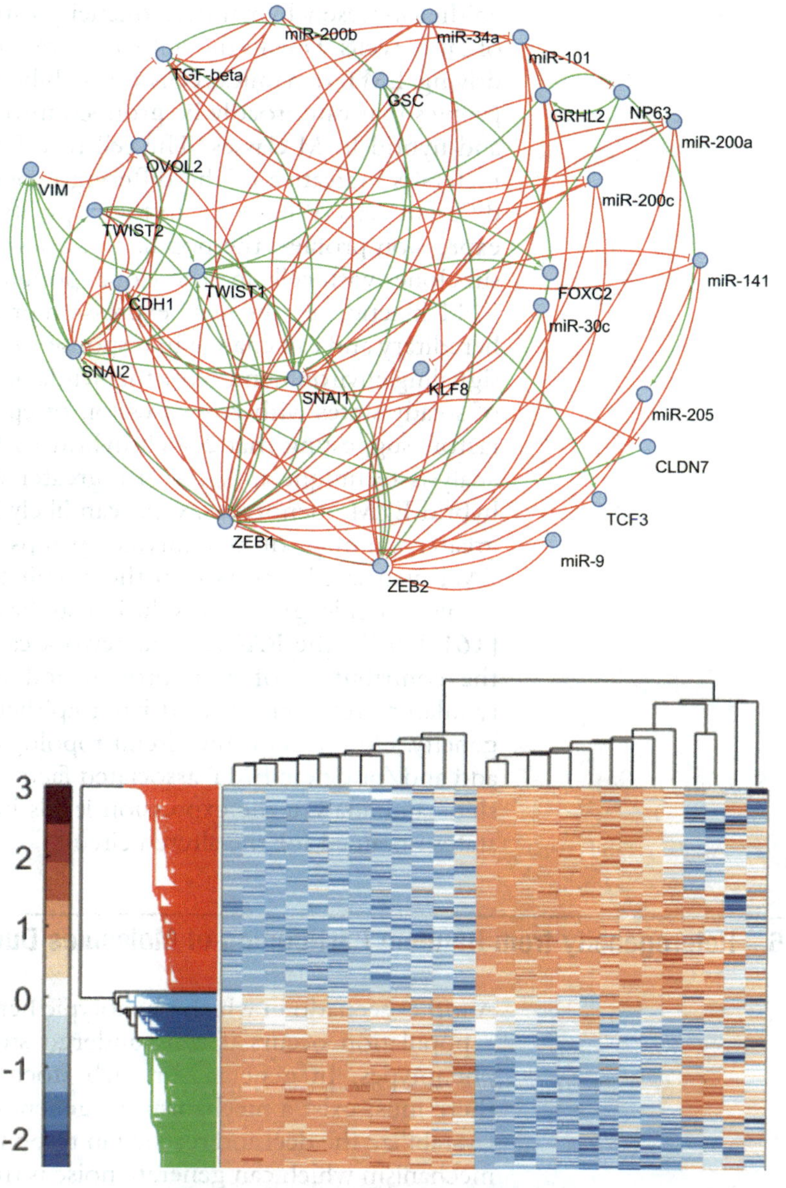

Fig. 3 The 26-node EMT topology (top). RACIPE was used to generate 5000 kinetic models corresponding to this topology. A total of 13,486 steady states were obtained via numerical integration of the ODEs in these kinetic models. Using hierarchical clustering, these steady states were grouped into four phenotypic classes (bottom)—epithelial (red), mesenchymal (green), and two hybrid E/M phenotypic classes (light blue and dark blue). Hierarchical clustering was carried out using the Z-scores of the log2 transformed expression levels

(group 4). Groups 2 and 3 thus co-express both epithelial and mesenchymal factors at intermediate levels.

Thus, analysis of a 26-node EMT circuit using the RACIPE framework demonstrates one mechanism by which epithelial-mesenchymal heterogeneity can emerge in a population of cancer cells. Due to the cell-to-cell variation of kinetic parameters driving EMT dynamics, cells can exhibit distinct gene expression profiles that can broadly be grouped into epithelial, mesenchymal, and hybrid E/M classes. The cell-to-cell variation in kinetic parameters is indicative of the differing exogenous signaling states in different cells. While cells in the population exhibit different gene expression profiles, the population does not consist of clones and subclones with cells in each clonal population exhibiting a specific EMT kinetic, i.e., the gene expression profile of a cell is not always hereditary and can change in response to changes in the exogenous signaling environment. Note that while our analysis reveals 2 groups of steady states with co-expression of epithelial and mesenchymal factors suggesting that 2 such hybrid E/M states exist, a different analysis technique may reveal a greater number distinct types of hybrid E/M phenotypes. Cells can likely be classified into an even greater number of phenotypic groups by incorporating other EMT-associated factors into the circuit topology [10, 11] which would provide greater resolution as has been reported recently [16]. Finally, the RACIPE framework can easily be used to probe the contribution of each protein and micro-RNA and of each regulatory relationship in driving epithelial-mesenchymal heterogeneity. One can edit the circuit topology file (extension .topo) to add and/or delete EMT-associated factors and regulatory relationships and analyze the expression levels in the collection of steady states obtained for the altered circuit.

5 Heterogeneity from Random Partitioning of Molecules During Cell Division

Another scenario in which phenotypic heterogeneity can emerge in a population occurs if cells undergo stochastic changes in their phenotypes. In general, for such stochastic changes to happen, there must exist a mechanism to generate noise and a mechanism to stabilize the decision reached in response to the noise [63]. One mechanism which can generate noise is the random partitioning of molecules (RNAs, proteins, etc.) in the parent cell among the daughter cells at the time of cell division [53, 54, 64]. This mechanism is likely to be a prominent source of noise in tumors wherein cells divide fast and uncontrollably. While phenotypic fluctuations in cells in response to noise are usually small and transient, the fluctuations can be amplified if the underlying response mechanism exhibits multi-stability, i.e., co-existence of multiple steady states. As described previously [9, 13], circuits which drive EMT and MET exhibit multi-stable behavior. Thus, random partitioning of EMT-associated factors during cancer cell division is likely to be a key contributor toward the emergence of epithelial-mesenchymal heterogeneity.

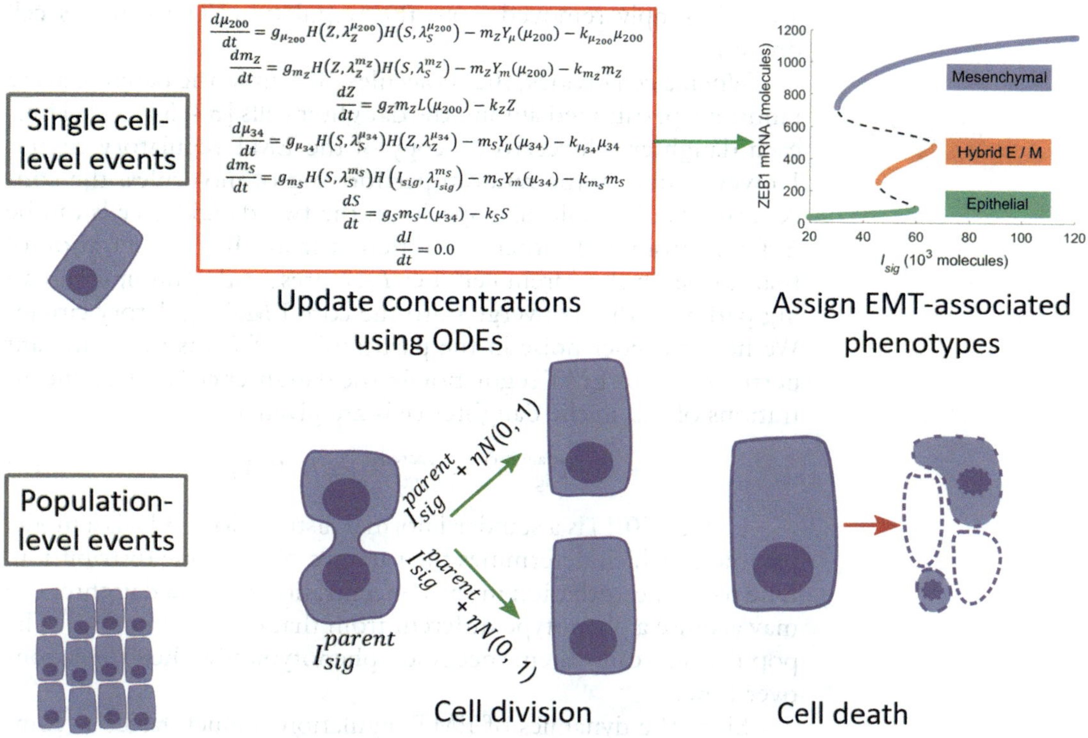

Fig. 4 A schematic representation of the model to investigate how epithelial-mesenchymal heterogeneity can arise from the random partitioning of proteins and RNAs during cell division. (Figure adapted from Tripathi et al. [65])

The schematic representation of a computational model that can be used to probe the role of this mechanism in the emergence of epithelial-mesenchymal heterogeneity is shown in Fig. 4. The model [65] builds upon the dynamics of the core regulatory circuit involving SNAIL, ZEB, miR-34a, and miR-200. These transcription factors and micro-RNAs together form a circuit that acts as a ternary switch, responding to the signaling pathways driving EMT and MET [9]. Stable steady states of this circuit can be mapped to different EMT-associated phenotypes—epithelial, mesenchymal, and hybrid E/M—on the basis of expression levels of ZEB (Fig. 4). To see the effect of random partitioning on the phenotypic composition of the population, we here consider a population of cancer cells with each cell carrying a copy of this EMT regulatory circuit. Since this regulatory circuit does not involve cell–cell communication, the dynamics of the regulatory circuit within each cell in the population can be simulated independent of other cells in the population. The dynamics of EMT regulation at the single-cell level are simulated using ordinary differential equations which have been described previously [9]. At the population level, there are two types of events that can take place. One is cell death during which a cell is simply removed from the population. The other is cell division.

When a cell divides, the molecules present in the parent cell are randomly partitioned among the daughter cells [53, 54, 64]. Thus, each daughter cell receives a copy of the EMT regulatory circuit. However, due to the random partitioning of molecules, the concentrations of a molecular species in the two daughter cells can be different from each other and different from the concentration of that species in the parent cell. Let I_{sig} represent the multiple signaling pathways that converge onto the core EMT regulatory circuit. We here consider noise in the partitioning of I_{sig} as the dominant perturbation to EMT regulation in the daughter cells. The concentrations of I_{sigs} in the daughter cells are given as:

$$I_{\mathrm{sig}}^{\mathrm{daughter}} = I_{\mathrm{sig}}^{\mathrm{parent}} + \eta N(0,1) \tag{5}$$

Here, $N(0,1)$ is a standard normal distribution and η is a model parameter which determines the variance of the noise distribution. Due to the perturbation in the concentration of I_{sig}, a daughter cell may acquire a phenotype different from that of the parent cell. The population can then become phenotypically heterogeneous over time.

Since the dynamics of EMT regulation is much faster as compared to the time scale at which cell division and cell death events take place, the model dynamics can be simulated in a multi-scale manner. Population-level dynamic, i.e., cell division and cell death, are simulated in a stochastic manner using Gillespie's algorithm [66]. Between two population-level events, the concentrations of RNAs and transcription factors within each cell are updated using ordinary differential equations. Previous studies have shown that different EMT-associated phenotypes can exhibit different rates of cell division [67–69]. However, one may consider a simpler case with equal division and death rates for all three cell types. In addition, to incorporate the effect of limited availability of nutrients in the tumor microenvironment, a logistic model of growth with a fixed carrying capacity can be used.

Dynamics of the model can be simulated as follows:

1. Choose an initial population size and randomly assign concentrations of molecules in the EMT regulatory circuit to different cells in the population. The concentrations are drawn from log-normal distributions such that the median concentration of each molecular species is within the range for which the regulatory circuit exhibits multi-stable dynamics.
2. Using Gillespie's algorithm [66], update the number of cells in the population. In the case of a cell death event, that cell is removed from the simulation and thus the population. In the

case of a cell division event, I_{sig} concentrations in the daughter cells are updated using Eq. 5.

3. At the end of the Gillespie update, the concentrations of molecules in each cell in the population are updated. Let Δt be the time interval between the last Gillespie update and the current one. Then, the concentrations of molecules can be updated by integrating the ordinary differential equations for the EMT regulatory circuit [9] over the time period Δt.

Computer code for simulating the model dynamics can be downloaded from GitHub (https://github.com/st35/cancer-EMT-heterogeneity-noise).

We simulated the model dynamics for populations with different initial phenotypic compositions. Figure 5 shows how epithelial-mesenchymal heterogeneity can emerge in a phenotypically homogeneous population over a period of 2 weeks. While epithelial and mesenchymal populations exhibit fairly stable phenotypic compositions, a hybrid E/M population can quickly give rise to a mixed population with both epithelial and mesenchymal cells. Such behavior has been confirmed in populations of mouse prostate cancer cells [17] and a comparison of experimental dynamics with the predictions from the model is shown in Fig. 5 (bottom panel).

The model thus shows that random partitioning of parent cell proteins and RNAs among the daughter cells can generate epithelial-mesenchymal heterogeneity in a population of cancer cells. Arising from cell division, this heterogeneity can emerge and be propagated from a small population, such as the one left after an anticancer regime. Note that the model proposed here is not sensitive to the choice of the core EMT/MET regulatory circuit. Any circuit topology can be used within the framework of this model as long as the circuit dynamics is multi-stable which is a key feature of EMT regulation.

6 Heterogeneity from Cell–Cell Communication Via Notch Signaling

In addition to the regulatory mechanism at the single-cell level, cell–cell communication also plays a major role in modulating EMT [6, 7]. Notch signaling [70, 71] is one such mechanism which operates via the binding of Notch, a transmembrane receptor, to a ligand expressed on the surface of a neighboring cell. This binding event triggers the cleavage of the Notch intracellular domain (NICD). NICD is then released into the cytoplasm where it can act as a transcriptional cofactor thereby promoting or inhibiting the expression of certain genes [70]. Notch signaling between neighboring cells can create varied spatial patterns in a population. The pattern type depends on the type of Notch ligands that are active in the population. NICD inhibits the expression of Delta ligands and

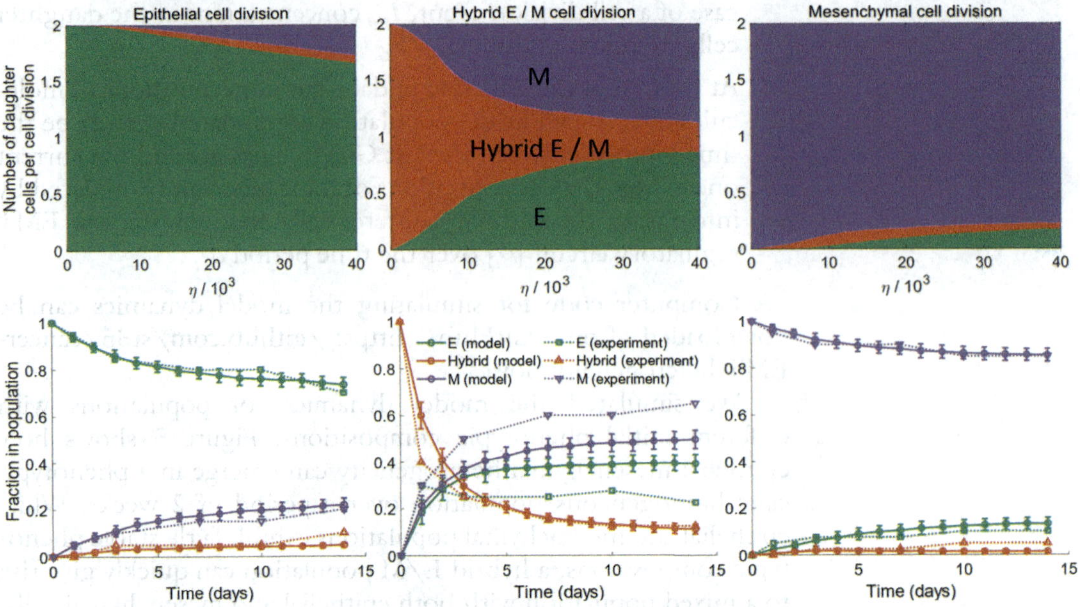

Fig. 5 (Top) Average number of epithelial, mesenchymal, and hybrid E/M daughter cells generated during the division of an epithelial cell (left), a hybrid E/M cell (middle), or a mesenchymal cell (right). Daughter cells can exhibit a phenotype distinct from that of the parent cell due to the random partitioning of I_{sig} during cell division. (Bottom) Change in the fraction of different phenotypes in a population of cancer cells when starting with a purely epithelial (left), a purely hybrid E/M (middle), or a purely mesenchymal population on day 0. Solid lines indicate the predictions from the proposed model. Dotted lines indicate the behavior for a population of mouse prostate cancer cells re-plotted from Ruscetti et al. [17]. (Figure adapted from Tripathi et al. [65])

promotes the expression of ligands of the Jagged family. Notch-Delta signaling leads to neighboring cells acquiring distinct phenotypes—the cell expressing high levels of the Notch receptor and low levels of Delta ligands acts as the "sender" cell while the neighboring cell with low levels of Notch expression and high expression levels of Delta ligands acts as the "receiver" cell [72] ("lateral inhibition"; Fig. 6 (top panel)). Notch-Jagged signaling, on the other hand, leads to neighboring cells acquiring the same phenotype which is characterized by the co-expression of Notch receptors and Jagged ligands [73] ("lateral induction"; Fig. 6 (bottom panel)).

The role of Notch signaling in EMT regulation arises from the coupling between the Notch signaling machinery and the core regulatory circuit that drives EMT (Fig. 7; top panel). miR-34 can posttranscriptionally inhibit the expression of Notch receptors and that of Delta ligands. miR-200 similarly inhibits the expression of Jagged ligands. Further, NICD promotes the expression of SNAIL, thereby acting as an EMT promoter [4]. Due to the cross talk between the Notch signaling and EMT circuits, the spatial patterns that emerge from Notch signaling translate into spatial patterning in the expression of epithelial and mesenchymal markers

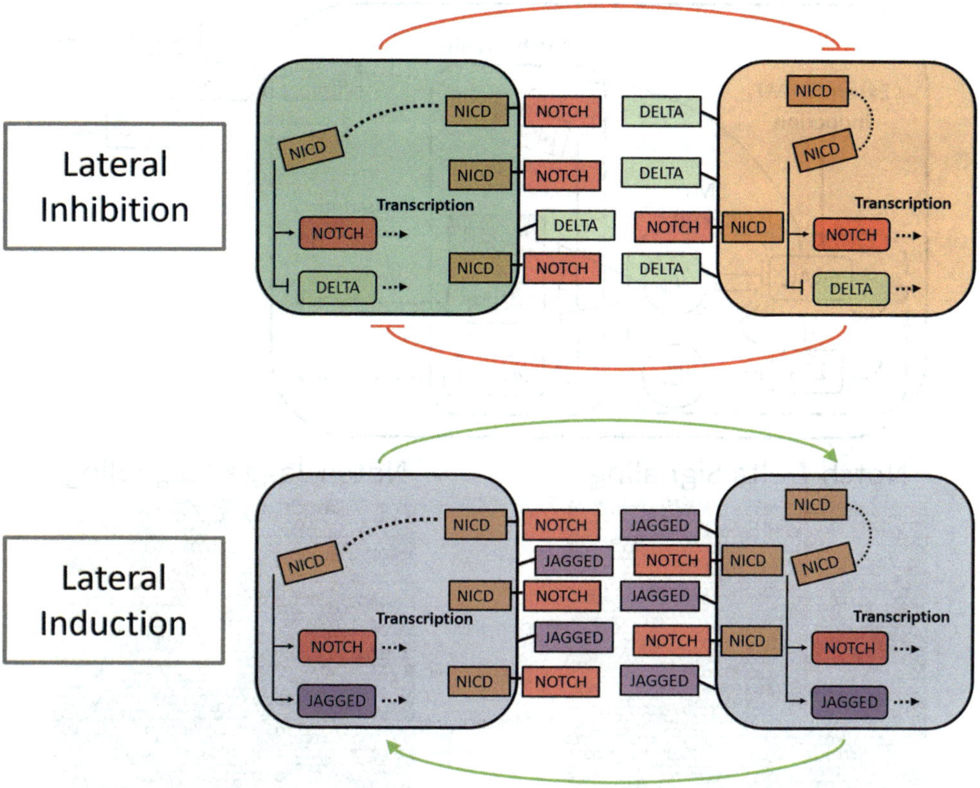

Fig. 6 Two types of Notch signaling-mediated coupling between neighboring cells. In the presence of Notch-Delta signaling (top), neighboring cells form a mutual inhibitory feedback loop causing them to exhibit distinct phenotypes. One of the cells acts as the receiver (green cell in the top panel) with high Notch, low Delta expression. The other cell acts as the sender (orange cell in the top panel) with low Notch, high Delta expression. On the other hand, in the presence of Notch-Jagged signaling (bottom), neighboring cells form a mutual excitatory feedback loop causing them to acquire the same phenotype. Each cell acts both as a sender and a receiver and both cells co-express Notch receptors and Jagged ligands

in a population of cells. In general, since NICD is an EMT promoter and Notch-Delta signaling leads to neighboring cells acquiring distinct phenotypes, Notch-Delta signaling leads to a spatial expression profile wherein hybrid E/M and mesenchymal cells are surrounded by epithelial cells. On the other hand, Notch-Jagged signaling can lead to the emergence of spatial clusters of hybrid E/M and mesenchymal cells due to the tendency of neighboring cells to acquire the same phenotype in the presence of in the presence of Notch-Jagged signaling (Fig. 7).

The spatial expression of epithelial and mesenchymal factors in a population in the presence of Notch signaling can be probed using ordinary differential equations to model the behavior of coupled Notch signaling and EMT circuits. The methodology differs from previous models of EMT regulation in that the dynamics of the circuit within each cell depends not only on the

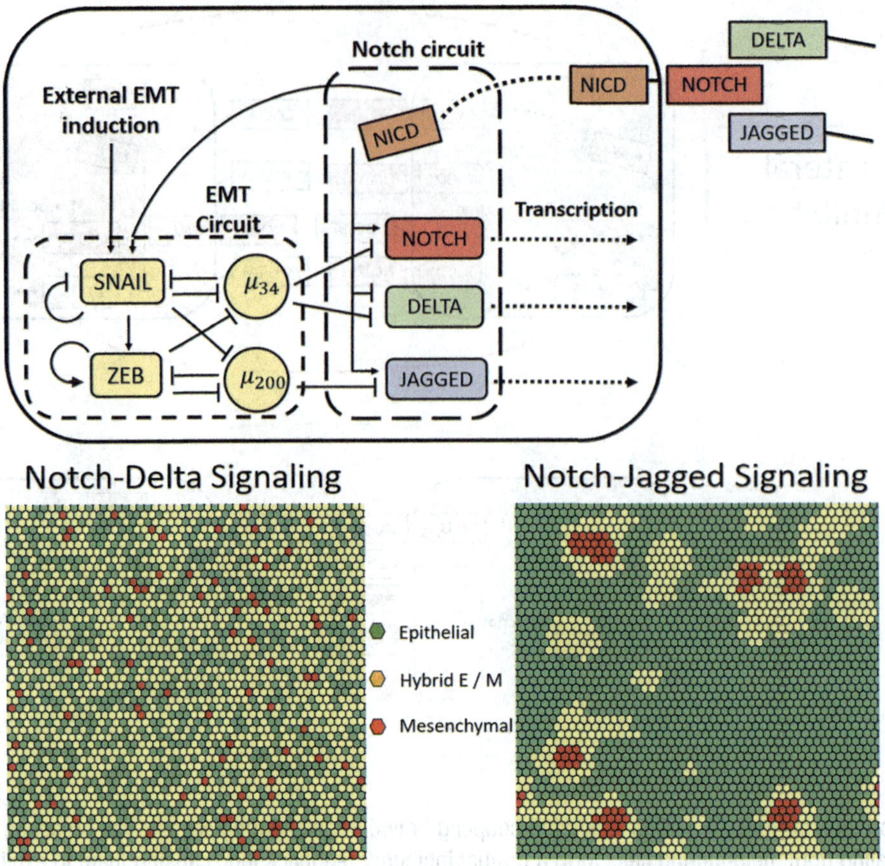

Fig. 7 (Top) Coupling between Notch-Delta-Jagged signaling and EMT regulation. (Bottom) Spatial heterogeneity in the expression of epithelial and mesenchymal markers in the presence of Notch-Delta signaling (left) and in the presence of Notch-Jagged signaling (right)

concentrations of molecules within the cell but also on the concentrations of molecules, particularly Notch receptors and ligands, on neighboring cells that are in direct contact with the given cell. Therefore, before simulating Notch signaling-mediated dynamics, one must choose a suitable spatial lattice wherein each lattice position is occupied by a single cell. This is essential in order to properly identify the neighboring cells for each cell in the population. We will not describe the mathematical form of the ordinary differential equations here since these equations have been described in detail previously [4].

Figure 7 (bottom panel) shows the spatial patterns that emerge via Notch signaling between cells occupying a hexagonal lattice wherein each cell communicates with six neighboring cells in the population. The results indicate that spatial epithelial-mesenchymal heterogeneity can emerge in a population of cancer cells due to the activity of the Notch signaling mechanism. Cells with differing expression levels of epithelial and mesenchymal factors can be spatially organized in distinct patterns in different contexts. While Notch-Delta signaling leads to a "salt-and-pepper" patterning wherein hybrid E/M and mesenchymal cells are surrounded by epithelial cells, Notch-Jagged signaling leads to the emergence of clusters of these cell types. The spatial organization of epithelial-mesenchymal heterogeneity is a distinguishing feature of this mechanism for emergence of heterogeneity. Neither cell-to-cell variation in kinetic parameters governing EMT regulation nor random partitioning of molecules during cell division can lead to such behavior. Spatial heterogeneity in the abundance of different phenotypes is a characteristic of tumors [74]. For example, mesenchymal cancer stem cells are abundant near the tumor-stroma boundary while cancer stem cells exhibiting a hybrid E/M phenotype tend to localize in the interior of the tumor [75]. The cell–cell communication-dependent mechanism for the generation of phenotypic heterogeneity described here can be used to understand and describe such features of the tumor microenvironment [76].

7 Modeling the Coupling Between EMT and Stemness in Cancer Cells

Across cancer types, subpopulations of tumor cells that exhibit stem-cell-like behavior, i.e., an increased capacity to repopulate tumors, have been observed [77]. These cancer stem cells (CSCs), often inherently resistant to anticancer therapies, can not only repopulate the tumor post-therapy but also re-create the intra-tumoral heterogeneity exhibited by the original tumor. The connection between EMT and the appearance of stem-cell-like properties in cancer cells has been studied for a long time. Initial studies argued that tumor cells must undergo a complete EMT in order to exhibit traits of CSCs [78, 79]. This proposition was consistent with the then prevalent perception of EMT as a binary process. Later studies showed that EMT/MET and cancer cell stemness are both highly dynamic processes. Cancer cells can exhibit hybrid E/M phenotypes and inter-convert between the different EMT-associated phenotypes. Similarly, cancer cells can switch between CSC and non-CSC phenotypic states, maintaining a dynamic equilibrium in a population of cancer cells [80–83]. Due to these developments, a more nuanced picture of the EMT-stemness connection has emerged wherein all EMT-associated phenotypes—epithelial, mesenchymal, and hybrid E/M—can exhibit stemness properties depending on the strength of the coupling between the modules regulating EMT and stemness.

In cancer cells, stemness is regulated by a 2-component decision-making circuit (Fig. 8) wherein LIN28 and let-7, a micro-RNA, form a mutual inhibitory loop. NF-κB activates the expression of both LIN28 and let-7 and thus acts as an input to this

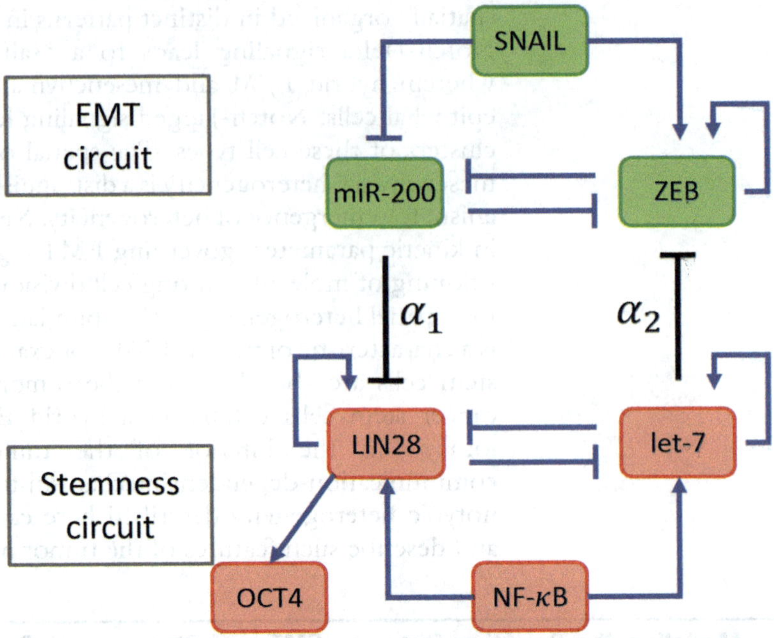

Fig. 8 Coupling between the circuits regulating EMT and stemness. The strength of coupling between the two circuits is governed by the parameters α_1 and α_2. α_1 is the maximum fold change in the rate of production of LIN28 that miR-200 can cause while α_2 is the maximum fold change in the rate of ZEB production that let-7 can cause. Since both coupling interactions are inhibitory, $0 \leq \alpha_1, \alpha_2 \leq 1$ with $\alpha_1, \alpha_2 \sim 1$ indicating weak coupling

regulatory module. Both LIN28 and let-7 can also activate their own expression. The dynamics of this regulatory circuit can be modeled using ODEs as has been done previously for the EMT regulatory circuit. These ODEs have been described in detail elsewhere [19]. The ODEs can be integrated numerically to obtain the steady state expression levels of LIN28 and let-7 for different concentrations of NF-κB. Three distinct phenotypes are evident from this analysis—high LIN28, low LIN28, and intermediate LIN28. LIN28 activates the expression of the pluripotency marker OCT4 [84] and the stem cell state is characterized by the expression of OCT4 within a range—both very low and very high levels of OCT4 expression lead to the loss of stemness [85–88]. Thus, only cells with such intermediate levels of OCT4 expression can acquire a cancer stem cell phenotype.

The stemness regulatory module couples with the EMT regulatory module via two micro-RNA mediated regulatory interactions. miR-200 inhibits the expression of LIN28 posttranscriptionally. Similarly, let-7 inhibits the expression of the EMT-driver ZEB (Fig. 8). These interactions can easily be included in the ODE-based models of EMT and stemness regulation to couple the two regulatory units [89]. Since both very low and very high levels of OCT4 expression lead to loss of stemness, to determine if a certain EMT-associated phenotype can acquire stemness, one can define a "stemness window"—range of expression levels of OCT4 for which a cell can acquire stemness. EMT-associated phenotypes that overlap with this stemness window can then acquire stemness. This overlap, and thus the set of EMT-associated phenotypes that can acquire stemness can be modulated by varying the strength of the coupling between the two regulatory units. This coupling is modeled via two parameters—α_1, the maximum fold change in the expression level of LIN28 that miR-200 can cause, and α_2, the maximum fold change in the expression level of ZEB that let-7 can cause. Since miR-200 inhibits LIN28 and let-7 inhibits ZEB, $0 \leq \alpha_1, \alpha_2 \leq 1$. A fold change close to 1 indicates weak coupling while a fold change close to 0 indicates strong coupling.

When there is no coupling between the two regulatory units ($\alpha_1 = 1, \alpha_2 = 1$), cells can exhibit three distinct phenotypes associated with the EMT circuit (epithelial, mesenchymal, and hybrid E/M) provided the concentration of the EMT-driver SNAIL is within the range for which the EMT circuit can exhibit tri-stability. Cells can further exhibit three distinct phenotypes corresponding to the stemness circuit (low LIN28, high LIN28, and intermediate LIN28). Thus, 9 (3×3) total phenotypes are possible. This number decreases when the strength of the coupling between the circuits is increased. Which of the EMT-associated phenotypes exist within the stemness window depends on the relative values of α_1 and α_2. When both α_1 and α_2 are close to 1 (weak coupling), all three EMT-associated phenotypes lie within the "stemness window" and thus can acquire stemness. Upon decreasing α_1, the stemness window shifts toward the mesenchymal phenotype. Epithelial cells can no longer acquire stemness in this scenario. When α_2 is decreased while keeping α_1 close to 1, the stemness window shifts toward the epithelial phenotype and mesenchymal cells cannot acquire stemness with such a coupling between the regulatory units. The different scenarios have been illustrated in Fig. 9.

The total number of phenotypes that may be exhibited by cells in a population will further depend on the concentration of SNAIL. For example, when the SNAIL concentration is very high, cells can only exhibit the mesenchymal phenotype. These mesenchymal cells can then acquire stemness provided the "stemness window" overlaps with the mesenchymal phenotype. Similarly, very low concentrations of SNAIL will lead to cells in the population exhibiting only the epithelial phenotype. In such a scenario, two distinct phenotypes may be acquired by tumor cells in the population—epithelial stem-like and epithelial non-stem-like. The number of phenotypes exhibited can further be tuned by varying the concentration of NF-κB which activates the expression of both LIN28 and

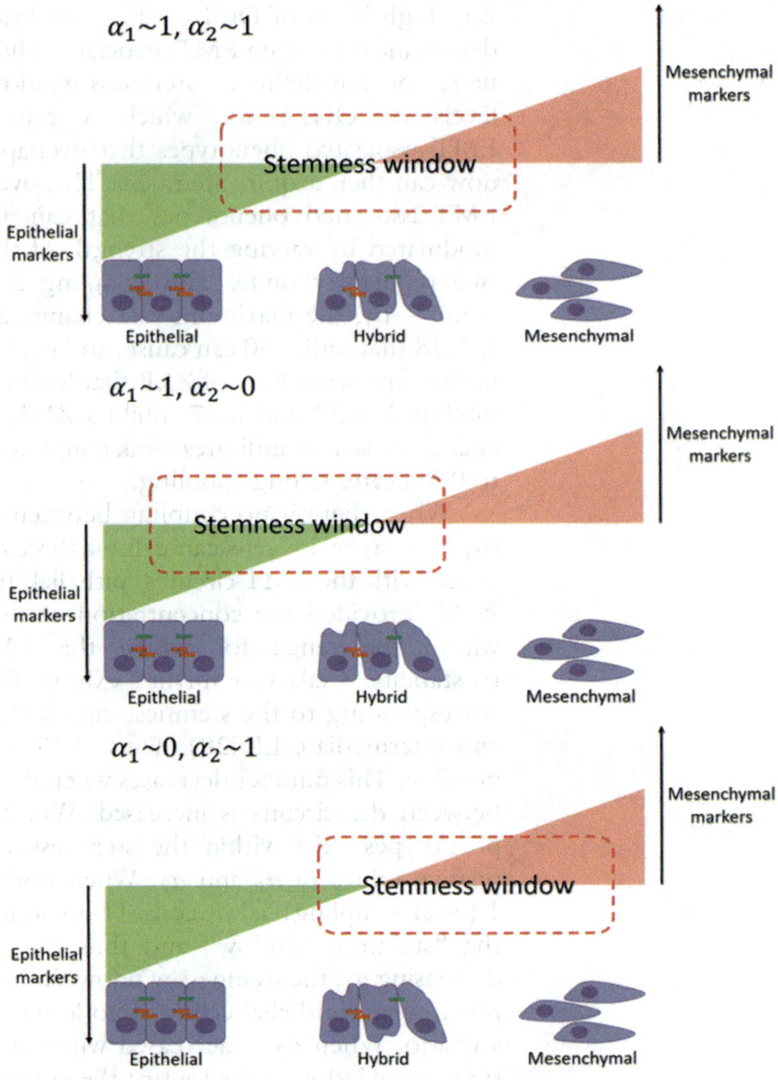

Fig. 9 The coupling parameters determine the overlap of the stemness window (expression of OCT4 within a range) with the spectrum of EMT-associated phenotypes. The overlap determines which of the phenotypes can acquire stemness. In the top panel, all three phenotypes can acquire stemness. In the middle panel, only epithelial and hybrid E/M phenotypes can acquire stemness. In the bottom panel, only hybrid E/M and mesenchymal phenotypes can acquire stemness

let-7. Very low or very high NF-κB concentrations, for example, will cause cells to lose their ability to acquire and maintain stemness due to very low and very high OCT4 expression levels, respectively.

The coupling of EMT and stemness regulatory modules thus allows for the existence of a myriad of phenotypes. Tumor cells in a population may exhibit all or some of these phenotypes depending on the signaling profile. Coupled with the spatial heterogeneity of signaling states within a tumoral mass, subpopulations exhibiting different phenotypic profiles can exist in different parts of the tumor. The EMT regulatory circuit in cancer cells is further coupled with other regulatory modules including the Notch signaling module. Such additional couplings can further increase the number of phenotypes that can be exhibited by cells in a population in a manner similar to the EMT-stemness coupling described above. Additionally, since Notch signaling leads to the emergence of spatial patterns in the distribution of different phenotypes, EMT-Notch-stemness coupling can lead to the localization of different stemness associated phenotypes in different parts of the tumor microenvironment [76].

8 Conclusion

Here, we have presented EMT from the lens of computational systems biology where the focus is on the emergent properties of the underlying regulatory network, instead of those of individual nodes in the network. We have highlighted various examples of how physics/engineering/mathematics driven approaches can reveal unprecedented insights into various aspects of EMT dynamics, such as multi-stability, reversibility/irreversibility, symmetry (or not) in EMT/MET, the effects of non-cell autonomous mechanisms in EMT/MET, and finally the connection of EMT/MET with other cellular traits such as stemness. The in silico models presented here have their own strengths, limitations, and assumptions, just as is the case with any in vitro, in vivo, or ex vivo model. The examples presented here emphasize how an iterative cross talk between mathematical modeling and experimental biology can help decode plasticity and heterogeneity in EMT/MET.

Acknowledgements

This work was supported by the National Science Foundation grant PHY- 1427654 and by the Ramanujan Fellowship awarded to M.K.J. by SERB, DST, Government of India (SB/S2/RJN-049/2018).

References

1. Jolly MK, Ware KE, Gilja S et al (2017) EMT and MET: necessary or permissive for metastasis? Mol Oncol 11:755–769. https://doi.org/10.1002/1878-0261.12083
2. Aiello NM, Kang Y (2019) Context-dependent EMT programs in cancer metastasis. J Exp Med 216:1016–1026. https://doi.org/10.1084/jem.20181827
3. Kumar S, Das A, Sen S (2014) Extracellular matrix density promotes EMT by weakening cell–cell adhesions. Mol BioSyst 10:838–850. https://doi.org/10.1039/C3MB70431A

4. Boareto M, Jolly MK, Goldman A et al (2016) Notch-Jagged signalling can give rise to clusters of cells exhibiting a hybrid epithelial/mesenchymal phenotype. J R Soc Interface 13. https://doi.org/10.1098/rsif.2015.1106
5. Li X, Jolly MK, George JT et al (2019) Computational modeling of the crosstalk between macrophage polarization and tumor cell plasticity in the tumor microenvironment. Front Oncol 9:10. https://doi.org/10.3389/fonc.2019.00010
6. Grosse-Wilde A, Kuestner RE, Skelton SM et al (2018) Loss of inter-cellular cooperation by complete epithelial-mesenchymal transition supports favorable outcomes in basal breast cancer patients. Oncotarget 9:20018–20033. https://doi.org/10.18632/oncotarget.25034
7. Neelakantan D, Zhou H, Oliphant MUJ et al (2017) EMT cells increase breast cancer metastasis via paracrine GLI activation in neighbouring tumour cells. Nat Commun 8:15773. https://doi.org/10.1038/ncomms15773
8. Nieto MA, Huang RY-J, Jackson RA, Thiery JP (2016) EMT: 2016. Cell 166:21–45. https://doi.org/10.1016/j.cell.2016.06.028
9. Lu M, Jolly MK, Levine H et al (2013) MicroRNA-based regulation of epithelial–hybrid–mesenchymal fate determination. Proc Natl Acad Sci 110:18144–18149. https://doi.org/10.1073/pnas.1318192110
10. Font-Clos F, Zapperi S, La Porta CAM (2018) Topography of epithelial–mesenchymal plasticity. Proc Natl Acad Sci 115:5902–5907. https://doi.org/10.1073/pnas.1722609115
11. Steinway SN, Zanudo JGT, Ding W et al (2014) Network modeling of TGF signaling in hepatocellular carcinoma epithelial-to-mesenchymal transition reveals joint sonic hedgehog and Wnt pathway activation. Cancer Res 74:5963–5977. https://doi.org/10.1158/0008-5472.CAN-14-0225
12. Celià-Terrassa T, Bastian C, Liu D et al (2018) Hysteresis control of epithelial-mesenchymal transition dynamics conveys a distinct program with enhanced metastatic ability. Nat Commun 9:5005. https://doi.org/10.1038/s41467-018-07538-7
13. Tian X-J, Zhang H, Xing J (2013) Coupled reversible and irreversible bistable switches underlying TGFβ-induced epithelial to mesenchymal transition. Biophys J 105:1079–1089. https://doi.org/10.1016/j.bpj.2013.07.011
14. Hong T, Watanabe K, Ta CH et al (2015) An Ovol2-Zeb1 mutual inhibitory circuit governs bidirectional and multi-step transition between epithelial and mesenchymal states. PLoS Comput Biol 11:e1004569. https://doi.org/10.1371/journal.pcbi.1004569
15. Jolly MK, Levine H (2017) Computational systems biology of epithelial-hybrid-mesenchymal transitions. Curr Opin Syst Biol 3:1–6. https://doi.org/10.1016/j.coisb.2017.02.004
16. Pastushenko I, Brisebarre A, Sifrim A et al (2018) Identification of the tumour transition states occurring during EMT. Nature 556:463–468. https://doi.org/10.1038/s41586-018-0040-3
17. Ruscetti M, Dadashian EL, Guo W et al (2016) HDAC inhibition impedes epithelial-mesenchymal plasticity and suppresses metastatic, castration-resistant prostate cancer. Oncogene 35:3781–3795. https://doi.org/10.1038/onc.2015.444
18. Karacosta LG, Anchang B, Ignatiadis N, et al (2019) Mapping lung cancer epithelial-mesenchymal transition states and trajectories with single-cell resolution. bioRxiv 570341. https://doi.org/10.1101/570341
19. Jolly MK, Huang B, Lu M et al (2014) Towards elucidating the connection between epithelial-mesenchymal transitions and stemness. J R Soc Interface 11:20140962. https://doi.org/10.1098/rsif.2014.0962
20. Grosse-Wilde A, d'Hérouël AF, McIntosh E et al (2015) Stemness of the hybrid epithelial/mesenchymal state in breast cancer and its association with poor survival. PLoS One 10:e0126522. https://doi.org/10.1371/journal.pone.0126522
21. Kröger C, Afeyan A, Mraz J et al (2019) Acquisition of a hybrid E/M state is essential for tumorigenicity of basal breast cancer cells. Proc Natl Acad Sci 116:7353–7362. https://doi.org/10.1073/pnas.1812876116
22. Bierie B, Pierce SE, Kroeger C et al (2017) Integrin-β4 identifies cancer stem cell-enriched populations of partially mesenchymal carcinoma cells. Proc Natl Acad Sci 114:E2337–E2346. https://doi.org/10.1073/pnas.1618298114
23. Drubin DG, Oster G (2010) Experimentalist meets theoretician: a tale of two scientific cultures. Mol Biol Cell 21:2099–2101. https://doi.org/10.1091/mbc.e10-02-0143
24. Igoshin O, Chen J, Xing J et al (2019) Biophysics at the coffee shop: lessons learned working with George Oster. Mol Biol Cell 30(16):1882
25. Albert R, Thakar J (2014) Boolean modeling: a logic-based dynamic approach for understanding signaling and regulatory networks and for making useful predictions. Wiley Interdiscip

Rev Syst Biol Med 6:353–369. https://doi.org/10.1002/wsbm.1273

26. Zhang J, Tian X-J, Chen Y-J et al (2018) Pathway crosstalk enables cells to interpret TGF-β duration. NPJ Syst Biol Appl 4:18. https://doi.org/10.1038/s41540-018-0060-5
27. Blagoev KB, Shukla K, Levine H (2013) We need theoretical physics approaches to study living systems. Phys Biol 10:040201. https://doi.org/10.1088/1478-3975/10/4/040201
28. Zhang J, Tian X-J, Zhang H et al (2014) TGF-β–induced epithelial-to-mesenchymal transition proceeds through stepwise activation of multiple feedback loops. Sci Signal 7:ra91. https://doi.org/10.1126/scisignal.2005304
29. Bustin SA, Huggett JF (2017) Reproducibility of biomedical research – the importance of editorial vigilance. Biomol Detect Quantif 11:1–3. https://doi.org/10.1016/j.bdq.2017.01.002
30. Fu Y, Glaros T, Zhu M et al (2012) Network topologies and dynamics leading to endotoxin tolerance and priming in innate immune cells. PLoS Comput Biol 8:1–14. https://doi.org/10.1371/journal.pcbi.1002526
31. Wang P, Song C, Zhang H et al (2014) Epigenetic state network approach for describing cell phenotypic transitions. Interface Focus 4:20130068. https://doi.org/10.1098/rsfs.2013.0068
32. Harris LA, Hogg JS, Tapia J-J et al (2016) BioNetGen 2.2: advances in rule-based modeling. Bioinformatics 32:3366–3368. https://doi.org/10.1093/bioinformatics/btw469
33. Box GEP (1976) Science and statistics. J Am Stat Assoc 71:791–799. https://doi.org/10.1080/01621459.1976.10480949
34. Jolly MK, Tripathi SC, Jia D et al (2016) Stability of the hybrid epithelial/mesenchymal phenotype. Oncotarget 7:27067–27084. https://doi.org/10.18632/oncotarget.8166
35. Jolly MK, Preca B-T, Tripathi SC et al (2018) Interconnected feedback loops among ESRP1, HAS2, and CD44 regulate epithelial-mesenchymal plasticity in cancer. APL Bioeng 2:031908. https://doi.org/10.1063/1.5024874
36. Meacham CE, Morrison SJ (2013) Tumour heterogeneity and cancer cell plasticity. Nature 501:328–337. https://doi.org/10.1038/nature12624
37. Lapidot T, Sirard C, Vormoor J et al (1994) A cell initiating human acute myeloid leukaemia after transplantation into SCID mice. Nature 367:645–648. https://doi.org/10.1038/367645a0
38. Al-Hajj M, Wicha MS, Benito-Hernandez A et al (2003) Prospective identification of tumorigenic breast cancer cells. Proc Natl Acad Sci 100:3983–3988. https://doi.org/10.1073/pnas.0530291100
39. O'Brien CA, Pollett A, Gallinger S, Dick JE (2007) A human colon cancer cell capable of initiating tumour growth in immunodeficient mice. Nature 445:106–110. https://doi.org/10.1038/nature05372
40. Ricci-Vitiani L, Lombardi DG, Pilozzi E et al (2007) Identification and expansion of human colon-cancer-initiating cells. Nature 445:111–115. https://doi.org/10.1038/nature05384
41. Singh SK, Hawkins C, Clarke ID et al (2004) Identification of human brain tumour initiating cells. Nature 432:396–401. https://doi.org/10.1038/nature03128
42. Collins AT, Berry PA, Hyde C et al (2005) Prospective identification of tumorigenic prostate cancer stem cells. Cancer Res 65:10946–10951. https://doi.org/10.1158/0008-5472.CAN-05-2018
43. Lehmann BD, Bauer JA, Chen X et al (2011) Identification of human triple-negative breast cancer subtypes and preclinical models for selection of targeted therapies. J Clin Invest 121:2750–2767. https://doi.org/10.1172/JCI45014
44. Calbo J, van Montfort E, Proost N et al (2011) A functional role for tumor cell heterogeneity in a mouse model of small cell lung cancer. Cancer Cell 19:244–256. https://doi.org/10.1016/j.ccr.2010.12.021
45. Stanta G, Bonin S (2018) Overview on clinical relevance of intra-tumor heterogeneity. Front Med 5:85. https://doi.org/10.3389/fmed.2018.00085
46. Hong D, Fritz AJ, Zaidi SK et al (2018) Epithelial-to-mesenchymal transition and cancer stem cells contribute to breast cancer heterogeneity. J Cell Physiol 233:9136–9144. https://doi.org/10.1002/jcp.26847
47. Kim JE, Leung EY, Baguley BC, Finlay GJ (2013) Heterogeneity of expression of epithelial–mesenchymal transition markers in melanocytes and melanoma cell lines. Front Genet 4:97. https://doi.org/10.3389/fgene.2013.00097
48. Pereira L, Mariadason JM, Hannan RD, Dhillon AS (2015) Implications of epithelial–mesenchymal plasticity for heterogeneity in colorectal cancer. Front Oncol 5:13. https://doi.org/10.3389/fonc.2015.00013
49. Stylianou N, Lehman ML, Wang C et al (2018) A molecular portrait of epithelial–mesenchymal

plasticity in prostate cancer associated with clinical outcome. Oncogene 38:913–934. https://doi.org/10.1038/s41388-018-0488-5

50. Creighton CJ, Li X, Landis M et al (2009) Residual breast cancers after conventional therapy display mesenchymal as well as tumor-initiating features. Proc Natl Acad Sci 106:13820–13825. https://doi.org/10.1073/pnas.0905718106
51. Tièche CC, Gao Y, Bührer ED et al (2018) Tumor initiation capacity and therapy resistance are differential features of EMT-related subpopulations in the NSCLC cell line A549. Neoplasia 21:185–196. https://doi.org/10.1016/j.neo.2018.09.008
52. Balaban NQ, Merrin J, Chait R et al (2004) Bacterial persistence as a phenotypic switch. Science 305:1622–1625. https://doi.org/10.1126/science.1099390
53. Huh D, Paulsson J (2011) Random partitioning of molecules at cell division. Proc Natl Acad Sci U S A 108:15004–15009. https://doi.org/10.1073/pnas.1013171108
54. Huh D, Paulsson J (2011) Non-genetic heterogeneity from stochastic partitioning at cell division. Nat Genet 43:95–100. https://doi.org/10.1038/ng.729
55. Boyer LA, Lee TI, Cole MF et al (2005) Core transcriptional regulatory circuitry in human embryonic stem cells. Cell 122:947–956. https://doi.org/10.1016/j.cell.2005.08.020
56. Moignard V, Macaulay IC, Swiers G et al (2013) Characterization of transcriptional networks in blood stem and progenitor cells using high-throughput single-cell gene expression analysis. Nat Cell Biol 15:363–372. https://doi.org/10.1038/ncb2709
57. Yan J, Wang H, Liu Y, Shao C (2008) Analysis of gene regulatory networks in the mammalian circadian rhythm. PLoS Comput Biol 4: e1000193. https://doi.org/10.1371/journal.pcbi.1000193
58. Zhang R, Lahens NF, Ballance HI et al (2014) A circadian gene expression atlas in mammals: implications for biology and medicine. Proc Natl Acad Sci U S A 111:16219–16224. https://doi.org/10.1073/pnas.1408886111
59. Milo R, Shen-Orr S, Itzkovitz S et al (2002) Network motifs: simple building blocks of complex networks. Science 298:824–827. https://doi.org/10.1126/science.298.5594.824
60. Huang B, Lu M, Jia D et al (2017) Interrogating the topological robustness of gene regulatory circuits by randomization. PLoS Comput Biol 13:e1005456. https://doi.org/10.1371/journal.pcbi.1005456
61. Lamouille S, Xu J, Derynck R (2014) Molecular mechanisms of epithelial–mesenchymal transition. Nat Rev Mol Cell Biol 15:178–196. https://doi.org/10.1038/nrm3758
62. Jia D, George JT, Tripathi SC et al (2019) Testing the gene expression classification of the EMT spectrum. Phys Biol 16:025002. https://doi.org/10.1088/1478-3975/aaf8d4
63. Losick R, Desplan C (2008) Stochasticity and cell fate. Science 320:65–68. https://doi.org/10.1126/science.1147888
64. Soltani M, Vargas-Garcia CA, Antunes D, Singh A (2016) Intercellular variability in protein levels from stochastic expression and noisy cell cycle processes. PLoS Comput Biol 12: e1004972. https://doi.org/10.1371/journal.pcbi.1004972
65. Tripathi S, Chakraborty P, Levine H, Jolly MK (2020) A mechanism for epithelial-mesenchymal heterogeneity in a population of cancer cells. PLoS Comput Biol 16(2):e1007619.
66. Gillespie DT (1977) Exact stochastic simulation of coupled chemical reactions. J Phys Chem 81:2340–2361. https://doi.org/10.1021/j100540a008
67. Lovisa S, LeBleu VS, Tampe B et al (2015) Epithelial-to-mesenchymal transition induces cell cycle arrest and parenchymal damage in renal fibrosis. Nat Med 21:998–1009. https://doi.org/10.1038/nm.3902
68. Vega S, Morales AV, Ocaña OH et al (2004) Snail blocks the cell cycle and confers resistance to cell death. Genes Dev 18:1131–1143. https://doi.org/10.1101/gad.294104
69. Handler J, Cullis J, Avanzi A et al (2018) Pre-neoplastic pancreas cells enter a partially mesenchymal state following transient TGF-β exposure. Oncogene 37:4334. https://doi.org/10.1038/s41388-018-0264-6
70. Bray SJ (2016) Notch signalling in context. Nat Rev Mol Cell Biol 17:722–735. https://doi.org/10.1038/nrm.2016.94
71. Siebel C, Lendahl U (2017) Notch signaling in development, tissue homeostasis, and disease. Physiol Rev 97:1235–1294. https://doi.org/10.1152/physrev.00005.2017
72. Shaya O, Sprinzak D (2011) From notch signaling to fine-grained patterning: modeling meets experiments. Curr Opin Genet Dev 21:732–739. https://doi.org/10.1016/j.gde.2011.07.007
73. Boareto M, Jolly MK, Lu M et al (2015) Jagged–Delta asymmetry in notch signaling can

give rise to a sender/receiver hybrid phenotype. Proc Natl Acad Sci 112:E402–E409. https://doi.org/10.1073/pnas.1416287112

74. Yuan Y (2016) Spatial heterogeneity in the tumor microenvironment. Cold Spring Harb Perspect Med 6:a026583. https://doi.org/10.1101/cshperspect.a026583
75. Liu S, Cong Y, Wang D et al (2014) Breast cancer stem cells transition between epithelial and mesenchymal states reflective of their normal counterparts. Stem Cell Rep 2:78–91. https://doi.org/10.1016/j.stemcr.2013.11.009
76. Bocci F, Gearhart-Serna L, Boareto M et al (2019) Toward understanding cancer stem cell heterogeneity in the tumor microenvironment. Proc Natl Acad Sci 116:148–157. https://doi.org/10.1073/pnas.1815345116
77. Nassar D, Blanpain C (2016) Cancer stem cells: basic concepts and therapeutic implications. Annu Rev Pathol Mech Dis 11:47–76. https://doi.org/10.1146/annurev-pathol-012615-044438
78. Mani SA, Guo W, Liao M-J et al (2008) The epithelial-mesenchymal transition generates cells with properties of stem cells. Cell 133:704–715. https://doi.org/10.1016/j.cell.2008.03.027
79. Morel A-P, Lièvre M, Thomas C et al (2008) Generation of breast cancer stem cells through epithelial-mesenchymal transition. PLoS One 3:e2888. https://doi.org/10.1371/journal.pone.0002888
80. Iliopoulos D, Hirsch HA, Wang G, Struhl K (2011) Inducible formation of breast cancer stem cells and their dynamic equilibrium with non-stem cancer cells via IL6 secretion. Proc Natl Acad Sci 108:1397–1402. https://doi.org/10.1073/pnas.1018898108
81. Yang G, Quan Y, Wang W et al (2012) Dynamic equilibrium between cancer stem cells and non-stem cancer cells in human SW620 and MCF-7 cancer cell populations. Br J Cancer 106:1512–1519. https://doi.org/10.1038/bjc.2012.126
82. Wang W, Quan Y, Fu Q et al (2014) Dynamics between cancer cell subpopulations reveals a model coordinating with both hierarchical and stochastic concepts. PLoS One 9:e84654. https://doi.org/10.1371/journal.pone.0084654
83. Gupta PB, Fillmore CM, Jiang G et al (2011) Stochastic state transitions give rise to phenotypic equilibrium in populations of cancer cells. Cell 146:633–644. https://doi.org/10.1016/j.cell.2011.07.026
84. Qiu C, Ma Y, Wang J et al (2010) Lin28-mediated post-transcriptional regulation of Oct4 expression in human embryonic stem cells. Nucleic Acids Res 38:1240–1248. https://doi.org/10.1093/nar/gkp1071
85. Niwa H, Miyazaki J, Smith AG (2000) Quantitative expression of Oct-3/4 defines differentiation, dedifferentiation or self-renewal of ES cells. Nat Genet 24:372–376. https://doi.org/10.1038/74199
86. Karwacki-Neisius V, Göke J, Osorno R et al (2013) Reduced Oct4 expression directs a robust pluripotent state with distinct signaling activity and increased enhancer occupancy by Oct4 and Nanog. Cell Stem Cell 12:531–545. https://doi.org/10.1016/j.stem.2013.04.023
87. Shu J, Wu C, Wu Y et al (2013) Induction of pluripotency in mouse somatic cells with lineage specifiers. Cell 153:963–975. https://doi.org/10.1016/j.cell.2013.05.001
88. Theunissen TW, van Oosten AL, Castelo-Branco G et al (2011) Nanog overcomes reprogramming barriers and induces pluripotency in minimal conditions. Curr Biol 21:65–71. https://doi.org/10.1016/j.cub.2010.11.074
89. Jolly MK, Jia D, Boareto M et al (2015) Coupling the modules of EMT and stemness: a tunable "stemness window" model. Oncotarget 6:25161–25174. https://doi.org/10.18632/oncotarget.4629

Chapter 29

Control of Cell Migration Using Optogenetics

Leo Valon and Simon de Beco

Abstract

Optogenetics uses light to manipulate protein localization or activity from subcellular to supra-cellular level with unprecedented spatiotemporal resolution. We used it to control the activity of the Cdc42 Rho GTPase, a major regulator of actin polymerization and cell polarity. In this chapter, we describe how to trigger and guide cell migration using optogenetics as a way to mimic EMT in an artificial yet highly controllable fashion.

Key words Optogenetics, Rho GTPase, Cdc42, Polarity, Migration, Lamellipodium, Microscopy

1 Introduction

Epithelial-to-Mesenchymal Transition (EMT) involves several phenotypic changes including loss of apicobasal polarity and cell–cell adhesion while cell migration and invasiveness are increased. Rho GTPases play a major role in these processes, and it is no surprise that they have been shown to be involved in Contact Inhibition of Locomotion (CIL) and EMT [1, 2]. Schematically, active Rac1 and Cdc42 are found at the cellular front where they drive actin polymerization resulting in the formation of lamellipodia and filopodia [3, 4]. Conversely, RhoA is mostly known to control myosin II activity resulting in the formation of stress fibers and retraction of the back of motile cells during mesenchymal migration [5]. Controlling Rho GTPases activity is thus of particular importance to study the mechanisms of cell migration during EMT.

Several methods have been developed in order to control Rho GTPases activity at the cellular and subcellular levels. The use of genetic mutations and drugs offers convenient methods but lacks the spatial and temporal precision needed to match the dynamics of these proteins as observed ex vivo and in vivo (typically, a few micrometers and tens of seconds, [6]). Microfluidics make it possible to create controlled gradients of chemoattractants that result in

Kyra Campbell and Eric Theveneau (eds.), *The Epithelial-to Mesenchymal Transition: Methods and Protocols*, Methods in Molecular Biology, vol. 2179, https://doi.org/10.1007/978-1-0716-0779-4_29,

intracellular gradients of Rho GTPase activity, thereby allowing to guide cell migration and to recapitulate EMT [7]. However, this is somehow technically difficult, especially if high spatial and fast temporal resolutions are required or if experiments are done in vivo. Recently, optogenetics emerged as a powerful tool to control signaling activities with an unprecedented spatiotemporal resolution [8–11]. By using light-sensitive protein domains derived from plants, it allows researchers to control the conformation or dimerization of synthetic polypeptides, thus making it possible to govern quantitatively the activity of proteins of interest in time and space with light [12, 13]. Rho GTPases are activated by Guanine Exchange Factors (GEFs) that promote the exchange from GDP to GTP. Optogenetic control of GEF localization at subcellular scale thus enables one to modulate locally the activity of Rho GTPases (Fig. 1) and to impose the direction and slope of intracellular gradients of these polarity signals in a quantitative fashion. One can then force cells to migrate and control the direction and speed of locomotion [14]. In this chapter, as an example of optogenetic manipulation of Rho GTPases, we explain how to manipulate quantitatively Cdc42 activity by controlling its GEF Intersectin1 catalytic domain (ITSN) at the subcellular level in order to guide cell migration (*see* **Notes 1** and **2**). Moreover, we detail some critical parameters for successful implementation and optimization of this technique.

2 Materials

2.1 Plasmids and Transfection

1. Two plasmids that encode the fusion proteins ITSN-CRY2-mCherry and CIBN-GFP-CAAX (optoGEF-Cdc42), with the expression controlled by strong eukaryotic promoter such as the CMV promoter. These plasmids can be shared upon request.
2. Transfection reagent: X-treme gene 9, or equivalent. X-treme gene 9 allows a strong expression of the transgenes in Hela cells, but can be replaced by any other transfection system adapted to the chosen cell type.

2.2 Cell Culture

1. HeLa cells.

   Alternately the CRY2/CIBN optogenetic system has been used by us and others on a wide range of cell types (MDCK cells, NIH 3 T3 fibroblasts, RPE1, etc.) without encountering specific issues.
2. Cell culture medium: DMEM (Dulbecco's modified Eagle's medium) supplemented with 10% fetal calf serum.

2.3 Cell Plating on Coverslips

1. A 6-well plate for cell culture.
2. Glass coverslips: 25 mm in diameter, 0.13–0.17 mm thickness.
3. Fibronectin solution: 20 μg/mL fibronectin from bovine plasma diluted in 100 mM NaHCO3 (pH 8.5).
4. A mounting holder for round 25 mm coverslips.

2.4 Stable Cell Line

Alternatively, to transient transfection, stable cell lines expressing the two optogenetic plasmids can be obtained and studied over months (*see* **Note 3**).

Use lentiviral infections to produce stable cell lines easily.

1. Viral particles containing the CIBN-GFP-CAAX and ITSN-CRY2-mCherry fusion genes.
2. A FACS-sorter to select cells with appropriate expression levels of the two partners of the dimerizer system (*see* **Note 4**) and to limit cell-to-cell variability.

2.5 Microscopy Setup for Optogenetics Imaging and Activation

The cellular response as well as the recruitment of ITSN-CRY2-mCherry to the plasma membrane are best imaged with TIRF microscopy, but can also be imaged with a spinning disk microscope, a fast scanning laser confocal microscope, or by epifluorescence. The TIRF setup allows fast imaging and accurate quantification of fluorescence signals at the membrane in contact with the coverslip without interference from the cytoplasmic signal.

1. A microscope with an atmosphere at 37 °C and 5% CO_2.
2. A high magnification objective (40–100×).
3. GFP and mCherry fluorescence filter sets.
4. Bright-field imaging possibility, ideally DIC (Differential Interference Contrast) with a red long-pass filter in the light path (cutoff wavelength above 550 nm) to avoid CRY2 activation by blue wavelengths contained in white light.
5. A FRAP module using a 488 nm laser to shine blue light in subcellular regions of the cells (*see* **Note 5**). Alternatively, any system allowing to illuminate specifically a Region of Interest within the field of view with blue light could also work (*see* **Note 6**).

3 Methods

By shining blue light at cells expressing the optoGEF-Cdc42 system, it changes the conformation of CRY2. This increases the affinity between CRY2 and CIBN, thus promoting the relocation of ITSN-CRY2-mCherry at the plasma membrane (where GFP-CIBN-CAAX is located). This allows ITSN to interact with the

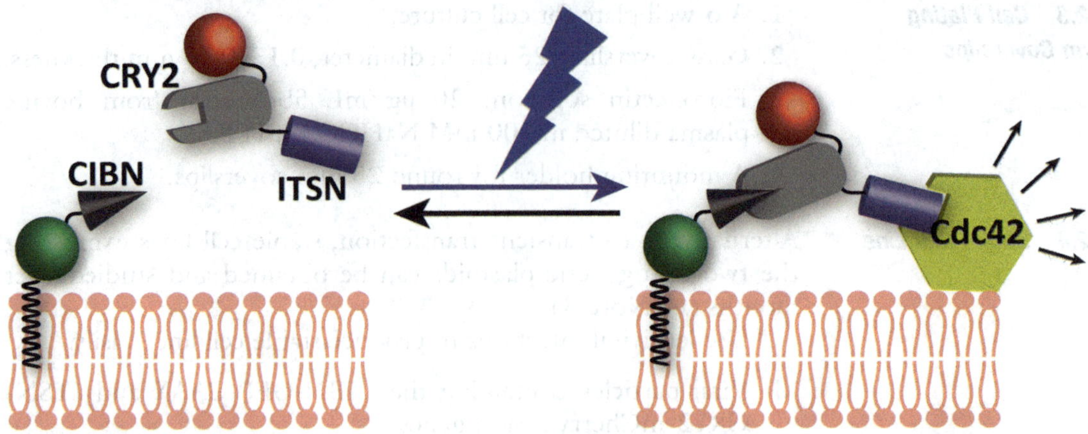

Fig. 1 Optogenetic activation of Cdc42 using the optoGEF-Cdc42 system. The optoGEF-Cdc42 system is composed of two components: CIBN-GFP-CAAX is inserted at the plasma membrane, whereas ITSN-CRY2-mCherry diffuses in the cytosol. Blue light increases the affinity between the two components, thereby inducing the relocation of ITSN-CRY2-mCherry to the plasma membrane where it can activate Cdc42

surrounding proteins such as Cdc42 (Fig. 1). When illuminating subcellular regions, it is thus possible to increase the proportion of activated Cdc42 in these regions, hence polarizing the cell.

3.1 Preparation of the Cell Samples

1. At D - 2 (2 days before the experiment), detach the HeLa cells and plate them at 25% confluency in 6-well plates.
2. At D - 1, transfect the cells with the plasmids coding for the two optogenetic partners (CIBN-GFP-CAAX and ITSN-CRY2-mCherry) using an equal amount of DNA for the two plasmids, according to the transfection reagent manufacturer's instructions.
3. Prepare the fibronectin-coated coverslips: Deposit a drop of 100 μL of 20 μg/mL fibronectin on a piece of parafilm, and place a coverslip on top of it. The droplet should be sandwiched between the parafilm and coverslip, and the whole glass surface should be covered by the solution. Incubate fibronectin at room temperature for 1 h.

3.2 Optogenetic Activations and Acquisition of Microscopy Images

1. 2–4 h before the experiment, detach the cells using Accutase (150 μL for 1–3 min) and plate them at the desired density (5–20%, 10-40.10^3 cells per round 25 mm coverslip when studying cell migration) on fibronectin-coated coverslips while maintaining the sample in low light conditions.
2. Mount the sample in a holder and place it on the microscope. Atmosphere should be kept at 37 °C and 5% CO_2.
3. Switch all sources of white or blue light off and wait 15 min to make sure all pre-existing CRY2-CIBN complexes are disassociated.

4. Use bright-field (with the red filter in the light path) and the mCherry signal from ITSN-CRY2-mCherry to find transfected cells with the desired shape and neighborhood crowding density (ideally no or few neighbors). Set up the TIRF so as to image only the signal from the membrane. Cells with a fairly high expression level of ITSN-CRY2-mCherry and maximal expression level of CIBN-GFP-CAAX should be chosen for the experiment to work (*see* **Note 4**), although it is not always possible to test the latter as visualization of CIBN-GFP-CAAX using blue light illumination will transiently activate Cdc42 in a uniform fashion throughout the cell.
5. Draw a region of interest (ROI) where blue light will be shined using the FRAP control software: target the cell edges, with a ROI ranging from 5 to 50 μm^2 approximately. For the cell to migrate in the direction dictated by the ROI, the ROI should encompass a few micrometers within the cytoplasm and a bigger area outside the cell.
6. Set up the imaging and activation frequency. One activation pulse every 10 s (100 ms pulses at approximately 10% power of a 488 nm, 100 mW laser, *see* **Note 7**) is sufficient for strong activation. In order to see the pre-activation and post-activation states, start these activation pulses from image number 20 on, and continue the movie after blue light illumination are stopped (*see* **Note 8**).
7. When all parameters are selected, launch the experiment. Recruitment of ITSN-CRY2-mCherry to the basal plasma membrane should be visible through a local increase of red fluorescence after a few seconds (*see* **Note 9**). If the microscope does not have auto-focus possibilities, make sure there is no drift of focus.
8. In the case of long experiments (>2 h), the cell starts to fully invade the ROI. It might then be necessary to adapt the activation ROI so that it remains localized at the cell border.

4 Notes

1. Many signaling pathways can be specifically targeted (activated or inactivated) using optogenetics. First, because many GTPases are activated by specific GEFs, activation can usually be achieved by recruiting one of these GEFs to the relevant membranes. Similarly, optogenetics-driven actuation (activation or inhibition) can be achieved by locally depleting the inhibitors or activators of the targeted pathway, respectively. This can be done by recruiting these regulators to some cellular compartments where they cannot activate their targets [15] or

by the induction of clusters where they are thereby inactive [16, 17]. As many membrane receptors are activated by ligand-induced dimerization, it is also possible to use optogenetics to mimic this activation [18, 19]. It is also possible to activate cytosolic kinases with optogenetics. Some of the constructs that have been designed to date are somehow involved in EMT, but the design of new constructs using these methods require an in-depth knowledge of the biochemical and structural properties of the targeted kinases [20, 21]. There are currently several optogenetic systems available to control the activity of Rho GTPases: photoactivatable Rac, making use of the LOV system [8], the CIBN/CRY2 system [9], the Phytochrome B/PIF system [10], the LOV/TULIP system [22], the iLID system [11]. None of these systems is perfect or intrinsically better than the others. One has to choose the best system for each experimental condition, and several parameters should be considered.

2. Spatiotemporal resolution of active Rho GTPase localization is dictated by the distribution of Rho GTPase activators. It is determined by the initial recruitment of the cytoplasmic partner to the membrane, the diffusion of the dimer at the plasma membrane (which increases the spreading of the activation) and the dissociation time of the dimer. The shorter the lifetime of the dimer is, the less it diffuses at the plasma membrane before dissociating. In practice, it is not possible to create subcellular activation smaller than 4–5 μm in the case of the first versions of CRY2/CIBN and TULIP systems [13]. So far, the recent iLID systems have the best spatial resolution (1–2 μm) as well as the PhyB/PIF (in which dissociation can be forced with far-red illumination) [10]. However, that comes with a price: as dissociation happens faster, it requires more frequent illumination to maintain a constant activation, hence more phototoxicity and photobleaching.

3. It is possible to achieve optogenetic activation of Cdc42 with the ITSN-CRY2-mCherry/CIBN-GFP-CAAX dimer through both transient transfections and stable cell lines. Transient transfections are faster to set up and will usually give rise to high expression levels of both components. However, their reproducibility is difficult to control and can be problematic if it is necessary to photo-activate a group of cells. Given the limitations enunciated before with regard to expression levels, stable cell lines are usually recommended for reproducibility and ease of use. In this case, it is highly recommended to FACS-sort the stable cell lines on the basis of their mCherry and GFP expression levels. However, we noticed that expression levels tend to be lower in stable cell lines, and that some cell lines tend to spontaneously repress either CIBN or CRY2,

in ways that sometimes make optogenetic experiments poorly favorable. If this is the case, other systems like iLID or TULIP could solve the problem.

4. Except for the photoactivatable Rac system, all the other systems are light-gated dimerizers at chemical equilibrium between the bound and free forms of the dimer. While the dissociation constant is unaffected by light, the association constant of the dimers is modified under light exposure. Thus, the dissociation constant at equilibrium, K_D, varies from high values in the dark to low value under light. For optogenetic activation to be efficient but only triggered when light is shone, cellular concentrations of the optogenetic components have to be higher than the K_D of the lit state and lower than the K_D of the dark state. Since it is usually difficult to control the level of expression in cellular systems, it is important to choose the optogenetic dimer with the largest difference between the K_Ds in the dark and in the lit state for the chosen cell line. The values of some of these constants have been measured, but not all [23]. In the case of the CRY2/CIBN system, this requirement translates into fairly high expression levels of CIBN and CRY2 for the system to work. However, both components should not be equally expressed: it is important that the concentration of CIBN-GFP-CAAX is in excess, so that it is not easily saturated with ITSN-CRY2-mCherry and that large amounts of the latter can be recruited to the plasma membrane. It is also important that ITSN-CRY2-mCherry is not expressed too strongly otherwise it can spontaneously form clusters in the cytoplasm. Even though the K_D in the illuminated state of the CRY2/CIBN couple has not been reported, it is sometimes difficult in some cell lines to reach expression levels that are high enough for its activation by illumination, suggesting this K_D is quite high. If this is the case, it can be worth testing a system with a lower K_D, like the iLID/SSPB nano couple.

5. All the optogenetic systems are not activated by the same wavelengths. Therefore, they are not all compatible with the same fluorophores. Mainly, the photoactivatable Rac, CRY2/CIBN, LOV/TULIP and iLID systems are activated by a wide range of blue light (400–530 nm approximately), while the Phytochrome B is activated by red light (centered around 650 nm). The wavelength of any other readout or additional fluorophore has thus to be compatible with this excitation spectrum as well as with potential fluorophores already attached to the system itself (ex: CIBN-GFP, CRY2-mCherry).

   Taking this into consideration, it limits the number of fluorophores that can be used without activating the dimerization. In the case of the use of the CRY2/CIBN system all wavelengths under 530 nm will activate it, which leaves only a few wavelengths free for imaging. Commonly, we keep CRY2 fused to mCherry to quantify the optogenetic perturbation. Additional biological readouts make use of fluorophores in the "far-red" domain (excitation ~640 nm, emission ~670 nm) such as iFP1 and upgraded iFP2 proteins [24].

However, in order to bypass this limitation regarding to wavelengths, two alternative strategies can be used. First, once the optogenetic effect is characterized, one can remove the mCherry protein fused to CRY2 and use it for another molecular reporter. Using this strategy, we managed to do optogenetic experiments while looking at two extra proteins tagged with mCherry and iFP2 respectively doing transient transfection with four plasmids. Second, it is possible to use the GFP channel to look at an extra readout. When the local activation is strong enough for the cytoplasm to be almost completely depleted in CRY2 protein, it is possible to acquire images in the GFP channel and visualize a GFP-fused protein without significantly modifying the CRY2 subcellular localization. This is at the condition that these blue illuminations are rare in time, with small exposure time and laser power. This second approach also requires to remove GFP from CIBN or to look at a protein fused to GFP with a localization or expression levels different than GFP-CIBN-CAAX.

6. Our approach to shine blue light using a FRAP module is not the only way to induce subcellular optogenetic activation. Other strategies are also commonly used in laboratories to illuminate restricted areas in a field of view. A very powerful option makes use of a Digital Micromirror Device (DMD). The light source required is not necessarily a laser but can also be a fluorescence lamp or LEDs, and it allows to shine patterns of light almost instantaneously. Moreover, with a DMD one can easily create complex patterns and gradients of light, which would need a strong implementation in the case of a FRAP module [14]. However, it also has drawbacks: in most cases, even if most of the light will be shone in the region of interest, a little amount of it will be shone everywhere in the field of view (<5%). Finally, a simple and inexpensive way to shine local patterns of blue light is to add an automatized square diaphragm with controlled aperture size in the light path of a spinning disk or epifluorescence microscope. By creating a channel for imaging at 488 nm in which the diaphragm is kept almost closed, it allows to shine squares of blue light only in the central part of the field of view [15]. This configuration is not versatile in terms of ROI shape but allows the control of exposure time and laser power, and is very convenient for preliminary tests on new optogenetic constructs or

cell lines. Without much implementation, it is also possible to do multiple cells activation at the same time (without recording different ROIs for each position). One has to place the ROI in a central area of the field of view, and then to select the different XY positions based on the position of the fixed ROI.

7. When starting using the CRY2-CIBN optogenetic system, one needs to know what is the amount of light required to activate the dimerization. In the case of the CRY2/CIBN dimer, a good reference value is to use similar light power and exposure times as the ones required to do GFP imaging without photobleaching. This should be sufficient to generate CRY2 recruitment to the membrane. Then, it is also possible to control quantitatively the level of activation. According to theory and experiments [13], two main parameters are important: the frequency of the activation pulses, and/or the light amount per pulse. With this approach, it is possible to create different levels of recruitment which can be quantified a posteriori. This approach remains correct as long as there is no saturation of the system, i.e., CIBN is in large excess at the membrane in comparison to CRY2 proteins, and that there is a large cytoplasmic CRY2 pool available for activation, in order to avoid the plateauing of the recruitment.

   Finally, it is also possible to create subcellular and/or multicellular gradients of activation from gradients of light, using a DMD for example [14]. This can be achieved in a straightforward manner as a linear gradient of light will create a linear recruitment of ITSN-CRY2-mCherry [13], at the condition that the system is not saturated.

8. When illumination is stopped after optogenetic activation is achieved, the optogenetic systems will revert to their initial state with timescales that depend on their biochemical properties. For example, the CRY2/CIBN couple, with its half-life time of 2 min, will be fully dissociated within 10 min. Hence, membrane-recruited ITSN will diffuse back to the cytoplasm and stop activating Cdc42. In our experiments with HeLa cells and NIH-3T3 cells, stopping blue illumination resulted in a quick loss of protrusions and a reshuffling of cell polarity—even after hour-long optogenetic activations. However, that might be dependent on cell types and experimental conditions.

9. It is sometimes necessary to quantify the recruitment of ITSN-CRY2-mCherry to the plasma membrane. Measure the mCherry fluorescence TIRF signal over time in the recruitment area. The mCherry signal is proportional to the concentration of ITSN proteins at the plasma membrane (Fig. 2). If fluorescence compensation is necessary, quantify the photobleaching

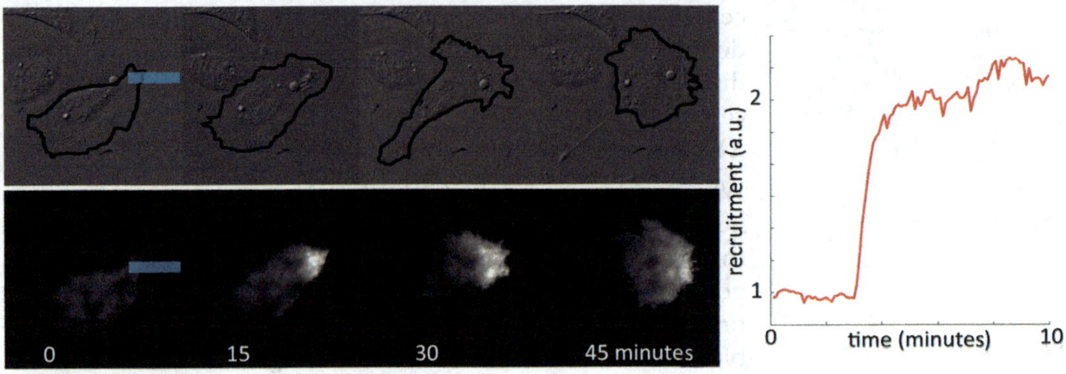

Fig. 2 Activation of optoGEF-Cdc42 and induction of migration of a HeLa cell. Left: Images are DIC (top) and mCherry in TIRF (bottom) images of a cell expressing the optoGEF-Cdc42 system and activated in the blue ROI as described in the protocol (one pulse of blue light every 10 s for 50 min, starting after 2.5 min without activation). Right: Quantification of the recruitment of mCherry fluorescence in the ROI during the first 10 min of the experiment

in neighboring transfected cells, or compare the evolution of fluorescence inside and outside the ROI in the targeted cell. In order to quantify the cell response in terms of migration, use the bright-field images and segment the cell contour to obtain the deformation and migration of the cells over time. It is also possible to run local morphometric analysis in order to quantify the amount of ruffling activity.

Acknowledgements

The authors would like to warmly thank Mathieu Coppey for fruitful discussions and critical reading of the manuscript. L.V. is supported by a post-doctoral grant "Aide au Retour en France" from the FRM (Fondation pour la Recherche Médicale, ARF20170938651) and by a Marie Skłodowska-Curie Fellowship.

References

1. Shoval I, Kalcheim C (2012) Antagonistic activities of Rho and Rac GTPases underlie the transition from neural crest delamination to migration. Dev Dyn 241:1155–1168. https://doi.org/10.1002/dvdy.23799
2. Scarpa E, Szabó A, Bibonne A, Theveneau E, Parsons M, Mayor R (2015) Cadherin switch during EMT in neural crest cells leads to contact inhibition of locomotion via repolarization of forces. Dev Cell 34:421–434. https://doi.org/10.1016/j.devcel.2015.06.012
3. Nobes CD, Hall A (1995) Rho, rac, and cdc42 GTPases regulate the assembly of multimolecular focal complexes associated with actin stress fibers, lamellipodia, and filopodia. Cell 81:53–62
4. Nobes CD, Hall A (1999) Rho GTPases control polarity, protrusion, and adhesion during cell movement. J Cell Biol 144:1235–1244
5. Burridge K, Wennerberg K (2004) Rho and Rac take center stage. Cell 116:167–179. https://doi.org/10.1016/S0092-8674(04)00003-0
6. Machacek M, Hodgson L, Welch C, Elliott H, Pertz O, Nalbant P, Abell A, Johnson GL, Hahn KM, Danuser G (2009) Coordination

of Rho GTPase activities during cell protrusion. Nature 461:99–103. https://doi.org/10.1038/nature08242

7. Te Kuo C, Liu HK, Huang GS, Chang CH, Chen CL, Chen KC, Huang RYJ, Lin CH, Lee H, Huang CS, Wo AM (2014) A spatio-temporally defined in vitro microenvironment for controllable signal delivery and drug screening. Analyst 139:4846–4854. https://doi.org/10.1039/c4an00936c
8. Wu YI, Frey D, Lungu OI, Jaehrig A, Schlichting I, Kuhlman B, Hahn KM (2009) A genetically encoded photoactivatable Rac controls the motility of living cells. Nature 461:104–108. https://doi.org/10.1038/nature08241
9. Kennedy MJ, Hughes RM, Peteya LA, Schwartz JW, Ehlers MD, Tucker CL (2010) Rapid blue-light-mediated induction of protein interactions in living cells. Nat Methods 7:973–975. https://doi.org/10.1038/nmeth.1524
10. Levskaya A, Weiner OD, Lim WA, Voigt CA (2010) Spatiotemporal control of cell signalling using a light-switchable protein interaction. Nature 461:1–5. https://doi.org/10.1038/nature08446
11. Guntas G, Hallett RA, Zimmerman SP, Williams T, Yumerefendi H, Bear JE, Kuhlman B (2015) Engineering an improved light-induced dimer (iLID) for controlling the localization and activity of signaling proteins. Proc Natl Acad Sci 112:112–117. https://doi.org/10.1073/pnas.1417910112
12. Toettcher JE, Voigt CA, Weiner OD, Lim WA (2011) The promise of optogenetics in cell biology: interrogating molecular circuits in space and time. Nat Methods 8:35–38. https://doi.org/10.1038/nmeth.f.326
13. Valon L, Etoc F, Remorino A, Di Pietro F, Morin X, Dahan M, Coppey M, di Pietro F, Morin X, Dahan M, Coppey M (2015) Predictive spatiotemporal manipulation of signaling perturbations using optogenetics. Biophys J 109:1785–1797. https://doi.org/10.1016/j.bpj.2015.08.042
14. de Beco S, Vaidžiulytė K, Manzi J, Dalier F, di Federico F, Cornilleau G, Dahan M, Coppey M, De Beco S, Manzi J, Dalier F, Cornilleau G, Dahan M, Coppey M (2018) Optogenetic dissection of Rac1 and Cdc42 gradient shaping. Nat Commun 9:1–13. https://doi.org/10.1038/s41467-018-07286-8
15. Valon L, Marín-Llauradó A, Wyatt T, Charras G, Trepat X, Marin-Llaurado A, Wyatt T, Charras G, Trepat X (2017) Optogenetic control of cellular forces and mechanotransduction. Nat Commun 8:14396. https://doi.org/10.1038/ncomms14396
16. Lee S, Park H, Kyung T, Kim NY, Kim S, Kim J, Do Heo W (2014) Reversible protein inactivation by optogenetic trapping in cells. Nat Methods 11:633–636. https://doi.org/10.1038/nmeth.2940
17. Taslimi A, Vrana JD, Chen D, Borinskaya S, Mayer BJ, Kennedy MJ, Tucker CL (2014) An optimized optogenetic clustering tool for probing protein interaction and function. Nat Commun 5:4925. https://doi.org/10.1038/ncomms5925
18. Grusch M, Schelch K, Riedler R, Reichhart E, Differ C, Berger W, Inglés-Prieto Á, Janovjak H (2014) Spatio-temporally precise activation of engineered receptor tyrosine kinases by light. EMBO J 33:1713–1726. https://doi.org/10.15252/embj.201387695
19. Bugaj LJ, Spelke DP, Mesuda CK, Varedi M, Kane RS, Schaffer DV (2015) Regulation of endogenous transmembrane receptors through optogenetic Cry2 clustering. Nat Commun 6:6898. https://doi.org/10.1038/ncomms7898
20. Zhou XX, Fan LZ, Li P, Lin MZ (2017) Optical control of cell signaling by single-chain photoswitchable kinases. Science 355:836–842
21. Leopold AV, Chernov KG, Verkhusha VV (2018) Optogenetically controlled protein kinases for regulation of cellular signaling. Chem Soc Rev 47:2454–2484. https://doi.org/10.1039/c7cs00404d
22. Strickland D, Lin Y, Wagner E, Hope CM, Zayner J, Antoniou C, Sosnick TR, Weiss EL, Glotzer M (2012) TULIPs: tunable, light-controlled interacting protein tags for cell biology. Nat Methods 9:379–384. https://doi.org/10.1038/nmeth.1904
23. Hallett RA, Zimmerman SP, Yumerefendi H, Bear JE, Kuhlman B (2016) Correlating in vitro and in vivo activities of light-inducible dimers: a cellular Optogenetics guide. ACS Synth Biol 5:53–64. https://doi.org/10.1021/acssynbio.5b00119
24. Filonov GS, Piatkevich KD, Ting L-M, Zhang J, Kim K, Verkhusha VV (2011) Bright and stable near-infrared fluorescent protein for in vivo imaging. Nat Biotechnol 29:757–761. https://doi.org/10.1038/nbt.1918

Index

A

Acidification 329, 336, 337
Adherens junctions (AJs) 7, 31, 43, 44, 66, 67, 76, 77, 120, 121, 128, 129, 149, 200, 304, 316
Adhesion
  cell-cell 15, 16, 49, 253, 257, 272, 276, 280, 346, 385, 415
  cell-matrix 15, 16, 43, 243, 258, 268, 272
Aggregates 46, 47, 253, 275–285, 297, 298

B

Bioenergetics 328, 329, 331, 333, 334, 336, 338

C

Cadherin
  E-Cadherin (CDH1) 44, 371
  N-Cadherin (CDH2) 371
Cancer
  associated fibroblasts 244
  breast 30, 50, 53, 244, 329, 354, 391, 392
  colorectal 161, 328, 391, 392
Carcinoma 14, 20, 37, 48, 52, 200, 244, 303, 315, 316, 328
Cell
  behaviour 21, 53, 80, 145, 174, 200, 228
  culture 44, 90, 190, 192, 229, 239, 244–247, 249, 265, 267, 270, 271, 290, 292, 293, 297, 331, 332, 338, 343, 355, 360–362, 416
  dissociation 36, 90, 267, 271, 307, 318, 320, 325
  dynamics 187, 278
  followers 96, 200, 244, 251
  leaders 96, 244, 251
  migrations 15, 53, 79–105, 173, 199–221, 228, 260, 266–267, 341, 354, 385, 415–424
  tracking 136, 266, 267, 271, 282
Cell sorting
  magnetic activated cell sorting 316, 321
Chromatin 51, 55, 80, 83–85, 87, 354, 369
Collective cell migrations 15, 199–221

D

Diffusion 145, 152, 155–157, 194, 310, 420

E

Electroporation 110, 136, 184, 186, 188, 190, 191, 194
Embryo
  chicken 107–114, 173–180, 184, 185, 188
  Drosophila 9, 35, 54, 65–77, 145–157, 199–221
  mouse 135–144, 200
  sea urchin 8, 9, 35
  Xenopus 259, 266, 268, 277, 279
  zebrafish 80
Endoderm 7, 30, 35, 65, 115, 135, 174, 177, 257, 264, 269, 271
Epiblast 45, 48, 135, 136, 139, 143, 184, 186, 191, 194, 257, 341
Epithelial cells 8, 15, 23, 35, 43–54, 65–67, 119, 157, 174, 227, 240, 275, 282, 290, 310, 317, 321, 337, 341–350, 354, 378, 402, 403, 405, 407
Epithelialization 185, 280, 282
Epithelial-mesenchymal transition (EMT)
  hybrid 22, 353, 385, 386, 389, 392, 397–399, 401, 403, 405, 407, 408
  partial 4, 5, 10, 14, 23, 24, 29–32, 47, 48, 50, 116, 131, 162, 200, 316, 322, 353
  pathological 21, 22, 30, 32, 44, 55
  physiological 21, 22, 30, 31, 183, 258, 354, 355, 363
Eversion 115–132

Kyra Campbell and Eric Theveneau (eds.), *The Epithelial-to Mesenchymal Transition: Methods and Protocols*, Methods in Molecular Biology, vol. 2179, https://doi.org/10.1007/978-1-0716-0779-4,

F

Fibrosis........23, 27, 29, 55
Fluorescence recovery after photobleaching (FRAP)........145–157, 417, 419, 422
Fluorescent in situ hybridization (FISH)........5, 353–382
Förster resonance energy transfer (FRET)........276, 289–292, 295, 296, 298

G

Gal4........84, 104, 117–120, 123, 125, 130, 162, 163
Gastrulation........7, 8, 11, 13, 35, 45, 48, 52, 65–68, 71, 76, 80, 135–144, 199, 200, 257, 271, 341
Genome........26, 47, 309, 328, 354, 358, 370
Glycolysis........328, 329, 331

H

Heterogeneity
  non-genetic........391, 392

I

Image
  analysis........90, 109, 137, 140, 163, 169, 178, 206, 210, 247, 251, 278, 282, 290, 295, 296, 375
  processing........92–97, 111, 113, 177, 193, 194
Imaginal disc........115–132
Immunofluorescence........345–347
Invasion........3, 7, 11, 24–26, 36, 37, 50, 161, 162, 199, 200, 228, 243–255, 304, 316, 329, 354, 385
In vitro........17, 36, 45, 52, 55, 108, 228, 231, 232, 239, 276, 289, 316, 342, 386, 391, 409
In vivo........8, 17, 36, 79–105, 107, 118, 120, 124, 126, 131, 145–157, 173, 183–195, 199–221, 228, 231, 258, 259, 268, 276, 303–312, 316, 386, 391, 409, 415

M

Mechanotransduction........55
Mesenchymal cells (MCs)........14, 30, 35, 43, 44, 46, 47, 65, 80, 135, 145, 174, 185, 186, 257, 276, 277, 316, 317, 320, 321, 341, 347, 401–403, 405, 407
Mesenchymal-epithelial transition (MET)........29–32, 35–37, 43–46, 65, 183–195, 275–285, 328, 341–350, 354, 376, 385, 393, 398, 401, 405, 409
Mesoderm........7, 46, 53, 65–77, 115, 135, 136, 148, 174–177, 185, 257, 262–264, 268–271, 279, 303, 304
Metabolism........24, 36, 328–338
Metastasis........3, 11, 13, 14, 25, 26, 31, 36, 37, 44, 48, 55, 65, 115, 161–169, 244, 258, 316, 328, 329, 342, 354
Micropatterns........55, 227–240
Mitochondria........328–331, 333, 334, 337, 338
Modeling
  computational........389, 392, 399
  mathematical........5, 155, 385–409
Morphogenesis........3, 11, 35, 47, 50, 52, 66, 107, 173, 174, 228, 239, 342
Mosaicism........139
Multi-photon microscopy........175, 179, 180
Multistability........409
Myosin........10, 17, 67, 69, 70, 76, 120, 121, 200, 218, 415

N

Neural crest
  chicken........107–114
  Xenopus........4, 257–270
  zebrafish........79–105
Neural tube (NT)........24, 49, 107, 109–112, 174–177, 183, 189, 191, 262, 263, 270

O

Optogenetics........415–423
Oxidative phosphorylation (OXPHOS)........328, 329, 331, 334
Oxygen........111, 147, 329, 330, 332, 334, 336–338, 378

P

Photoconversion........90, 101, 105
Plasticity........4, 14, 22, 24, 25, 32, 37, 43, 48, 258, 315, 328–334, 385–409
Polarity
  apico-basal........4, 35, 257, 258, 415
  front-rear........35, 258
Post-transcriptional regulation........5
Protrusions........15, 116, 200, 220, 423

R

Respiration........329, 334, 336, 338
Rho GTPases........17, 415, 416, 420
RNAi........115, 116, 118–120, 123, 124, 130, 131

S

Secondary neurulation (SN)........183–195

Single cell 4, 5, 11, 14–16, 26, 80, 87, 107, 135–144, 232, 237, 247, 266, 268, 271, 272, 303–312, 325, 389, 395, 399, 401, 404
Somites 35, 46, 80, 110, 111, 173–180, 189
Spheroids 243–255, 328

T

Tamoxifen 84, 90, 99, 139, 142, 143
Time-lapse imaging 91, 92, 111, 180, 184, 187, 190, 192, 193, 195, 205, 206, 209–212, 249, 251, 260, 263, 265, 271, 272
Tissue engineering 227–240
Transcription factors (TFs) 4, 5, 9, 10, 14, 20, 36, 37, 47, 258, 303, 311, 312, 318, 363, 393, 394, 396, 399, 400

U

UAS 84, 87, 99, 104, 117, 119, 120, 124, 130

W

Whole embryo culture 108

9781071607787